INTRODUCTION TO SPACE:

The Science of Spaceflight

THIRD EDITION

INTRODUCTION TO SPACE:

The Science of Spaceflight

THIRD EDITION

by
Thomas D. Damon
Professor Emeritus
Pikes Peak Community College

with Foreword by
Edward G. Gibson

KRIEGER PUBLISHING COMPANY
MALABAR, FLORIDA
2001

Original Edition 1989
Second Edition 1995
Third Edition 2001

Printed and Published by
KRIEGER PUBLISHING COMPANY
KRIEGER DRIVE
MALABAR, FLORIDA 32950

Library of Congress Cataloging-in-Publication Data

Damon, Thomas.
Introduction to space : the science of spaceflight / by Thomas D. Damon ; with foreword by Edward G. Gibson.—3rd ed.
p. cm.
Includes bibliographical references and index.
ISBN 0-89464-065-8 (alk. paper)—ISBN 0-89464-066-6 (pbk. : alk. paper)
1. Space flight. I. Title.

TL791 .D36 2001
629.4′1—dc21 00-059950

10 9 8 7 6 5 4 3 2

To James, Jacob, and Maxx . . . may they one day walk on Mars.

Series editors
Edwin F. Strother, Ph.D.
Donald M. Waltz

Contents

CHAPTER 4
The Space Environment 59

CHAPTER 5
Satellites 81

CHAPTER 6
Remote Sensing 103

CHAPTER 7
Astronomy from Space 119

Preface to the Third Edition

The technology of reaching space has not changed much in the last 30 years. The Space Shuttle was designed and built in the 1970s. The booster rockets now in use are descendants of the 1950s intercontinental ballistic missiles. Design of the International Space Station (ISS) now under construction began in the early 1980s. Landings on Mars still fail as often as they succeed.

Using these "old" technologies, however, the application of orbiting satellites to everyday activities has moved forward rapidly since the second edition of this book was published five years ago. Space applications have matured, are in everyday use, and have become big business. Where television programs were once sent via satellite to local stations for broadcasting, now they are beamed directly from satellites to homes. The global positioning system (GPS), once the exclusive domain of the military, has become a must for aviators and archaeologists, sailors and surveyors, map makers and mountaineers. Although telephone companies have been using satellites to carry phone conversations both intercontinental and intracontinental, an individual can now access a satellite directly with a handheld telephone. High resolution images from Earth observing satellites, once available only to the intelligence community, are finding applications from agriculture to zoology. Exciting things have been happening in space-based astronomy also. As one astronomer states, just as explorers discovered the Earth in the 16th century, we are now discovering the universe. Powerful new telescopes in orbit are answering age-old questions about stars and galaxies, and posing new questions as well.

This third edition updates these activities and describes where we hope to go in the next few decades: The Space Shuttle will be replaced by space planes that will make almost daily trips to orbit carrying astronauts, equipment, and perhaps tourists. Humans will return to the Moon and land on Mars. Home computers will connect to the Internet directly through high-speed satellite data links. The Space Station orbiting Earth will be permanently inhabited. Some of our energy needs may be met by satellites converting sunshine into electricity and beaming it to Earth. We may have definitive knowledge of other intelligent species elsewhere in the universe.

The organization of the book is unchanged, that is, the subject of each chapter is the same as in the second edition. There are many new or updated illustrations and twice as many color pages. The MATHBOXES have been a widely acclaimed technique for introducing mathematics to an essentially nonmathematical book and are therefore continued in this edition. All measurements are now presented in both common units and in the international System of Units (SI), commonly called metric units. SI values carry the same number of significant figures as the original number. In some cases, the number of significant figures is not obvious and is assumed from the context.

At the end of the 20th century we look back to its beginning and smile at the Wright brothers' primitive airplane, the first horseless carriages, the Morse telegraph, Marconi's first radio transmission, Lowell's canals on Mars, Edison's first lightbulb. At the end of the 21st century people will look back to its beginning and smile at our primitive efforts to leave the confines of Earth's gravity. Who can predict what kind of vehicle they will drive, how fast, and how far. Even time travel and faster-than-light travel are no longer considered impossible. The possibilities are mind-boggling.

Preface to the Second Edition

Five years have passed since this book was first published. A lot has happened in space research and exploration in those five years. The Space Shuttle is back in business. The Hubble Space Telescope was launched, found to be flawed, and repaired on orbit. Venus and the Moon have been completely mapped, one by radar and the other by multispectral sensors. Voyager left the Solar System. New spacecraft are bound for the Sun, Jupiter, and Saturn, and a few have failed. Space defense resources came into play in the Persian Gulf War, employed as they had never been before in actual combat. The Global Positioning System is completed. Direct broadcast satellites are about to become a reality. Biomedical research in space is leading to a better understanding of how the human body functions in orbit and on Earth. And more. . . .

Remarkable changes have taken place in world politics and economics as well, changes thought to be impossible in 1988. The Soviet Union disintegrated and the 45-year Cold War came to an end. Russian space scientists and engineers are joining with their counterparts in the United States, seeking to work together on cooperative projects. The possibility that all spacefaring nations will collaborate in building and operating an international space station is not as far-fetched as it seemed only five years ago. Can international colonization of the Moon and human exploration of Mars be far behind? Successful cooperation in the space station could be a model for successful cooperation in other endeavors, both on Earth and in space. It could be the cornerstone for joining all peoples of Earth into permanent peaceful coexistence.

This second edition has been updated with the latest information in all aspects of space research and exploration. A special effort has been made to include space activities of all nations, past, present, and future. Space defense is no longer treated as a separate subject; the information in the previous Chapter 7 has been distributed to other appropriate chapters throughout the book. On the other hand, the discussion of astronomy from space has been removed from Chapter 6 to a separate chapter, Chapter 7 in this edition. Otherwise, the organization of the book is essentially the same as it was in the first edition.

Some of the MATHBOXES have been revised and five new ones have been added. There are numerous new or revised illustrations, more of them in color. For example, with printouts from new computer software, the orbit diagrams in Chapter 3 are clearer and more readable. Several new discussion questions have been added and the additional reading lists are updated and expanded.

I would like to acknowledge receipt of a grant from the Pikes Peak Community College International Education Committee to enable me to expand on international space activities in my course and, consequently, in this book. Special thanks to John Bally of the University of Colorado at Boulder for the remarkable Space Telescope image of the Orion Nebula. Special thanks also to my dear wife, Anita, for her encouragement, patience, and understanding during the past six months.

Preface to the First Edition

Most of what the general public "knows" about space has been learned from watching *Star Trek, Star Wars,* and the evening news. Unfortunately, these sources of "information" often lead to misconceptions about the laws of nature with regard to what is scientifically and technically possible. For example, in a classroom discussion concerning the hazards of rocket debris and dead satellites that orbit the Earth, one student suggested that the Space Shuttle could simply fly around picking up the pieces of space junk, drawing them into the cargo bay with a tractor beam. The student seriously believed that the Star Trek tractor beam really exists.

Several colleges and universities around the country offer courses in space science and technology, heavy in math and oriented toward upper division or graduate students in science and engineering. Because of the high interest in space activities in Colorado Springs, we at Pikes Peak Community College felt the need for a general science, general education course primarily for liberal arts students who would like to know, among other things, how spacecraft get into orbit, why astronauts float, what satellites are doing up there, and if permanent colonies can be built in space.

We searched for an appropriate textbook, but none was available. Considerable effort has gone into space education curriculum development at the grade school and junior high level, but ours may be the only course of its kind for nonspecialist college students and adults. At first we assembled a collection of NASA pamphlets, government publications, and contractor reports, along with a few freshly written sections on basic science. As the course became more popular, the need for a more formal textbook became obvious. This book is the result.

I have tried to avoid writing a trivia reference book. This is, I hope, a book of scientific and technical substance, not just a collection of interesting facts. On the other hand, I have avoided any "heavy" math, but have included a number of MATHBOXES for those students and classes that can handle simple algebra. MATHBOXES are set apart from the rest of text and may be omitted without disrupting the continuity of the book.

The subjects included here are those that have been of greatest interest to my students. A brief history is summarized in Chapter 1. The basic science of propulsion, orbital mechanics, and the space environment is covered in Chapters 2, 3, and 4. These topics are fundamental to understanding the rest of the book. The next three chapters discuss unmanned spacecraft with particular emphasis on remote sensing. Chapters 8, 9, and 10 deal with man in space, particularly the Space Shuttle, our only manned space program. The final three chapters project into the future with solid facts and some speculation about humankind's ventures into the final frontier.

The thirteen chapters can be covered at the average rate of one per week during a 15 week semester, with time left for tests and other subjects. Some topics, particularly Chapters 2 and 3, may require more than a week; some others can be covered more quickly.

I must acknowledge the help of many people in this project. Particular thanks go to Dr. Mario Iona, Professor Emeritus of Physics at Denver University, and to Dr. William Bennington, PPCC biologist, for their review and critique of certain chapters. The numerous individuals and groups that provided background information, artwork, and photographs are mentioned by organization, if not personally, in the captions of the illustrations. This book could never have been completed without their assistance. Special thanks go to Mary Roberts, editor, whose words of encouragement always came when they were most needed, and to Robert Krieger for his willingness to venture into the unknown.

Foreword

This time we've really gone and done it!

Regardless of the means—feet, horse, boat, covered wagon, car, or plane—we humans have always felt compelled to push outward, to take that next step, to surge into the space around us, to explode into every fresh frontier that offers growth and challenge. But this time, with the technology of the rocket, we've pierced gravity's cocoon, stepped outward and come face to face with our ultimate frontier, an insatiable frontier—the remainder of our universe.

True, so far we've made just a small, tentative step, maybe even just a shuffle. Yet, make no mistake about it, our first step into space is as important as those first steps we've made onto land, across mountains, over oceans, and into air. For we've just made that first step that must precede the strides and dashes to planets, the marathons between stars and that matter-of-fact galactic travel that someday will fill eons yet to come. No longer just observers of the heavens, we can now leave Earth and in only minutes become participants.

Intimidating? No doubt about it.

For after looking back at Earth from the Moon, we've responded by pulling back to regroup, to sort out the motives, methods, and timing of our next steps. Before beginning Tom Damon's clear, concise, and exciting presentation of the science of spaceflight, let's raise some key issues.

In every venture into a new frontier, we humans always benefit by proceeding in the same orderly manner: explore, develop, and use. Is space to be the same?

Yes. With manned and unmanned spacecraft we've explored the edge of our new frontier—low Earth orbit—and have now developed the most logical means to use it—space stations. The United States is planning to end its visitor status in space with its Space Station, as the Soviet Union has already done, and to reap the benefits of a permanent presence.

In the short term, Space Station provides a Space Research Center for science and technology, as well as for the development of new commercial products and industries. Occupancy in low Earth orbit provides researchers and entrepreneurs with an environment unequaled in any earthbound facility: near zero gravity, near perfect vacuum, and a vantage point from which to study Earth as well as the undistorted image of the universe.

In the long term, Space Station will be an assembly base, a stepping stone, a central gateway in an evolving infrastructure that enables our future expeditions to the Moon, Mars, and farther out. Behind our expeditions, colonies will spread permanent human presence throughout the Solar System. We can feel confident that with time, modest upgrades to our technology, and our permanent presence on space stations, this next outward surge will occur.

But let's be "prudent and economical" about this new frontier. Let's just develop it remotely from the ground or dart back and forth to it now and then and avoid the expense of a space hotel. We don't really have to stay on-board Space Station day after day, do we?

Yes! In earthbound research and industrial facilities, our human perception, logic, creativity, and dexterity are, without question, unique and indispensable. The value of this hands-on direction and support is undiminished by the relocation of our facilities into space. In truth, it is our continuous presence that permits us to most profitably operate, repair, maintain, and upgrade every facility and gives us the time and experience to evolve every concept, design, instrument, and operation to its fullest potential.

Should we do anything more now than give space token attention?

Yes! The time for space is now. At any time in the past, only a few nations have had the strength and opportunity to shape human history. Not so today. Many nations are strong and have access to space. Those nations who do shape our future and, indeed, those who remain free of military or economic domination will be those who understand that response to challenge enhances strength and that, from this time onward, strength in space is closely tied to strength on Earth. Our decision can be delayed, our future cannot.

But didn't we say that once we've left gravity's cocoon behind we'd be ready for "our entire universe"? That's a bit pretentious. What about the distances and the technology required to cover them? Sure, new frontiers have always shocked us with their remoteness, until technology advancements reduce this to inconvenience. But beyond our Solar System loom invincible expanses that take years to travel, even at light speed. Can interstellar distance also be overcome?

Yes! But here our confidence requires more than a direct extrapolation of today's technology and time; it also requires a faith inspired by the history of human achievement and an awareness of what shaped it—humanity's expanding intellect and incessant drive.

Lastly, let's go back to the basic premise: ". . . compelled to push outward, to take that next step, to surge into the space around us. . . ." All froth and no substance? Do we need to allow anything other than economics, politics, or other concrete intellectual considerations to enter our deliberations on space?

Once again—yes! In the exploration and development of previous frontiers, we humans have always been obsessed with lists of scientific objectives and practical benefits. Yet when we look back, we realize that our rationalizations, often even our imaginations, fell far short of predicting the frontier's true importance. We really accepted the challenge for more basic reasons.

So it is now with our presence in space. We may underestimate its importance to us and our descendants as well as our basic motivation. Data, information, pictures, or other nourishments of the mind, by themselves, ring hollow. They can only thicken our intellectual shell, only toughen our educated hide. Inside us, as always, the essential spark resides, alive and well with all the primordial drives and desires, itches and urges. In time, we must break the physical limits of our turf and go there, to see and feel the new territory up close, to take physical measure of it with our own person before it can become a true part of our world. To deny this drive is to deny our humanity.

Tom Damon provides for you, the reader, a lucid understanding of the science and technology of entering space, what we are finding, and some of our plans for the future. Enjoy it! It is an extraordinary treatment of an exciting subject, an outstanding introduction to mankind's greatest, open-ended adventure.

EDWARD G. GIBSON
ASTRONAUT, SKYLAB 4
1989

Chapter 1

History of Spaceflight

When the history of this time is written, what will be considered the most important of the daily multitude of momentous events to characterize this era? When those future historians chronicle our most significant contribution, I believe they will say that this was the time when humans took their first tentative steps out of the cradle, when they left Earth to seek greener pastures, to start anew, to try again. Space is the new and final frontier.

Americans have already impressed footprints in the dust of the Moon, put robots on the surface of Mars, sent spacecraft to fly past the other planets and moons in our Sun's family, and sent people commuting to work in orbit. The former Soviet Union has also sent robot explorers to the far reaches of the Solar System and has built a permanently manned space station where cosmonauts have set endurance records of more than 1 year living in space. Only 66 years, less than an average lifetime, took us from the Wright brothers' first 120 foot airplane flight in 1903 to a walk on the Moon in 1969.

The exploration and colonization of other worlds will be the most exciting undertaking of all time. It is inevitable, as inevitable as the exploration and colonization of America.

That is what spaceflight and this book are all about.

Rocketry: The Means of Getting There

The history of spaceflight begins with the history of rocketry; without rockets there would be no spaceflight.

The history of rocketry begins in China. The Chinese invented gunpowder which exploded violently when ignited in an enclosed space. However, if gunpowder were ignited in the open, it just burned rapidly. By putting the gunpowder into a chamber with one end open so the gases formed by the combustion could escape, the whole thing took off "like a rocket" instead of exploding. Indeed, it was a rocket. As early as 1212, Chinese armies fired rockets carrying flaming materials into Mongol encampments to start fires.

About the year 1500, according to legend, a Chinaman named Wan Hu made the first attempt to build a rocket powered vehicle. He attached 47 rockets to some sort of cart and at a given signal 47 coolies lit the 47 rockets simultaneously. In the explosion that followed the entire vehicle disappeared in a cloud of smoke and Wan Hu was never seen in this world again.

Artillery rockets carrying bombs became highly developed in Europe by 1800. They had been used on a large scale against British troops in India, and in response, Sir William Congreve developed the modern stick-rocket in 1805. These new weapons of war were used by most world armies of that time. The British used them against Napoleon in Europe and against the newly organized United States of America. In 1814 during the 25 hour long British bombardment of Baltimore, Francis Scott Key, watching the action from a ship in Baltimore harbor, wrote a poem later to become our national anthem. "And the rocket's red glare . . . gave proof through the night that our flag was still there."

Science fiction writers have long written stories about people travelling to far-off planets. Propulsion systems included flocks of birds, balloons, and just closing your eyes and wishing. A cannon was the choice of Jules Verne who gave us *From the Earth to the Moon and Around the Moon* in the mid-1800s.

Russian schoolteacher, Constantin Tsiolkovski, was a space dreamer. He first published some of his ideas in 1883 and in 1903 published a treatise "Exploration of the Universe With Rocket Propelled Vehicles." By the time of his death in 1935 he had considered every aspect of spaceflight. Although he was considered mad by many during his lifetime, a monument to his memory now stands in Moscow.

Robert Goddard was the first to build successful liquid-fueled rockets. Using gasoline and liquid oxygen for propellants and a blowtorch for an igniter, he launched for the first successful flight from a farm in Auburn, Massachesetts, on March 16, 1926. It lasted less than 3 seconds at a maximum speed of 60 miles per hour (100 km/h), covered a distance of 184 feet (56 m), and reached an altitude of 41 feet (12.5 m). The rocket weighed less than 6 pounds (2.5 kg) and carried about 5 pounds (2.3 kg) of propellant.

Goddard was a quiet person and, when he suggested that rockets would be the way to leave Earth and fly to the Moon and Mars, he met with such ridicule that he never sought public attention again. *The New York Times* said Goddard did not have the knowledge of a high school student. The *Times* printed a retraction 49 years later when Apollo 11 landed on the Moon. In 1930, with financial backing from Charles Lindbergh and from the Guggenheim Foundation, Goddard moved to New Mexico where he continued his experiments until 1941.

At Goddard's request, his papers describing interplanetary flight were held secret by the Smithsonian Institution until 1970, twenty-five years after his death. Four papers covered nearly all the basic science and engineering principles that were needed for flight to the planets. He held 214 patents for his many inventions. In 1960 the U.S. government agreed that the large military rocket engines used for intercontinental ballistic missiles infringed on Goddard's patents and paid his widow a million dollars in damages. She gave half of it to the Guggenheim Foundation.

In Europe, the story was different. Unlike Goddard's bad public experience in the United States, Germans were highly interested in rocketry and the prospect of space travel. Rockets were being built by German amateur rocket societies, especially Verein fur Raumschiffahrt (Society for Space Travel). Hermann Oberth in 1923 published a book, *The Rocket into Interplanetary Space*. In Russia, Tsiolkovski's works were eagerly read and rocket societies were actively experimenting in the 1920s and 1930s. In England, the British Interplanetary Society did serious studies on all aspects of spaceflight. One of its early presidents was Arthur C. Clarke who later wrote *The Sentinel*, a short story which was made into the movie *2001: A Space Odyssey* by Stanley Kubrick.

The technology needed to actually fly into space and return safely was the outgrowth of several technologies that developed in mid-twentieth century. The German military, limited by the World War I Treaty of Versailles, became interested in the rockets being built by amateur rocket societies. Military research with rocket-powered artillery led to the first ballistic missile in 1943 and culminated in the V-2 "vengeance weapon" that carried warheads from Peenemünde, along the Baltic coast of Germany, to England in the closing days of World War II. It was a fearful weapon because there was no sound, no forewarning before the warhead hit and exploded. The V-2 rocket boosted the 1 ton warhead to 3,500 miles per hour (5600 km/h) and burned out. The warhead continued on a ballistic trajectory to a range of 200 miles (320 km) and fell on its target. By contrast, there was some warning of an attack by airplanes and by the earlier guided missiles whose engines continued running until they reached their targets. Fortunately for the Allies, the V-2s came too late in the war to have any effect on the outcome.

While the Russian army was advancing across Germany at the end of the war, the German rocket scientists and engineers from Peenemünde sought out the American Army and surrendered. Some were captured by the Russians and taken to the Soviet Union. The Americans confiscated some of the V-2 rockets and took them along with Wernher von Braun and co-workers to White Sands, New Mexico, where they continued their research and development work. When asked about the design of their V-2, the Germans said to ask Robert Goddard; it had been copied from a rocket Goddard flew in 1939.

By 1948 the German engineers in the United States had flown a two-stage rocket to an altitude of 244 miles. Von Braun dreamed of space travel. He knew that the work he and his colleagues were doing for the Army could lead to rockets which could fly into orbit, but any time the subject came up it was vetoed. The Army was not in the space business. Nonetheless, without the military rocket development programs, the advances in rocketry necessary for spaceflight would not have come about as they did.

While the U.S. Army was interested in artillery, the U.S. Air Force was working toward the development of very long range rocket-powered missiles. Both the United States and the Soviet Union saw rockets as a means to deliver nuclear weapons against the enemy. Successful development of intercontinental ballistic missiles (ICBMs) was dependent on three things: the perfection of reliable high powered rocket engines, guidance systems, and reentry vehicles. The problem of reentry into Earth's atmosphere was particularly important. Without protection form the searing heat of atmospheric friction, the nuclear weapons would melt and vaporize before reaching their targets.

At about the same time, jet aircraft were being perfected and experiments with rocket-powered aircraft were beginning (Figure 1.1). These craft were able to fly to the upper fringes of the atmosphere, where a human could not survive without special protective equipment, the forerunners of the modern space suit and life support equipment.

The center of aircraft research was at Edwards Air Force Base in the desert northeast of Los Angeles. There, in the Bell X-1 rocket plane, Chuck Yeager became the first man to fly faster than sound in spite of broken ribs suffered in a fall from a horse a few days before (which he didn't tell the doctor). The X-1, shaped like a bullet, was powered by four liquid rockets and was carried to high altitude strapped under a B-29.

Further experiments with rocket airplanes led to the X-15 which flew 199 flights up to six times the speed of sound. Neil Armstrong, later to become the first man to walk on the Moon, was an X-15 test pilot.

Sputnik: Dawn of the Space Age

On October 4, 1957, the United States was shocked when the Soviet Union launched the first satellite into orbit. Sputnik I

Figure 1.1 A research rocket plane is dropped from a B-29 as a jet fighter chase plane follows. *Courtesy of NASA; artist William F. Phillips.*

was a small sphere carrying a radio transmitter which "beeped" a signal for all the world to hear of the supremacy of Russian technology. A month later, on November 3, Sputnik II was orbited carrying the space-faring dog, Laika.

In November 1957, President Eisenhower went on television and showed the first man-made object that had survived the heat of reentry and was recovered from orbit. The solution to the reentry problem was one of the main keys to successful manned spaceflight as well as successful ballistic missiles. Eisenhower emphasized that we were preparing to launch a satellite but that our major effort focussed on the development of ballistic missiles, not on space. In 1958 the president established a civilian space program and created NASA, the National Aeronautics and Space Administration, to oversee it.

American incentive was spurred and, after several spectacular failures of the Vanguard project, Explorer 1 was successfully launched on January 31, 1958. It was a 30-pound (14 kg) steel cylinder containing two detectors for micrometeoroids and one for high energy particle radiation. Their information was transmitted to the ground until May 23. Although small and simple, Explorer 1 made a major discovery, the Van Allen radiation belts around Earth.

Man in Space: A Russian

The Soviets also beat the Americans in putting a man into space. Yuri Gagarin orbited Earth in a Russian spacecraft on April 12, 1961. Soviet preeminence in these space spectaculars can be attributed to their very large rockets which had been built to carry their heavy hydrogen bombs. In the United States a breakthrough in nuclear weapons technology by Dr. Edward Teller made possible the construction of relatively lightweight warheads. Thus, American ICBM booster rockets were not large enough to carry heavy loads into orbit.

Mercury: Man in a Can

The first U.S. manned spaceflight program was called Mercury. Flights of chimpanzees preceded manned flights; Ham was the first passenger in the Mercury capsule on January 31, 1961. The first astronauts were all military test pilots, some of them from the rocket aircraft research at Edwards AFB. Alan Shepard became the first American in space when he flew a 15 minute 22 second suborbital flight on May 5, 1961 (Figure 1.2). He reached an altitude of 116.5 miles and a maximum speed of 5180 miles per hour (8330 km/h). He landed in the Atlantic Ocean, 302 miles from his starting point at Cape Canaveral, Florida.

Figure 1.2 Mercury capsule, Freedom 7, sits atop the Redstone booster rocket to carry Alan Shepard on the first American flight into space. The towerlike structure at the top is equipped with small rockets which would pull the capsule away from the booster in case of an emergency. The rocket tower was jettisoned if everything was operating properly. *Courtesy of NASA*.

Twenty days later President Kennedy stated the national goal of placing a man on the Moon and returning him safely to Earth before the end of the decade. Nearly a year went by before the first American orbited Earth. John Glenn made a three orbit flight on February 20, 1962.

The Mercury capsule, shown in Figure 1.3, was crowded with only one person inside. The astronaut barely had room to turn his head and move his arms. It measured only 6 feet 10 inches (2.1 m) long and 6 feet 2-1/2 inches (1.9 m) in diameter at its base. An escape tower was attached to the top to pull the capsule off the booster rocket in case of an emergency. It was jettisoned if not needed. Two different rockets carried the Mercury capsule: Redstone, a modified Army missile, and Atlas, an Air Force ballistic missile.

A retro-rocket motor slowed the vehicle to take it out of orbit and return it to Earth. Attached to the blunt end of the capsule was a heat shield which absorbed the 3000°F (1600°C) heat generated by atmospheric friction during reentry. The capsule landed in the ocean, its descent slowed by parachutes, and was recovered by ships waiting at the site. Each capsule was used only once. Six successful manned flights were made; the final one flew 22 orbits in more than 34 hours on May 15 and 16, 1963.

Gemini: Twins in Orbit

Named for the constellation of stars which includes the twins Castor and Pollux, the Gemini spacecraft, Figure 1.4, was a two man capsule with a volume about the size of the front part of a compact car. It measured 19 feet long and 10 feet in diameter and had the same basic design as the Mercury capsule. Equipment and expendables that were not needed for the return to Earth were put outside and left behind, providing more room in the cabin. As with Mercury, the capsule was designed for one-time use and landed at sea. Although development of a capability for landing the spacecraft on land was begun in the United States, it was never completed. The Soviets, on the other hand, recovered their manned spacecraft on land right from the beginning.

Titan, a modified Air Force ballistic missile, carried Gemini spacecraft to orbit. Ten flights were made in 1965 and 1966. In those cramped quarters, one of the flights spent just a few hours short of two weeks in orbit, a world record at that time. Gemini astronauts learned how to maneuver, change orbit, and rendezvous and dock with other spacecraft. Gemini VI launched on December 15, 1965, and accomplished the first space rendezvous with Gemini VII which had been launched on December 4, seen in Figure 1.5. They flew together for over 5 hours at distances of 1 foot to 295 feet (0.3 m to 90 m).

The first space walk was on June 5, 1965, when Edward White opened the hatch and floated into space, attached to Gemini IV by a 23-foot (7-m) tether line.

Apollo: To the Moon

Apollo was the project that took men to the Moon and back. To meet President Kennedy's challenge, dozens of technological breakthroughs had to be made. The spacecraft carried

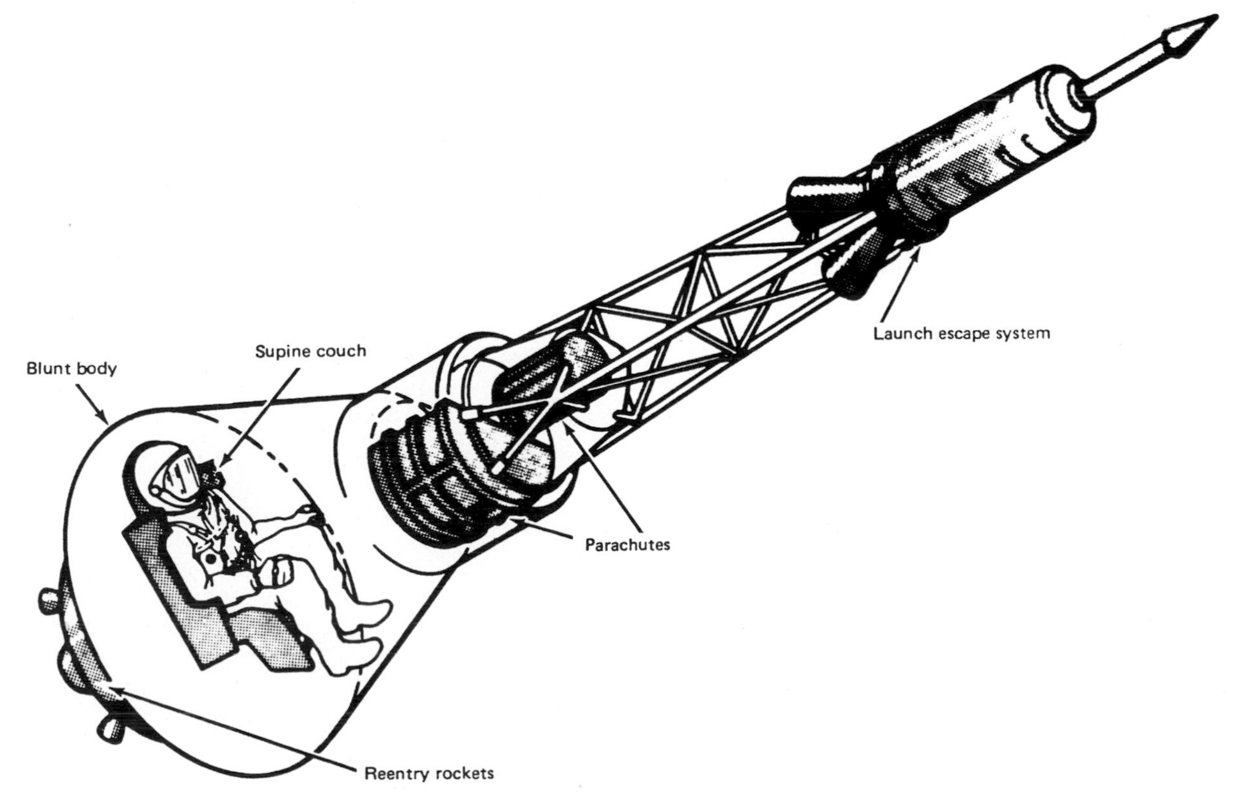

Figure 1.3 Diagram of Mercury capsule. *Courtesy of NASA.*

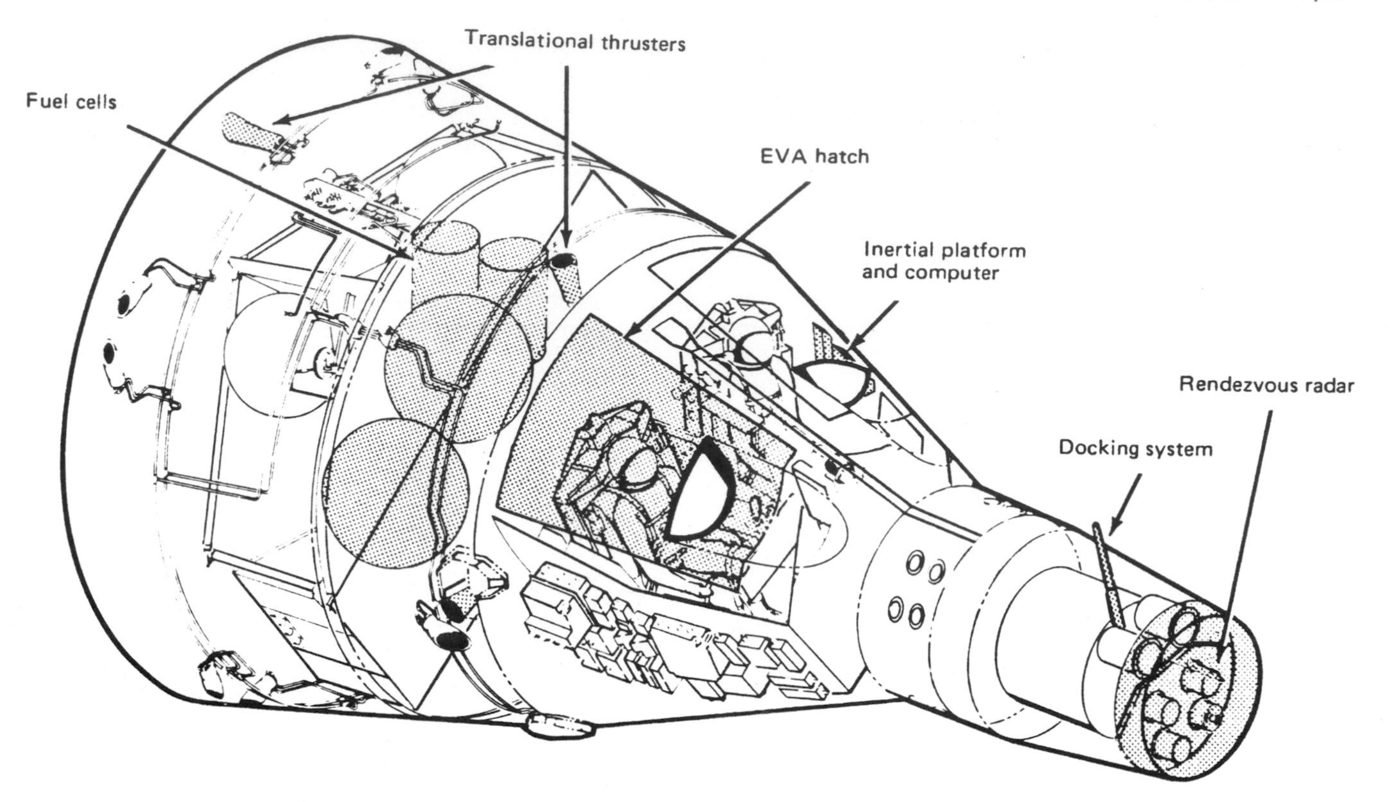

Figure 1.4 The Gemini spacecraft had a crew of two. Fuel cells generated electricity to power the craft. Astronauts could exit the vehicle on a "space walk" through the EVA hatch. Thrusters allowed the spacecraft to maneuver, rendezvous, and dock with other spacecraft. *Courtesy of NASA.*

three men in a volume about the size of the inside of a minivan. It contained a shirtsleeve environment; that is, the astronauts could take off their space suits and move around a bit. They also had hot water aboard to prepare meals.

The huge Saturn V rocket which launched the Apollo spacecraft on its trajectory to the Moon is shown in Figure 1.6 with the previous manned space vehicles drawn to the same scale. Saturn V stood 363 feet (111 m) high and weighed 6.5 million pounds (3.0 million kg). It burned liquid oxygen and kerosene, 15 tons (14 metric tonnes) per second at liftoff. The second and third stages used liquid hydrogen and liquid oxygen propellants.

The Apollo spacecraft itself is shown in Figure 1.7 compared to Gemini. The cone-shaped command module in the center carried the astronauts. It was 10 feet 7 inches (3.23 m) high and 12 feet 10 inches (3.91m) in diameter. The cylindrical service module at the top carried electrical equipment, oxygen tanks, and the rocket engine for leaving lunar orbit to return to Earth. The spider-like 23-foot (7.0-m) tall lunar module (LM) at the bottom was the vehicle that descended from lunar orbit carrying two of the three astronauts to the surface of the Moon (Figure 1.8). Because it didn't have to operate in the atmosphere (the Moon has none) it was not necessary to streamline it; it was actually the first true manned space vehicle. When the activity on the surface was finished, the lunar module rejoined the command and service

Figure 1.5 The Gemini VII spacecraft as seen through the window of Gemini VI during the first space rendezvous. *Courtesy of NASA.*

Drawings indicate relative sizes of launch vehicles.

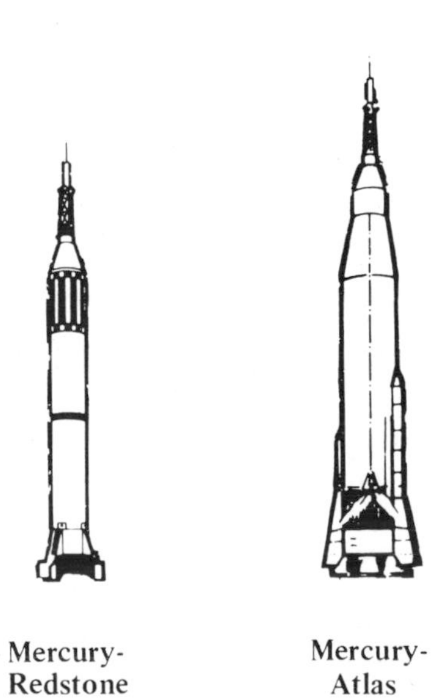

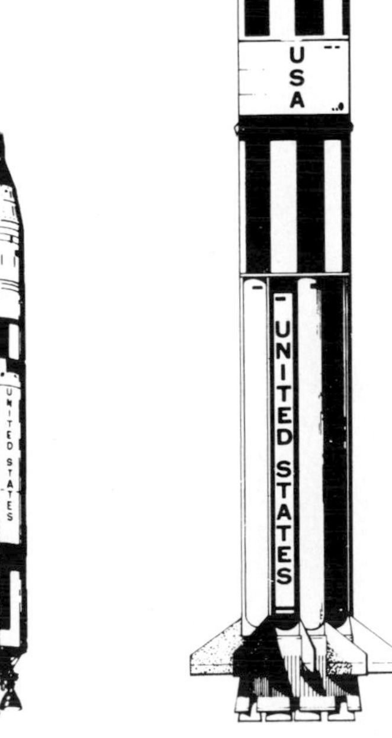

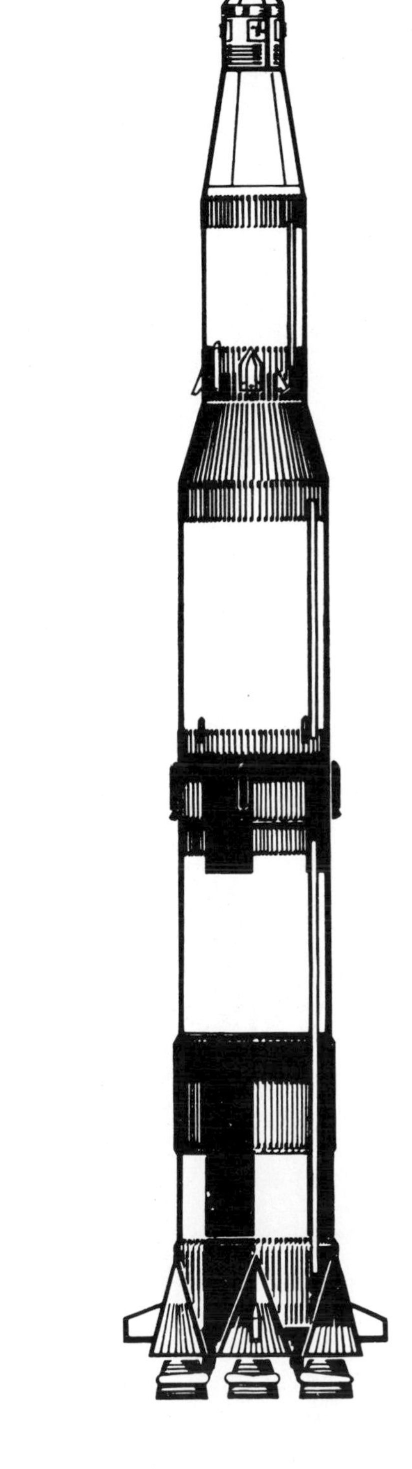

Figure 1.6 Comparison of manned launch vehicles. *Courtesy of NASA.*

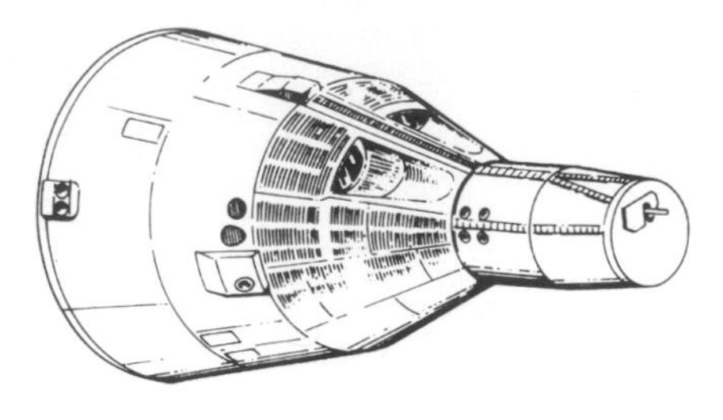

Gemini

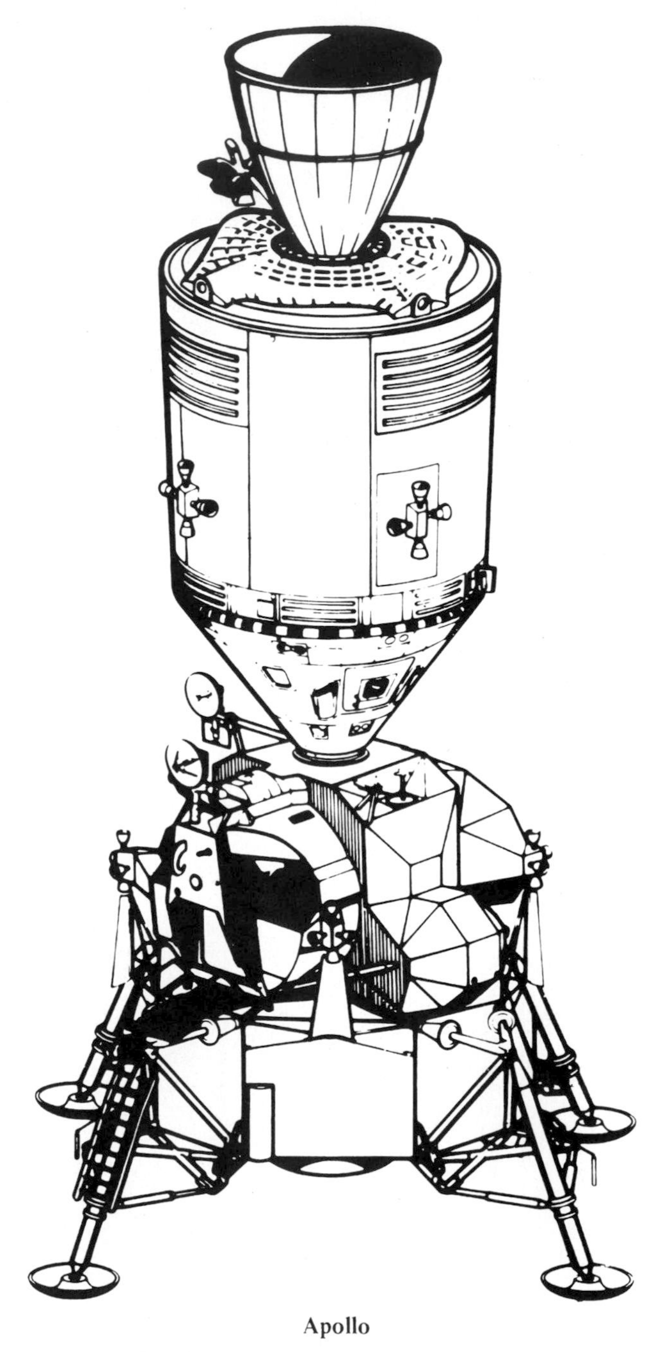

Apollo

Drawings indicate relative sizes of spacecraft.

Figure 1.7 Apollo and Gemini spacecraft drawn to the same scale. *Courtesy of NASA.*

modules, still in lunar orbit, for the return trip home. Only the command module with its three main crew landed back on Earth. As with the Gemini and Mercury, Apollo landed in the ocean (Figure 1.9). Nothing, not even the command module, was reused.

For safety, lunar landings had to be made on smooth, level terrain, often uninteresting geologically. An electric powered car, the lunar rover shown in Figures 1.8, 1.10, and 1.11., was carried to the Moon on the last three flights so the astronauts had transportation while on the surface. The rover allowed them to drive up to 6 miles (10 km) to hills, cliffs, and craters to take photos and bring back samples. Each of the four wheels had its own electric motor so that if one or two failed, the rover could still operate. Power was supplied by silver-zinc batteries. The astronauts covered a total of almost 60 miles (100 km) on the three flights. The rovers were left behind along with just about everything else that was taken to the surface.

Disaster struck during the early days of the Apollo program. Three astronauts lost their lives in a launch pad fire on January 27, 1967, during preliminary checkout of the Apollo. During an actual flight, the command module contained an atmosphere of pure oxygen at a pressure of 5 pounds per square inch (34 kPa), about one-third normal atmospheric pressure. At that pressure things burn as they would in a normal oxygen-nitrogen atmosphere. But during the checkout the command module was pressurized to near normal sea level pressure with pure oxygen. Under these conditions materials burn explosively. A 2-1/2 month investigation resulted in a design for a fireproof spacecraft. Other changes were made, delaying the program about a year and a half, but the goal set by President Kennedy was met. Two lunar landings were made before the decade of the 60s was up.

Apollo 11 landed on the Moon July 20, 1969. Neil Armstrong's words "one small step for man, one giant leap for mankind" are part of history. A plaque was left behind, engraved with the words "Here men from the planet Earth first set foot upon the moon July 1969 A.D. We came in peace for all mankind."

Apollo 13 nearly ended in disaster when on April 13, 1970, the third day out, an explosion in one of the oxygen bottles blew a hole in the side of the service module and started a leak in a second oxygen bottle. Without oxygen, the fuel cells could not generate electricity and the command module was disabled. The three-man crew squeezed into the two-man LM which became their lifeboat. There was no way to turn back; they had to continue on around the Moon, without landing, and back to Earth. The LM was designed to support two men for 2 days; but it would take about 6 days to get back to Earth. The temperature in the LM dropped to 38°F (3°C). Most of the food was dehydrated and required hot water to rehydrate; but no water was available. By careful minimal consumption of food, water, oxygen, and electricity, and by using the LM rockets for course adjustment, they managed to limp home. The three men lost a total of 31.5

Figure 1.8 An Apollo landing site. The lunar module is at the center and a lunar rover at the right. Because there is no air on the Moon, the flagstaff was fitted with a horizontal bar to keep the flag "flying." *Courtesy of NASA.*

pounds (14.3 kg), but did not lose their lives. In that respect, the mission was a success; in spite of a serious problem the astronauts, with advice from the ground, used their limited equipment and supplies with great ingenuity and creativity to make it back.

Apollo was not only an exciting and spectacular adventure, it was also a scientific, technological, and political success. From 1969 to 1972 three manned flights were made around the Moon (Apollo 8, 10, and 13) and six landings were made (Apollo 11, 12, 14, 15, 16, 17). The locations are

Figure 1.9 Three parachutes gently lower an Apollo spacecraft into the sea. *Courtesy of NASA.*

Figure 1.10 An astronaut and his rover on the moon. *Courtesy of NASA.*

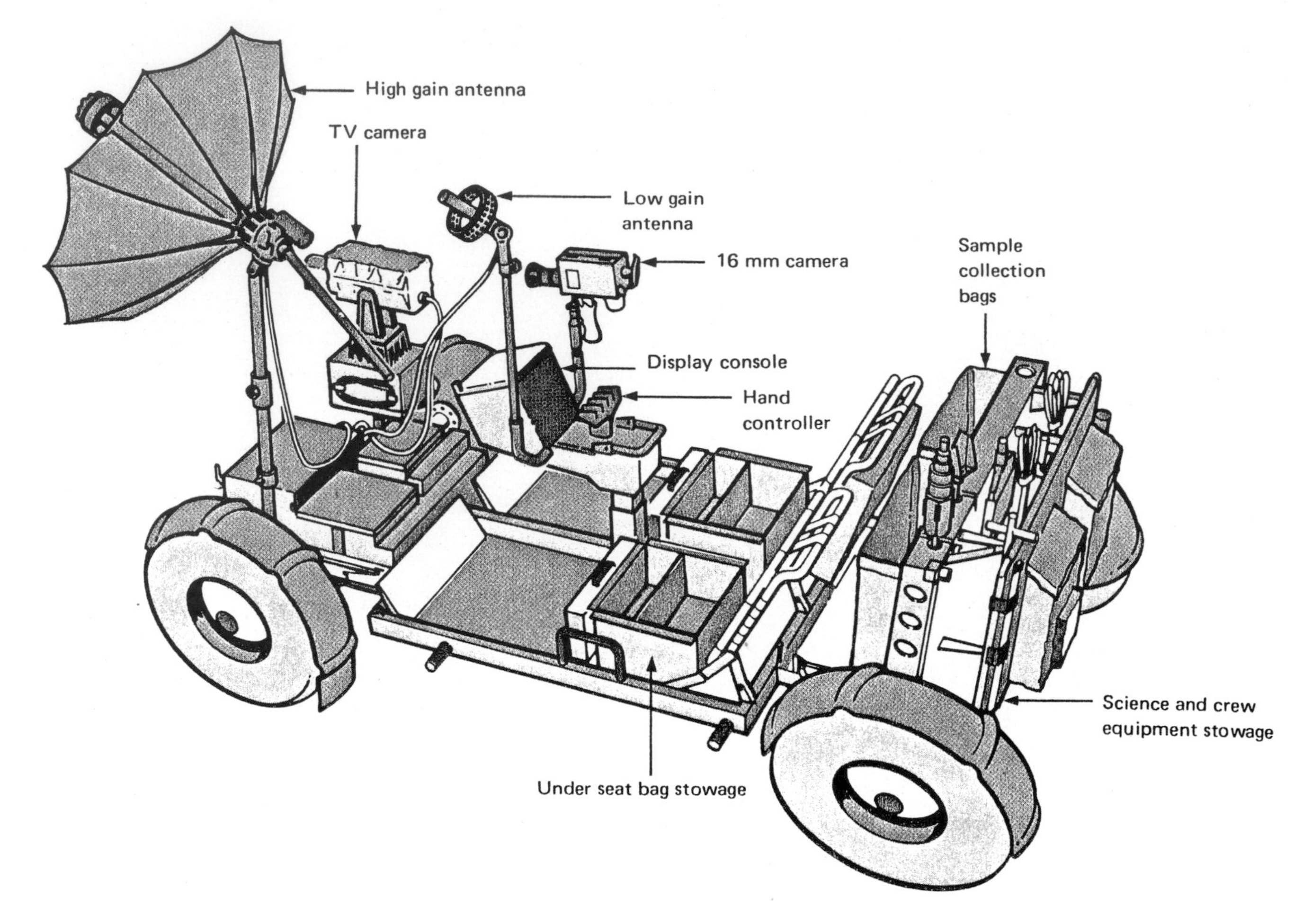

Figure 1.11 Diagram of the lunar rover. *Courtesy of NASA.*

shown in Figure 1.12. Ten landings were originally planned but three were deleted because of NASA budget cuts. Twelve American astronauts left footprints in the lunar soil during 166 man-hours of surface exploration. About 850 pounds (390 kg) of soil and rocks were brought to Earth and nearly 100 experiments were carried out in orbit and on the surface.

As a result of the Apollo lunar exploration scientists have a pretty good understanding of the history of the Moon from its formation at the same time Earth formed, through a period of severe impacts by asteroids and meteoroids, to the present quiet. They also know the composition of lunar rock and soil so when a colony is established on the Moon we know what kinds of raw materials are available for construction and mining. This is discussed in Chapter 12.

Equally important were knowledge and technology gained from the research and development needed to make Apollo succeed: building huge reliable rockets, developing life support systems, refining orbital mechanics, miniaturizing electronic devices, and advancing computer technology. This kept America preeminent in space for the following decade.

The Soviet Union planned manned flights to the Moon also, but because of a number of booster failures, never succeeded in putting a man on the Moon. The first stage booster consisted of 30 rocket engines all burning simultaneously. Apparently the problem was in maintaining reliability and stability while keeping the thrust of the 30 engines balanced. By comparison, the Saturn V first stage had a cluster of five engines.

The Soviets did have a very comprehensive and successful unmanned lunar exploration consisting of 24 flights including the first soft landing on the Moon; the first flight around the Moon with photos of the back side which never faces the Earth; two Lunokhod roving vehicles which landed on the Moon, drove about for more than a year examining the soil and sending back 100,000 photographs of the lunar surface; and several flights which returned samples of the lunar soil to Earth.

Soyuz-Salyut: Soviet Space Stations

Meanwhile, the Soviets concentrated on manned flights in low Earth orbit. The Soyuz spacecraft was the workhorse of their manned space activities. The Soyuz spacecraft consists of three segments which can be seen in Figure 1.13. On one end a module carries equipment and instruments and two so-

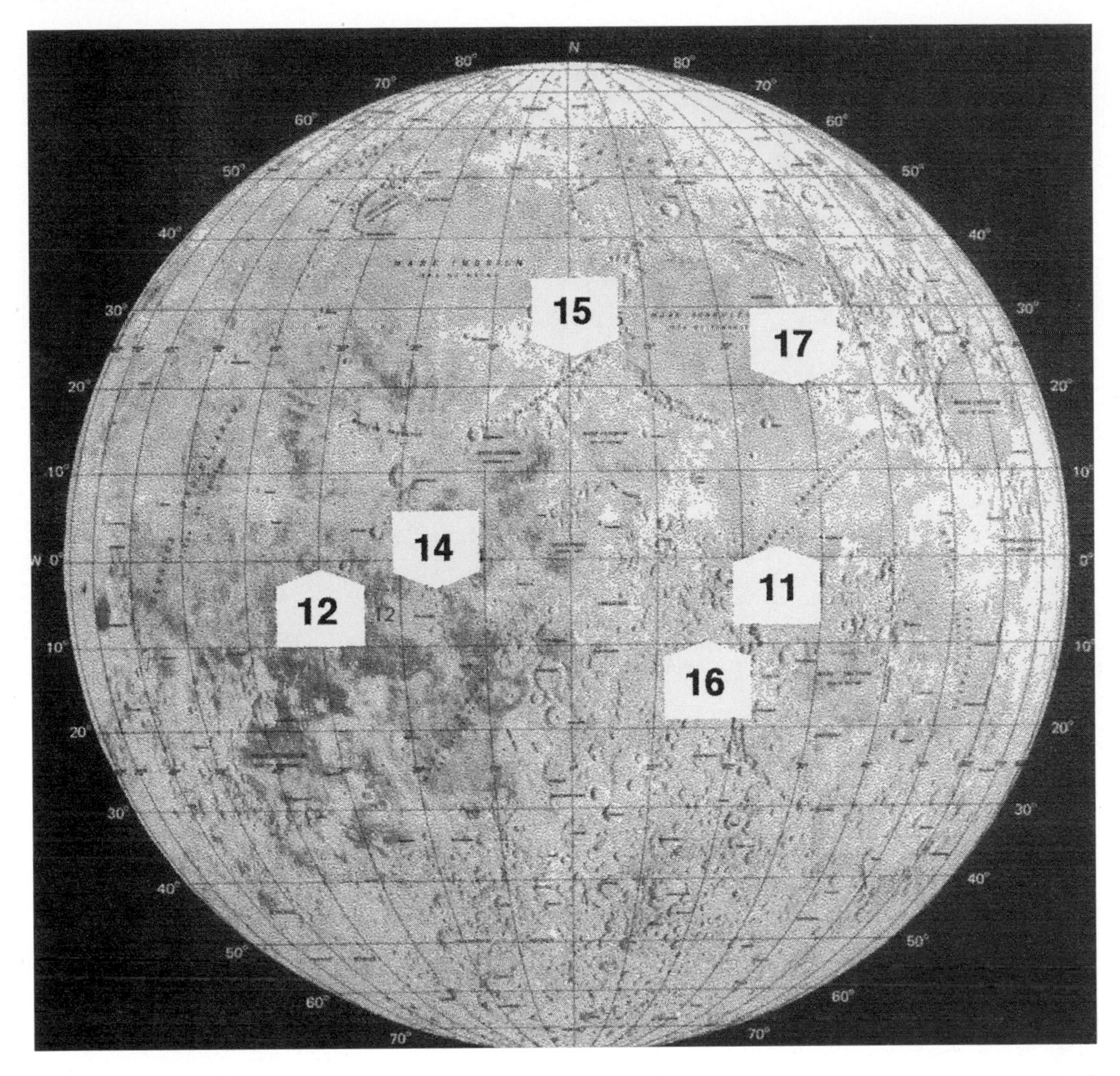

Figure 1.12 The six Apollo landing sites. The Soviet Union put two unmanned spacecraft on the Moon, Luna 16 and 20, which picked up samples of lunar soil and returned to Earth. *Courtesy of NASA.*

lar panels extend from its opposite sides. On the other end is a spherical-shaped living and laboratory module. The dome-shaped segment in the center is the reentry vehicle which carries the crew back to Earth. The crew consists of up to three cosmonauts. While Mercury, Gemini, and Apollo spacecraft landed in the ocean, Soyuz returns to dry land. Small retro-rockets fire just before touchdown to bring the craft to a gentle landing. At least 60 Soyuz flights were made since 1965.

The seven Salyut space stations were launched beginning in 1971. At first, they were placed in low orbits which decayed in just a few months. Later models remained in orbit for years. Salyut's main section was about 40 feet (12 m) long by 14 feet (4.2 m) in diameter and weighed just over 20 tons (18,000 kg). It enclosed a laboratory and living section, an equipment-propulsion-control section, and an airlock equipped for the Soyuz vehicles to dock. Soyuz acted as a shuttle craft to bring crews to and from Salyut.

Figure 1.13 Artist's concept of the hookup of Apollo command and service modules (left) with Soviet Soyuz spacecraft on the right. *Courtesy of NASA.*

Figure 1.14 The Skylab space station, photographed by the crew as they departed for Earth in their Apollo capsule. Skylab was damaged during launch and had to be repaired on orbit by the astronauts. Solar cells cover the X-shaped structure and the panels to the lower right; the left solar cell array is missing. A sunshield was improvised over the top of the workshop where the micrometeoroid shield was torn off. *Courtesy of NASA.*

Skylab: U.S. Space Station

Built from the empty third stage of a Saturn V rocket, the 100 ton Skylab was about the size of a small three bedroom house. See Figure 1.14. It was launched on May 14, 1973, into a 270-mile orbit (434 km). In 1973 and 1974 Skylab was manned by three crews of three men. Crews and supplies were ferried to the space station using Apollo vehicles. The crew would rendezvous and dock with Skylab and enter the space station; the Apollo command module and service module would remain attached until it was time to go home. The last Skylab mission still holds the record for the longest American spaceflight: 84 days. The crew included Edward Gibson, William Pogue, and Gerald Carr.

Much research in solar physics, space physics, earth science, and human biology was done. It gave the United States its best experience in long duration space flight, especially the effects of weightlessness on the human body. Medical research was given top priority. Many interesting problems arose: How do you weigh a weightless person? Answer: the period of oscillation of a seat attached to springs is proportional to the person's weight. How do you eat, sleep, and take a shower? And the most asked question, "How do you go to the bathroom in space?" the title of a fascinating book by William Pogue. These and other questions will be discussed in later chapters.

Apollo-Soyuz

The Apollo program ended in July 1975 with history's first international spaceflight when an Apollo spacecraft linked up with a Russian two-man Soyuz spacecraft. The Apollo capsule was equipped with a docking device which allowed it to connect to the Soyuz. An artist's concept of the hookup is shown in Figure 1.13. For nearly 2 days the astronauts and cosmonauts moved between the two vehicles, did joint experiments, and displayed good fellowship. Although 28 scientific experiments were conducted, this event was more of a political achievement than a scientific or technological breakthrough.

Unmanned Spacecraft

Meanwhile, satellites were invading the space surrounding the Earth. By way of definition, a *spacecraft* is a self-con-

Figure 1.15 An artist's concept of a spacecraft passing Saturn with the Sun and other planets in the background. *Courtesy of NASA.*

tained vehicle designed for spaceflight. A spacecraft is called a *satellite* when it is in a closed orbit around some celestial body such as the Earth or a planet.

The first spacecraft were research satellites which studied the space environment. Instrumented satellites have measured the content of near-Earth space, electrons, protons, atmospheric gases, the solar wind, radiation from the Sun, aurora, the magnetic field, and micrometeorites. Dozens of communications satellites were launched to relay telephone and television signals from one side of the Earth to the other. Other orbiting satellites looked down at the Earth, measured its size and shape, took its picture in visible light and invisible infrared, observed and reported weather, and watched for rocket launches and signs of military activity. A series of satellites prepared the way for a global navigation system.

Spacecraft have been sent to all the planets of our Solar System except Pluto. See Figure 1.15. In 1962, Mariner 2 was the first to fly close to another planet, Venus. Mariner 4 flew by Mars in 1965 and two Viking spacecraft landed on the red planet in 1976. The Voyager spacecraft sent back spectacular pictures of the giant gas planets, Jupiter, Saturn, Uranus, and Neptune, from 1979 to 1989. Mariner 10 flew by Venus in 1974 and by Mercury three times in 1974 and 1975. Magellan, in orbit around Venus, made radar maps of the surface through the dense clouds that blanket that planet. Others are still on their way to their targets. Galileo is on its way to Jupiter for a closer examination of the largest planet and its moons, Cassini will be a return mission to the Saturnian system, and Ulysses is a mission to study the polar regions of the Sun. A flyby of Pluto is planned for early in the 21st century.

Astronomical instruments have also been sent into orbit enabling astronomers to view the sky unhindered by the Earth's atmosphere.

We will examine the specifics of unmanned satellites and spacecraft in the following chapters.

Space Shuttle

The Space Shuttle, Figure 1.16, is a reusable spacecraft that takes off like a rocket, flies in orbit like a spaceship, and returns to Earth like an airplane. The emphasis is on reuse in order to reduce the cost of transportation into orbit. The Shuttle is the mainstay of the U.S. manned space program. We will discuss it in detail in following chapters. The Soviets built what appears to be a duplicate of the Space Shuttle orbiter which was flown on suborbital flights only.

Figure 1.16 Space Shuttle lifts off. This and the Russian Soyuz are the two vehicles capable of carrying people into space. *Courtesy of NASA.*

DISCUSSION QUESTIONS

1. What do you suppose were some of the technical problems with the Apollo-Soyuz rendezvous in space?
2. Why couldn't an automobile gasoline engine have been used on the lunar rover?
3. What would have been some of the considerations in choosing the landing sites on the Moon?
4. Why did the returning manned U.S. spacecraft land in the oceans?
5. Do you think that exploration of space should continue, or should the government concentrate its resources on other more earthly problems?

ADDITIONAL READING

Bainbridge, William S. *The Spaceflight Revolution.* Krieger Publishing Co., 1983.

Baker, D. *The History of Manned Space Flight.* Crown Publishers, 1982.

Benford, Timothy B., and Brian Wilkes. *The Space Program Quiz and Fact Book.* Harper and Row, 1985.

Bilstein, Roger E. *Stages to Saturn.* NASA SP-4206, 1996. A technological history of the Apollo/Saturn launch vehicles. From the NASA History Series.

Caprara, Giovanni. *The Complete Encyclopedia of Space Satellites.* Portland House, Crown Publishers, 1986.

Every military and civil satellite of the world from 1957 to 1986.

Cortright, Edgar M., editor. *Apollo Expeditions to the Moon,* NASA SP-350, Government Printing Office, 1975.

Dawson, Virginia. *Engines and Innovation.* NASA SP-4306, 1991. Lewis Laboratory and American propulsion technology. From the NASA History Series.

Dunne, James A., and Eric Burgess. *The Voyage of Mariner 10.* NASA SP-424, Government Printing Office, 1978. Complete well-illustrated story of the Mariner 10 mission to Venus and Mercury.

Fimmel, Richard O., et al. *Pioneer Venus.* NASA SP-461, Government Printing Office, 1983. Complete well-illustrated story of the "assault" on Venus in December 1978 by ten U.S. and Soviet spacecraft.

Hart, Douglas. *The Encyclopedia of Soviet Spacecraft.* Exeter Books, 1987. Describes all Soviet space programs from Sputnik I to 1987. Well illustrated.

Kerrod, Robin. *The Illustrated History of Man in Space.* Mallard Press, 1989. Picture book.

Logsdon, John M., and Alain Dupas. "Was the Race to the Moon Real?" *Scientific American,* June 1994.

Miller, Ron. *The Dream Machines: An illustrated history of the spaceship in art, science and literature.* Krieger Publishing Co., 1993.

NASA. The First 25 Years, 1958–1983. Government Printing Office, 1983.

Pogue, William R. *How Do You Go to the Bathroom in Space?* Tom Doherty Associates, 1985. What it's like to live in space, written by a Skylab astronaut.

Taylor, G. Jeffrey. "The Scientific Legacy of Apollo." *Scientific American,* July 1994.

Wilson, Andrew, ed. *Interavia Space Directory, 94–95.* Jane's Information Group, Inc., 1994.

Yenne, Bill. *The Pictorial History of NASA.* Gallery Books, 1989.

NOTES

Chapter 2

Propulsion

Watching a space vehicle lift off the ground and head for space is one of the most thrilling spectacles you can imagine. The engines roar and vibrate, shaking the ground beneath you; even 3 1/2 miles away, a lifting-off Space Shuttle shakes your shirt, lungs, and pant legs. Exhaust trails across the sky until it disappears in the distance, followed by silence and you know it is somewhere up there.

When it comes to space travel, the rocket is the means of getting there. At least, at the present time it is. Therefore, we first study the principles of rocketry, especially the chemical rocket engine. In the future there may be more "exotic" propulsion systems in use; we will look at them as well.

Basic Principles of Rocketry

How does a rocket ship get off the ground? You have probably heard the terms *thrust, force,* and *acceleration.* These words have precise physical meaning but are often misunderstood and misused. To understand them, you need to understand a little physics. In this matter, we will follow Albert Einstein's advice, "Everything should be made as simple as possible, but no simpler." That is, we will explain the laws of motion as clearly as possible without oversimplification.

Mass and Weight

First, we must distinguish between mass and weight. The *mass* of an object is related to the quantity of matter that goes to make up the object, the number of protons, electrons and neutrons, plus the energy that binds them together. An object's mass is the same wherever it is in the universe. You have seen pictures of astronauts floating around the cabin of their spaceship. It is said that they are "weightless." But they are not "massless." They still have the same mass, the same number of particles of matter, that they had when they were on Earth.

The *weight* of an object, on the other hand, depends on gravity. Weight is the force that the object exerts downward as a result of the gravitational attraction between its mass and Earth's mass. The farther you are from the center of the Earth, the less is the force of gravity. You would weigh less on top of Pikes Peak than you would in Houston, but not very much less.

A metric note: In the metric system, the unit of mass is the *kilogram* and the unit of weight or force is the *newton.* In the English system, the pound is the unit of both weight or force and mass. This can be confusing. When it is necessary to differentiate between the two in technical writing, the terms *pound-force* and *pound-mass* are used. Actually it is not correct to say that you weigh 154 pounds (70 kg). The two are equated in everyday use because a mass of 70 kilograms weighs 154 pounds *on Earth.* It would not be true on the Moon or other planets where gravity is different. There you would still have 70 kilograms of mass, but your weight would be something else. Therefore, because weight is a force, we will use newtons for weight when we want to be precisely correct.

If you weigh 150 pounds (667 N) in Houston, you would weigh 149.8 pounds (666 N) on Pikes Peak. At 100 miles (160 km) altitude you would weigh 114 pounds. The decrease in weight is about 5 percent. MATHBOX 2.1 shows how these calculations are made.

If gravity were zero, then weight would be zero. But gravitation is a natural and universal force of attraction between *any* two masses. It does decrease with distance, but even in deep space, far from Earth, there is a force present due to the gravity of the planets, the Sun and the stars. Gravity is never zero—it cannot be turned off.

Scales are devices which measure the force due to gravity. When you stand on a scale it tells your weight, that is, how much force you are exerting toward the center of the Earth. If it reads 150 pounds (667 N), then you are pushing on the Earth (through the scale) with a force of 150 pounds (667 N). You can feel that force on your feet as you stand on the scale. If you were on the Moon your weight force would be only about one-sixth of what it is on the Earth, that is 25 pounds instead of 150 pounds (111 N instead of 667 N). A scale would show your reduced weight. The gravitational force be-

MATHBOX 2.1

Change of Weight with Altitude

The weight of an object decreases as it moves farther from Earth. Because the force of gravity, which determines weight, decreases with the square of the distance from the center of Earth, the formula is

$$W = \frac{R^2}{(R + r)^2} w$$

where R is the radius of Earth (3960 miles; 6370 km), r is the altitude of the object above the surface of Earth, w is the weight of the object at the surface and W is the weight of the object at altitude r.

Example: If an astronaut weighs 120 pounds on Earth, how much does she weigh at an altitude of 100 miles?

$$W = \frac{(3960)^2}{(3960 + 100)^2} 120 = 114 \text{ pounds}$$

In a metric example, a 55 kg astronaut at an altitude of 160 km would weigh 52 kg. Check the arithmetic yourself.

tween you and the Moon is smaller because the Moon is so much smaller and less massive than the Earth. So weight, unlike mass, depends on where in the universe you are.

Another metric note: Metric scales are commonly marked in kilograms. If your metric scale says you "weigh" 66 kilograms here on Earth and you take it to the Moon it will read 11 kilograms. Of course, that is incorrect. Kilograms are units of mass and your mass is the same on the Moon as it was on Earth. Metric scales are calibrated for use only on Earth.

Incidentally, the type of scale you use can further confuse the issue. Figure 2.1 shows that the ball weighs 4 ounces. On this type of scale, weights are put into the right pan until the scale balances. The two downward forces must then be the same, that is, the ball must weigh the same as the weights. Question: Where was the picture taken? It could have been taken on Earth, but when we take our experiment to the Moon, the picture would be the same. Both objects weigh only one-sixth of their Earth weight, but both still have the same mass. This type of scale actually compares the masses of the two objects. If they have the same mass, the scale is in balance.

Acceleration

Acceleration describes a *change* of motion of an object. When accelerated, a motionless object will begin to move; an object already in motion will change its speed or its direction or both.

You feel the effect in a car when you press the accelerator to increase your speed. You are pushed back against the seat while your speed is changing, that is, while you are accelerating. When you once again reach a steady speed, when your acceleration is zero, you no longer feel the push against the back of the seat.

When you accelerate vertically you get a sensation that your weight changes. An example is when you begin to descend in an elevator. As the elevator accelerates downward, that is, as it starts moving from a stop, you feel momentarily lighter. Indeed, if you were standing on a scale, it would in-

Figure 2.1 A pan balance compares masses.

dicate that you weigh less. Similarly, as the elevator accelerates upward, you feel momentarily heavier, and a scale would register an increase in weight. Notice that your weight seems different to you only while the elevator is changing speed, that is, accelerating, either upward or downward. As long as the elevator's speed is constant, you feel nothing unusual. Of course, it isn't a change in gravity that makes you feel that way; the force of gravity inside an elevator is no different than the force of gravity outside the elevator. It is the brief acceleration that gives the momentary sensation of a change in weight.

Forces

A *force* is simply defined as a push or a pull. When a force is applied to an object, the object will accelerate unless the force is cancelled out by other forces. If Jim gives you a shove on your right arm, you move to the left. If Mike shoves on your left arm, you move to the right. If both push at the same time with equal but opposite forces, you do not move. (You may wish they would quit, but in the interest of science you sit there and contemplate the laws of physics.) The sum of all forces acting on an object is called the *net force*. Two equal forces acting in opposite directions cancel each other out. They add to a net force of zero.

Sitting on your chair, you exert a force downward, your weight. Why do you not accelerate downward if you are exerting a force downward? There must be some balancing force such that the net force is zero. Fortunately, the interconnections between the molecules in a solid substance form a somewhat flexible lattice network. The lattice that makes up the chair "stretches," similar to the way a spring stretches, and it exerts an upward force on your bottom-side equal to your downward weight force due to gravity. The net force is zero and you remain still. Similarly, when you stand on the floor, the floor exerts an upward force to balance the downward force of gravity. If you push on a wall, the wall pushes back, and so on.

But suppose you sat down on a shoe box. The molecular lattice of the shoe box probably could not exert an upward force equal to your weight without coming apart. The downward weight force exceeds the upward force from the box; the net force is not zero so you accelerate down to the floor. If the net force is not zero, then the forces are said to be unbalanced.

Figure 2.2 illustrates the principle. The molecular lattice of the yardstick stretches, applying an upward force to the weight, trying to hold together, until the downward weight exceeds what the molecules can hold and the yardstick breaks.

Isaac Newton

All of this was summarized by Isaac Newton (1642–1727) in a book titled *Philosophiae Naturalis Principia Mathematica,* usually shortened to just *Principia,* published on July 5, 1687.

Newton's first law of motion says that an object at rest will remain at rest unless acted on by an unbalanced force. If the net force is zero, the object does not move. If the net force is not zero, the object will accelerate in the direction of the net force. A second part to this law states that an object in motion will continue in straight line motion at constant speed unless it is acted on by an unbalanced force. Newton's first law is sometimes called the law of inertia because it points out that a force must be applied to change the motion of a body, that is, to accelerate it.

Newton's second law of motion tells us how to calculate the acceleration of an object: simply divide the net unbalanced force by the mass of the object. His *third law* explains the principle of rocket engines which we will soon encounter.

(And may the net force be with you!)

Freefall

Now suppose that you are back in the elevator. The floor of the elevator is pushing up against the bottoms of your feet with a force equal to your weight. The net force is zero and everything is OK. But what happens if the cable breaks? The floor of the elevator falls away under you and no longer exerts an upward force on your feet. Gravity is accelerating you and the elevator downward at the same rate; you are both in free fall. If you could manage to put a scale under your feet, it would register zero because it too is in free fall accelerating downward at the same rate. The gravitational force of the Earth still pulls on you, but because you, the scale, and the elevator are all in free fall together, you are weightless.

Bouncing on a trampoline, jumping from a diving board, and sky diving all give similar sensations. Because you have

Figure 2.2 The molecules of wood in a yardstick apply an upward force to balance the downward weight force of the object. If the weight exceeds the force that holds the lattice together, then it comes apart.

no support, no upward force to balance your downward weight force, you are in free fall.

This is precisely the situation with the astronauts orbiting the Earth in their space ship. The spacecraft, astronauts and all, are in free fall toward the Earth. Therefore they have the feeling of weightlessness. More about this in Chapter 3 when we discuss orbits.

Liftoff!

Consider the Space Shuttle in Figure 2.3 sitting on the launch platform attached to its external tank and booster rockets. Fully loaded and ready to go it weighs about 4.5 million pounds (20 million newtons). That is, it is exerting a 4.5 million pound (20 million newton) force downward on the launch platform. The launch platform, a structure of concrete and steel, is able to support the spacecraft; it can exert an upward force of 4.5 million pounds (20 million newtons) without breaking apart. The forces are in balance, the net force is zero, and the spacecraft just sits there. Now, how do you make it accelerate, preferably upward?

Think for a moment about a more manageable problem. The book in Figure 2.4a is lying on the table and you want to

Figure 2.3 The Space Shuttle poised on the launch platform. The structure of the platform can support the downward weight force of the Shuttle. *Courtesy of NASA.*

a

b

c

Figure 2.4 Forces in balance (a and b) and unbalanced (c).

put it up on a shelf. The book weighs 1 pound; gravity, acting on the mass of the book, is exerting a 1 pound force downward on the table. The table is exerting a 1 pound force upward on the book so the net force on the book is zero and it just sits there. Now pick it up in your hand (Figure 2.4b) and hold it stationary. Like the table, you are exerting a one pound upward force to balance the downward weight force. It does not move because the forces are balanced and the net force is zero.

If you tighten your muscles and exert an upward force greater than the 1 pound downward weight force, the net force on the book is no longer zero. The net unbalanced upward force causes the book to accelerate upward (Figure 2.4c).

So it is with the Shuttle on the launch pad. All we need to start it accelerating upward is to apply an upward force greater than its weight. Where do you find a force greater than 4.5 million pounds? Rocket engines! As the rocket engines fire and the Shuttle lifts off, the launch pad no longer provides support. The engines must provide the necessary 4.5 million pounds (20 million newtons) of upward force, and then some. To accelerate upward, there must be an unbalanced force, that is, a net upward force, greater than the weight of the shuttle.

When discussing the propulsion capabilities of rocket engines, we use the word *thrust* rather than *force*. A rocket engine rated at 100,000 pounds thrust, therefore, can, at most, lift 100,000 pounds off the ground. This does not mean, however, that the engine could put 100,000 pounds into orbit. In addition to lifting the spacecraft, the rocket engine must lift itself and its fuel. As fuel is burned and exhausted out the nozzle, the vehicle becomes lighter and requires less force to keep it accelerating.

The interaction of these and other variables becomes a complex mathematical problem. However, we can make a simple first estimate of the acceleration of the Space Shuttle as it lifts off the launch pad. *Newton's second law of motion* states that to calculate the acceleration of an object we divide the net force by the mass of the object. MATHBOX 2.2 shows the calculations.

The conclusion from this calculation is that the Shuttle accelerates at about 12 miles per hour per second (about 19 km/h/sec); that is, after 1 second it is going 12 miles per hour (19 km/h), after 2 seconds its velocity is 24 miles per hour (38 km/h), after 5 seconds it will be going 60 miles per hour (95 km/h), and so on. At the end of 5 minutes, 300 seconds, it will be moving at a speed of 3,600 miles per hour (5700 km/h). To go into orbit, the spacecraft must be travelling at more than 17,000 miles per hour (27,000 km/h).

In violation of Einstein's advice, we oversimplified this calculation. During the flight, none of the numbers that we used remain constant. They all change as the Shuttle heads for orbit. Its mass decreases as fuel is burned and thrown overboard. Parts of the vehicle detach and drop away, further decreasing the mass. The thrust of the rocket engines is made to vary during the flight, as we will see. As the vehicle pitches over to horizontal its thrust no longer is directed directly against gravity. Also, we did not include atmospheric frictional drag. Air drag increases as the vehicle increases speed but then decreases as it reaches higher altitudes where the air is less dense. In addition, the force of gravity decreases as the spacecraft moves away from the Earth. These factors are significant and must be taken into account in a "real-world" calculation.

Using computers, engineers make a step-by-step, second-by-second calculation, taking into account the changing mass, forces, and gravity. The result of their calculation is a predicted flight profile showing where the Shuttle should be, how fast it should be going, and in what direction it should be heading. During a flight, then, the actual position and speed are compared to the predicted values to see if everything is running OK.

Reaction Engines

A rocket is a *reaction engine*. It is basically a very simple device, consisting of a chamber in which propellants (solid, liquid, or gas) are accelerated to high speed and expelled through the nozzle at the open end. In reaction to the motion of the propellant, the engine is accelerated in the opposite direction.

A toy balloon is a simple example of this principle. When you blow up the balloon, the air inside is contained under pressure. When you release it, the air is expelled from the nozzle and the balloon speeds off in the opposite direction. In Figure 2.5a the balloon is tied shut. The air inside pushes outward while the stretch of the balloon rubber pushes in. The force in each direction is balanced by a force in the opposite direction. A is balanced by B, C is balanced by D, etc. The net force is zero and the balloon does not accelerate.

However, in Figure 2.5b, the nozzle of the balloon is open so air escapes. The force at C is no longer balanced by an equal and opposite force at D; therefore the balloon accelerates in the direction of the arrow at C.

The motion of balloons and rockets also was explained by Newton. His *third law of motion* states that for every action there is an equal and opposite reaction. The acceleration of the propellants out the engine in one direction causes an acceleration of the engine in the opposite direction.

See Figure 2.6. Because there is an opening in the rear, the forces of the gases in the combustion chamber forward and rearward are unbalanced. Forces on the sides are balanced by forces on the opposite sides. But in the fore-aft direction there is a net force forward. The thrust of the rocket is, then, the unbalanced force on the forward wall of the combustion chamber which accelerates the vehicle forward. The thrust of any reaction engine is determined primarily by the amount of propellant expelled out the nozzle and how fast the propellent is

MATHBOX 2.2

Newton's Second Law of Motion Applied

The Space Shuttle sitting fully loaded on the launch pad weighs 4,511,200 pounds (mass of 2,050,500 kg). It has three main engines and two solid rocket boosters. Rated at 393,800 pounds of thrust each (1,751,600 N), the three main engines apply a total force of 1,181,400 pounds (5,254,800 N). The two solid boosters provide 2,900,000 pounds (12,900,000 N) of thrust each. Total thrust from the five engines at takeoff, then, is 6,981,400 pounds (31,053,300 N)!

Newton's second law states that the acceleration of an object is equal to the net force acting on it divided by its mass. Mathematically,

$$a = \frac{\Sigma F}{m}$$

where the symbol, Σ, read "sum of," means the sum of all the forces acting, that is, the net force.

To solve this in the English system, we must first find the mass of the Shuttle on the launch pad. In the English system, the mass of an object is found by dividing its weight by g, the acceleration of gravity, which is 32 feet per second per second at the surface of the Earth. For our present problem,

$$\text{mass} = \frac{\text{weight}}{g} = \frac{4{,}511{,}200 \text{ lb}}{32.2 \text{ ft/s}^2} = 140{,}100 \text{ pounds-mass}$$

Pounds-mass is a unit of mass in the English system of measurement. Now, apply Newton's second law. In the case of the Shuttle, the downward force is its weight, 4.5112 million pounds and the upward force is the total thrust of the rockets, about 6.9814 million pounds. The net force is, then,

$$6{,}981{,}400 - 4{,}511{,}200 = 2{,}470{,}200 \text{ pounds upward}$$

See Figure 2.2.1.

$$a = \frac{\Sigma F}{m} = \frac{2{,}470{,}200 \text{ pounds}}{140{,}100 \text{ pounds-mass}} = 17.6 \text{ feet/s}^2 = 12 \text{ mi/hr/s}$$

This means that for each second that passes at liftoff, the Shuttle increases its speed by 12

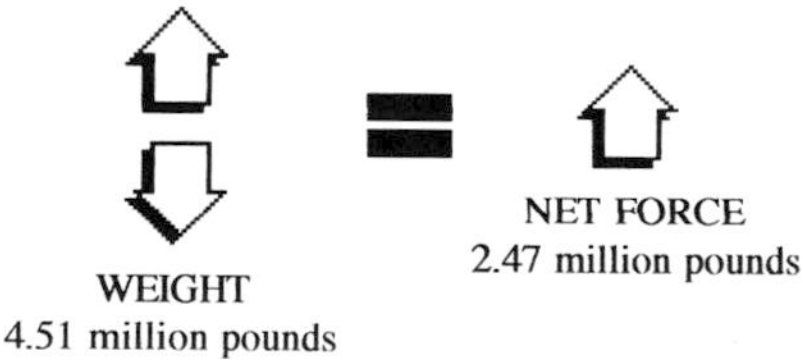

Figure 2.2.1.

miles per hour or 17.6 feet per second.

To do this calculation in metric units, we must first find the weight force of the vehicle on the launch pad in newtons. We do that by multiplying the mass in kilograms by the acceleration of gravity in metric units, 9.8 meters per second per second.

$$w = mg = 2{,}050{,}500 \text{ kg} \times 9.8 \text{ m/s}^2 = 20{,}094{,}900 \text{ N}$$

$$\text{Net force} = 31{,}053{,}300 \text{ N} - 20{,}094{,}900 \text{ N} = 10{,}958{,}400 \text{ N}$$

$$a = \frac{\Sigma F}{m} = \frac{10{,}958{,}400 \text{ N}}{2{,}050{,}500 \text{ kg}} = 5.3 \text{ m/s}^2$$

This is the same answer we got using English units. (Check it yourself.)

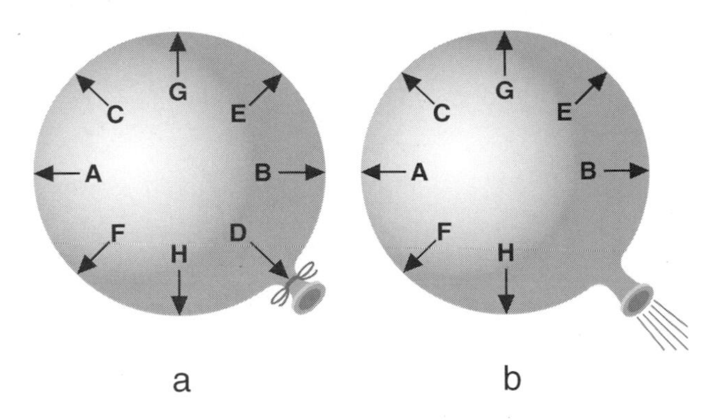

Figure 2.5 Forces in and out of balance in a balloon.

moving. The product of mass times speed is called *momentum*. We will encounter this quantity again later. To increase thrust, simply throw more propellant out the nozzle at higher speed.

A jet engine is also a reaction engine. Air is taken into the front and compressed; fuel is burned so the hot gases expand and are exhausted out the rear (Figure 2.7). The action of the exhaust jet of hot gas produces a reaction which drives the engine forward (along with whatever is attached to it). The exhaust, as it moves through to the rear, also drives the turbine that drives the compressor at the front.

A rocket engine is a reaction engine also, but it does not intake air. It carries its own oxygen supply in addition to the fuel. Therefore, it can operate in space where there is no air.

Chemical Rockets

Combustion is a chemical process in which a fuel oxidizes, that is, combines with oxygen. For example, wood burns in air as in a campfire. An automobile's internal combustion engine burns (oxidizes) gasoline and a jet engine burns jet fuel. In these two vehicles the oxygen supply comes from the surrounding air. A chemical rocket engine carries both fuel and oxidizer and does not depend on outside air to support the combustion.

An upper stage engine of the Saturn V Moon rocket is shown in Figure 2.8. In the combustion chamber of a rocket engine, Figure 2.9a, the confined gases from the burning fuel build up tremendous *pressure,* exerting force against the walls of the chamber. Pressure is the force divided by the area

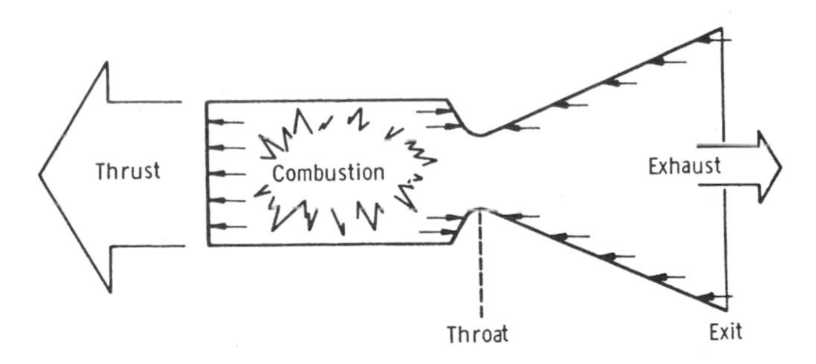

Figure 2.6 Rocket engine, a reaction engine. *Courtesy of NASA.*

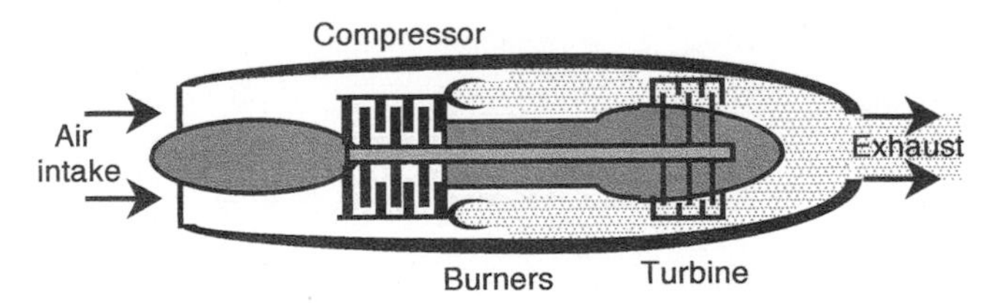

Figure 2.7 Cross section of a turbojet engine.

over which the force is applied. The pressure is highest in the combustion chamber, maintained by the high temperature of the burning propellants in the chamber. On the way out, the exhaust gases pass through the throat, a constriction or reduced area which increases their speed. Once through the throat, the gases expand rapidly in the nozzle, accelerating further and reducing their pressure. The sideways expansion of the gases applies pressure to the nozzle walls producing the thrust. Figure 2.9a is a diagram of the commonly used bell-shaped nozzle.

The direction of the thrust is opposite the direction of the moving exhaust. Therefore the direction the rocket moves can be changed by changing the direction of the exhaust. In

Figure 2.8 One of the second stage engines of the Saturn V Moon rocket.

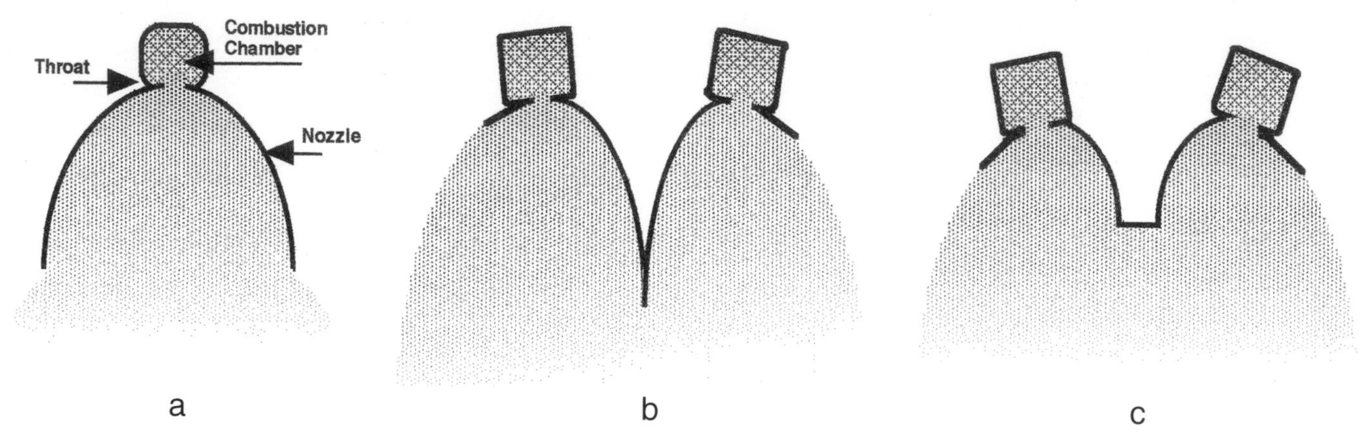

Figure 2.9 a. A bell nozzle. b. An aerospike nozzle. c. A truncated aerospike.

most rockets the engine is mounted on a device called a gimbal which allows the nozzle to be swiveled to one side or another, changing the direction of the exhaust flow. This process is called gimballing.

For the engine to be most efficient, the pressure in the exhaust gases must decrease to the pressure of the atmosphere outside by the time it reaches the exit end of the nozzle. At that point the velocity of the gases would then be maximum and the engine is most efficient. If the outside air pressure were greater than the pressure of the exhaust gas, it would inhibit the flow of gases outward. If the exhaust pressure were greater, some of the energy of the hot gas would be wasted, producing no useful thrust. In the vacuum of space, of course, the outside pressure is zero. The curvature of a bell-shaped nozzle determines what the pressure will be at the exit end, so a nozzle designed for a booster rocket operating in the atmosphere is shaped different from one designed to operate in space.

A different kind of nozzle, called a *plug nozzle* or an *aerospike* is sketched in Figure 2.9. Imagine a bell nozzle cut in half, top to bottom, and the two halves switched places and reconnected (Figure 2.9b). The reconnection point becomes a spiked-shaped protrusion in the exhaust flow. Each of the halves has its own combustion chamber. Now visualize this in three dimensions. If the combustion chamber is shaped like a hollow ring with the burning taking place inside, then the spike takes the shape of a cone with a curved side. Looking at the end of this engine you would see a ring of flame with a plug in the center.

What is the purpose of this odd configuration? As discussed above, the most efficient engine is designed to allow complete expansion of the exhaust gases. On one side of the aerospike nozzle, the exhaust expands and applies pressure against the spike producing the thrust. On the other side the exhaust expands uninhibited so it does not retard the outflow of the gas coming through the throat.

A variation of the aerospike is shown in Figure 2.9c. The point of the spike is cut off leaving a flat base against which the exhaust applies pressure to add to the thrust. Another design called a linear aerospike has two rows of combustion chambers with the exhaust expanding against a long curved surface. Figure 2.10 is a diagram of this engine. New space planes under development, replacements for the aging fleet of Space Shuttles, will probably use this type of engine. They will be discussed in Chapter 8.

It must be pointed out emphatically that a rocket does not obtain its forward motion by pushing against the outside air. If that were the case, it would not operate in the near vacuum of space. In fact, the outside air only hinders the motion of the rocket, reducing its efficiency in two ways. First, the gases expanding out the nozzle are retarded in their expansion by the surrounding atmosphere. Equally important, the atmosphere causes a frictional drag which reduces the forward acceleration. A rocket operating in a vacuum is significantly more efficient than one operating in the atmosphere. That is one of the major reasons for launching straight up, to get out of the dense lower atmosphere as quickly as possible.

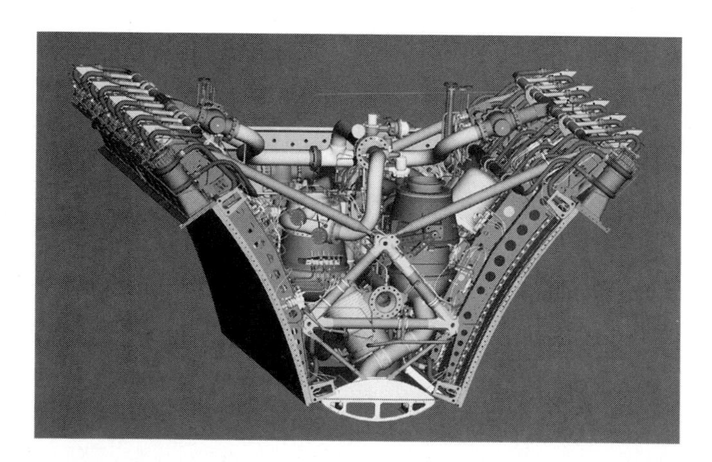

Figure 2.10 A linear aerospike engine. Two rows of five combustion chambers each are located on the front end (top of the picture). Their exhaust expands against the two curved ramps. The engine is 90 inches (2.3 m) from front to rear (top to bottom in the picture). *Courtesy of Boeing.*

To go into orbit a spacecraft must be accelerated to a speed of over 17,000 miles per hour (27,000 km/h), requiring a lot of fuel producing high thrust for a long time. In fact, the weight of fuel and oxidizer in the Shuttle at liftoff is about 3.8 million pounds (about 1.7 million kilograms of mass), 86 percent of the total weight! It is all thrown overboard to produce the thrust. The orbiter, which goes into orbit, is only 5 percent of the total. The remaining 95 percent is propellant, engines, and tanks which drop off along the way. Research and development efforts are underway to reduce or eliminate this expensive and wasteful way to put things into orbit. Later in this chapter we will discuss some of these efforts.

Specific impulse is a number which indicates the effectiveness of a rocket fuel. It may be thought of as the time that 1 pound of fuel will burn while it is producing 1 pound of thrust. The specific impulse of some typical fuel-oxidizer combinations is shown in Table 2.1. A typical solid propellant has a specific impulse of 300 seconds; that means that 1 pound of propellant will produce 1 pound of thrust for 300 seconds. (The same thing is true in metric units if you substitute newtons for pounds.) A higher number indicates a more effective thrust-producing fuel. Various fuel-oxidizer combinations will be discussed further in the following sections.

Solid Propellant Rockets

The first rockets, built by the Chinese, were propelled by a mixture of solids similar to black gunpowder: potassium nitrate, sulfur, and charcoal. These three ingredients burn together without the need for air. Oxygen for combustion is supplied by the oxidizer, potassium nitrate. (". . . ate" at the end of a chemical name generally indicates that oxygen atoms are contained in the molecule.) When combined in certain proportions and ignited in a closed space, these three ingredients will explode. The problem is to control the explosion, to slow it down so the energy is released over a longer period of time rather than a quick bang. Varying the percentage of charcoal provides a control over the speed of the combustion; increasing the charcoal content slows the burning.

Products of the combustion in the confines of the rocket engine include carbon dioxide, nitrogen, sulfur dioxide, and oxides of nitrogen. These gases, heated by the combustion process, are expelled at high speed out the rocket nozzle to provide the forward thrust. Control of thrust can be achieved by varying the fuel-oxidizer ratio. As mentioned before, increasing the proportion of charcoal in black powder slows the burning.

Look for a moment at the flight of a model rocket shown in Figure 2.11. Some model rocket engines are constructed with low charcoal content at the start of the burn for maximum thrust (see Figure 2.12). Then the charcoal is increased to yield a slow burn of constant thrust. After a delay period while the rocket coasts to maximum altitude, an explosive charge at the top of the engine ignites to blow off the nose cone and eject the parachute for descent. This can be seen in the graph of thrust versus time shown in Figure 2.12.

Modern solid propellants include potassium perchlorate (oxidizer) with asphalt (fuel), ammonium perchlorate (oxidizer) with aluminum powder (fuel), and nitrocellulose with nitroglycerine (both oxidizers and fuel). The ingredients are mixed with a binder to hold them in a desired shape, a stabilizer to keep them from decomposing, and sometimes a catalyst to speed up the reaction. The mixture is molded into a *grain* of the desired shape and size to fit into the outer casing or shell which provides structural integrity.

TABLE 2.1 Specific Impulse of Various Propellants (fuel-oxidizer) (Note: The specific impulse depends on the engine design.)

Propellant	Specific Impulse
Typical solid fuels, e.g., gunpowder	250–350 seconds
Kerosene-LOX	300–380 seconds
Liquid hydrogen-LOX	390–460 seconds
Liquid hydrogen-fluorine	400–480 seconds
Hydrogen heated by nuclear reactor	800–1000 seconds
Hydrogen gas heated by electric arc	1500–2500 seconds
Solar thermal engine	800 seconds
Ion engine	2000–10,000 seconds
Matter-antimatter	100,000 seconds
Solar sail (no propellants, no mass is expelled)	Infinite

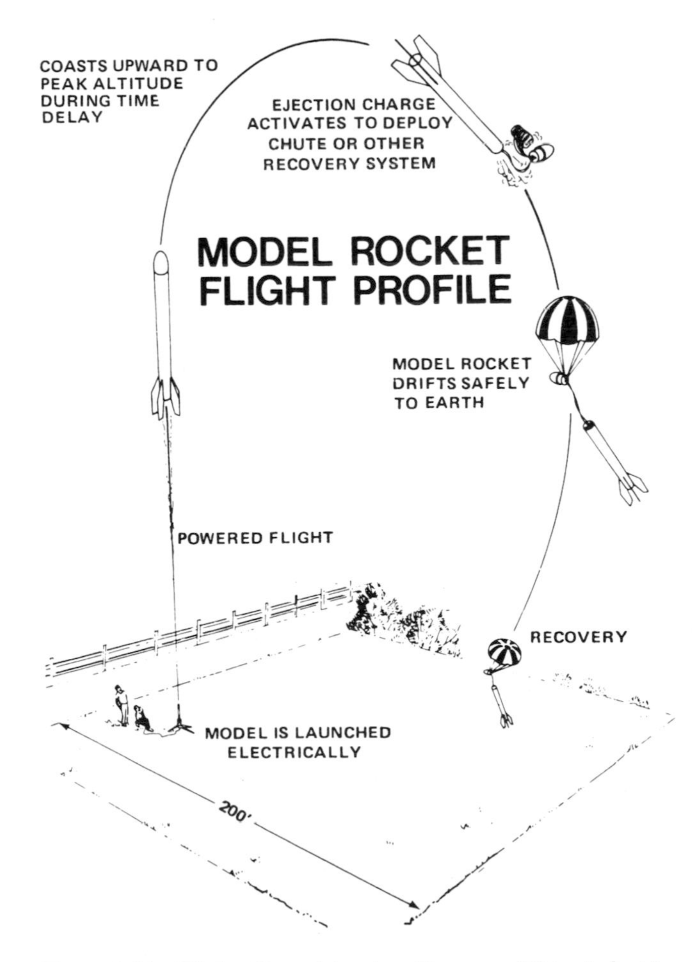

Figure 2.11 Flight of a model rocket. *Courtesy of Estes Industries.*

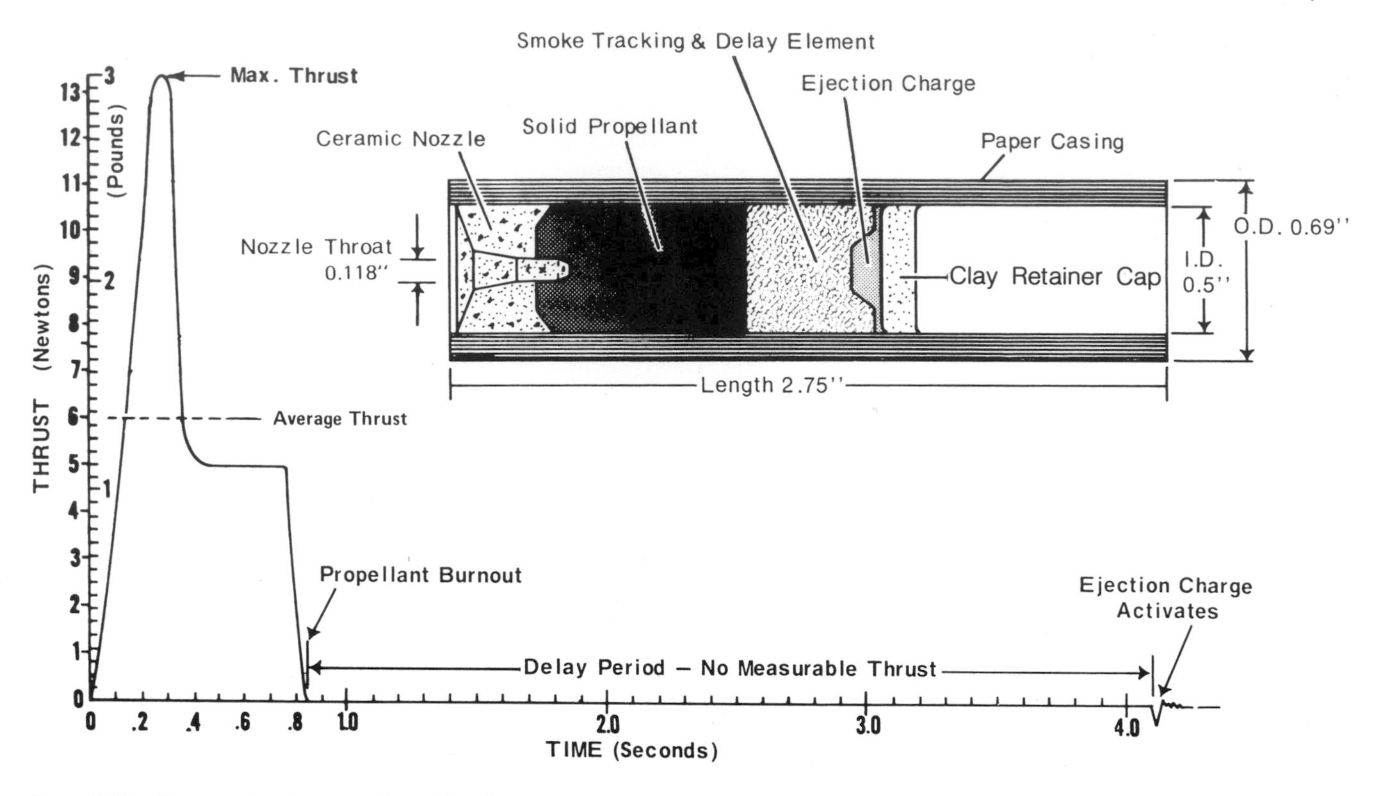

Figure 2.12 Cross section diagram of model rocket engine and its time-thrust curve. Thrust is given in both pounds and newtons (metric units of force). Note that the thrust rises quickly to maximum in about 0.3 second, and that the thrust is constant at about 1.1 pounds (5 newtons) from 0.4 to 0.8 second. The nose cone is blown off and the parachute is ejected about 4 seconds after ignition. *Courtesy of Estes Industries.*

A grain of solid propellant burns on its exposed surface and produces hot gases. Besides varying the fuel-oxidizer mixture, thrust can also be controlled by the way the grain is shaped. Thrust depends on the rate of burning, that is, how fast the hot gas is being produced and exhausted through the nozzle. That, in turn, depends on the exposed surface area. Refer to Figure 2.13. A cylindrical-shaped grain which completely fills the casing burns like a cigarette from one end to the other. Because the burning surface area remains constant throughout the burn, the thrust is constant, a *neutral burn.* A problem that must be considered when designing a rocket with a cylindrical grain is that it becomes top-heavy as the bottom burns away.

Some grains are molded with holes of various shapes and sizes running down the center, called *perforations,* in which the burning takes place. See Figure 2.14. As the burn progresses, the surface area increases and so does the thrust, a *progressive burn.* Yet another type of grain is made smaller in diameter than the outer casing and held centered in the casing by metal spiders. See Figure 2.15. In this type, the outer surface of the grain is the burning surface. Because this surface area decreases as the burn progresses, the thrust decreases with time, a *regressive burn.*

Figure 2.16 shows several grain cross sections. The end-burning cylinder has a smaller burning area than those of the same size but with perforations; therefore, its thrust is lower and constant. A star-shaped perforation has a large burning area producing a high thrust for a shorter time; perhaps surprisingly, it has a neutral burn. The multiperforated grain is progressive-regressive with a large increasing exposed area at first, then a decreasing area as the tubes merge and the burn comes to an end. Try to figure out what kind of burn is produced by the cruciform grain; consider step by step how the burning area is changing.

An *inhibitor* is coated over the surfaces of the grain that are not supposed to burn, the ends of those with perforations for example. A simple but effective inhibitor is latex paint.

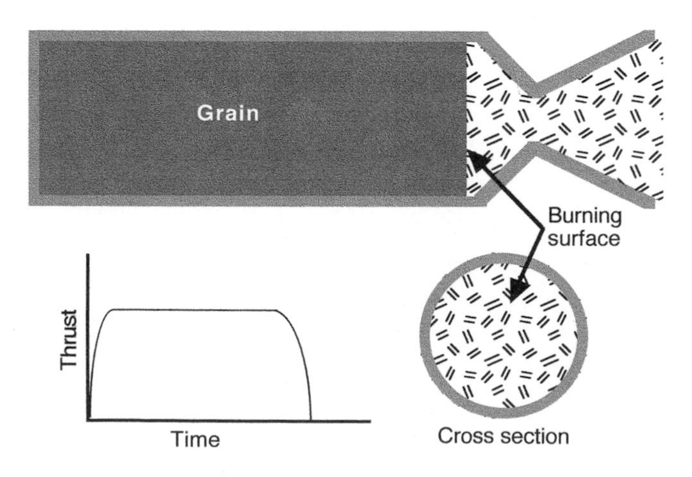

Figure 2.13 A neutral burn, cylindrical grain of solid propellant.

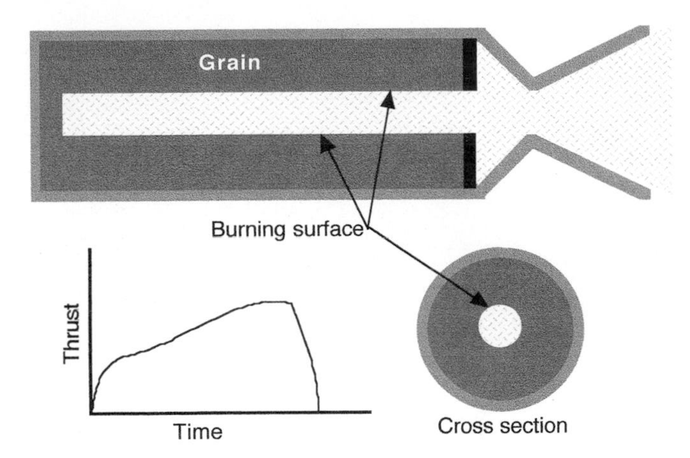

Figure 2.14 A progressive burn, bored cylinder grain of solid propellant

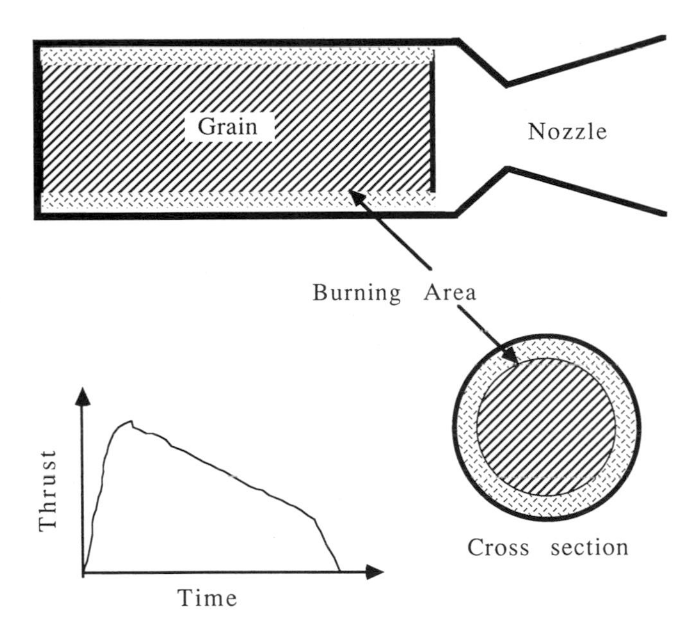

Figure 2.15 A regressive burn, supported cylinder grain.

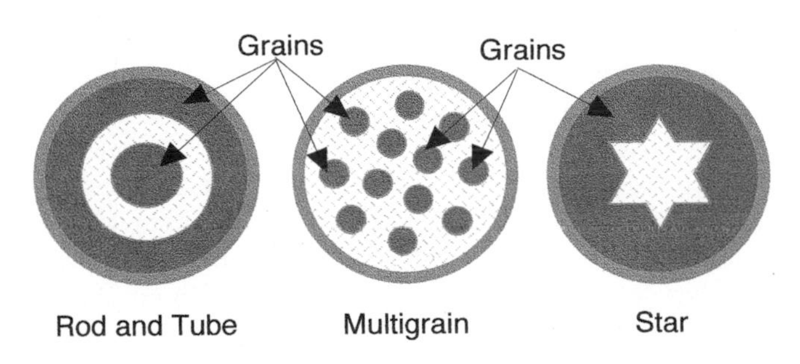

Figure 2.16 Cross sections of three types of solid propellant grains.

A major limitation of a solid rocket is that once it is ignited it cannot be turned off and restarted. In an emergency, however, thrust can be terminated by blowing off a top cap (which effectively destroys the rocket engine) and allowing the gases to escape from both ends, i.e., an unbalanced force no longer exists.

Solid Propellant Boosters

Solid propellant rockets are often used as booster rockets. They provide a desired amount of thrust for a specific amount of time and are discarded when they burn out. There are a number of advantages to solid rockets. They are simple and easy to construct, relatively low in cost, highly reliable, easy to store, and can be made ready to fire in a short time. The Space Shuttle uses two solid rocket boosters (SRBs) which can be seen in Figure 2.3, one on either side of the spacecraft. Standing 149 feet (45.5 m) tall and over 12 feet (3.7 m) in diameter, these are the largest SRBs ever built and the first to be used for manned spaceflight. They are listed in Table 2.2 along with liquid rocket engines that have been used for manned spaceflight. SRBs will be described in more detail in Chapter 8.

Very large solid propellant rockets must be constructed in several segments rather than in one piece. A one-piece grain the size of the Shuttle SRB would likely crack while hardening and curing. Furthermore, it would be nearly impossible to move such a large rocket in one piece from the manufacturer in Utah to the launch site in Florida without damage. On arriving at Kennedy Space Center, the segments making up the SRB are stacked and connected together.

TABLE 2.2 U.S. Manned Spaceflight Rocket Engines

Vehicle	Fuel	Oxidizer	Thrust
Mercury-Redstone	Alcohol	LOX	78,000 lb (350,000 N)
Mercury-Atlas	RP-1	LOX	365,000 lb (1,620,000 N)
Gemini-Titan			
First stage	Aerozine 50	Nitrogen tetroxide	430,000 lb (1,900,000 N)
Second stage	Aerozine 50	Nitrogen tetroxide	100,000 lb (450,000 N)
Saturn 1B			
First stage	RP-1	LOX	1,600,000 lb (7,100,000 N)
Second stage	Liquid hydrogen	LOX	200,000 lb (890,000 N)
Saturn V			
First stage	RP-1	LOX	7,760,000 lb (34,500,000 N)
Second stage	Liquid hydrogen	LOX	1,150,000 lb (5,120,000 N)
Third stage	Liquid hydrogen	LOX	230,000 lb (1,023,000 N)
Space Shuttle			
Main engine	Liquid hydrogen	LOX	394,000 lb (1,750,000 N)
Solid rocket booster	Aluminum powder	Ammonium perchlorate	2,900,000 lb (12,900,000 N)

The joints between segments must be tightly sealed against the enormous pressure of the combustion taking place inside the rocket to prevent hot gases from leaking through. A leak would melt the rocket casing, reduce the internal pressure and thrust of the motor, and would produce unwanted thrust in a direction opposite the leaking gas.

The top of each segment has a U-shaped groove, called a *clevis,* into which the bottom of the next segment, called a *tang,* fits. Figure 2.17 is a simplified diagram of the original Space Shuttle solid rocket booster joint. Because metal against metal does not seal well, two flexible, rubber-like seals, *O-rings,* are set into slots cut into the clevis. In addition, zinc chromate putty fills the space between the two segments. When the booster ignited, the high pressure gas inside is supposed to push against the putty to compress the air ahead of it and force the O-ring into the gap between the tang and clevis making a leakproof seal. This seal must take place in less than one second or the hot gas will blow by the O-ring.

That is exactly what caused the *Challenger* accident in January 1986. After 24 successful Shuttle flights, one booster failed catastrophically. It was a cold morning in Florida and the O-rings were cold and stiff. One did not seal properly and the hot gas leak caused the accident. The puff of smoke in

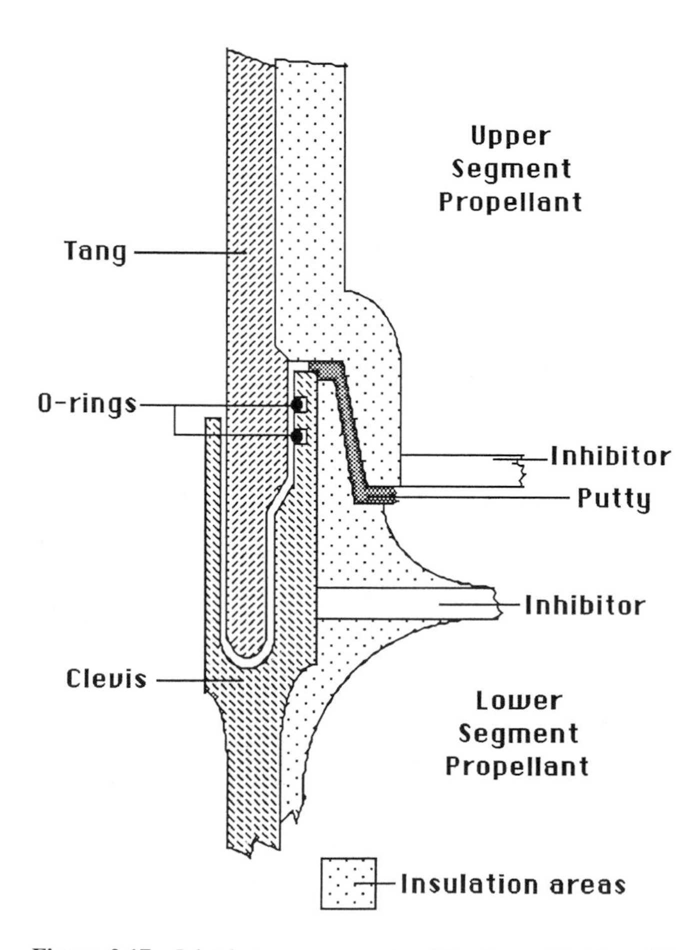

Figure 2.17 Joint between segments of the Space Shuttle solid rocket booster.

Figure 2.18 The puff of smoke when the *Challenger* boosters ignited shows where the seal failed and hot gases leaked through the joint. *Presidential Commission Report, NASA photo.*

Figure 2.18 occurred at the point where the O-ring did not seal. First the putty came through the gap, then came the hot gases from the burning propellant which caused the destruction of the external tank and the orbiter. After the accident, the joint was redesigned to incorporate a third O-ring, the zinc chromate putty was replaced by bonded insulation, and electric heaters were incorporated into each joint to keep the O-rings warm and flexible even in cold weather.

Liquid Propellant Rockets

Solid propellant rockets have been around a long time; the Chinese recorded their use before A.D. 1200. But it wasn't until March 16, 1926, in Auburn, Massachusetts, that the first successful liquid fuel rocket was flown. Engineered by Robert Goddard, the pioneer flight lasted 2.5 seconds, reached an altitude of 41 feet (12.5 m), a range of 184 feet (56 m) and a speed of 60 miles per hour (km/h). The rocket weighed less than 6 pounds (2.7 kg) and carried about 5 pounds (2.3 kg) of propellant.

The development of liquid propellant rockets was not easy. Two formidable problems had to be overcome: the pumping of extremely cold (*cryogenic*) liquid propellants into the combustion chamber and the high temperature of the burning fuel. The propellants used first by Goddard were gasoline for the fuel and liquid oxygen (LOX) as the oxidizer. An assistant ignited the rocket with a blowtorch.

Handling the gasoline was not particularly difficult. There had been considerable experience with automobile engines to draw on. It could be stored in steel tanks, easily pumped, and rubber gaskets could be used for seals. Liquid oxygen (LOX) was another matter. It boils at −297°F (−182°C) and must be kept colder than that or it boils away and returns to a gas. Such cold substances are called *cryogenic*. Metal valves and pipes contract and crack at that low temperature. Rubber freezes and shatters. Atmospheric water vapor freezes on and in everything. Lubricants solidify, oils congeal, and pumps freeze up. On the other hand, when LOX and gasoline burn, temperatures in the combustion chamber reach 5560°F (3070°C). Steel melts at that temperature. The hot gases flowing out the combustion chamber at high speed can melt through the throat of the nozzle where the heat transfer from the exhaust gas is greatest. All these engineering problems and many more took time, talent, and money to solve.

Liquid propellant rockets have been developed to a high degree of performance and reliability. A much simplified diagram of a liquid propellant rocket is shown in Figure 2.19. The fuel and oxidizer are carried in separate tanks and are pumped into the engine's combustion chamber by pumps powered by a turbine engine. Some of the fuel and oxidizer is burned in a gas generator to produce the pressure needed to operate the turbine.

Table 2.2 shows the propellant combinations used in American rockets for manned spaceflight. RP-1 is a kerosene-like fuel. Note that liquid oxygen (LOX) is most often used as the oxidizer. *Oxidation* refers to a type of chemical reaction, not necessarily involving oxygen. In that context, there are oxidizers that are chemically more energetic than oxygen, e.g., fluorine. However, fluorine is extremely corrosive, and toxic. When a reaction occurs between fluorine and hydrogen, the exhaust includes hydrofluoric acid which reacts with most common substances. In a laboratory hydrofluoric acid must be stored in wax bottles because it dissolves glass. Thus, LOX is the most widely used oxidizer because it is safer and easier to handle, and produces less hazardous exhaust products. Fluorine has been used only in experimental rockets.

Nitrogen tetroxide was used in Titan missiles because, unlike LOX, it is easily stored. Aerozine 50 and nitrogen tetroxide are *hypergolic*, that is, they react chemically on contact, like Alka Selzer in a glass of water. No igniter is needed. These features are most desirable for a military rocket which remains in "storage" for its entire life (we hope) yet must be ready to go instantly if necessary. Hypergolic propellants are also used in rocket engines and thrusters used for rotating and maneuvering while in orbit because they can be turned on and off repeatedly with high reliability.

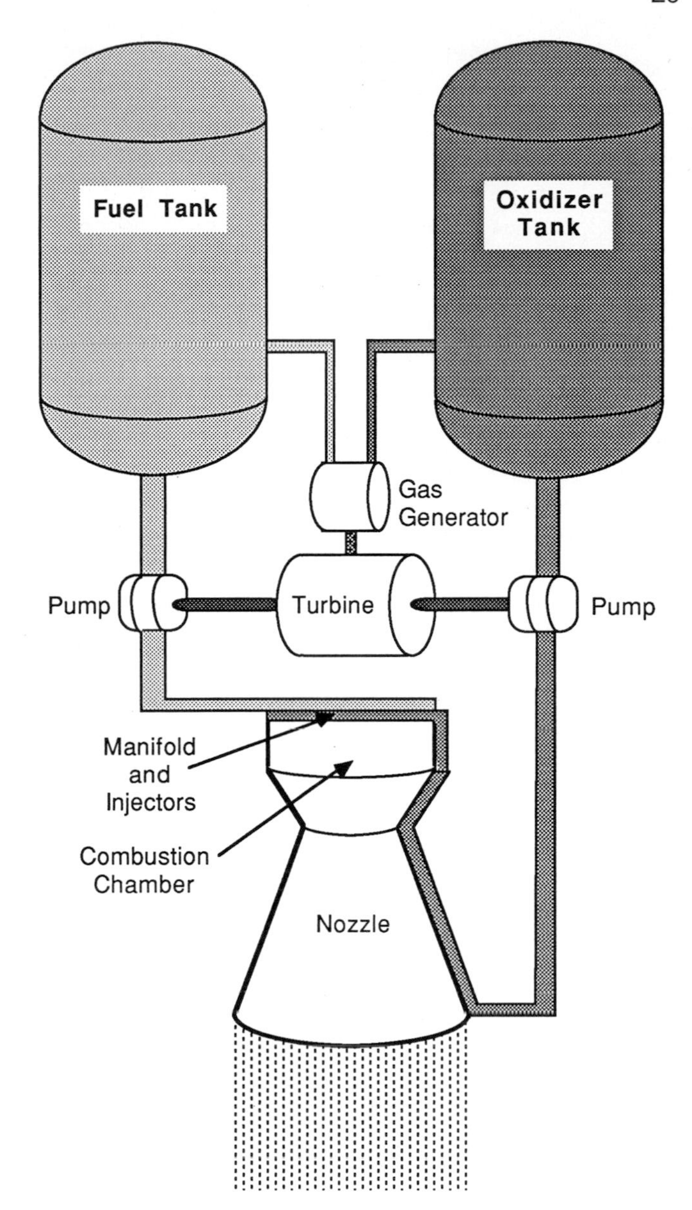

Figure 2.19 Simplified diagram of liquid propellant rocket engine.

Fuel and oxidizer enter the combustion chamber through *injectors* where they are ignited and burn. The primary purpose of the injectors is to break up the propellants into fine streams or droplets and spray them into the combustion chamber. They perform a function similar to automobile carburetor and injectors which mix fuel from the gas tank with an incoming stream of air. Injectors for rocket engines come in a variety of designs, some of them quite complex. Early injectors were nothing more than metal plates punched with fine holes at the forward end of the combustion chamber. Later models looked something like shower heads or garden hose nozzles. Figure 2.20 shows some types and arrangements of injectors.

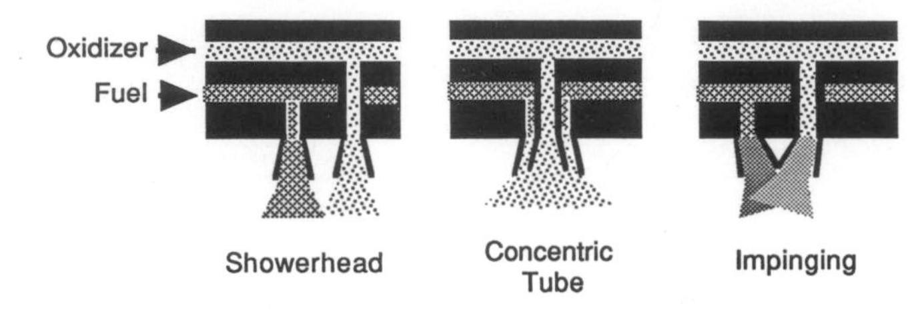

Figure 2.20 Liquid propellant rocket injectors that feed fuel and oxidizer into the combustion chamber.

It is important that the burning take place in the center of the chamber so the side walls and injectors do not overheat and melt. It is also important that the propellants be thoroughly mixed in the proper proportions. The fineness of the spray depends on the size of the injector holes and on the difference in pressure between the pipes and the combustion chamber. Smaller holes and greater pressure difference produce a finer spray.

The fuel is ignited by any of a number of devices, commonly a spark plug similar in principle to an automobile spark plug. Once burning begins, the temperature rises rapidly, fuel ignites as it reaches the burning temperature and the igniter is no longer needed to keep the combustion going. As mentioned above, hypergolic propellants do not need igniters; they react on contact.

The high pressure produced by the burning fuel in the combustion chamber opposes the propellants spraying into the chamber through the injectors. Therefore, the propellant pumps must produce an even higher pressure than the combustion produces. Designing and building reliable high-velocity, high-pressure pumps to pump liquid oxygen at near −300°F (−184°C) was not an easy task.

To keep the combustion chamber from melting, one common solution is to run the cold LOX feed lines through pipes or a cooling jacket around the combustion chamber. This arrangement can be seen in Figure 2.19. An automobile engine has a similar arrangement where a water-antifreeze mixture circulates in a jacket around the cylinders to absorb the heat, then flows through the radiator where it loses the heat to the cool air passing through. In a rocket engine the cold propellant circulates around the combustion chamber and nozzle. This accomplishes two things: it cools the engine and preheats the propellant at the same time. It is important that the fuel and oxidizer be at nearly the same temperature. For example, if cold LOX and RP-1 are brought into contact, the RP-1 would freeze solid.

Thrust of a liquid rocket engine can be controlled by varying the rate at which the propellants are pumped into the combustion chamber, thereby varying the rate at which exhaust gases are produced. A throttle works on the same principle as the accelerator pedal on your car. Also, some liquid rocket engines have been designed so they can be be turned off and back on again as often as needed, using hypergolic fuels for reliable ignition.

The Space Shuttle has three liquid rocket engines at the rear of the orbiter, visible in Figure 2.21. Propellants are fed to them from the large external tank between the solid rocket boosters. Pumps on the Space Shuttle carry 64,000 gallons of LOX and liquid hydrogen per minute at −423°F (−253°C) to the engines through 17-inch (43-cm) pipes, an engineering problem of the highest order! They are probably the most complex rocket engines ever built.

When hydrogen burns in oxygen the result is water (H2O). The cloud of exhaust coming from the three main engines in

Figure 2.21 Space Shuttle on its way to orbit. Notice the exhaust from the three main engines is barely visible. They burn liquid hydrogen and liquid oxygen and the exhaust is pure water. By comparison, the solid rocket booster exhaust is a mixture of many different chemical compounds from the combustion of the aluminum fuel, ammonium perchlorate oxidizer, and other materials used in making the grain. *Courtesy of NASA.*

Figure 2.21 is a real cloud, nothing put pure water. The Shuttle has several other small liquid propellant rocket engines and thrusters which will be discussed in Chapter 8.

Hybrid Engines

A *hybrid rocket* attempts to combine the advantages of solid and liquid propellants into one engine. Some experimental hybrid engines have been tested in an attempt to exploit the advantages of both. One design that appears to be particularly promising uses a grain of solid fuel without an oxidizer mixed in; LOX is stored in a separate tank and fed into the perforation of the fuel grain as needed for combustion. The engine can be throttled to a certain extent by varying the LOX flow, or even shut down completely by cutting off the LOX supply and restarted when desired. Hybrid engines of this type have been tested producing 10,000 pounds (44,000 N) of thrust and 250,000 pound (1.1 million newton) engines have been designed.

Another proposed hybrid engine would burn solid propellant at low temperature to generate a gas which is then further burned with liquid oxygen to produce the high energy exhaust gas. In effect, the solid propellant "smolders" like a fire that has insufficient air; the "smoke" becomes more completely burned when the oxygen is added. This engine could be throttled by controlling the flow of LOX and could be immediately shut down in case of emergency.

Unconventional Reaction Engines

Any device which throws mass overboard in one direction in order to accelerate in the opposite direction qualifies as a reaction engine. In a chemical rocket, the fuel and oxidizer serve two purposes: first, by chemically combining in the process of combustion they release heat energy which causes them to expand and, second, after burning they *are* the particles which are expelled to produce the thrust.

In other reaction engines, the propellants and the energy source may be independent of one another. The heat to expand and accelerate the propellant may be produced by some means other than combustion, or the propellants may be accelerated by some means other than heating them. This allows for a wider choice of propellants because a chemical process, combustion, is not required of them. If the energy source and the propellant are independent of each other, the efficiency of the engine can be maximized by using a heat source that produces the highest possible temperature and selecting a propellant with a low molecular weight.

Many of the engines we will discuss produce low thrust, not enough to overcome gravity and lift themselves off the ground; they must be carried to orbit by a booster rocket. However, they are designed to run for hours or days, rather than minutes like a chemical rocket. Even a small amount of thrust, producing a small acceleration, given a long running time, will gradually bring the vehicle to high speed. Thus these types of engines are most useful as upper stages for transferring to higher orbits or to escape the Earth on an interplanetary or interstellar trajectory.

Ion Engine

An ion engine is a reaction engine, but its principle of operation is quite different from that of a chemical engine. In a chemical engine the propellant is burned to increase the pressure and expel the exhaust particles at high speed. In an ion engine the particles (ions) are accelerated by electric and magnetic fields. Such engines are now being used on interplanetary spacecraft and communications satellites.

The principle is similar to that of your television picture tube. The phosphor on the inside of the face of the tube glows when electrons strike it. A stream of electrons, accelerated and steered to the proper spots on the screen by electric and magnetic fields, creates the picture in the phosphor.

An atom consists of a positively charged nucleus with negatively charged electrons in orbits around it. Normally the number of orbiting electrons equals the number of positive charges in the nucleus. Thus, an atom is electrically neutral. However, electrons can be separated from the atom by collisions with other particles, by applying an electric current, by heating a metal cathode, or by other means. With an electron missing, the atom bears a net positive charge and is called a positive ion. Positively charged particles repel one another; so do negatively charged particles. But a positively charged particle and a negatively charged particle attract each other. As with other forces, the force of attraction or repulsion causes the particles to accelerate toward or away from one another.

Charged particles also accelerate in an electric field; positive ions accelerate in one direction while negative electrons accelerate in the opposite direction. By ionizing a propellant and introducing the ions into an electric field, the propellant can be accelerated out the nozzle at very high speeds to produce thrust. To be practical, the propellant used in an ion engine must be easy to vaporize and ionize. In general, metals are easy to ionize, although many have high boiling temperatures. Possible propellants include mercury, potassium, lithium, sodium, and cesium. Of these, cesium has the lowest boiling point, 1238°F (670°C).

A simplified diagram of one type of ion engine is shown in Figure 2.22. The propellant is heated to boiling and the vapor moves to the ionizing plates composed of a heated metal such as platinum, iridium, nickel, or tungsten, among others. When a propellant atom strikes the plate, an electron is knocked off, ionizing the atom. The plates must be arranged in such a way as to assure that nearly 100 percent of the atoms collide and are ionized. The temperature of the plate must be hot enough that the atoms have enough energy to bounce off without sticking.

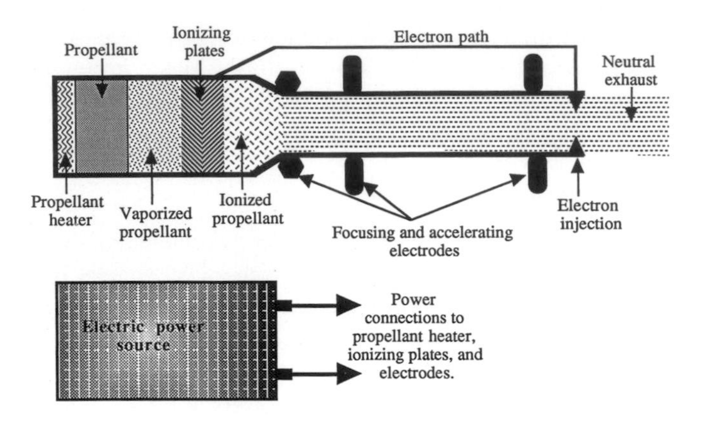

Figure 2.22 Diagram of an ion engine.

At this point, the ions are moving toward the exhaust end, but also have some random motion toward the sides. When they enter the electric field, focusing electrodes straighten out their motion and they are accelerated out the engine.

There is a major difficulty: if positive ions are continually being thrown overboard as exhaust, the electrons remaining behind will build up a strong negative charge on the engine. A negatively charged engine will attract the positively charged ions it is trying to expel! A solution to this dilemma is to feed the electrons into a wire like an electric current and insert them into the ion stream at the exit. There the electrons and ions are attracted to each other and they recombine into neutral atoms but continue to fly out the nozzle as a good propellant must. Because of this electrical charging problem, experimental ion engine chambers cannot be scaled up in size to make a larger more powerful rocket engine. Instead, just as increasing the number of cylinders in an automobile engine increases its power, a more powerful ion engine is constructed of several chambers all expelling exhaust in the same direction.

It is possible to get very high values of specific impulse in an ion engine, 2,000 to 10,000 seconds compared to 250 to 500 seconds in a chemical engine. The specific impulse of an ion engine is so great because exhaust speeds are 12 to 30 miles per second (20 to 48 km/s) and with such a high exhaust speed, it produces more thrust with less propellant mass. However, the rate at which mass is expelled from the nozzle is very low. Because of that, the thrust of an ion engine is very low, less than a pound (fraction of a newton); it must be carried to orbit on a conventional rocket and then started up. If you are in no hurry, the small thrust from an ion engine, continuously applied, will eventually accelerate the vehicle to high speed. Theoretically, a spacecraft with a small but constant acceleration can eventually approach the speed of light.

Being an electric engine, an electric power supply is required. It is the most massive part of the engine, as much as 90 percent of the engine's total mass. A nuclear electric generator or solar cells are the two most obvious choices. Exhaust velocity depends on the voltage applied to the accelerating electrodes, which may be as high as 100,000 volts. The largest power supplies available for use in space produce less than 100 kilowatts which limits the power of the engine. Power supplies in the megawatt range are needed for a manned trip to Mars. Of course, the power supply can also provide electricity to the rest of the vehicle, its sensors, transmitters, and other equipment.

Small ion engines are now being used in applications where time is not of the essence. Ion thrusters on communications satellites adjust their orientation, maintain their orbit, and keep them in the proper location in orbit. One engine was started and restarted more than 300 times over a period of 10 years, demonstrating that ion engine thrusters are very practical for long missions. Deep Space 1, an interplanetary probe, uses an ion engine with an exhaust speed of nearly 70,000 miles per hour (31 km/s). It made a close pass by an asteroid using only 1/10 the amount of fuel a conventional liquid rocket would have used to do the same job.

Nuclear Thermal Rocket

Figure 2.23 is a diagram of a nuclear thermal powered rocket. A nuclear fission reactor replaces the combustion chamber of a chemical rocket. Instead of burning fuel and oxidizer, the reactor heats the propellant to increase its temperature. This builds up the pressure in the reactor chamber and the propellant particles accelerate out the nozzle to produce thrust. There is no oxidizer and there are very few moving parts. Exhaust temperatures of 4000°F (2200°C) have been achieved in solid core reactors; theoretically, 9500°F (5200°C) is achievable in liquid core reactors and nearly 18,000°F (9700°C) in gas core reactors.

The most efficient engine uses the lowest mass exhaust particles at the highest possible temperature. Thus, hydrogen is the propellant of choice. As with a chemical rocket, the liquid hydrogen is pumped from the storage tank by turbopumps. First it is circulated around the nozzle and structural parts of the reactor; this cools the engine and prewarms the

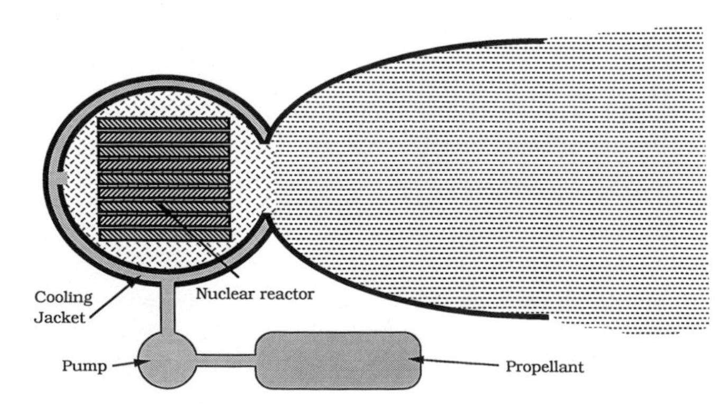

Figure 2.23 Diagram of a nuclear thermal rocket engine.

hydrogen before it enters the reactor core where it is heated further and ejected at high speed through the nozzle. In one design, about 3 percent of the hot hydrogen is bled off at the nozzle and routed to drive the turbopumps. It then enters the core and joins the main propellant flow.

Nuclear rockets were undergoing serious development from 1955 to the 1970s. Project NERVA (nuclear engine for rocket vehicle application) was ready to test fire an upper stage prototype engine on a ballistic trajectory when the project was abandoned in favor of the Space Shuttle. Antinuclear activists may have influenced the decision, also. The NERVA engine was ground tested in 1972 using uranium carbide, UC_2, imbedded in graphite as the reactor fuel with hydrogen as the propellant. Exhaust temperatures ranged up to 4000°F (2200°C). The engine weighed 5100 pounds (mass of 11,250 kg) without the shielding and with the shielding its thrust was three times its weight. The Phoebus-2a was the most powerful nuclear rocket engine in the NERVA series; it was the prototype for a 250,000-pound thrust engine under development at the time the project came to an end. Figure 2.24 is a sketch of a proposed nuclear powered vehicle.

Interest in the further development of nuclear thermal rockets increased again in the 1990s, especially for a manned Mars mission, to inject the spacecraft onto an interplanetary trajectory to Mars and return it to Earth. Since the demise of the NERVA program, there have been numerous engineering advances in chemical rocketry which could be applied to greatly improve the performance of nuclear thermal rockets. Specifically, the turbopumps developed for the Space Shuttle main engines deliver nearly twice the mass flow rate as those used on the NERVA engine, yet they weigh only two-thirds as much. The Shuttle nozzles weigh only half as much as the NERVA nozzles but can operate under much higher chamber pressure, higher exhaust temperature, and can shed excess heat faster. It is likely that these and other new engineering developments will soon produce a practical nuclear thermal rocket with a thrust of five times its weight with shielding, and engine weight below 14,000 lb (mass below 6400 kg) plus shielding and propellant.

One more recent design uses a bed of tiny pellets of uranium reactor fuel which makes possible a high rate of heat transfer to the hydrogen propellant as it flows around and through the pellets. To prevent a chemical reaction with the hydrogen, the pellets would be coated with zirconium-carbide.

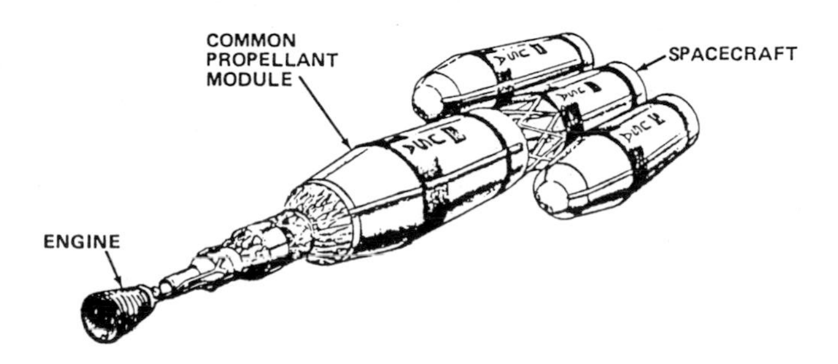

Figure 2.24 Sketch of vehicle using NERVA engine. *Courtesy of NASA.*

Obvious problems with operating a nuclear reactor are shielding people and equipment from the radiation in the core; starting and re-starting in space; and designing structural materials for containing the heat and radiation, especially with the possibility of a core meltdown. Another concern is an accident during launch; the reactor would have to be capable of surviving an impact or explosion without rupturing and scattering radioactive material. Because of these concerns, a nuclear rocket would probably only be used for upper stages, in high orbits, or on interplanetary missions to avoid the possibility of a crash landing on Earth.

Radioisotope Heat Cycle Engine

Radioactive elements, such as polonium or plutonium, emit large quantities of heat as they decay. If a quantity of the material is sealed in a container, the hot container could be the heat source for raising the temperature and pressure of a propellant flowing around it. As with other reaction engines, thrust is produced as the heated propellant is accelerated through a nozzle.

Radioisotope heat generators are now used on interplanetary spacecraft not for propulsion, but to generate electricity. Plutonium is encased in small metal spheres imbedded in graphite. The hot graphite then heats a gas which drives an electric generator. Electric power for most of the spacecraft headed for the outer Solar System has been supplied by such generators, some for more than 20 years. At such great distances from the Sun, photovoltaic (solar) cells do not produce much electricity.

In a propulsion application, the heat produced by the decay of the radioisotope accelerates the spacecraft instead of generating electricity.

Thermal Electric Propulsion

Propellants can be heated in other ways; use of an electric heater or electric arc seems simple and obvious. A diagram of a possible configuration is shown in Figure 2.25 in which an electric arc is placed in the "combustion" chamber. An electric arc consists simply of a voltage applied across the gap between two electrodes. Carbon is commonly used. The two electrodes are touched together momentarily to complete the circuit so an electric current flows vaporizing some of the carbon. Then they are slowly pulled apart to produce a gap filled with hot ionized air and carbon molecules; current continues to flow across the gap producing an arc, which emits both heat and light. As the carbon rods vaporize, the gap would widen and the current would stop; therefore, a small motor drives the rods toward each other at the same rate at which they vaporize to keep the arc a constant length. Temperatures in the arc may reach 35,000°F to 90,000°F (20,000 to 50,000°C).

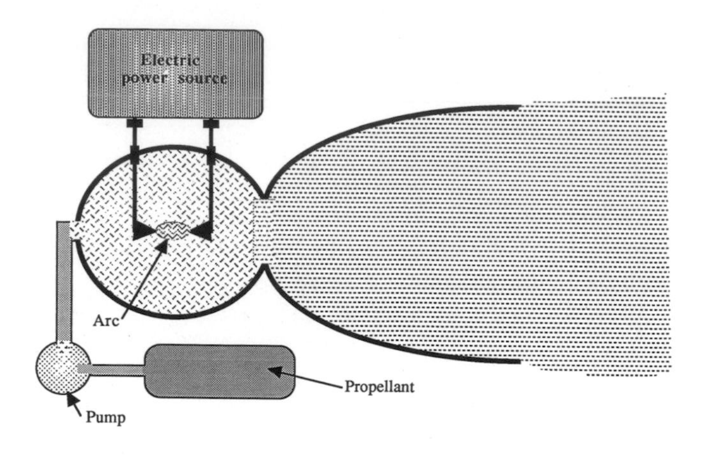

Figure 2.25 Diagram of a thermal electric arc engine.

Electric arcs are commonly used for high intensity lights in motion picture projectors, on movie sets, and sometimes for stage lighting. Welding can be accomplished using an arc. The metal to be welded is one conductor; the other electrode is a metal rod or wire. The arc between the rod and the metal to be welded heats, melts, and fuses the joint.

To be used as a rocket, propellant is pumped into and through the arc at a relatively slow rate so its temperature will rise to perhaps 18,000°F (10,000°C). If the temperature goes much above that, some of the energy is lost when it goes to ionizing atoms or dissociating molecules. The propellant then expands through a nozzle as in any other reaction engine. Using a low molecular weight propellant, such as hydrogen, a theoretical specific impulse of 1500 seconds could be achieved. A thrust of less than 25 pounds (100 N) is the best that can be expected because the mass flow rate must be kept low. If the propellant is flowing too fast, it does not have time to heat up to the required temperature.

Heating elements, such as are found in an industrial furnace or kiln, are also a possible source of energy for heating propellants. Temperatures of 2700°F (1500°C) may be reached in a kiln.

Electricity could be supplied from any of several sources. A radioisotope electric generator is perhaps the most practical, but a large array of photovoltaic cells would work near Earth where the intensity of sunlight is sufficient to supply the required current. Beyond the orbit of Mars, the sunlight is so weak that solar cells are not practical.

These are very low thrust engines, and, because of the massive electric power supply required, they cannot lift themselves off the ground. However, they are useful for low thrust applications such as satellite pointing. Several satellites have flown using this type of engine for thrusters.

Solar Thermal Engine

Another way to heat a propellant is to use solar energy directly. Flat plate solar collectors have long been used to heat buildings and water in homes. Some industrial applications have used curved mirrors, similar to large dish satellite antennas, to concentrate the solar energy to a small spot of extremely high temperature. An inflatable concentrating collector, 98 feet (30 m) in diameter, has been under development for a second stage rocket to boost satellites to higher orbits.

Solar power above the Earth's atmosphere is 0.127 kilowatt per square foot (1.365 kW/m^2) of area pointed directly at the Sun. A collector 98 feet (30 m) in diameter would intercept about 960 kilowatts continuously. A pair of such collectors could raise the temperature in the engine to about 4600°F (2500°C). Coils of tubing at the hot spot would carry the hydrogen propellent to be heated and expanded out a nozzle generating thrust. A metal with a high melting point is needed for the coils. Tungsten melts at about 6100°F (3400°C). A specific impulse of more than 800 seconds has been theoretically calculated.

The engine would produce very small thrust, and would be useful only as an upper stage. If the reflectors are kept pointed at the Sun the engine operates continuously. A 15 ton (13.6 metric ton) payload could be transferred from a 200 mile low Earth orbit to a 22,000 mile orbit in 15 to 20 days. (A metric ton is 1000 kilograms.)

Laser Propulsion

Yet another possible method of heating a propellant would use a ground-based laser. Small rockets, 100 pounds (50 kg) or so, would be given an initial upward velocity by some sort of a compressed gas gun; then the laser beam would be directed through the nozzle of the moving rocket into what might be called a combustion chamber. There the laser energy vaporizes and heats the solid propellant so it accelerates out the nozzle to provide thrust.

One proposal suggests that the small packages could be launched as often as one every 10 minutes. While each package is small, the frequent launch capacity would allow a large total mass to be put into orbit in a day's time. This would be particularly useful for resupplying a space station or sending materials to a construction site.

Most any low molecular weight propellant that vaporizes easily could be used. Ice has been suggested; it vaporizes at only 212°F (100°C). One pulse from the laser would vaporize a quantity of the ice and a second more powerful pulse would heat it to a high temperature.

The major advantage of such an arrangement is its low weight—the energy source remains on the ground. The major disadvantage is that it would require several megawatts of electrical power, as much as is consumed by a good-sized city.

Far-out Propulsion Concepts

The most expensive part of spaceflight is getting that first hundred miles off Earth into a low orbit, mostly because, except for the Space Shuttle, the vehicles are used once and dropped into the ocean. There must be a better way. The gasoline and diesel engines with their pollution and dependence on oil, sometimes from unfriendly sources, are the standard for automobile and truck use. Cars could run on electricity, hydrogen, natural gas, or a number of other fuels, but the automobile and oil industries are geared up to build gasoline-propelled cars and after almost a century of building them that way, it would be difficult and expensive to change. There is little incentive to do so. Just as we could build cars using a different means of propulsion, we could build other types of engines to take us off this planet, but here too, the bias toward standard proven chemical rockets is dominant. It is difficult to convince people to attempt something new when you have something that works, especially if hundreds of millions of dollars are at stake.

Computers are another example. If it hadn't been for Steve Jobs and his innovative development of the personal home computer, computers would still be relegated to office, research, and engineering functions. But with the advent of the Apple computer, born in a garage, a whole new industry came into being and the firmly established computer manufacturers were forced to enter the market. Now a small box on a desk top has more memory, greater speed, and more sophisticated input and output devices than a room-sized computer of the 1970s.

It is unlikely that a spaceship capable of carrying a crew to the Mars could be developed in a garage, but there are many innovative people who have ideas worth exploring. Are nuclear rockets, solar sails, electric engines, warp drives, wormholes, antigravity devices, antimatter engines, magnetic levitation, and quantum fluctuations just technobabble? The answers are no, no, no, maybe, maybe, yes, no, no, and who knows. In 1998 NASA established a new Institute for Advance Concepts to explore some of the promising new ideas in Earth science, space science, human exploration, aeronautics, and space transportation. The ground rules are simple: the proposal must be relevant to NASA's mission and must not violate known physical laws.

Plasma Pulse Engine

A plasma is a mixture of positively charged ions, electrons, and neutral atoms. An electric arc passing through a plasma is affected by a magnetic field which accelerates it through the engine and out the exit end producing thrust. High specific impulse is achievable, but much of the energy imparted to the plasma is lost by collisions against the walls of the engine causing heating and possible damage. Pulsing the arc intermittently allows some time for cooling between pulses, but reduces the efficiency of the engine. As with ion engines, a large electric power source is needed. Test engines have been flown.

Gun Launchers, or Shades of Jules Verne!

In the 1865 science fiction story, *From the Earth To the Moon,* Jules Verne used a gun to launch a projectile "spacecraft" carrying a crew and its equipment to the Moon! The gun was a vertical shaft, 790 feet (240 m) long, dug into the Earth. The projectile was about 10 feet (3 m) in diameter and weighed 30,000 pounds (13,600 kg mass). What Verne failed to consider was the acceleration required to get his spacecraft going fast enough to escape Earth's gravity. When a projectile leaves the barrel of the gun it has the highest speed it will have; from then on it slows down because of gravity and frictional drag as it travels through the atmosphere. To escape Earth it must have a starting speed of about 7.5 miles per second (12 km/s) to overcome gravity. The calculation in MATHBOX 2.3 shows that the acceleration would be 30,000 g, making everything weigh 30,000 times its normal weight.

New technologies now under development may make Verne's gun launcher a reality, at least to place objects in orbit. When it reaches orbital altitude a payload must have sufficient speed to stay in orbit, about 5 miles/sec or 8 km/sec, not as much as is needed to escape Earth. Therefore, the gun's muzzle speed must be 8 kilometers per second plus whatever it will lose to atmospheric friction on the way up. Total required muzzle speed may be 10 kilometers per second. For a gun that is 0.9 mile (1.5 km) long, the acceleration would be about 3300 g's. People could not survive that, but materials and equipment could if they were properly packaged.

Proposed designs could launch up to 4 tons (3.6 metric tons) into low Earth orbit, and several dozen launches per day could probably be achieved. There is a need for such launches. Delivery of material for construction of the Space Station and other large structures, resupply of the Space Station after it is completed, and placement of constellations of small communications satellites are a few potential uses. As it is now, the Space Shuttle carries construction materials and fresh supplies to the Space Station and satellites are generally launched by throwaway rockets.

Gun launch from the Moon would be much easier because gravity on the surface of the Moon is only one-sixth Earth gravity and because the Moon has no atmosphere, there would be no drag. Once the gun was operational, the cost of a launch would be only a small fraction of the cost of fuel for rocket launches. Transporting ore from a mining colony, for example, may be a practical application. We will discuss this further in Chapter 12.

Solar Sails

Sailors use the momentum of the wind to propel their boats through the water. Why not use the solar wind to propel a spacecraft through space? Solar wind particles, mostly protons, electrons, and alpha particles, flow continuously from the Sun with great speed, of the order of 250 miles (400 km) per second in the vicinity of Earth. But it cannot be used to propel a spacecraft because the solar wind has too little mass to provide the necessary force. There are only about 80 to 160 particles per cubic inch (5 to 10 particles per cubic centimeter), not enough to substantially accelerate a spacecraft.

The momentum of the light from the Sun, however, is thousands of times greater than that of the solar wind. Light

MATHBOX 2.3

Jules Verne's Gun

To calculate the acceleration required to reach escape speed from Earth to the Moon with Jules Verne's gun we use the following equation:

$$v^2 = v_0^2 + 2ax$$

Because it starts from rest, v_0 is zero and v is the speed the projectile has when it leaves the gun. It accelerates only while it is in the barrel of the gun, a length $x = 790$ feet $= 240$ meters. It must achieve escape speed, $v = 7.5$ miles/second $\times$ 5280 feet per mile $= 39{,}000$ feet/second $= 12{,}000$ meters/second in that distance. Solve for a and enter those values in the equation in common units and in metric units:

$$a = \frac{v^2}{2x}$$

$$a = \frac{(39{,}400 \text{ ft/s})^2}{2(790 \text{ ft})} = 980{,}000 \text{ft/s}^2 = 186 \text{ miles/s}^2$$

$$a = \frac{(12{,}000 \text{ m/s})^2}{2(240 \text{ m})} = 300{,}000 \text{ m/s}^2$$

That is about 30,000 times the acceleration of gravity, 30,000 g. A person inside the projectile would weigh 30,000 times normal weight and would be squished into a puddle at the bottom of the spacecraft!

Because the length of the barrel is in the denominator, the acceleration would be less if the barrel were longer. By a similar calculation, a gun 0.9 mile (1.5 km) long would need an acceleration of about 3300 g to reach orbital speed, 6 miles (10 km) per second. (Try this calculation yourself.) Properly packaged material could withstand this acceleration.

pressure can be used to accelerate a sail to high velocity. In fact, the pressure of sunlight had to be considered in planning the Viking missions to Mars. If it had not been included in the calculations, the spacecraft would have missed Mars by as much as 9000 miles (15,000 km).

Light waves transfer energy from place to place in "bundles" called photons. Photons have many of the characteristics of particles, but are not particles in the common sense of the word. They have no mass at rest but because they travel at the speed of light, they have momentum. The sail is practically standing still compared to the speed of the photons. When any object collides with another and bounces off, momentum is transferred twice: once when the object strikes and once when it moves away. The first is like the cue ball hitting another ball on the pool table and sending it off toward a pocket. The second, when the photon reflects off the sail, is the same as exhaust leaving a rocket nozzle and transferring momentum to the rocket. Remember Newton's third law. A highly reflective mirror-like sail would reflect photons of sunlight and the momentum of the photons would be doubly transferred to the sail. See Figure 2.26.

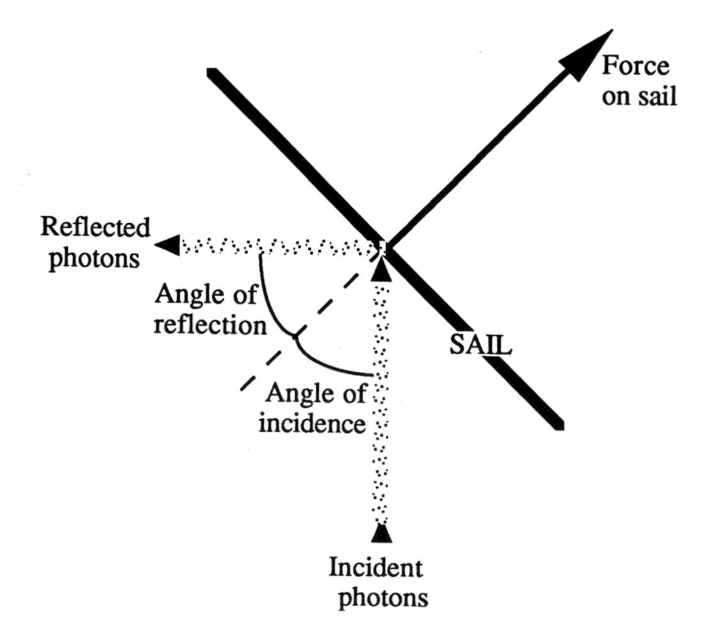

Figure 2.26 Solar sailing. The angle of the incident photons (measured from the line perpendicular to the sail) is equal to the angle of the reflected photons bouncing off the sail. Therefore, the thrust is perpendicular to the sail.

A chemical rocket burns for only a few minutes to boost a spacecraft into orbit; after the rocket burns out, the spacecraft just coasts. On the other hand, a solar sail thrusts continuously as long as the Sun is shining on it. The pressure is very small, only 0.02 billionths of a pound per square foot (9 millionths of a newton per square meter), less than the weight of an ant. But the continuous flow of photons from the Sun steadily increases its speed. As it speeds up, it climbs to higher and higher altitudes; i.e., it spirals outward from the Earth. It could not take off from the ground; to do so its thrust must exceed its weight, so it would have to be launched into orbit on a chemical rocket.

In near Earth space, three forces are acting on the solar sail: light pressure away from the Sun, gravity toward the Sun, and gravity toward the Earth. To move the way you want it to, the sum of the three forces must be a net force in the direction you want to go. Just as a sailor on Earth adjusts his sail at an angle to the wind, a solar sail is angled in the light stream from the Sun to control the direction of the light pressure. In one design the sail is kept properly oriented by small triangular vanes attached to the corners of the main sail. Small motors rotate the vanes so they catch the photons, applying a small force to turn the main sail toward the streaming sunlight. Just as a sailboat tacks into the wind, tacking can steer a solar sail toward the Sun if desired.

The forces due to gravity must be kept as small as possible. Remember that gravity is related to mass so the sail must be made of very lightweight material, thinner than plastic wrap. Very strong, lightweight materials less than 100 atoms thick are becoming available. A solar sail is flimsy and fragile on Earth, but stretched out in the near vacuum of space it is quite sturdy. Of course, it must be rolled up and folded for launching. Designing a device to automatically unfold and unfurl the sail when it reaches orbit is not an easy problem. It must be streatched out tight without wrinkles and without tearing. A Russian cosmonaut, while leaving the Mir space station in 1993, was the first to successfully deploy such a sail, a 65-foot (20-m) disk. The packaged sail was attached to a motor that made it spin so the sail drew itself out of the package to its full extent. In an American design, stainless steel struts would be attached to the accordian-folded sail, flattened and rolled up on a spool. Pressurized nitrogen gas would stretch it out and orient it toward the sunlight.

In 1977 and 1978, a team of researchers at NASA Jet Propulsion Laboratory (JPL) proposed constructing a solar sail spacecraft to rendezvous with Halley's comet. One design called for a huge 2600-foot (800-m) square sail, nine times the length of a football field. Another design would have looked like helicopter rotors consisting of 12 blades, each 26 feet (8 m) wide and 3.75 miles (6.25 km) long for an area of about 6.5 million square feet (600,000 m^2). NASA abandoned the concept in favor of developing an ion engine. Although it never sailed, the researchers showed that solar sailing is entirely feasible and, indeed, can outperform rocket engines for long flights.

Since then, experimental solar sails have been designed in several countries and a race to Mars was planned for 1992 to commemorate the 500th anniversary of Columbus's sailing trip to North America. Unmanned race vehicles from the United States, Europe, Japan, and Russia were to be launched into Earth orbit on a chemical rocket but would use only solar sail propulsion for the trip to Mars. When the participants ran into financial trouble, the race was delayed two years and scaled back to a lunar voyage. The winner would be the first one to transmit a picture of the far side of the Moon.

The proposed United States entry was to consist of a square sail, 180 feet (55 m) on a side and would weigh only about 65 pounds (mass of 180 kg). Unfurled, its area would be about 32,000 square feet (3000 m^2). Struts would be made of two thin stainless steel strips, curved and welded together to form a tube.

The 1990 British design called for a circular sail, 820 feet (250 m) in diameter supported by 36 carbon fiber ribs, and weighing less than 660 pounds (mass of 300 kg). It would be packaged in a spiral fold for unfurling when it gets to orbit. The designer claimed it would eventually reach 6 miles (10 km) per second and could make a trip all the way to Mars in only 200 to 300 days depending on how the planets are aligned. It, too, ran into financial trouble.

The race never happened. However, new practical uses for solar sails have surfaced. For example, solar radiation pressure forces satellites out of orbit and small thrusters must be fired from time to time to bring them back into position. The GOES-8 weather satellite has an attached 10 foot (3 m) solar sail to exactly compensate for the radiation pressure against its solar cell array. Thus the thrusters do not have to be fired so often and less thruster propellant must be carried in the spacecraft.

Orion Nuclear Rocket

The Interstellar Orion Rocket would use nuclear bombs to propel a spacecraft through interstellar space. Small bombs would be released out the back of the vehicle and at a safe distance would explode. The hot expanding gas of the explosion would push against a large plate at the rear of the vehicle to drive it forward. Space is already full of high energy particles and a few more would not bother anything. (It seems like a good way to get rid of the world's huge stockpile of nuclear weapons!)

Antimatter Propellants

The matter that we encounter in everyday life on Earth consists of positive protons and negative electrons. But there also exists antimatter with negative protons, called antiprotons, and positive electrons, called positrons. It cannot exist very long in our part of the galaxy, however, because oppositely charged particles of matter and of antimatter attract one another and when they collide they disappear in a gamma ray, a puff of energy. The energy released is a hundred million times that of the same amount of the best chemical propellants. Theoretically the specific impulse could be as high as 100,000 seconds.

Antimatter is the stuff of science fiction, the propellant for the Starships *Enterprise*. But antimatter is real and it can be manufactured in high energy particle accelerators found in physics research laboratories. Only a few hundred-thousandths of an ounce (a few milligrams) of antimatter coming in contact with regular matter would produce enough energy to heat several tons of propellant in a rocket engine producing several hundred thousand pounds of thrust. One-thousandth of an ounce (35 milligrams) of antimatter could produce the same energy as all the solid and liquid propellants in the Space Shuttle. Cost to produce and store the antimatter is estimated at a million dollars per milligram.

The problem is keeping it out of contact with our kind of matter until it is to be used. It can be stored in the lab in the particle accelerator. However, the unsolved problem is moving it and storing it in the spacecraft until flight time. The *Enterprise's* "antimatter containment field" is in the distant future.

DISCUSSION QUESTIONS

1. If an astronaut weighs 120 pounds at an altitude of 100 miles above the Earth, then why is she weightless in orbit?
2. Explain what is meant by miles per hour per second or feet per second per second as units of acceleration.
3. Why does a toy balloon fly an erratic path when it is released, but a rocket ship flies straight?
4. How does the turbojet engine on an airliner work? Is it a reaction engine?
5. Why is the engine sticking out so far behind the spacecraft in Figure 2.24?
6. What precautions would have to be taken when using a nuclear-powered rocket on the way to orbit? In orbit? On a trip carrying a crew to Mars?

ADDITIONAL READING

Bilstein, Roger E. *Stages to Saturn*. NASA SP-4206, Government Printing Office, 1996. Technological history of the Apollo/Saturn launch vehicles.

Corliss, William R. *Propulsion Systems for Space Flight*. McGraw Hill, 1960. An early textbook on the subject.

Dawson, Virginia. *Engines and Innovation*. NASA SP-4306, Government Printing Office, 1991. History of the Lewis Laboratory, now named the Glenn Research Center, from aircraft engines to rockets.

Estes Industries. *The Classic Collection, Model Rocketry*. Estes Industries. Nonmathematical technical notes and reports.

Miller, Ron. *The Dream Machines*. Krieger Publishing Co., 1993. Pictorial history of the spaceship in art, science and literature, 360 B.C. to the present.

Muolo, Michael J., et al. *Space Handbook (AU-18)*. Air University Press, December 1993. Two volumes aith algebra level mathematics.

NASA. *Exploring in Aerospace Rocketry*. Government Printing Office, 1971. Algebra level mathematics.

Newgard, John J., and Myron Levoy. "Nuclear Rockets." *Scientific American,* May 1959. Early work, before nuclear rockets were shelved.

Noordung, Hermann. *The Problem of Space Travel, The Rocket Motor*. NASA SP-4026, Government Printing Office, 1995. Translation of a historic 1929 German book on rockets.

Presidential Commission on the Space Shuttle Challenger Accident. *Report to the President*. Government Printing Office, 1986. Thorough investigation and report.

Stine, G. Harry. *Handbook of Model Rocketry*. Follett Publishing Co., 1976. Minimal mathematics.

Sutton, George P. *Rocket Propulsion Elements*. John Wiley and Sons, 1986. Engineering textbook.

Williamson, Mark. *Dictionary of Space Technology*. Adam Hilger, 1990. Comprehensive.

Wright, Jerome L. *Space Sailing*. Gordon and Breach Science Publishers, 1992. Technical, non-mathematical, comprehensive, and readable account of solar sails.

NOTES

Chapter 3

Orbits

Getting into Orbit

Throw a ball straight up into the air (Figure 3.1). It slows down as it rises, stops at the top of its path, and comes back down to you, speeding up as it falls. Why? Once the ball leaves your hand, you have no more control over it—it is on its own. There must be some other force affecting its motion: gravity, of course. While it is moving away from Earth it slows down; as it falls back to Earth it speeds up.

Throw the ball harder and it goes higher before stopping and returning. If you could throw it hard enough, it would keep on going, slowing down but never stopping, never returning; it escapes the gravity of Earth. How hard is hard enough? From the surface of Earth, greater than 7 miles per second would do it. But in this experiment we have only succeeded in sending our ball out to escape Earth forever. How do we get it into an orbit around the Earth?

Let us try something else. Shoot a cannonball from a cannon. It follows a curved path as it falls to Earth (Figure 3.2). Eventually it hits the ground. Fire it with a little more oomph and it travels farther before hitting the ground.

Newton's first law of motion states that an object in motion will continue to move in a straight line unless an external force acts on it. Once the cannonball leaves the cannon, it is no longer affected by the cannon. The external force is gravity, which causes the normally straight line path to curve back to Earth.

In a larger perspective, consider that Earth is actually spherical. The cannonball travels part way around Earth before it strikes the ground (Figure 3.3). However, if it were possible to give the cannonball enough speed, it would never strike the ground. It would continue to fall in its curved path toward Earth, but would never quite get there. In other words, it would be in orbit. Isaac Newton knew this back in the 1600s; a diagram similar to Figure 3.3 appears in his notes.

There are two practical problems with this scheme for orbiting a cannonball. First, the speed necessary to follow a path which does not intersect Earth is over 17,000 miles per

Figure 3.1 When Jimmie throws a ball up, it always returns.

Figure 3.2 Can a cannon fire a satellite into orbit?

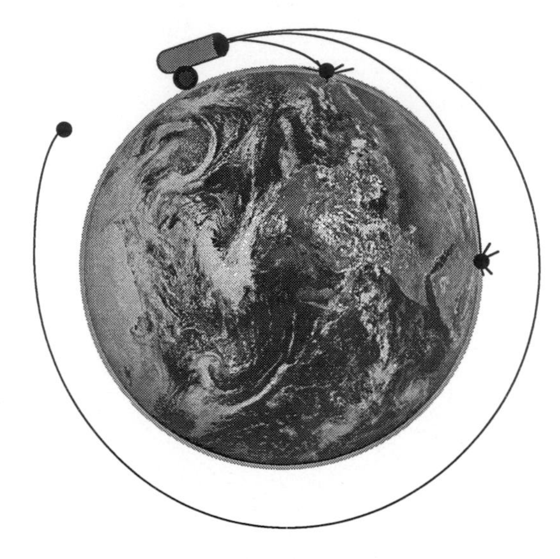

Figure 3.3 Putting a cannonball into orbit

hour. No cannon has yet been built which can impart that speed to a cannonball. Second, even if the cannonball could be shot at 17,000 miles an hour, the friction with the air would heat it to the point of melting, vaporizing, and disintegrating. Did you ever hold your hand out the window of a car moving at 60 miles an hour? Imagine what it must be like at 17,000 miles an hour!

As mentioned in Chapter 2, it may be feasible to use a "cannon" to launch a vehicle to high altitude where a rocket engine would ignite and carry it to orbit. Research is now being done to determine whether it would be practical on Earth. It would work on the Moon, however, for two reasons. First, the Moon has no atmosphere. Second, because the Moon has a smaller mass than Earth, the required orbital speed is much less. Just fire the cannonball with enough horizontal speed so its curved path does not intersect the Moon. It might be advisable to launch from the top of the highest mountain so the orbit is high enough to clear all obstacles.

To launch a spacecraft into Earth orbit, it is first necessary to get it up out of the densest part of the atmosphere, then give it sufficient horizontal velocity so that its curved path toward Earth does not actually intersect the ground. It is a common misconception that satellites are launched straight up into orbit. The primary reason for launching vertically is to carry the spacecraft through the dense part of the atmosphere at a lower speed to keep it from burning up or breaking up. Then it is pitched over horizontally and accelerated to orbital speed (Figure 3.4). Without sufficient horizontal velocity it will simply fall back to Earth (Figure 3.5); it is that motion parallel to the surface of the Earth that keeps it in orbit.

Ellipses

Earth and the other planets are in orbit around the Sun in just the same way as satellites are in orbit around Earth. In the late 1500s, during a lifetime of study of the motion of the planets, Johannes Kepler discovered laws which describe their motion around the Sun. His initial assumption, like many other natural philosophers of the time, was that the orbits of the planets were circles. The circle was considered a perfect mathematical shape and if the universe was perfect, the orbits should be perfect. Kepler, after years of studying the precise observations made in the observatory of the Danish nobleman, Tycho Brahe, finally came to the conclusion that they were ellipses, not circles. Kepler was disappointed. He expected the universe to have been created with perfection. This important discovery applies as well to any object in orbit around another body, including spacecraft in orbit around Earth.

An *ellipse* is an oval shape as shown in Figure 3.6. You can draw an ellipse by using two thumb tacks and a piece of

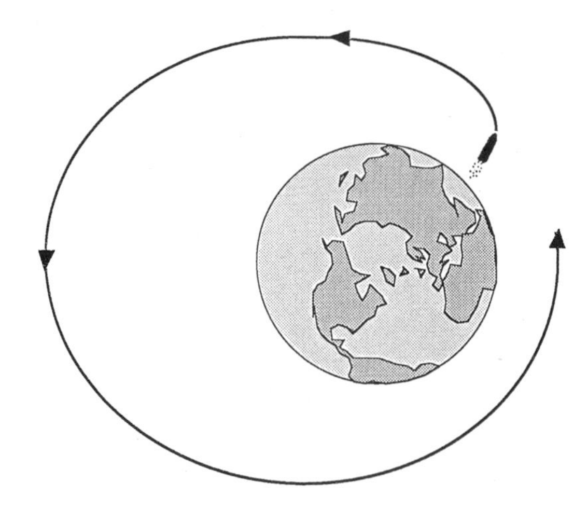

Figure 3.4 Launch up out of the densest atmosphere, then pitch over and accelerate to orbital speed.

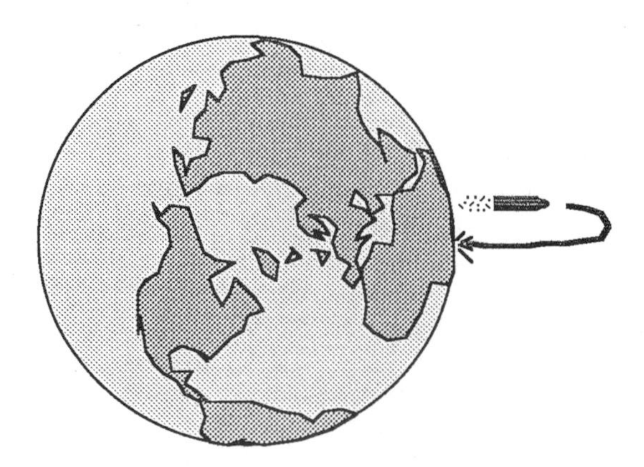

Figure 3.5 Launch something straight up and it comes straight down.

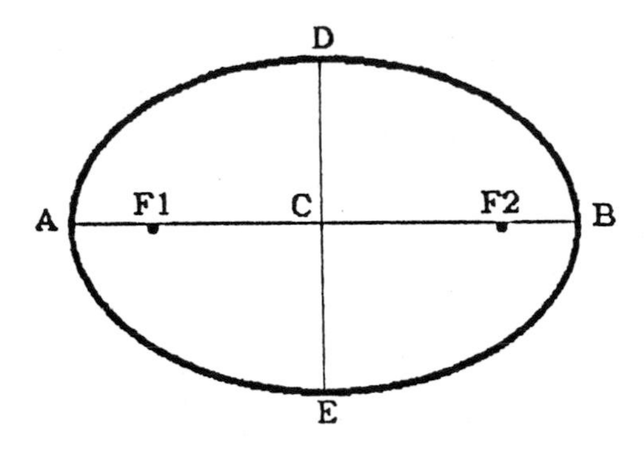

Figure 3.6 The ellipse. A to B is the major axis; D to E is the minor axis. The center is at C and the two focal points (foci) are at F1 and F2.

string. See Figure 3.7. Stick the thumbtacks into a board; they will become the two *focal points* or *foci*. Tie the string into a loop and hang it loosely around the thumbtacks; then insert your pencil into the loop and draw it up tightly. As you move the pencil around the thumbtacks while keeping the string taut, it will trace out an ellipse. Use a shorter piece of string or set the thumbtacks far apart for an elongated ellipse. Use a longer piece of string or set the thumbtacks close together for a near-circle. Set them on top of one another and you will get a circle, i.e., use only one thumbtack at one focal point.

The *eccentricity* of an ellipse is a number between zero and one which tells just how elongated the oval is. It is calculated by dividing the distance between the foci by the length of the major axis as shown in MATHBOX 3.1. If the ellipse is a very elongated oval, then the eccentricity is close to one. At the other extreme, the eccentricity of a circle is zero.

Kepler's first law of planetary motion states that planets move in elliptical orbits with the Sun at one focus. The eccentricity of the Earth's orbit is very small, 0.017, indicating that the orbit is nearly circular. In fact, if you draw it to scale on a piece of paper, it looks like a perfect circle. Pluto's orbit has the greatest eccentricity of any of the planets, 0.248.

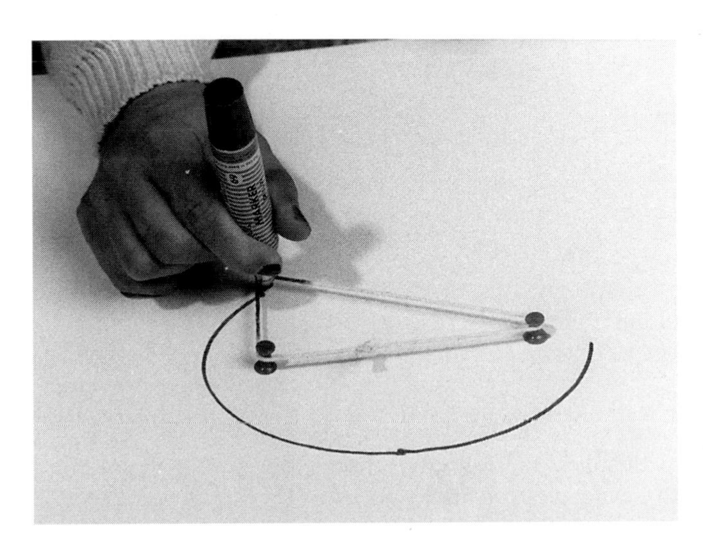

Figure 3.7 Drawing an ellipse.

Satellite Orbits

Satellites in orbit around Earth also follow Kepler's first law, moving in elliptical paths with the center of Earth at one of the foci. The satellite's speed, direction, and distance from Earth at the instant the engines burn out determine the size and shape of the orbit. The only force acting is gravity, and left on its own, the satellite will follow its elliptical orbit and return to the point where the rocket engines shut down.

The point of closest approach to Earth is called *perigee* and the point on the orbit farthest from Earth is *apogee*. (*Geos* is the Greek word for Earth.) In Figure 3.8, Earth is at F1. Notice that perigee and apogee are on opposite ends of the major axis. When referring to an orbit, the major axis is called the *line of apsides*. For an orbit around the Sun, the nearest and farthest points to the Sun are called *perihelion* and *aphelion* respectively. (The Greek word for Sun is *helios*.) The orbit is fixed in space with the spacecraft moving on it and Earth rotating under it.

The speed of a spacecraft varies from point to point in an elliptical orbit depending on how far it is from Earth. The farther out it is, the slower it travels. Thus, a spacecraft's speed is greatest at perigee and slowest at apogee. As a consequence, its speed is increasing as it moves from apogee to perigee and it is slowing down as it moves from perigee back to apogee. This makes intuitive sense. A ball thrown upward, moving away from Earth, slows down; falling back, moving toward Earth, it speeds up. Of course, if the spacecraft's orbit is a circle then its distance from Earth does not change, perigee and apogee are undefined, and its speed is constant. MATHBOX 3.2 shows how to calculate the speed of a spacecraft in circular orbit.

Kepler recognized these facts in his studies of planetary motion around the Sun. His second law states that equal areas are swept out in equal times. At first glance, that statement seems unrelated to our previous discussion about speed, but look at Figure 3.9. In a certain time, say 10 minutes, the spacecraft moves from point A to point B near perigee, and the line from the spacecraft to the focus at the center of Earth sweeps out the area A1. Since it moves slower at apogee, the distance covered in 10 minutes will be only from point C to point D, sweeping out area A2. Kepler's second law states that A1 must equal A2. The fact that the speed of an orbiting spacecraft depends on its distance from Earth is a most important consideration in orbital maneuvering and rendezvous.

The *inclination* of an orbit is the angle the orbit makes with the plane of the equator (Figure 3.10). An orbit directly over Earth's equator has an inclination of zero degrees; an orbit that passes over both poles has an inclination of 90 degrees. The angle is measured counterclockwise from the equator at the ascending node, the place where the spacecraft crosses the equator heading north. Thus, it is possible to have an orbit with inclination greater than 90 degrees as illustrated in Figure 3.11. The *period* of an orbit is the time it takes the

MATHBOX 3.1

Eccentricity

Let us represent the distance between the foci of an ellipse as f and the length of the major axis from A to B as d. The definition of eccentricity, e, is

$$e = f/d$$

In the ellipse of Figure 3.1.1, f is 2 inches and d is 5 inches. Then its eccentricity is

$$e = 2/5 = 0.4$$

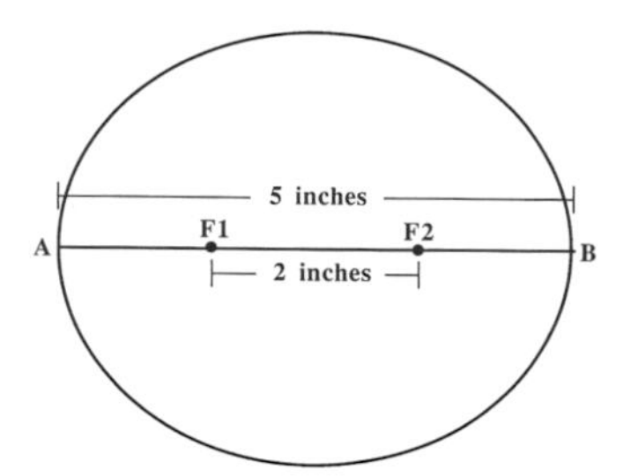

Figure 3.1.1 Ellipse, eccentricity of 0.4.

In Figure 3.1.2, $f = 4$ inches and $d = 5$ inches. Its eccentricity is

$$e = 4/5 = 0.8$$

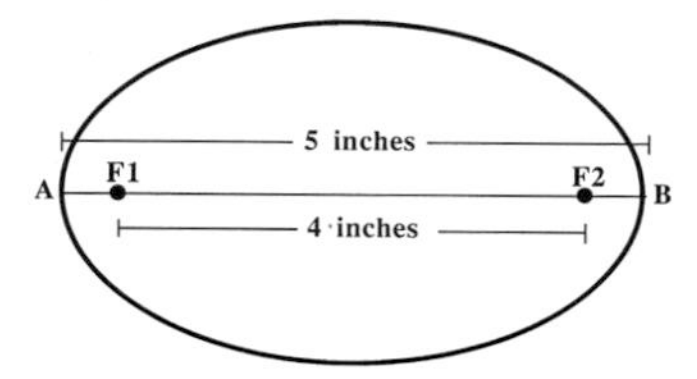

Figure 3.1.2 Ellipse, eccentricity of 0.8.

Finally, the third ellipse in Figure 3.1.3 is really a circle with a diameter of 3 inches. The two foci are coincident so $f = 0$, and d is the diameter. The eccentricity of the circle is zero since

$$e = 0/3 = 0$$

(Zero divided by anything is zero)

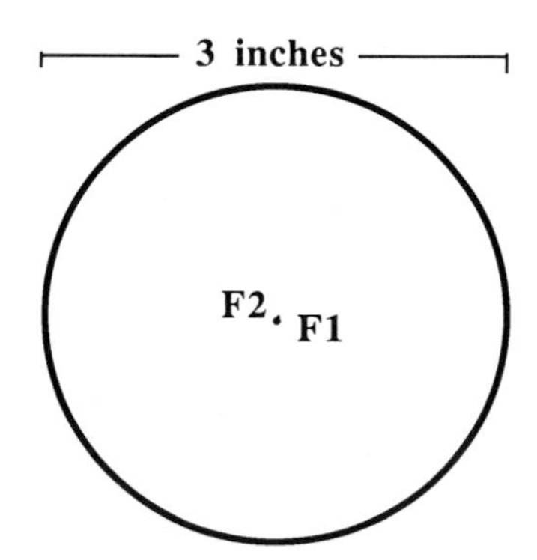

Figure 3.1.3 Circle, eccentricity of zero.

Note that if the two foci are close together, the distance between them is very small, then the eccentricity is a very small number. In the extreme case where there is only one focus, the ellipse becomes a circle.

On the other hand, if the foci are far apart the ellipse is elongated and the eccentricity is a larger number. In an extremely elongated ellipse, the distance between the foci is nearly the same as the length of the major axis of the ellipse and the eccentricity is nearly 1.

So the eccentricity of an ellipse is a number with values greater than zero but less than one; it indicates how elongated the ellipse is.

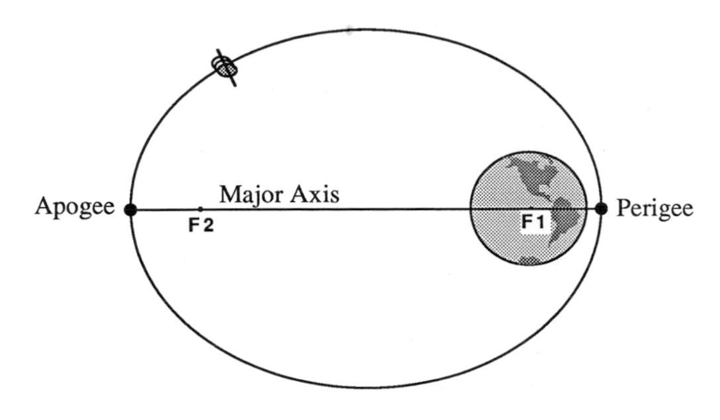

Figure 3.8 An elliptical orbit around Earth.

spacecraft to make one complete orbit. MATHBOX 3.3 tells how to calculate the period of a circular orbit.

The *ground trace* of a spacecraft is the track it makes over the surface of the Earth. The inclination is the farthest north and south latitude that the spacecraft reaches on its ground trace. Ground traces, as we shall see, often make unexpected and unusual patterns when drawn on a flat map of Earth.

Useful Orbits

Described here are several particularly useful orbits.

1. *Low Earth orbit, 250 miles (400 kilometers) above the surface, nearly circular, period about 90 minutes.* This is a

MATHBOX 3.2

Speed in Circular Orbit

According to Newton, an object will move in a straight line unless some force acts on it. Therefore, a force must be acting on a spacecraft to keep it moving in a circular orbit. That force, directed toward the focus of the orbit, is gravity. Any centrally directed force is called a *centripetal force*. For circular motion its magnitude is given by

$$F_c = \frac{mv^2}{r}$$

where m is the mass of the object, v is its speed, and r is its distance from the center of the circle.

Newton's law of gravitation tells us the magnitude of the force of gravity:

$$F_g = \frac{GmM}{r^2}$$

where M is the mass of the central body (Earth, Sun, etc.) and G is the universal gravitation constant, a number which is the same everywhere in the universe. The m and r are the same as in the previous equation. The gravitational force is the centripetal force for an object in orbit. Therefore, for a circular orbit

$$\frac{mv^2}{r} = \frac{GMmM}{r^2}$$

If we solve this for v we get the circular orbital velocity

$$v = \sqrt{\frac{GM}{r}}$$

Now, G times M is a constant for any particular central body. For Earth it is 1.24×10^{12} if r is given in miles and the speed v is in miles per hour. Let us make the calculation for an orbit with an altitude of 200 miles. Because r in the formula is the distance to the center of the orbit, we must add the radius of the Earth, about 4000 miles (6400 km). Then

$$v = \sqrt{\frac{1.24 \times 10^{12}}{4200}} = 17{,}180 \text{ miles per hour}$$

Try this calculation using metric units. If r and v are in meters and meters per second, then GM is 3.9860×10^{14}. Also, calculate v for orbits at other altitudes. Remember, this equation is valid only for circular orbits.

G times M for the Sun is 4.13×10^{17} when r is in miles and v is in miles per hour. In metric units, GM is 1.3286×10^{20}. How fast is the Earth moving, assuming the orbit to be circular?

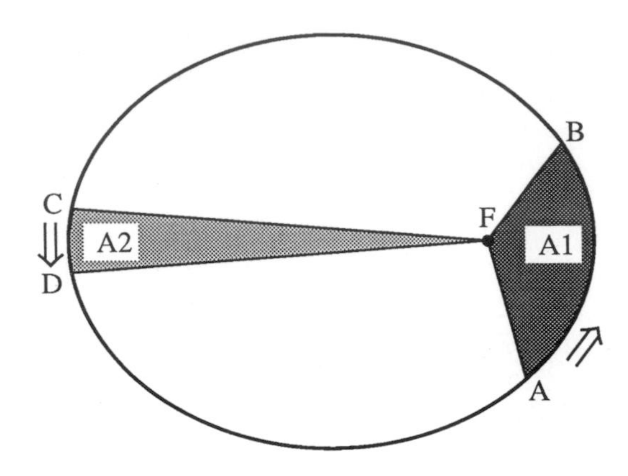

Figure 3.9 Kepler's second law of motion of orbiting bodies.

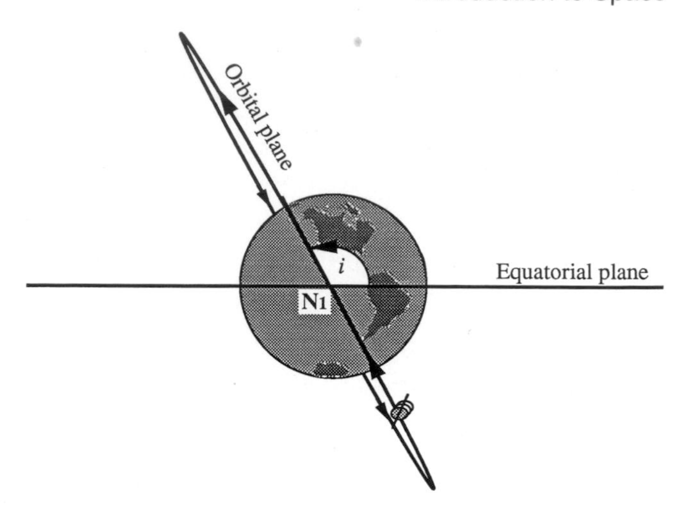

Figure 3.11 An orbit with an inclination angle (i) greater than 90 degrees.

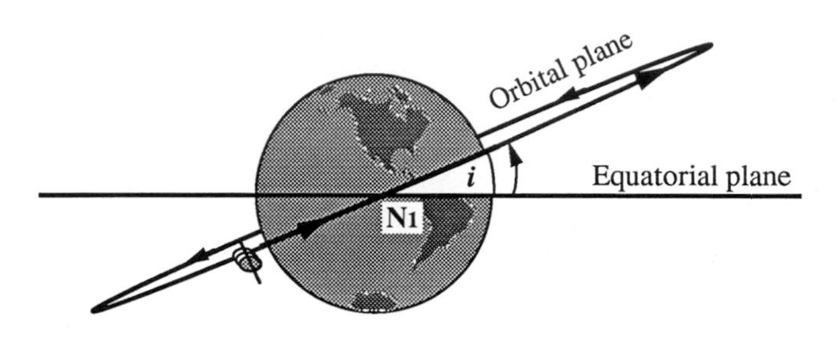

Figure 3.10 An orbit with an inclination angle (i) less than 90 degrees. N_1 marks the place where the spacecraft crosses the equator moving from the southern hemisphere into the northern hemisphere, called the ascending node.

typical orbit for the Space Shuttle, space stations, surveillance satellites, scientific research satellites, etc. When drawn to scale as in Figure 3.12 it appears remarkably low, just skimming the surface. The ground trace looks like a sine wave that does not repeat itself. About 1.2 orbits are shown in Figure 3.13. Starting northeast of Australia, the satellite circles Earth to a point northwest of Australia. In space, the orbit closes on itself; but when we view the path on a flat map

MATHBOX 3.3

Period of a Circular Orbit

Kepler also had a third law which related the orbital period of a planet, its "year", to its distance from the Sun. In words, it says the square of the period is proportional to the cube of the distance from the Sun. It wasn't until Newton, however, that the constant of proportionality was found. The formula is

$$P = \sqrt{\frac{4\pi^2 r^3}{GM}}$$

where r is the radius of the orbit and P is the period.

Kepler's third law also applies to satellites in circular orbit around the Earth. Remember that the radius of the orbit must be measured to the center of the Earth.

Now, GM is a constant (see MATHBOX 3.2), and π is also a constant (3.14159). So for Earth satellites in circular orbit the formula becomes

$$P = 5.64 \times 10^{-6} \sqrt{r^3}$$

if r is in miles and P is in hours. If r is in kilometers

$$P = 2.76 \times 10^{-6} \sqrt{r^3}$$

For a satellite in a 200 mile (320 km) orbit above the Earth, r is 4200 miles (6700 km). The period turns out to be 1.5 hours. Of course, it is the same in either set of units.

You can use the same formula to find the period of a planet given its distance from the Sun, or vice versa, but you must use GM for the Sun. That number was given in MATHBOX 3.2. Try finding the orbital period of the Earth about the Sun.

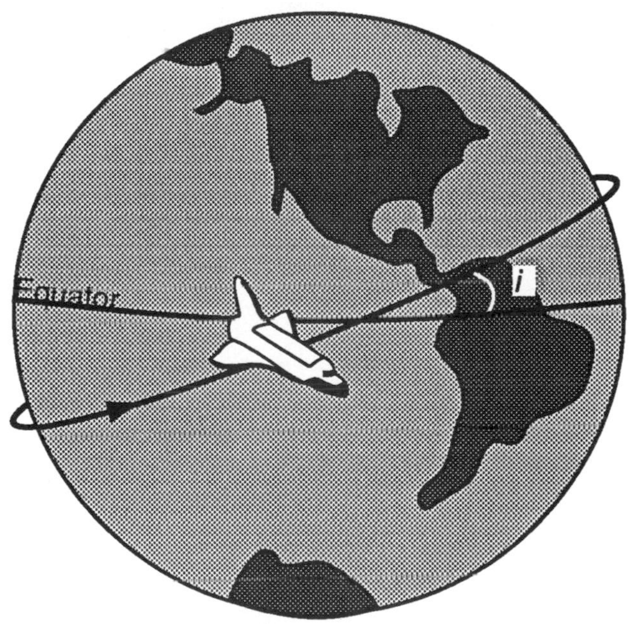

Figure 3.12 Low Earth orbit, circular, drawn to scale. Altitude is 250 miles (400 km). Top view is from over the north pole: lower view is from above the equator.

it does not appear to do so because the Earth is rotating under the satellite orbit. In the 90 minutes that the satellite takes to complete the orbit, the Earth rotates 22.5 degrees of longitude, about 1500 miles (2400 km) at the equator. During those 90 minutes the flat Earth map shifts to the east by that amount, so the satellite does not pass over the point where it began, but 22.5 degrees of longitude farther west. Each point on the ground trace is 22.5 degrees farther west than the comparable point on the previous orbit. Notice that the inclination of the orbit is 28 degrees and the satellite reaches to 28 degrees latitude north and south of the equator. With the combined motion of the spacecraft and rotation of the Earth, it eventually passes over all areas of the Earth between 28 degrees north latitude and 28 degrees south latitude. Because of its low inclination, it does not overfly areas to the far north or south.

2. *Polar orbit, inclination near 90 degrees, circular.* In contrast to the low inclination orbit, a polar orbit passes over the entire surface of the Earth in just a few days. Such an orbit at an altitude of 500 miles and period of about 100 minutes is drawn to scale in Figure 3.14. The ground trace for a 90 degree orbit is shown in Figure 3.15. Because it passes over all areas of the Earth including the polar regions, a satellite in polar orbit is useful for observing and mapping weather, ice pack, structures of glaciers, and Earth resources.

3. *Sun-synchronous orbit, inclination greater than 90 degrees.* Figure 3.16 is the ground trace for an orbit with a 98.2 degree inclination. Notice that it reaches 81.8 degrees north and south latitudes. The westward drift of this orbit just keeps pace with the Sun so that the satellite crosses the equator at the same local time each pass. For a given inclination there is only one altitude at which the orbit is Sun-synchronous. For a 98.2 degree inclination, the altitude is 438 miles (705 km). In Chapter 6 we will see the value of this orbit for Earth and weather observing.

4. *Geosynchronous orbit, 22,400 miles above the surface, circular, zero degree inclination, over the equator.* This orbit is drawn to scale in Figure 3.17. The satellite completes one orbit in 23 hours and 56 minutes, the same time as it takes Earth to rotate once on its axis. (The motion of the Earth in its orbit around the Sun makes up the other 4 minutes.) Therefore, from a vantage point on Earth, the satellite appears to remain fixed in the sky. Consequently, the orbit is sometimes called a *geostationary orbit.* The ground trace, Figure 3.18, is simply one spot on the equator. This orbit is particularly useful for communications satellites because the antennas can be pointed at that fixed spot. From that altitude the satellite transmitter can cover nearly one-third of the Earth; it cannot reach far north countries, however. The geostationary orbit is also valuable for warning of missile attacks because the satellite sensors can continuously watch large areas of Earth.

There are some interesting variations on the geosynchronous orbit. If the orbit has an inclination other than zero, the satellite's ground trace is a line extending north and south of the equator to the latitude equal to the inclination of the orbit. If the orbit is not circular, then, because the satellite speeds up at perigee and slows down approaching apogee, its ground trace is an east-west line along the equator. If the orbit is both noncircular and inclined, the ground trace may be an oval or a figure eight. See Figure 3.19.

5. *High eccentricity, with perigee at about 300 miles and apogee at about 25,000 miles above Earth.* The orbit in Figures 3.20 and 3.21 has an inclination of 64 degrees, an eccentricity of 0.69, and a period of 11 hours and 58 minutes.

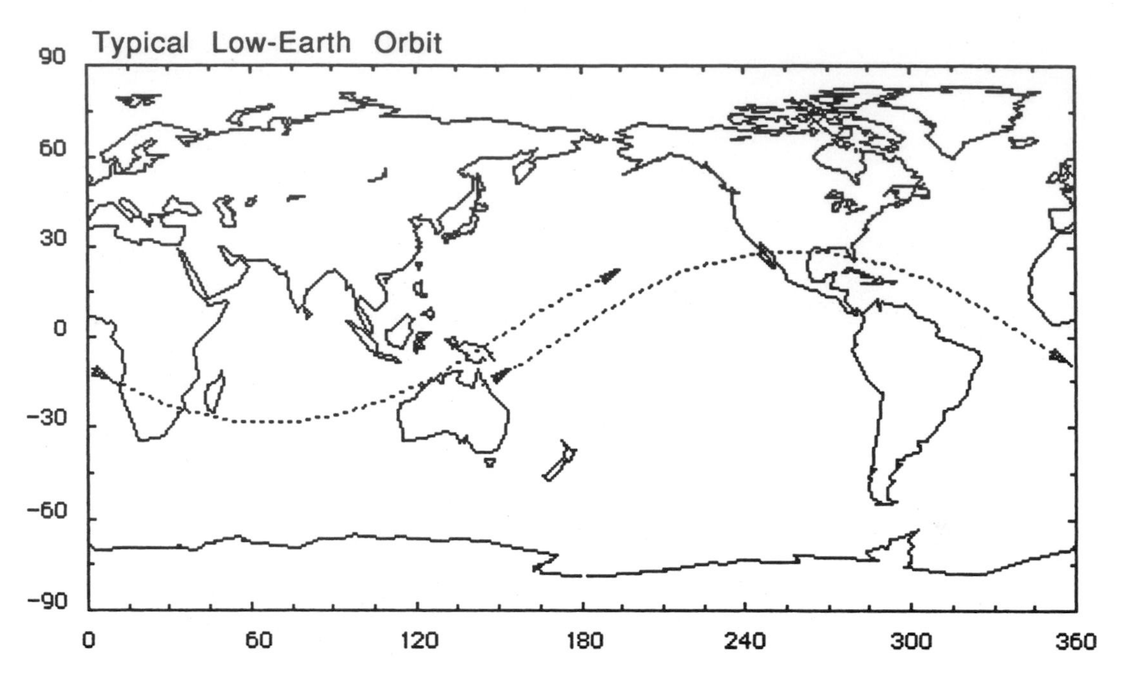

Figure 3.13 Ground trace of the low Earth orbit shown in Figure 3.12. The illustration shows 1.2 orbits, 110 minutes. Latitude is shown on the left; degrees south of the equator are negative. Longitude is measured eastward from Greenwich, England; negative numbers are longitude westward from Greenwich. Dots are about 1 minute apart; because it is a circular orbit traveling at a constant speed, the dots are equally spaced.

The dots on the ground trace are at five minute intervals. Notice that they are far apart in the southern hemisphere, indicating that the satellite is traveling at high speed, while they are so close together in the northern hemisphere that they form a continuous line. Because apogee is near geosynchronous altitude, the satellite moves very slowly, hovering over one point on Earth for a long time, then moves very rapidly past perigee to return to apogee and hover again. This is the orbit of a Soviet Molniya communications satellite. It remains nearly stationary for more than 8 hours at apogee over northern Siberia; then it speeds up, drops down past perigee and returns to apogee in less than 4 hours. Meanwhile, Earth will have rotated half a turn so apogee is over Canada. Twelve hours later it is back over Siberia. This orbit is particularly useful for communications in the far north where geosynchronous satellites over the equator cannot be seen.

Orbital Perturbations

So far, our discussion of orbits has been somewhat theoretical. Theoretically, once in orbit, no engines are required to keep the satellite going. In the real world, orbits are never perfect ellipses or circles. The gravitational attraction of Earth is the dominant factor in shaping the orbit. But the Earth is not a perfect sphere; it bulges at the equator and is flattened at the poles. This causes the line of apsides to drift westward, a perturbation called *regression*.

The gravity of the Moon and Sun may affect high altitude satellites by changing their inclination or their altitude of perigee. The effect is small because the Sun and Moon are so

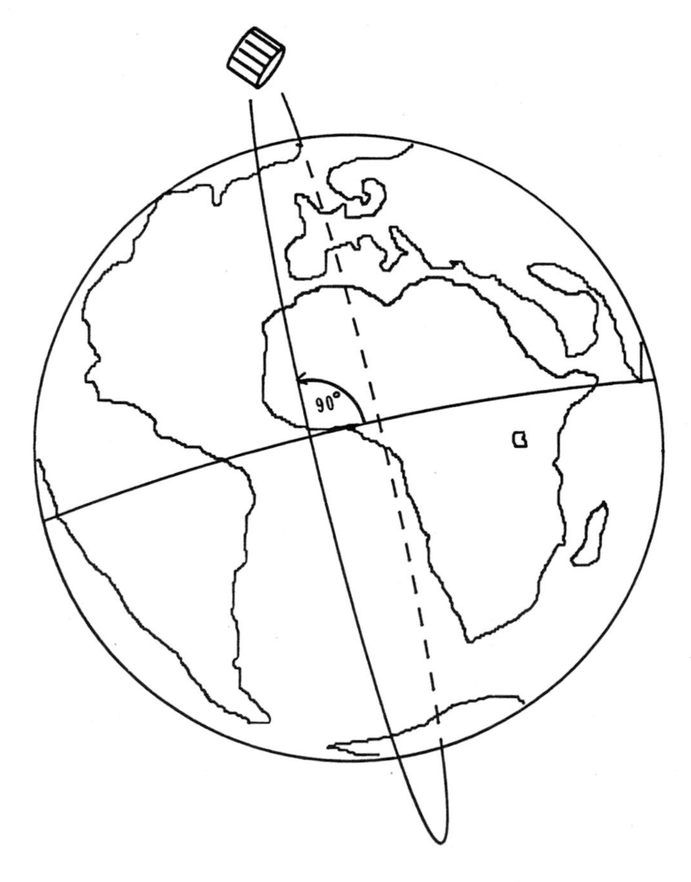

Figure 3.14 Sketch of a circular polar orbit, inclination of 90 degrees, altitude of 500 miles (800 km), period about 100 minutes.

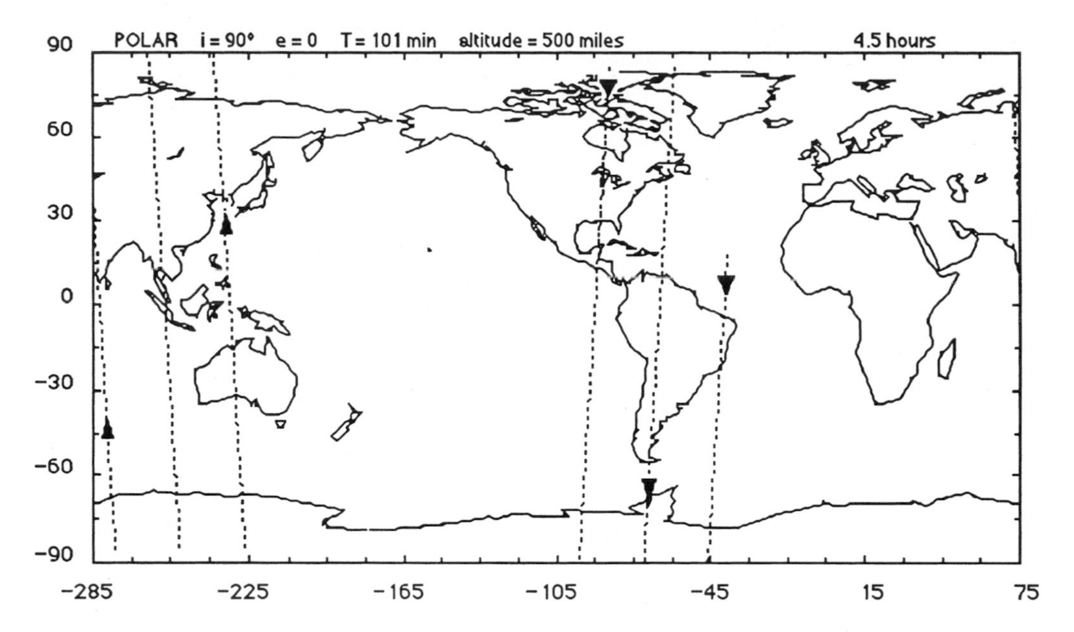

Figure 3.15 Ground trace of the orbit shown in Figure 3.14. Although it is actually heading due south, the spacecraft drifts south-southwest on the map because the Earth it rotating under it. Similarly, it drifts north-northwest on the map when it is heading due north. The trace shows 2½ orbits.

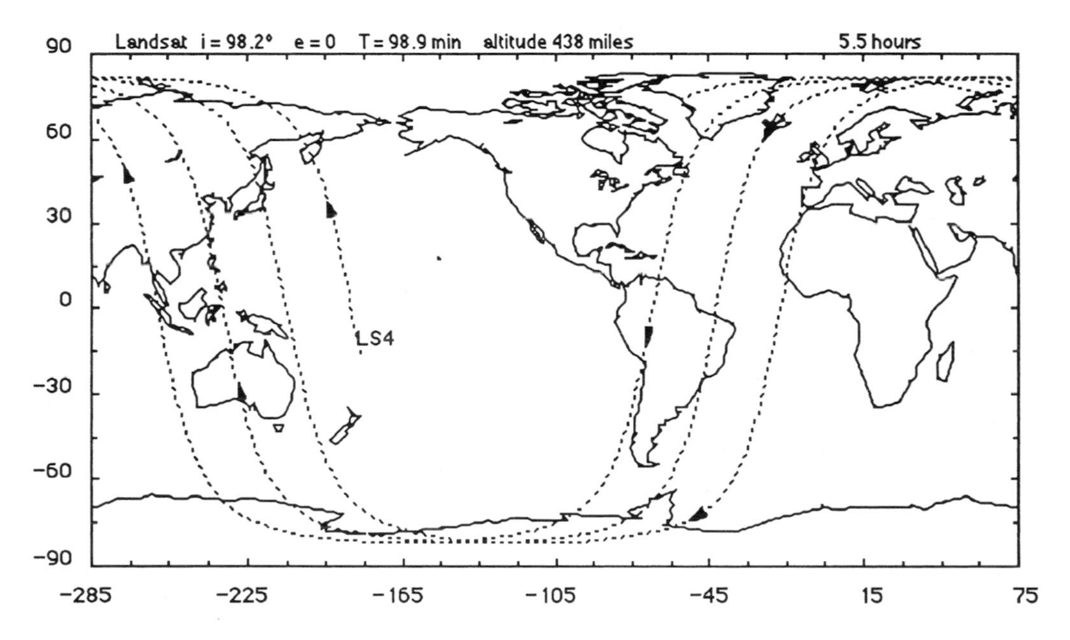

Figure 3.16 A Sun-synchronous orbit at 438 miles (705 km) altitude, inclination 98.2 degrees. It passes over the equator at the same *local* time on each orbit. The trace shows 3 1/3 orbits.

far away, but can become significant as it accumulates over a long period of time. Another small perturbation is due the radiation pressure of the sunlight as was described in the section on solar sails in Chapter 2.

Atmospheric drag, friction with the air, causes the most significant perturbation in the orbit of low-flying satellites. The effect is most pronounced at perigee where the spacecraft is closest to Earth. Each time the spacecraft passes perigee, the drag reduces its energy slightly so it cannot make it out as far as its previous apogee. On each orbit, then,

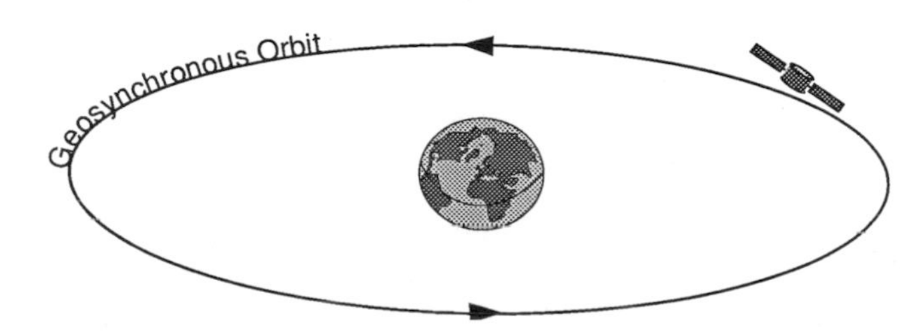

Figure 3.17 Geosynchronous orbit drawn to scale. The circular orbit is at an altitude of 22,300 miles over the equator. Period is 23 hours 56 minutes.

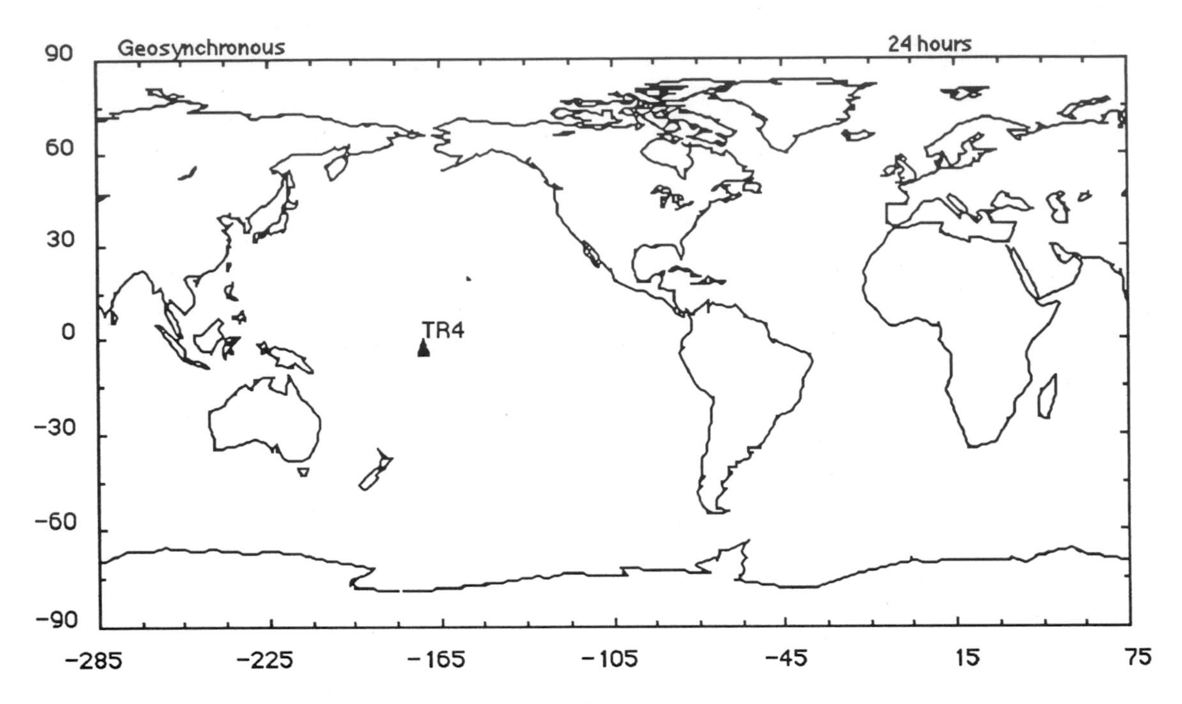

Figure 3.18 The ground trace of a geosynchronous satellite over a spot in the Pacific Ocean is simply a dot on the equator.

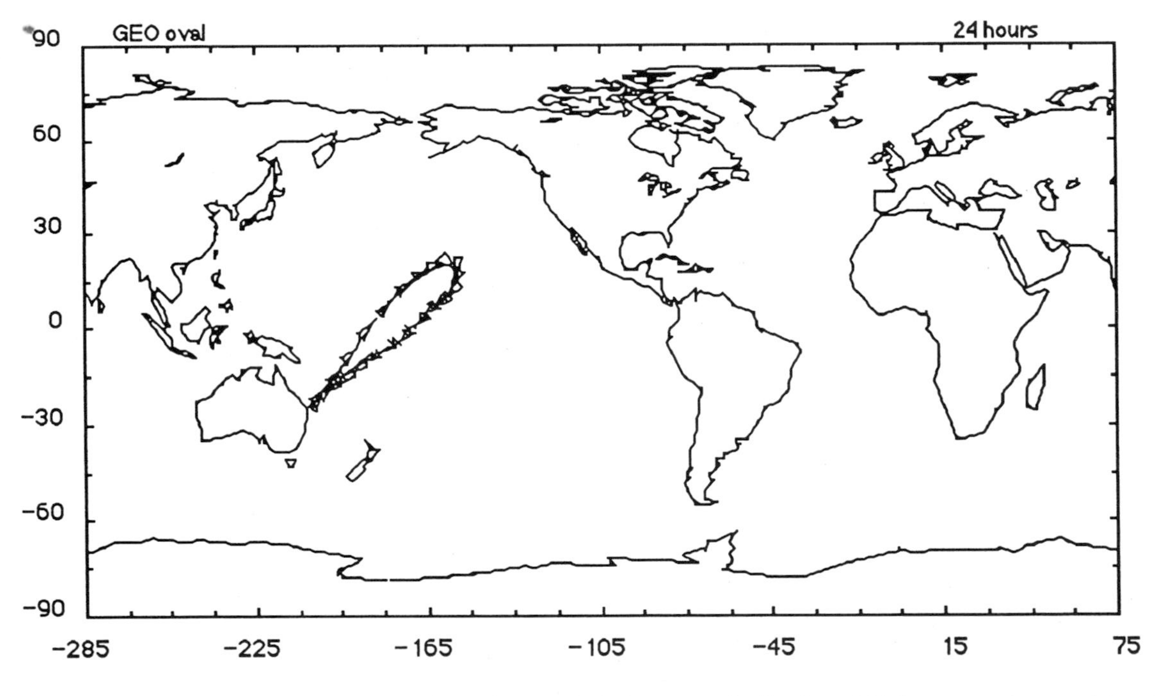

Figure 3.19 Oval-shaped ground trace of a satellite with a period of 1 day, an inclination of 20 degrees, and eccentricity of 0.2.

apogee is lowered slightly while perigee remains at about the same altitude. See Figure 3.22. The eccentricity decreases as the orbit becomes more circular. After a period of time apogee lowers into the denser air so that the spacecraft experiences continuous frictional drag. Finally it cannot remain in orbit and it falls back to Earth, usually burning up in the atmosphere on its way down. The lowest of the satellite orbits decay rather rapidly over a period of a few weeks or months. Atmospheric drag varies seasonally, and with disturbances on the Sun and in the space environment, a subject to be covered in the next chapter.

Orbital Maneuvering

Once in space, the orbit of a satellite can be modified or changed by applying thrust in the proper direction and at the

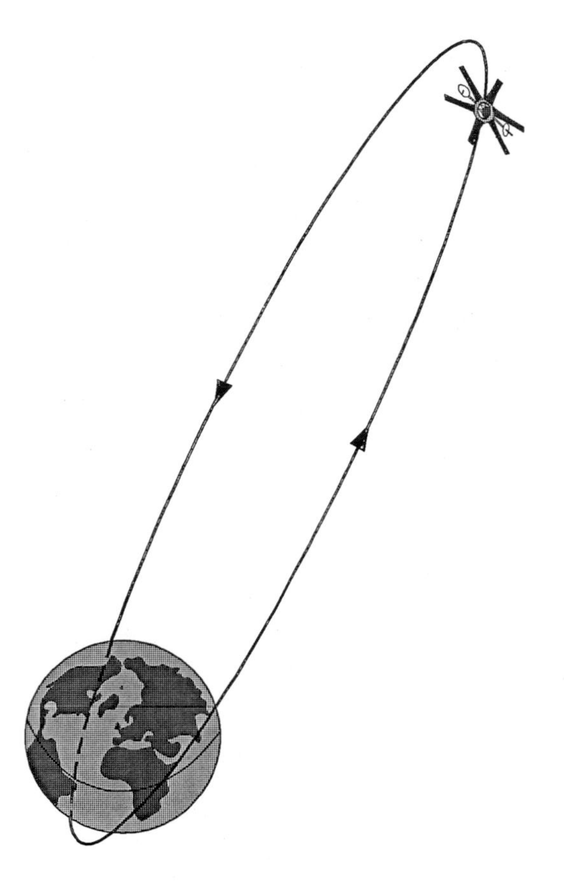

Figure 3.20 Molniya orbit drawn to scale. Perigee is at about 300 miles in the southern hemisphere. Apogee is at about 25,000 miles in the northern hemisphere.

proper time to change the speed and energy of the spacecraft. In general, the orbit will increase in size if energy is increased by firing the engines normally. If energy is decreased by turning around and firing the engines in the direction of motion, called *retrofiring,* then the orbit decreases in size.

Changing Eccentricity

Look at Figure 3.23. If the rockets are fired at perigee the satellite's speed and energy increase. It will go into a larger orbit, farther out beyond its original apogee. The orbit has become more eccentric. Refer now to Figure 3.24. Suppose the satellite is at apogee when the engines are ignited. Once again, it transfers to a larger orbit because it has increased energy, but note that this time the orbit becomes less elliptical. If just the proper amount of energy is added, the orbit becomes a circle. This technique, called *apogee kick,* is used to circularize the orbits.

Note, also, that firing the engines anywhere in a circular orbit will increase the eccentricity of the orbit. *Any* change from circular increases the eccentricity!

Hohmann Transfer

Next, suppose a satellite is to be raised to a higher altitude by transferring it from one circular orbit, position 1 in Figure 3.25, to another higher circular orbit, position 4. The move is accomplished by igniting the engines twice. First, when the satellite is at point 2, the rockets are fired to increase its speed and put it into an elliptical transfer orbit with apogee at 3, the

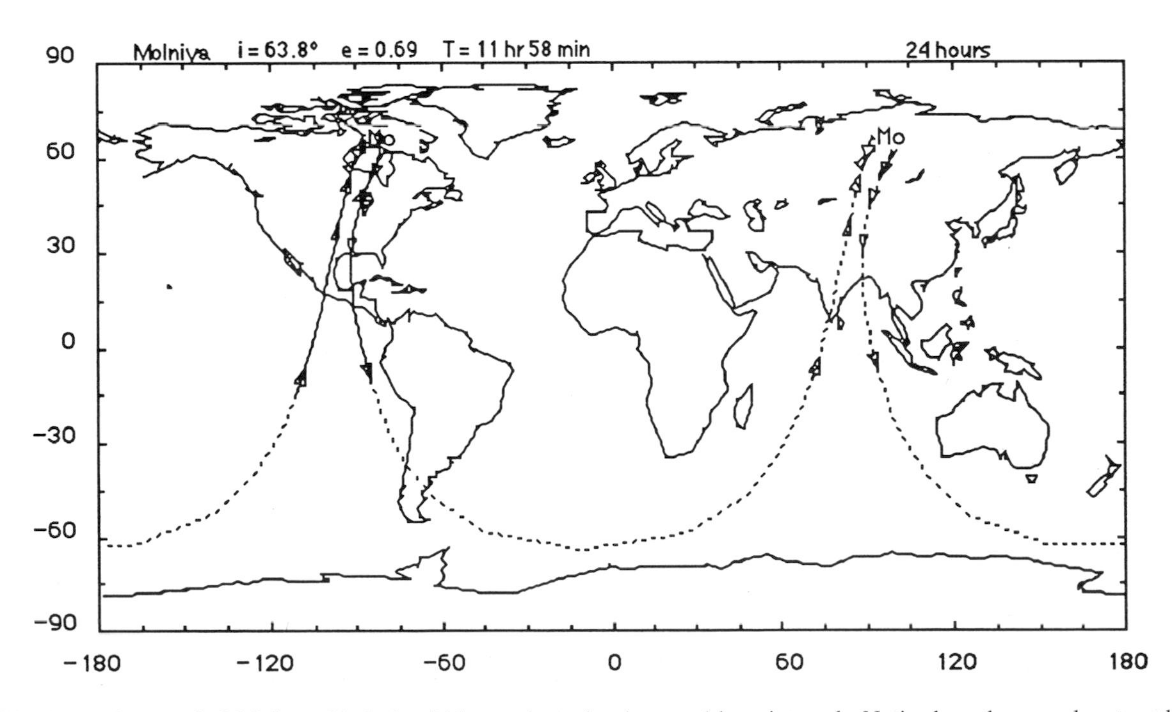

Figure 3.21 Ground trace of a Molniya orbit during 24 hours. Arrowheads are at 1 hour intervals. Notice how they are close together in the northern hemisphere, showing a slow speed near apogee. In the southern hemisphere the arrowheads are far apart indicating a higher speed near perigee. Thus, the satellite hovers for 9 to 10 hours alternately over Siberia and over Canada and spends only about 2 hours in the southern hemisphere.

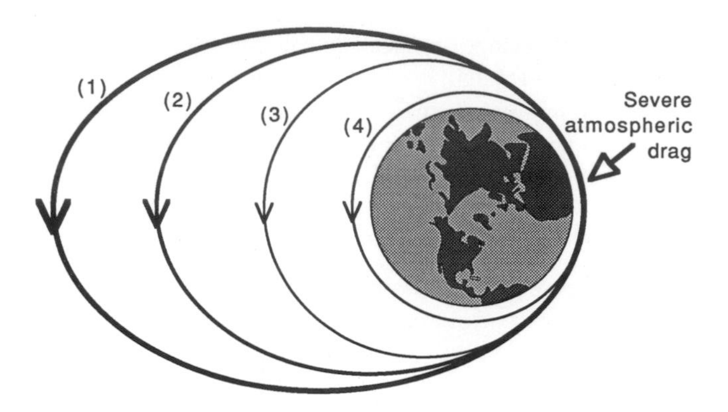

Figure 3.22 Atmospheric drag circularizes an elliptical orbit. The orbit starts at (1). Atmospheric drag at perigee causes apogee to become lower with each orbit (2) and (3) until it becomes nearly circular (4) and burns in.

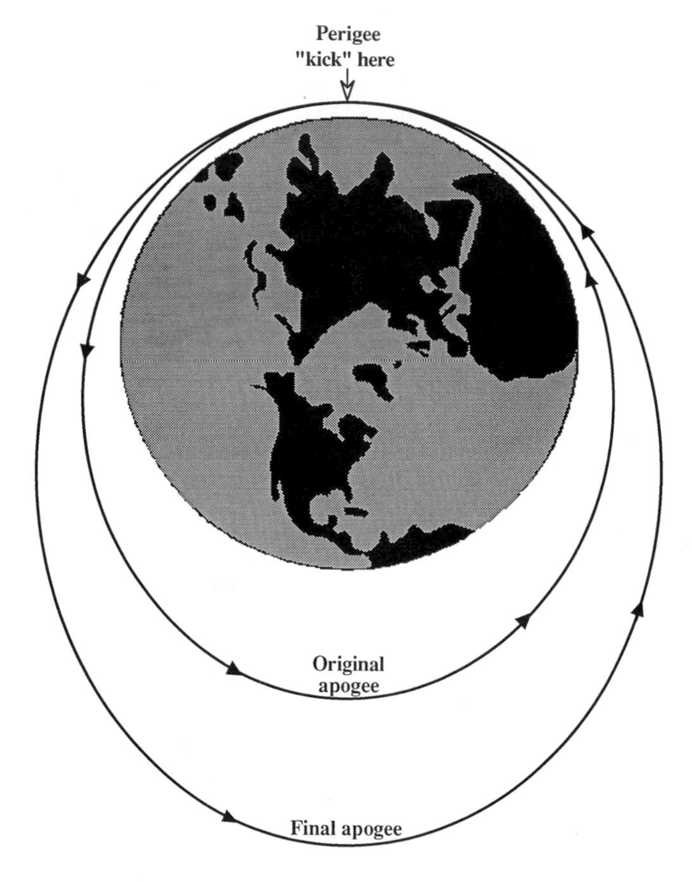

Figure 3.23 Firing the engine at perigee increases the altitude of apogee and the eccentricity.

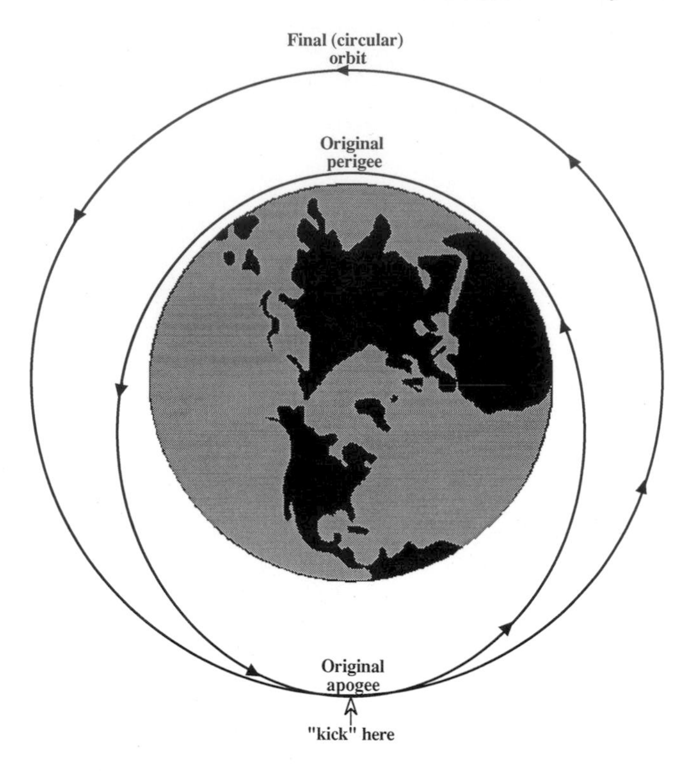

Figure 3.24 Firing the engines at apogee increases the size of the orbit. Just the right kick will circularize the orbit.

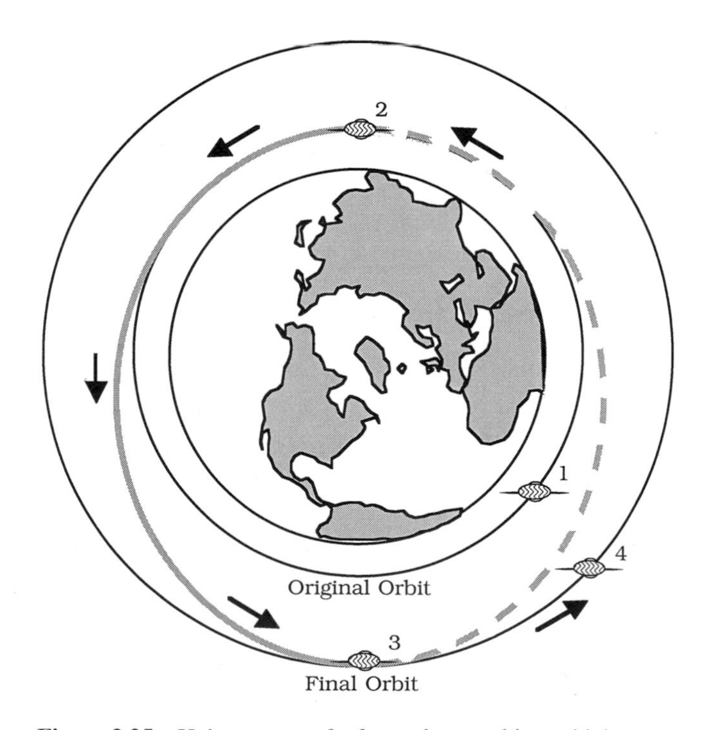

Figure 3.25 Hohmann transfer from a lower orbit to a higher one.

desired final altitude. When it reaches point 3 the engines are ignited once more to circularize the orbit at that altitude. This maneuver is called a *Hohmann transfer,* named after the German who, in 1925, first came up with the idea.

An alternative method of moving from one circular orbit to another is shown in Figure 3.26. Engines are fired at position 2, but with a greater increase in energy than for a Hohmann transfer. The elliptical transfer orbit is therefore larger and intersects the desired circular orbit at 3. When the satellite reaches 3 the engines are ignited again, but this time greater energy is required because the direction as well as the speed must be changed. The satellite has to turn a corner. This method is faster than the Hohmann transfer, but requires more fuel. The Hohmann method uses the least fuel but requires more time to complete.

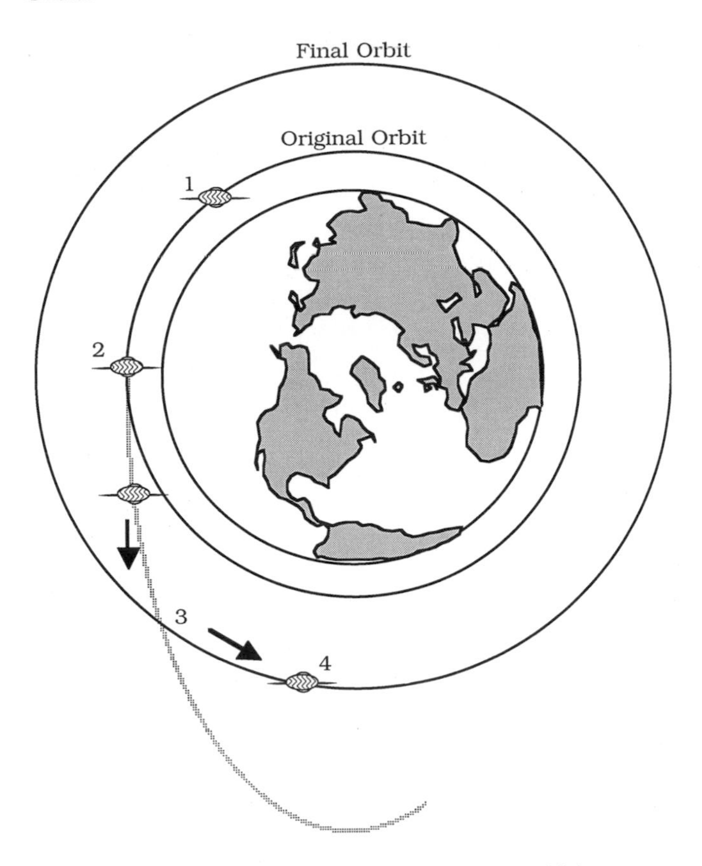

Figure 3.26 Fast transfer from a lower orbit to a higher one.

Moving from a larger, higher altitude orbit to a lower, smaller orbit may be accomplished by a Hohmann transfer, also. Retrofiring the rockets moves the satellite into an elliptical transfer orbit. Then a second retrofiring at the perigee of the transfer ellipse circularizes the orbit at the desired altitude. Again, there are faster ways of doing it, but the Hohmann technique requires the least fuel.

Rendezvous

How can one spacecraft catch up and rendezvous with another? Sounds like a simple problem. Luke Skywalker and Han Solo had no problem in the *Star Wars* movies. Just point at the Empire tie fighter or cut inside its turn and "step on the gas." That is the way jet fighter pilots do it!

In the real world of space it is not so easy. A jet fighter is supported by its wings moving through the air and the air in turn produces a frictional drag on the aircraft. By increasing the drag more on one side than the other the plane turns in the direction of the greater drag. This is accomplished with ailerons, elevators, and rudders. But in space there is no air to "push against."

The first attempt at orbital rendezvous was in June 1965, when Jim McDivitt and Ed White tried to maneuver the Gemini IV spacecraft next to the second stage rocket which was in the same orbit. But McDivitt's experience as a pilot was of no use to him in this situation.

The first successful orbital rendezvous came in December 1965, when Wally Schirra and Tom Stafford in Gemini VI came within 6 feet of Gemini VII carrying Frank Borman and Jim Lovell and flew side by side for over 5 hours. Figure 1.5 is a photograph of their success.

Imagine you are in orbit in a spaceship and want to join up with a space station just ahead of you. You are both in the same orbit and therefore going at the same speed. You aim toward the space station, fire your engines to catch up, and discover that you fall farther behind! Why?

Look again at Figure 3.26. Suppose your spaceship is in the inner orbit and you are at perigee. Fire your engines and you increase your speed and your energy. Therefore, you will not remain in that orbit, but will move to the larger, higher orbit. To further complicate the situation, because you have transferred to a higher orbit, you move farther from Earth and consequently you slow down, even though you just fired your engines to speed up.

Rendezvous with another spacecraft, then, is rather tricky. Increasing your energy to catch up will move you to a higher orbit and you will instead fall behind. Imagine the flying saucer at A in Figure 3.27 wants to rendezvous with the mother ship now located at B. The saucer ignites its engines to move to the elliptical transfer orbit. The proper increase in energy will put it into a transfer orbit which touches the mother ship's orbit at R, the apogee of the transfer orbit. When it reaches R it again fires its engines, an apogee kick, to transfer to the mother ship's orbit. This is actually a Hohmann transfer.

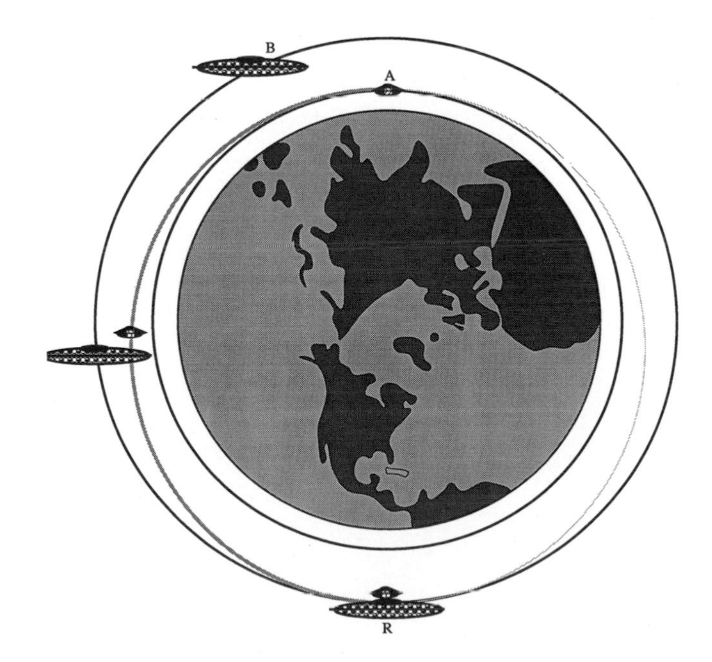

Figure 3.27 The flying saucer at A decides to rendezvous with the mother ship located at B. In a lower orbit, the saucer is moving faster than the mother ship in the higher orbit. Timing is critical so they both reach the rendezvous point R at the same time. See text.

The tricky part is to time the engine ignitions in such a way that the flying saucer and the mother ship both reach the rendezvous point, R, at the same time. Because the mother ship is in a higher orbit, it is moving at a slower speed. Therefore, the saucer must wait until the mother ship is closer to the rendezvous point because the saucer is traveling at a higher speed. Calculating the exact time is quite involved and we will not go into it here.

Other Maneuvers

It is also possible to change the inclination of an orbit, change the location of perigee, rendezvous with spacecraft in different orbital planes, and perform other maneuvers. Each requires the application of the proper amount of thrust in the proper direction and at the proper time.

Deorbit

Reentry and landing back on Earth are accomplished by turning the spacecraft around and firing the engines. This maneuver applies thrust in the opposite direction to "put on the brakes," to reduce the satellite's energy, transferring it to a lower orbit which intersects Earth, increasing speed as it falls. In Figure 3.28, retrofiring at the place shown transfers the spacecraft to the more elliptical orbit which intersects Earth at the place of splashdown. Timing and length of burn must be precise so the new orbit intersects Earth at the desired location for landing. The atmospheric drag during reentry distorts the orbit from a true ellipse which must be taken into account also.

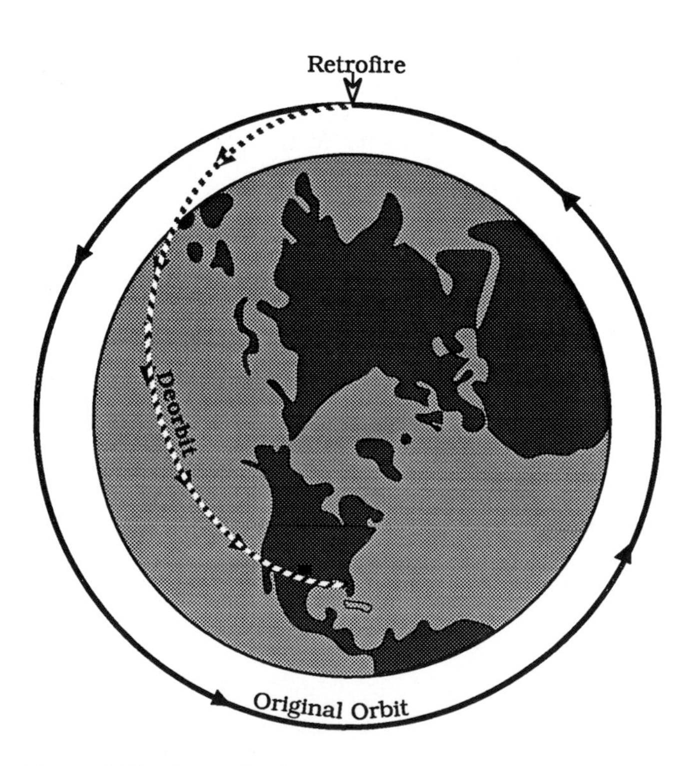

Figure 3.28 Space Shuttle retrofires over the Indian Ocean to deorbit and land in Florida.

The Lagrangian Points

There are points in space where the gravitational attraction of Sun and Earth combine in such a way that a small object placed at that point will just stay there. These places are called *libration points* or *Lagrangian points* after Joseph Lagrange the French mathematician who discovered them theoretically. The L point locations are shown in Figure 3.29. To say "it will just stay there" is misleading. Actually, an L point is really a rather large area with room for many spacecraft.

L1, L2, and L3 are on the Earth-Sun line. L1 is located about 1.5 million miles (2.4 million km) from Earth on the sunny side, while L2 lies the same distance from the Earth on the dark side. L3 lies on the opposite side of the Sun. These three points are unstable. If displaced perpendicular to the Earth-Sun line, the object would return to its position. However, if displaced toward Earth or the Sun, it would continue to drift away from the L point.

The other two points lie in Earth's orbit. L4 lies 60 degrees ahead of the Earth and L5 lies 60 degrees behind the Earth. Thus the Sun, Earth, and the L point form an equilateral triangle with the L point at one corner and the Sun and Earth at the other two. A spacecraft at the L point is the same distance from the Earth as it is from the Sun, 93 million miles (150 million km). These are stable points; placed at one of these locations, the spacecraft would remain with no further expenditure of energy. Spacecraft at L4 or L5 would remain essentially fixed in the Earth's sky like a geosynchronous satellite.

The L1 point, nearly a million miles closer to the Sun, is now occupied by the Advanced Composition Explorer ACE and the Solar and Heliospheric Observatory (SOHO) spacecraft that monitor the Sun and the solar wind. By remaining

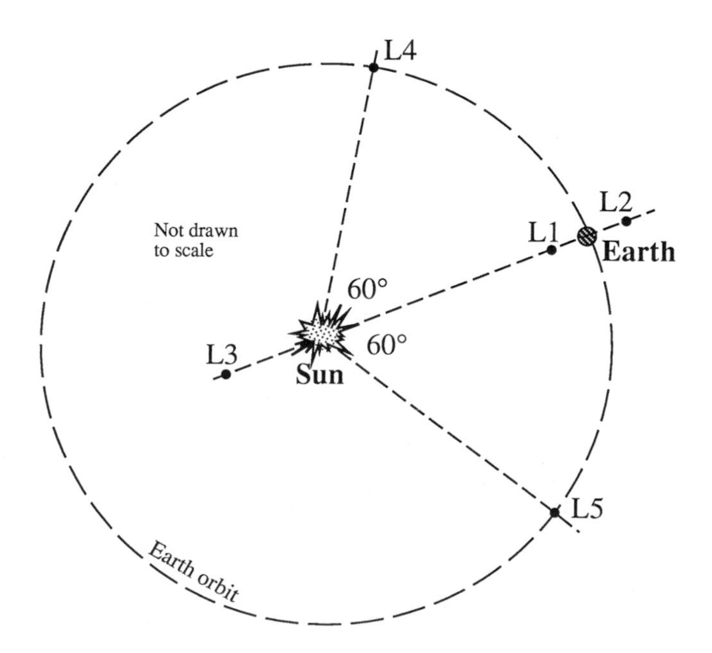

Figure 3.29 The Lagrangian points in the Earth-Sun system.

locked in at L1 on the Earth-Sun line, ACE and SOHO make the trip around the Sun in the same time that Earth does, 1 year. This seems to violate Kepler's third law which says that any object in a circular orbit has a period that depends only on the radius of its orbit, i.e., its distance from the center. (Sample calculations were done in MATHBOX 3.2 and MATHBOX 3.3.) Here we have spacecraft moving around the Sun in an orbit whose radius is a million miles smaller than Earth's orbit; yet it has the same orbital period as Earth, 1 year. With a smaller orbit, the period should be shorter and they should move faster than Earth. The answer to this apparent contradiction is that Kepler formulated his laws based on two bodies, the Sun and a planet. Here we have three: a spacecraft, Sun, and Earth. At L1, Earth's gravity pulls on the spacecraft just the right amount to slow them down so they keep pace with the Earth. They are in a non-Keplerian orbit.

A number of other spacecraft are destined for the libration points in the next few years. L1 is an excellent spot for those studying the Sun. Geophysicists and solar astronomers would like to have a spacecraft like ACE closer to the Sun than L1. As was pointed out above, if an object is displaced away from L1 toward the Sun, solar gravity will pull it in. However, radiation pressure on a solar sail attached to the spacecraft would exert a force away from the Sun that just balances solar gravity, keeping the spacecraft in place. Geostorms, a spacecraft equipped with a 230 foot (70 m) solar sail, is being designed to do just that. It would find a balance point between gravity and radiation pressure about 1.5 million miles (2.4 million kilometers) from L1 toward the Sun. It may be launched by 2005.

L2, on the other hand, is always in Earth's shadow. This would be an ideal place for an astronomical telescope where the Earth would act as a shield to protect the instruments from damaging sunlight. The L points could also be stepping stones for trips beyond the Moon.

Similar Lagrangian points lie in the Earth-Moon system. In Figure 3.29, just put Earth in the place of the Sun and the Moon in the place of the Earth and you see the same arrangement of the points. Unstable L1, L2, and L3 are on the Earth-Moon line. Stable L4 and L5 lie in the Moon's orbit, and would make good locations for space colonies as will be described in Chapter 11.

In fact, any two body system has similar points. The Trojan asteroids lie at L4 and the Greek asteroids are at L5 in the Sun-Jupiter system. They are asteroids that in their wanderings happened to drop into those locations and remain there.

Trajectory to the Moon

The Moon is in orbit around the Earth just like a spacecraft, so going to the Moon is similar to a rendezvous maneuver. There is one important difference, however; the Moon is so large that its gravitational attraction will control the orbit of the spacecraft as it draws near.

First consider a Hohmann transfer. The rockets are ignited to inject the spacecraft from a low Earth orbit to an elliptical transfer orbit with apogee at the Moon. As the spacecraft heads for apogee it slows down, as usual. But as it approaches the Moon and comes under lunar gravitational influence, it deviates from its transfer ellipse and falls toward the Moon with increasing speed. One of two things may happen: it hits the Moon or it misses but because of its speed it curves around the Moon and escapes. Continuing on, it is then in a different elliptical orbit around Earth. Figure 3.30 shows this situation. If our goal was to land on the Moon, we did something wrong. We succeeded in our rendezvous, but our speed was too great. The solution is to slow down at the proper moment so the spacecraft is captured by the Moon and goes into a lunar orbit.

The near miss, described above and in Figure 3.30, that accelerates the spacecraft and sends it on a curved trajectory past the Moon can be used to advantage. Called a *gravity assist,* it changes the speed and direction of flight of the spacecraft, a maneuver that would otherwise require large amounts of fuel. Voyager used a near miss past Jupiter to send it on its way to Saturn and the outer Solar System. Ulysses received a gravity assist from Jupiter to direct it over the poles of the Sun. Gravity assists were also used to send Cassini past Venus and Earth to Saturn, and Galileo on a complex tour of the moons of Jupiter. We will examine this subject again in later chapters.

Figure 3.31 is the Apollo mission profile, not drawn to scale. First, the Saturn V rocket carried the command module, service module, lunar module, and a booster rocket into a low parking orbit. After everything was checked out and

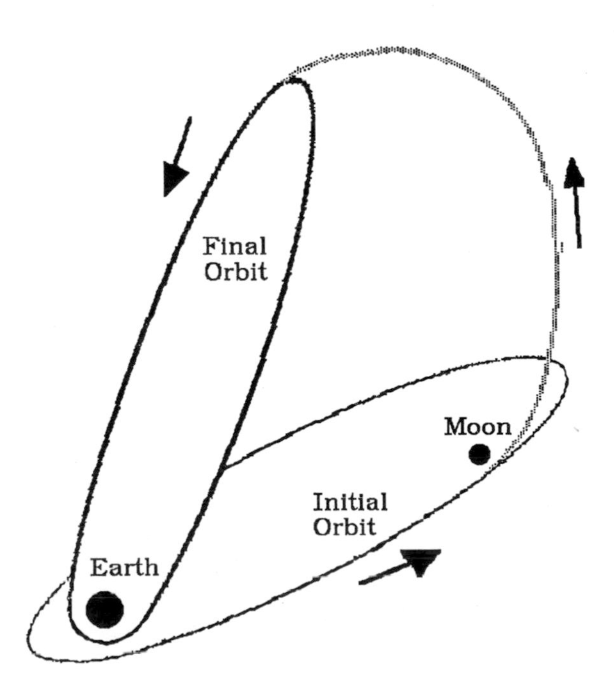

Figure 3.30 Trajectory past the Moon transfers the spacecraft to a new orbit around Earth.

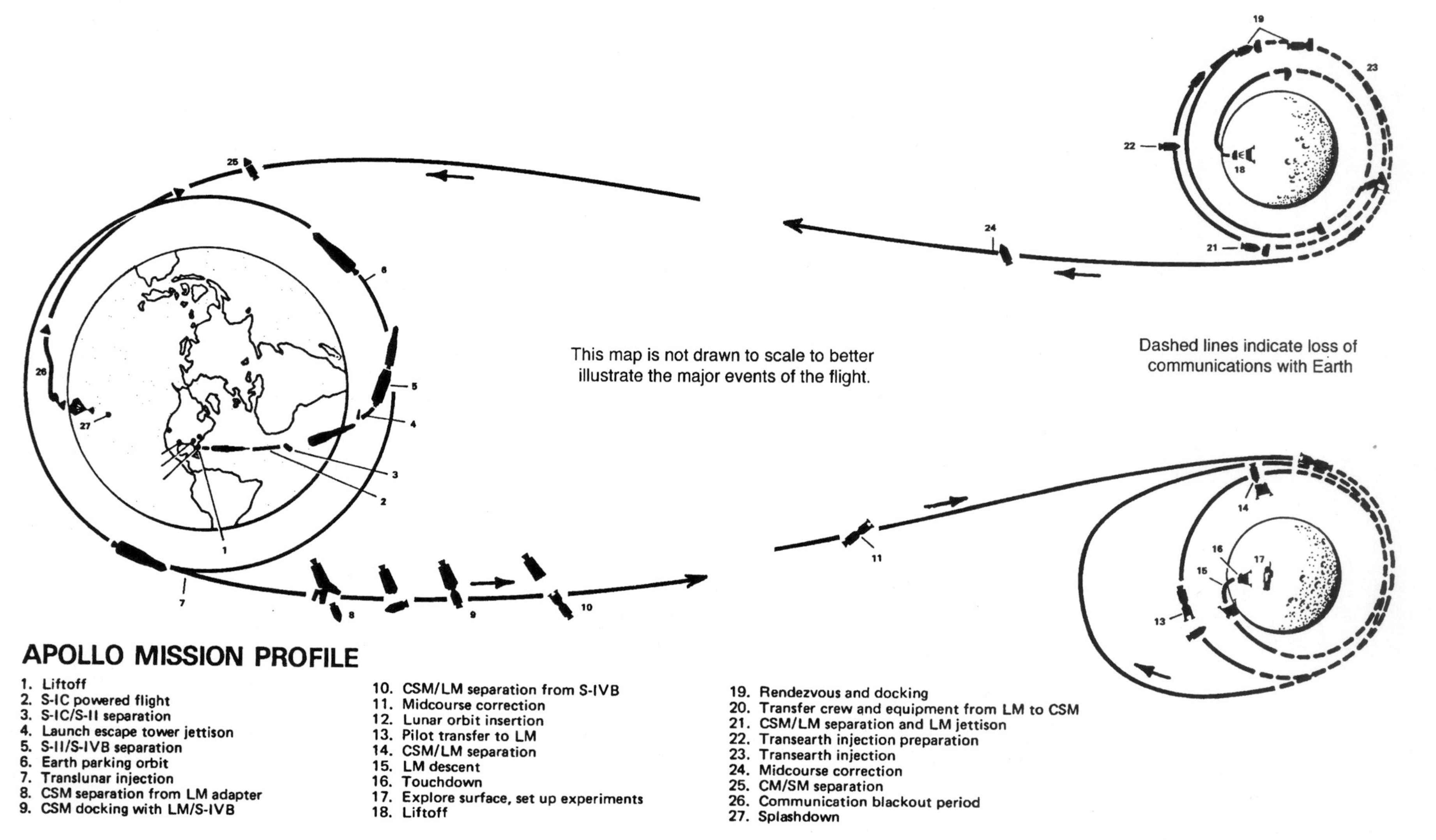

Figure 3.31 Apollo mission to the Moon. *Courtesy of NASA.*

the vehicle was at the correct position in orbit, the booster rocket fired to inject it into the transfer orbit to the Moon (lower path in Figure 3.31). If there was any error in the transfer orbit, engines were fired again for a midcourse correction. The vehicle coasted for nearly 3 days and as it approached the Moon, the gravitational attraction caused it to leave the transfer orbit and curve around the Moon. Retrofiring the rockets slowed the vehicle and injected it into an elliptical lunar orbit. Another rocket burn circularized that orbit. The lunar module with two astronauts separated from the command and service module and its rockets retrofired to take it out of orbit to a landing. As it approached the surface, rockets were fired to slow its descent and set it gently onto the lunar soil. After surface activities were finished, the astronauts again fired the lunar module rockets to take it back to orbit and a rendezvous with the command and service modules where the third astronaut was waiting (upper part of Figure 3.31). The lunar module was then separated and left behind. Engines were fired once more to inject the command and service modules into a transfer orbit back to Earth. A midcourse correction was made if necessary. As Earth gravity took over, the spacecraft deviated from the transfer orbit for a direct entry into the atmosphere. Only the command module with the three astronauts splashed down in the ocean.

Apollo 13 had a problem 56 hours into the flight when an oxygen tank exploded in the service module. A safety feature was built into the Apollo trajectory calculations, however. Their speed was sufficient to reach the Moon, but just short of escape speed from Earth's gravity. (MATHBOX 3.4 shows how to calculate escape speed from Earth or any other celestial body.) Thus, even if they missed the Moon, they would remain in orbit around the Earth as was shown in Figure 3.30. However, they still had to fire the rockets to bring them down to an ocean landing or they would have remained in Earth orbit "forever." The lunar module became their lifeboat for surviving the 6 day journey with minimal food, water, heat, and electricity, all of which depended on the lost oxygen. The lunar module descent engine, designed for landing on the Moon and returning to lunar orbit, was used to adjust their course back to Earth.

These were the most complicated manned spaceflights ever made. Count how many times rocket engines had to be fired! Each firing had to be made at precisely the right moment for precisely the correct period of time in exactly the proper direction. Computing those times and directions applied principles of orbital mechanics that had never before been used in manned flight. It was a tremendous accomplishment; nothing comparable has been achieved since. As a side benefit, the need to make the complex calculations prompted the advancement of computer science.

To Another Planet

Flying to another planet is similar to completing a rendezvous with a space station. The difference is that in near Earth orbits the gravitational attraction of Earth controls the

MATHBOX 3.4

Escape Speed

Throw a baseball up and it slows, stops, and comes back down. How fast do you have to throw it so it keeps going never to return? That speed is called escape speed. It is different for every planet or other central body and it can be calculated by the formula

$$v = \sqrt{\frac{2\,GM}{r}}$$

where r is the distance of the starting point from the center of the central body and GM is the gravitation constant for the central body as explained in MATHBOX 3.2.

Let us find the escape speed from the surface of the Earth. GM for Earth is 1.24×10^{12} if r is given in miles and v is in miles per hour. At the surface we are about 4000 miles (6400 km) from the center, therefore,

$$v = \sqrt{\frac{(2)\,(1.24 \times 10^{12}}{4000}} = 24{,}900 \text{ mi/hr}$$

In metric units, GM for Earth is 3.9860×10^{14} if r and v are in meters and meters per second.

$$v = \sqrt{\frac{(2)\,(3.986 \times 10^{14})}{6{,}400{,}000}} = 11{,}200 \text{ m/s}$$

We can calculate escape speed at any distance from any central body provided we know its mass so that we can calculate GM for that body. Again, refer to MATHBOX 3.2.

orbit, while in interplanetary space the Sun is in control and toward the end of the flight the gravitational attraction of the target planet takes over. So we can examine the trajectory to a planet by separately considering these three segments.

Imagine a flight to Mars as depicted in Figure 3.32. The Sun is at the focus of the orbits of both Mars and Earth. From our point of view high above the north pole, Earth and Mars are both moving counterclockwise in their orbits. The spacecraft is first injected into a low Earth orbit. At just the proper time the engines are ignited and the spacecraft is transferred to the dashed trajectory at a speed greater than escape velocity from Earth. This new trajectory is not a closed ellipse around Earth, but is an open curve called a *hyperbola.* It is curved at first but approaches a straight line as it gets farther from Earth. About a million miles out from Earth the gravity of the Sun becomes dominant and the spacecraft is in an elliptical orbit around the Sun.

We can look at the path of the vehicle from two points of view: with respect to the Earth or with respect to the Sun. With respect to Earth it is on an escape trajectory, never to return. But with respect to the Sun it is in an elliptical orbit, just like a planet. Actually, the spacecraft was orbiting the Sun even when it was on the launch pad, since the Earth and everything on it is in orbit around the Sun.

When we fire the vehicle into the hyperbolic escape trajectory, we send it off in the same direction that the Earth is moving. It is then traveling faster than the Earth, which transfers it to an orbit larger than Earth's orbit toward the outer planets. This elliptical transfer orbit is like a Hohmann transfer except at the two ends where the gravity of Earth and Mars control the motion.

If the spacecraft is to land on Mars, another maneuver is necessary. Just as in the flight to the Moon, as the spacecraft approaches Mars it falls in toward the planet and its speed increases. Retrofiring the engines reduces its speed so it enters an orbit around Mars. Without this braking maneuver the vehicle would be deflected somewhat by Mars gravity and continue on a modified orbit around the Sun. Actually, if the vehicle missed the planet completely, it would just continue to orbit the Sun.

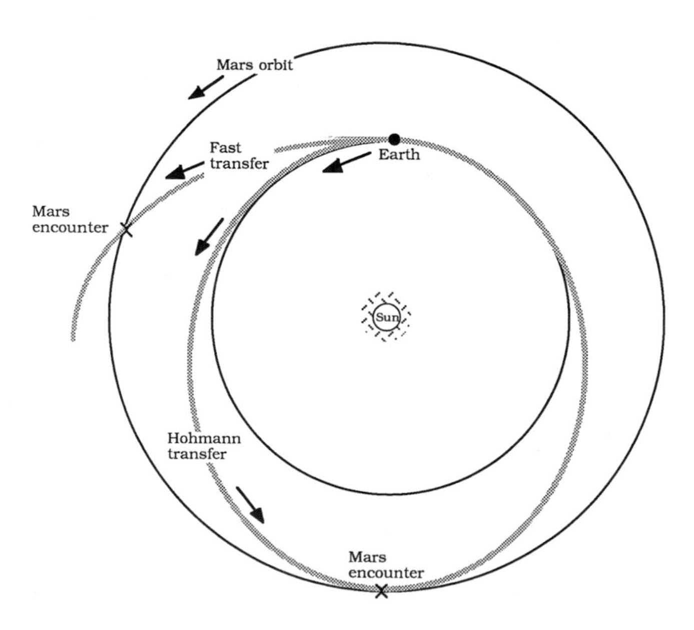

Figure 3.32 Transfer orbits to Mars.

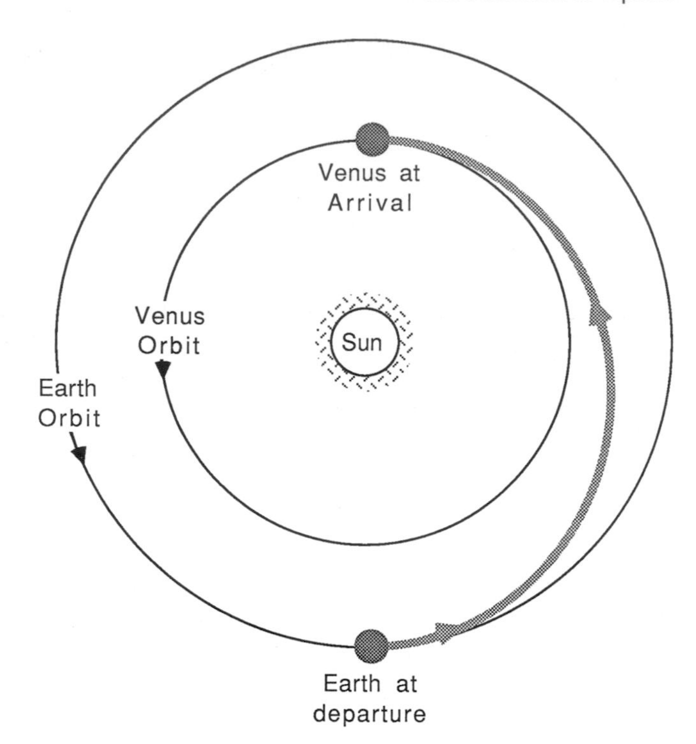

Figure 3.33 Hohmann transfer to Venus.

A method of braking that does not require fuel is *aerobraking*. A space vehicle approaching a planet can be steered on a near miss trajectory, close enough to skim the top of the atmosphere without hitting the planet. The craft slows down by the air drag and is captured into a long elliptical orbit around the planet. After a number of passes the orbit becomes more circular (Figure 3.22) and a burst of the rockets can pull perigee up out of the drag. A sturdy, well-insulated spacecraft is needed because the heat and stress of entering the atmosphere at high speed could damage or destroy it. This technique was used for circularizing the orbit of the Magellan spacecraft around Venus.

Going to Venus is similar to going to Mars except that Venus's orbit lies inside Earth's orbit. Therefore, the vehicle must be accelerated to escape velocity on a hyperbolic trajectory from Earth, but in the direction opposite to the direction of Earth's motion. Its speed with respect to the Sun is then less than Earth's speed and it moves into a smaller transfer orbit toward the inner planets. See Figure 3.33.

Many spacecraft have left Earth for encounters with other planets. We will discuss some of them further in later chapters.

DISCUSSION QUESTIONS

1. According to Newton, an object accelerates (changes velocity) only as long as a force is applied. When a rocket engine shuts down there is no more thrust. Why then does a spacecraft change speed in an elliptical orbit? Why doesn't it change speed in a circular orbit?
2. If Earth rotated on its axis once in 12 hours instead of once in 24 hours, would a geosynchronous orbit be higher or lower?
3. Describe the ground trace of a satellite orbiting at an altitude of 40,000 miles.
4. If a spacecraft is 10 miles directly above a space station, how would you get it down for rendezvous and docking?
5. What are the advantages and disadvantages of a Hohmann transfer to carry a human crew to Mars? a cargo ship to Mars?
6. How are the motions of satellites in orbit around Earth and electrons in orbit around the nucleus of an atom the same? How are they different?

ADDITIONAL READING

American Astronautical Society. *The Journal of the Astronautical Sciences.* Quarterly publication of the AAS.

Bate, Roger R., Donald D. Mueller, and Jerry E. White. *Fundamentals of Astrodynamics.* over Publications, 1971. Calculus level mathematics.

Cortright, Edgar M., editor. *Apollo Expeditions to the Moon.* NASA SP-350, Government Printing Office, 1975. Detailed, well-illustrated history of the Apollo program..

Melbourne, William G. "Navigation between the Planets." *Scientific American,* June 1976.

Mickelwait, Aubrey B., et al. "Interplanetary Navigation." *Scientific American,* March 1960.

Muolo, Michael J. *Space Handbook, An Analyst's Guide, Volume Two.* Air University Press, Government Printing Office, December 1993. Mostly descriptive with some algebra.

Prussing, John E., and Bruce A. Conway. *Orbital Mechanics.* Oxford University Press, 1993. Vector calculus level mathematics.

Thomson, William T. *Introduction to Space Dynamics.* Dover Publications, 1986. Calculus level mathematics.

NOTES

Chapter 4

The Space Environment

When a spacecraft leaves Earth it enters a rather unfriendly place. At the surface of Earth we live in a warm comfortable environment with air to breathe, water to drink, land to walk on. Space offers none of these. We often hear and use the term "empty space." True enough, the content of space is pretty sparse when compared to our earthly environment. The atmosphere surrounding us contains about 2 pounds of air in each cubic yard (about 1 kg/m^3) at a pressure of 14.7 pounds per square inch (101 kilopascals, kPa). At 150 miles altitude a cubic yard contains only a millionth of a pound at a pressure of 0.07 pound per square inch (0.5 kPa), a better vacuum than can be produced by laboratory equipment on Earth.

Space is far from being truly empty. In the space environment around Earth we find not only atoms and other particles, many of them electrically charged, but also electromagnetic energy in the form of waves: x-rays, ultraviolet, gamma rays, visible light, infrared, radio waves, and microwaves. Cosmic rays which are really particles, not rays, whiz past at speeds approaching the speed of light. In addition, comets leave behind gases and pieces of debris during their excursions through the Solar System, and meteoroids, which are small pieces of dust and debris, abound in space. Meteoroids and comets are thought to be objects left over from the formation of the Solar System. Man-made debris from spaceflight activities, from flecks of paint to burned-out rockets and satellites, is scattered about in low Earth orbits.

The electromagnetic radiant energy and the pieces of matter that are present in space could damage a spacecraft and could be lethal to an unprotected human. This chapter discusses these components of the space environment and their potential hazard to spaceflight.

The Sun

The Sun is the source of all the particles and waves in the space environment, except for comets, meteoroids, and man-made debris. Some cosmic rays come from the Sun, but most come from outside the Solar System; indeed, some of them may come from outside our Galaxy. To understand the contents of the environment in which spacecraft and astronauts fly, we must first understand the structure and activity of the Sun. We also need to become familiar with the electromagnetic spectrum. Electromagnetic waves will be of interest to us several times as we progress through this book.

The Sun is a star. We can see that it is round and through telescopes we can observe features on its surface. Other stars appear as just points of light even with the largest telescopes because they are so far away. The average distance from the Sun to Earth is about 93 million miles (150 million km). Light from the Sun, traveling at approximately 186,000 miles per second (300,000 km/s), takes 8.3 minutes to travel that distance. On the other hand, light from the next nearest star, Alpha Centauri, the brightest star in the constellation Centaurus, takes 4.3 years to reach us.

The Sun is a sphere about 865,000 miles (1.4 million km) in diameter, 110 times the diameter of Earth and over a million times the volume of Earth. It is a medium-sized star. Like other stars, it is composed mostly of hydrogen with some helium and traces of other elements. Many stars are much larger; some giants are 400 times the diameter of our Sun.

Earth orbits the Sun with a period of 1 year; that is the definition of a year. The orbit is nearly circular with an eccentricity of 0.017, a perihelion of 91.3 million miles (148 million km), and aphelion of 94.4 million miles (153 million km).

Atoms

Before going farther, it is important to understand the structure of an atom. As long ago as 400 BC Greek philosophers believed that if an object were divided into smaller and smaller parts, it would eventually get down to a smallest piece that could not be divided any further. That smallest piece of matter was called an *atom*. Twenty-three hundred years later scientists discovered that atoms could be split further into three more elementary particles. *Electrons* are the

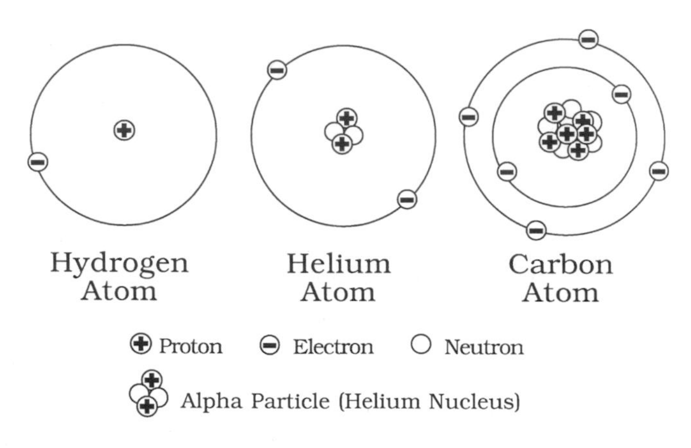

Figure 4.1 Structure of atoms.

smallest and they carry a negative electrical charge. *Protons* are about 1800 times more massive than electrons and they carry a positive charge. *Neutrons* have about the same mass as protons but are electrically neutral.

All atoms in the universe are made up of these three elementary particles. See Figure 4.1. Each *element* such as oxygen, iron, gold, lead, hydrogen, and carbon has a unique combination of protons, neutrons, and electrons, different from all other elements. The interior of an atom is something like a miniature Solar System, with a nucleus composed of protons and neutrons located at the center and electrons in orbit around it. The simplest possible atom would be a one-proton nucleus and one electron in orbit; that describes an atom of hydrogen. In scientific "shorthand" this is written ^{1}H, indicating one particle in the nucleus. The next more complex atom, helium, ^{4}He, has two protons and two neutrons in the nucleus with two electrons in orbit. Gold, ^{197}Au, has 79 protons and 118 neutrons with 79 electrons in orbit. The number of neutrons may vary. For example, carbon has six protons and six, seven, or eight neutrons in the nucleus (^{12}C, ^{13}C, or ^{14}C). These are called *isotopes* of carbon. The number of protons determines which element it is and the number of neutrons determines which isotope it is.

Notice that in a normal atom the number of protons and the number of electrons are equal. Thus, with an equal number of positive and negative charges, the atom is electrically neutral. Under some circumstances one or more electrons may be removed from an atom. We say that it is *ionized*. If all the electrons are stripped away, nothing is left but the nucleus. A hydrogen nucleus is simply a proton; a helium nucleus, two protons and two neutrons, is given a special name, an *alpha particle*.

A Nuclear Furnace

Now back to the Sun. The Sun shines by its own light; the Moon, planets, and comets shine by reflected sunlight. The source of the Sun's energy is a nuclear "furnace" in its center, powered by a process called *nuclear fusion*. Deep in the Sun's interior, hydrogen is being converted to helium with a release of energy. To be more specific, four protons, hydrogen nuclei, fuse together to form an alpha particle. In this fusion process two of the original protons have become neutrons; that is, they lost their positive charge. If you add together the mass of the four protons, you find that they total more than the mass of an alpha particle: about 0.7 percent of the mass has disappeared in the fusion. According to Albert Einstein's famous equation, $E = mc^2$, the lost mass is converted to energy. See MATHBOX 4.1 for an actual calculation. The mass-converted-to-energy is released in the form of a *gamma ray,* an electromagnetic wave.

Fusion requires temperatures in excess of 18 million degrees Fahrenheit (10 million degrees Celsius) and a pressure of a trillion tons per square inch. Such temperatures and pressures exist in the cores of stars. Once fusion begins, the heat produced keeps the process going until the hydrogen fuel runs out or the pressure drops. The necessary high pressure is maintained by the weight of the overlying mass of the star held together by gravity.

Nuclear fusion is the same process that takes place when a thermonuclear hydrogen bomb explodes. In a thermonuclear bomb, the fusion of a quantity of hydrogen into helium takes place in a fraction of a second and the energy is released as an enormous explosion causing a massive rearrangement of things in the immediate vicinity. It is a star on Earth for an instant. If it were possible to control the fusion, to slow it down so the energy would be released over a long period of time, it could be used beneficially, for example, to boil water to run steam electric generators.

The temperature at the center of the Sun is about 28 million degrees Fahrenheit (15 million degrees Celsius) and fusion takes place in the core which extends about one-fourth of way out from the center. There, hydrogen is fusing to form helium at a rate of nearly 5 million tons a second. Fusion in the Sun started about 4.5 billion years ago when the pressure of the overlying layers of gas compressed the core and raised the temperature sufficiently for it to turn on. The heat produced by the fusion then caused the layers to expand outward again until the inward contraction from gravity balanced the outward expansion from the heat. The Sun has been stable since then. With the bountiful supply of hydrogen available, it will continue to remain stable for another 5.5 billion years. We do not have to worry that it will blink out on us, at least not this weekend. This calculation is also shown in MATHBOX 4.1.

Our Sun, then, is a middle-aged star, about halfway through its lifetime. Temperatures and pressures are great enough to support the fusion in only about the central 25 percent of the Sun. After the hydrogen in that region is consumed, the nuclear furnace will decrease in temperature and the outer layers of the Sun will press inward causing the temperature to rise once more. When it reaches a temperature even higher than before, the helium will begin to fuse into

MATHBOX 4.1

Nuclear Fusion

When four protons combine into one alpha particle some mass is "lost," converted to energy.

The mass of atomic particles is measured in atomic mass units (amu). A proton has a mass of 1.0078, so four protons have a mass of 4.0312 amu. An alpha particle has a mass of 4.0026.

$$4.0312 - 4.0026 = 0.0286 \text{ amu converted to energy}$$

$$\frac{0.0286}{4.0312} = .0071 = 0.71\% \text{ of the mass is converted to energy}$$

One atomic mass unit is 1.7×10^{-26} kilograms. So 0.0286 amu is 4.8×10^{-29} kilograms. The energy produced by this mass is found from Einstein's famous equation

$$E = mc^2$$

where c is the speed of light, 3×10^8 meters per second.

$$E = (4.8 \times 10^{-29} \text{ kg})\ (3 \times 10^8 \text{ m/s})^2 = 4.3 \times 10^{-12} \text{ joules}$$

Now, that is not very much energy per fusion. It must be going on at a tremendous rate in the core for the Sun to be radiating energy away at the rate of 3.8×10^{26} joules per second.

$$\frac{3.8 \times 10^{26} \text{ j/s}}{4.3 \times 10^{-12} \text{ j/fusion}} = 8.9 \times 10^{37} \text{ fusions per second}$$

Let us go one step further. We found that 4.8×10^{-29} kg are converted to energy in each fusion. So the Sun must be losing

$$(4.8 \times 10^{-29} \text{ kg/fusion})\ (8.9 \times 10^{37} \text{ fusion/s}) = 4.3 \times 10^9 \text{ kg/s}$$

In more familiar units, 4.3 billion kilograms per second is about 4.7 million tons per second of matter being converted to energy. At that rate, how long can the Sun last? The Sun has a mass of 2×10^{30} kg. About the inner 10%, 2×10^{29} kg, is involved in the fusion, and 0.71% of it will be converted to energy before the core goes out.

$$\frac{(2 \times 10^{29} \text{ kg})\ (.0071)}{4.3 \times 10^9 \text{ kg/s}} = 3.3 \times 10^{17} \text{ seconds} = 10 \text{ billion years}$$

Since the Sun is about 4.5 billion years old, it still has enough fuel to run for another 5.5 billion years.

---

heavier nuclei: beryllium, oxygen, carbon, and through several other steps to all the elements up to iron in the chart of the elements. Finally, the core will go out and the Sun will step through various phases until it slowly fades and dies. A star much larger than the Sun will end its life in a spectacular explosion called a supernova during which the elements heavier than iron are manufactured. It is thought that only hydrogen and a little helium were created in the Big Bang that started the universe going. All the other elements were manufactured in the cores of stars and distributed through space in such supernova explosions.

The life cycle of stars is a fascinating study in nuclear physics and is the subject of intense investigation by astronomers. Here we are only concerned with our Sun at its present stage of evolution.

Surface Features

Figure 4.2 is a diagram of the surface and a cross section of the Sun. The *photosphere* is the visible surface of the Sun. Above the photosphere is the *chromosphere,* a region that can best be described as an atmosphere. Next is the *corona,*

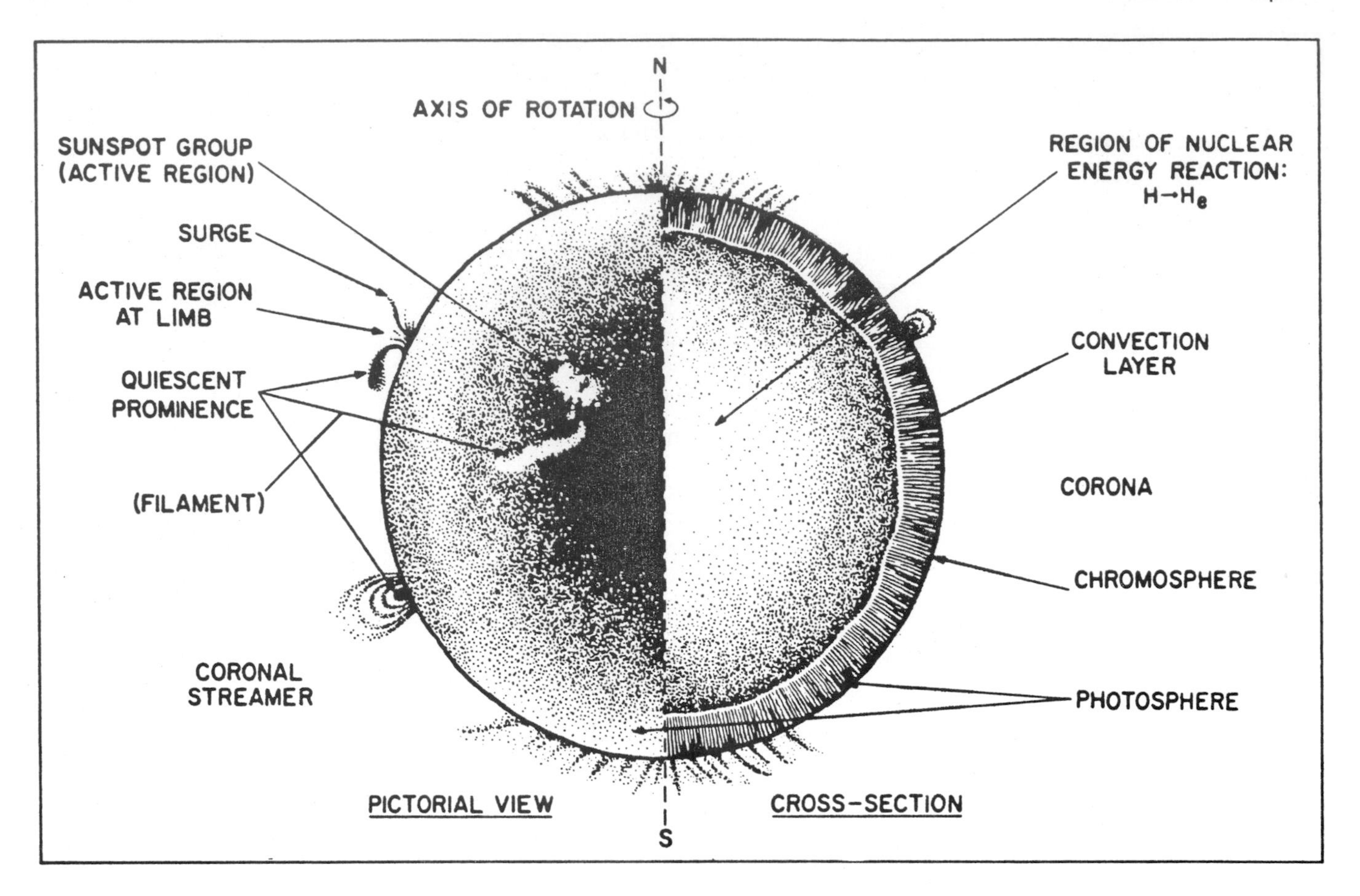

Figure 4.2 Diagram of the Sun. The left side of the diagram shows the surface features; the right half is a cross section of the interior. *Courtesy of the U.S. Air Force.*

the upper atmosphere of the Sun, which extends out to the planets. It may truly be said that Earth lies in the solar atmosphere. The corona is the hottest part of the solar atmosphere. Its temperature reaches several million degrees compared to about 11,000°F (6000°C) in the photosphere. Just why the corona is hotter than the lower regions of the Sun's atmosphere is one of the mysteries of solar physics. It is probably related to solar magnetic fields transporting energy upward from below.

The *core* is the nuclear furnace where the energy is produced. Gamma rays produced in the core radiate outward through a *radiation zone* as photons, bundles of electromagnetic energy. In free space all electromagnetic waves travel at the same speed, 186,000 miles per second (300,000 km/s). However, in the dense interior of the Sun they travel much more slowly, so slowly that it takes more than a million years for the gamma rays to reach the *convection zone,* the outer 30 percent of the solar radius. En route, they lose energy to their surroundings. When they reach the convection zone, the electromagnetic energy is absorbed by the gases and converted to a form of heat energy. The hot gases rise upward while cooler gases from above sink into the deeper parts where they are heated and rise again. Convective cells may be about 20,000 miles (30,000 km) in diameter in the convection zone, but they break up into smaller cells when they reach the photosphere. As a result, the photosphere appears granular and is in constant motion. Individual cells are 600 to 1000 miles (1000 to 1600 km) across and last about 25 minutes.

These continuously circulating convection cells are very similar to the ones that produce cumulus clouds in Earth's atmosphere. Air in contact with the warm ground is heated from below and rises, bubble-like, to higher altitudes, expanding and cooling as it goes. Meanwhile cooler air sinks toward the surface where it is warmed and rises. Clouds form in the upward moving air if sufficient water vapor is present. This continuous convective cell circulation can also be observed in a pot of water on the stove. Water heated by contact with the hot bottom of the pot can be seen to rise while cooler water sinks around the edges. If you look closely you can see the shimmering in the water. The heat is thus distributed by the circulating water.

Similarly, in the convective zone of the Sun, convective cells of circulating hydrogen carry the heat energy from the zone of radiation to the photosphere where it flows into space, mostly as light, infrared, and ultraviolet. It is indeed fortunate for us that the lethal gamma rays do not reach the surface of the Sun and radiate into space. If they did life would be impossible on Earth or anywhere else in the Solar System.

The Sun rotates on its axis just as Earth and other celestial bodies do. The photosphere and the convection zone rotate together, but because the Sun is a fluid body, it rotates differentially, that is, different latitudes rotate at different speeds. The polar regions take about 32 days to make a complete rotation, while the equatorial regions rotate once in about 25 days. About one-third of the way down, at the bottom of the convection zone, the rotation becomes more uniform and the interior rotates more like a rigid body. That leads to complicated flow patterns in the upper layers of the Sun and distortion of its magnetic fields, leading to hot spots, cooler spots, and explosions called flares. We will discuss flares in more detail later in this chapter.

Sunspots

Dark spots are often visible in the Sun's photosphere (Figure 4.3). Although they appear tiny from our viewpoint, smaller *sunspots* may be about the size of Earth and largest ones are bigger than a hundred Earths. They vary in number from day to day and year to year. Sunspots were observed by Galileo and others in 1610 shortly after he first pointed a telescope at the heavens. Records of their number have been kept since that time, although the records are quite erratic for the first two hundred years after Galileo saw them. Actually observations of sunspots had been recorded 2000 years earlier in China. On several occasions, when the Sun was low on the horizon behind a thin veil of clouds or fog, spots were large enough to be seen with the naked eye.

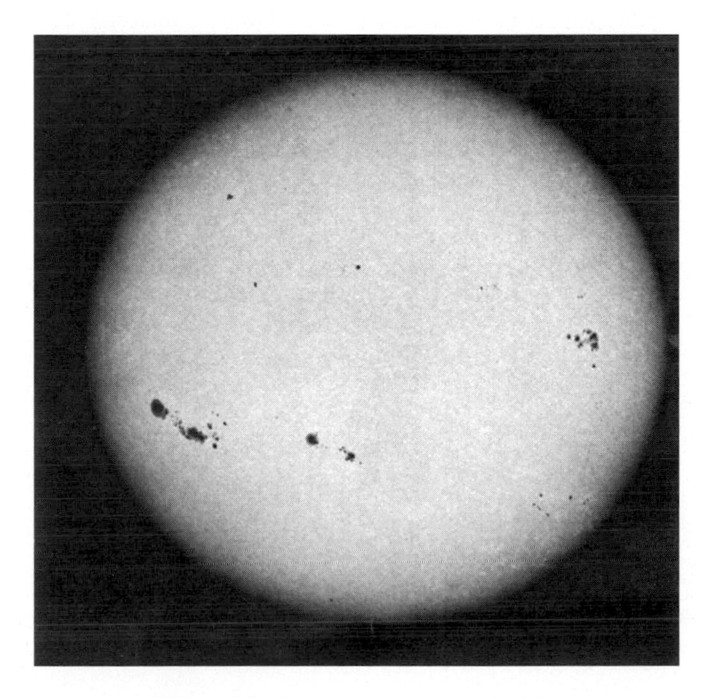

Figure 4.3 Sunspots. *Courtesy of NASA.*

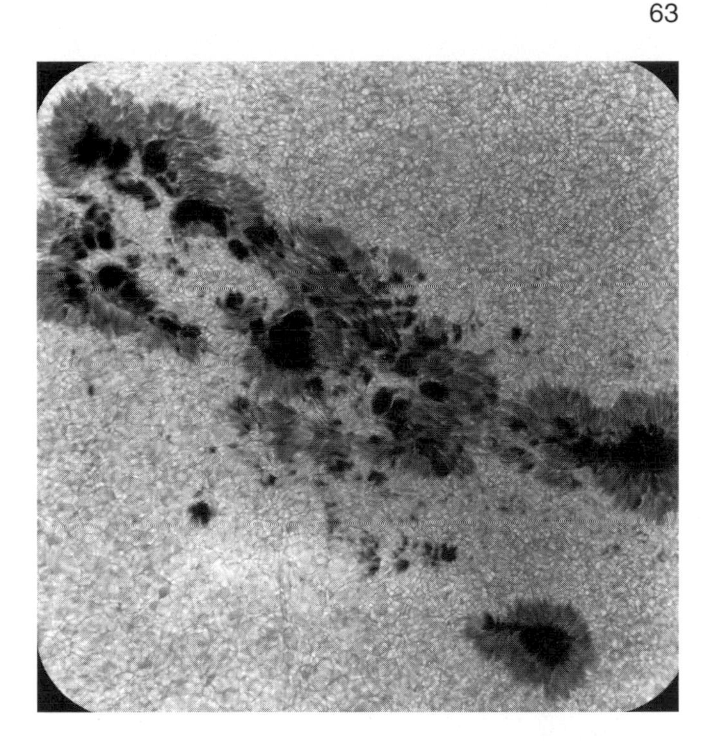

Figure 4.4 Closeup photo of a sunspot group. *Courtesy of National Solar Observatory, Association of Universities for Research in Astronomy, Inc.*

Sunspots may be observed with any telescope equipped with a device for reducing the light intensity by at least 98 percent. Looking through a telescope without some sort of light-reducing device will cause permanent eye damage. You have perhaps used a magnifying glass to focus the Sun onto a sheet of paper and watched the paper burst into flames. That is some indication of what happens to the retina of an eye focused on the Sun; smoldering retinas are painful! Focusing the image of the Sun through the telescope onto a white card held a couple of feet behind the eyepiece is an easy way to view sunspots without damaging your eyes. Figure 4.4 is a close-up photo of a group of sunspots.

Sunspots are "cool" regions in the photosphere. The coolest temperatures in sunspots are about 7500°F (4200°C). Although 7500°F (4200°C) is not very cool, the spots appear black against the hotter photosphere surroundings where the temperature is nearly 11,000°F (6000°C). If we could see a sunspot against the cold blackness of space it would appear bright. The cause of the lower temperature is involved with the magnetic fields of the Sun. Whenever an electrically charged particle is in motion, it is accompanied by a magnetic field. The convection cells in the Sun are mass motions of charged particles, mostly negative electrons and positive protons. Therefore, magnetic fields are present. When complex flow patterns occur in the convection zone, the magnetic fields become entwined and tangled. This prevents the usual rather smooth flow of heat from the interior to the photosphere. Consequently, with the heat flow interrupted, the region becomes cooler. The differential rotation of the Sun also contributes to the complexity of the heat flow. This simpli-

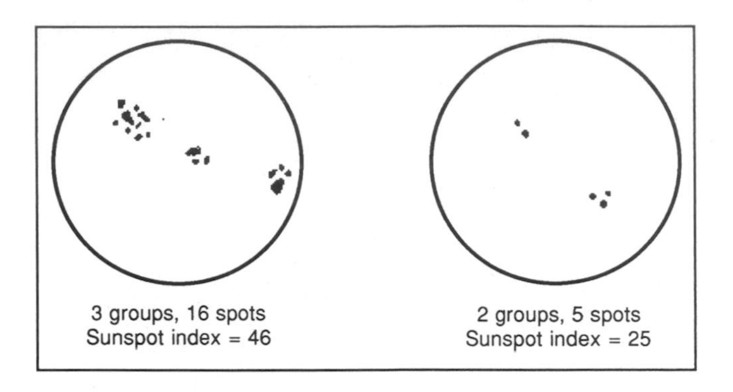

Figure 4.5 Counting sunspots.

fied explanation may not be completely precise. However, complex magnetic field structures are observed in sunspot regions and there is undoubtedly some relationship between them.

Frequently sunspots occur in pairs with opposite polarities, one a magnetic north pole and one a magnetic south pole. The magnetic field lines connect between the two in a loop extending upward into the solar atmosphere. Other sunspots occur singly; their magnetic field lines extend disconnected into space.

The Sunspot Cycle

For the past two centuries daily observations of sunspots have been tabulated and studied. Spots occur in groups with as many as 100 spots or as few as one to a group. A standard method of counting sunspots originated in the 1800s. Because it was thought that the existence of a group might be significant, groups of sunspots were given extra value. Therefore, the sunspot index for the day is equal to ten times the number of groups plus the number of spots. An example is shown in Figure 4.5.

When sunspot indices are averaged over 12 months and then plotted on a graph extending over many years, a cyclic pattern shows up with a period of about 11 years. Figure 4.6 shows such a graph from 1600 to the present. The most recent maximum occurred in 1989 when the running average sunspot number reached 157.6, the third highest on record, followed by a minimum in 1996 of 8.6. The highest maximum ever recorded was 201.3 which occurred in March 1958. The next maximum should occur in 2000 or 2001, but we will not know until at least a year after the peak because the data are averaged over a 12-month period to eliminate any short period fluctuations.

As the Sun rotates, the sunspots appear to move westward across the disk when viewed from Earth. As pairs of sunspot traverse the disk, all of the leading spots will have the same polarity during an entire sunspot cycle. During the following cycle, all the leading spots will have the opposite polarity. For example, if the leading spots are north poles during the present cycle, they will all be south poles during the next cycle. Thus a new cycle begins when the leading spots change polarity; using this criterion, the first spots of the current cycle were seen in August 1995. Because of this reversal in polarity, some scientists define a magnetic sunspot cycle as 22 years long.

Few spots were seen in the 60 years from 1645 to 1705, a period of time called the Maunder Minimum, named after the scientist who first studied it. Indeed they were so rare that many people doubted their existence; some thought Galileo was seeing spots before his eyes from staring at the Sun through his telescope.

There have been many attempts to relate sunspots to events on Earth, particularly weather and climate. For example, the late 17th century, when sunspots were rare, was also a period of abnormal cold, sometimes referred to as the "Little Ice Age." The Thames River in London froze over in win-

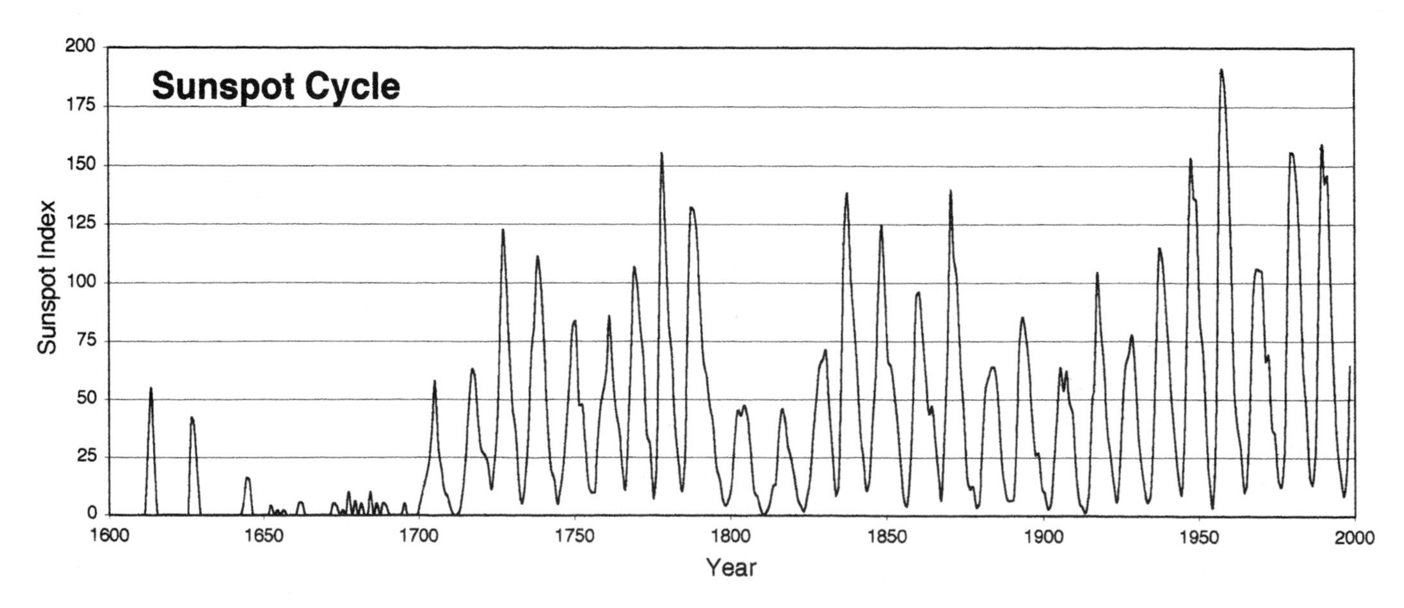

Figure 4.6 The 11-year sunspot cycle from 1600 to 1999.

ter which has never happened since. Coincidence? Solar energy output is lowest when sunspots are at a minimum.

Statistical correlations have been found between the sunspot cycle and many other things: droughts on the high plains of the United States, the stock market, the length of women's skirts. But no cause-effect relationships have been found. That is, given that the stock market varies with the sunspot cycle, what is the cause? "Coincidence" is the only logical response.

Sunspots by themselves do not do anything significant to the space environment but, as we will see, they often accompany other activity on the Sun that may adversely affect spaceflight.

The Chromosphere

Some telescopes are equipped with a special filter that allows only one wavelength of light to pass through, the red light of hydrogen-alpha. The main source of hydrogen-alpha light is the chromosphere, the lower atmosphere of the Sun. Figure 4.7 is a photograph taken through a hydrogen-alpha filter that reveals many interesting features. First, bright areas that look like clouds are hot regions called *plages,* the French word for beaches. They do have the appearance of a white sandy beach, although I wouldn't recommend them for a vacation spot.

The long black streaks are called *filaments*. They protrude upward into the corona and are cooler than the photosphere. Therefore they appear black the same as sunspots do. If a filament is on the *limb,* the edge of the solar disk, it appears

Figure 4.7 The Sun in hydrogen-alpha light. The north pole is at the top, south at the bottom, east on the left side, and west on the right. Many filaments and plages are scattered over the surface. Four large spots and some smaller ones are visible. A small prominence can be seen on the left (east) side. Little activity takes place at the equator. *Courtesy of National Solar Observatory, Association of Universities for Research in Astronomy, Inc.*

Figure 4.8 Loop prominences on the edge of the Sun. To take this photo, a disk was placed in the telescope to block out the light from the surface of the Sun. Otherwise, the nebulous features on the edge would not be visible. *Courtesy of National Solar Observatory, Association of Universities for Research in Astronomy, Inc.*

bright against the blackness of space and is called a *prominence* because of the way it sticks out from the limb (Figure 4.8). Prominences appear in many different shapes with fanciful names: loops, arches, hedgerows. They can be seen to grow, dissipate, change shape, oscillate, and sometimes explode into space. They are generally associated with *active regions,* near plages and sunspots.

Solar Energy Output

The Sun emits huge quantities of energy in the form of electromagnetic radiation, mostly light and ultraviolet, and energetic particles, mostly electrons and protons. Usually the energy flows in a continuous steady stream, but at times it comes in great bursts. This energy expands into the Solar System in all directions. If we were able to capture all the energy released in all directions by the Sun in just one second, we would have enough to power the entire United States for 9 million years!

The Solar Constant

The total energy reaching Earth's orbit from the Sun, called the "solar constant," is 1143 watts per square yard (1367

W/m^2). For many years it was thought that the solar constant was exactly constant. Measurements made on Earth, through the atmosphere, were not precise enough to show any variability. However, satellite measurements above Earth's atmosphere show that the solar constant varies as much as 0.5 percent over a period of several weeks and its averaged value increases and decreases by about 0.2 percent in an 11-year cycle. We will discuss more about this 11-year cycle later in this chapter.

There is also some indication that the intensity of solar radiation may have increased by 0.5 watt per square meter between 1988 and 1998. That seems insignificant until you realize that it equals about 70 times the amount of energy produced by all the nations on Earth each year. Yet, if the Sun has increased its energy output by that amount, it would account for only a small portion of the global warming that has been occurring on Earth, less than 0.1°C.

Theoretically, there should be a relationship between the diameter of the Sun and its energy output. Some recent satellite measurements suggest that the diameter of the Sun changes, increasing and decreasing some 400 miles (700 km) with a period of about 90 years. Besides satellite data, the diameter of the Sun can be determined from the duration of total eclipses, the time Mercury takes to cross the Sun when it is along the line of sight from Earth, and the time it takes the Sun to cross a reference line in the southern sky. A study of old records seems to support the 90-year periodicity. If this is true, it would not be unusual among stars. More than 23,000 variable stars have been catalogued. The diameter of Delta Cephei varies by 2 million miles (3 million kilometers) and nearly doubles its brightness in only 5.4 days. Others vary with periods from hours to years.

Electromagnetic Waves

Most of the Sun's energy is radiated as visible light (41 percent of the total), infrared (52 percent), and ultraviolet (7 percent). Smaller quantities of x-rays, radio waves, and microwaves are also emitted. Light waves emitted by the Sun are quite constant, however ultraviolet radiation varies considerably over a period of years, while x-rays and radio waves come in great bursts.

X-rays, gamma rays, ultraviolet, radio waves, television waves, and light are really all manifestations of the same thing, *electromagnetic waves.* See Figure 4.9 (Plate 1). The only difference among them is their wavelength. As the name suggests, they have an electrical and a magnetic component. They are created when an electrically charged particle, such as an electron, oscillates or accelerates. For example, they are produced in atoms as the electrons change orbit around the nucleus, escape from the nucleus (ionization), or are captured by a nucleus (recombination).

Certainly you have seen water waves or ripples in a pond. Water waves and sound waves must have a medium in which to travel. Sound waves travel in air, through liquids (you can hear sounds when you are swimming), or through solids like walls and windows. But electromagnetic waves do not require a medium. Therefore, they can travel through empty space. They all travel at the same speed, the speed of light, 186,000 miles (300,000 km) per second.

The wavelength of a wave is the distance from one crest to the next crest, or from one trough to the next. Figure 4.9 (Plate 1) shows various electromagnetic waves in order of wavelength from the shortest to the longest. Wavelengths are given in metric units. A kilometer is about 0.6 mile. It takes about 25,000 micrometers to make an inch; a grain of sand is about 300 micrometers in diameter. A micrometer is sometimes called a micron.

The length of a radio wave, measured from crest to crest or trough to trough, ranges from a tenth of a meter to many kilometers (a few inches to many miles). Microwave oven waves are about 1 centimeter (0.4 inch) long. Infrared waves range from a micrometer (millionth of a meter) to several hundred micrometers in length. Light waves, on the other hand, measure from 0.4 micrometer to 0.7 micrometer. Ultraviolet, x-rays, and gamma rays are even shorter. Gamma rays and x-rays come from the most violent, high energy events in the universe. MATHBOX 4.2 shows the relationship between wavelength and frequency.

Solar Wind, Coronal Holes, Coronal Mass Ejections

In addition to electromagnetic waves, a steady stream of particles, mostly electrons, protons, and alpha particles, flows outward from the Sun in what has been named the *solar wind.* The solar wind is actually an extension of the solar corona blowing outward past the planets, filling the Solar System with charged particles. Usually they speed past Earth at about 250 miles per second (400 km/s) with a density of about 1600 particles per cubic inch (100 per cubic centimeter). However, sometimes the solar wind increases in density and speeds as high as 500 miles per second (800 km/s) during periods of high solar activity. Exposure to the solar wind is like being exposed to low level radioactive material. Dosage does not usually reach a dangerous level.

In areas of the Sun where single sunspots have open magnetic field lines, usually closer to the poles, the electrons and protons have an easy escape route into the Solar System. In these regions, called *coronal holes,* the speed of the solar wind is higher and steadier as the particles exit the surface of the Sun. In other regions of the lower corona, most often nearer the equator, pairs of sunspots with their looping magnet fields trap and hold the plasma. These features and coronal holes can be seen clearly in the x-ray picture of the Sun, Figure 4.10 (Plate 1).

As more electrons and protons are carried upward from the convective zone, the plasma cloud expands like a balloon, stretching the magnetic field lines and twisting them into tight corkscrews. Eventually the magnetic field can no longer

MATHBOX 4.2

Waves: Speed, Wavelength, and Frequency

The frequency of oscillation of a wave in hertz (oscillations per second) multiplied by its wavelength gives the wave's speed. The equation is

$$v = f\lambda.$$

The Greek letter, λ (lambda), stands for wavelength.

Electromagnetic waves all travel at the same speed, the speed of light, 9.8×10^8 feet per second or 3.0×10^8 meters per second. So knowing the frequency we can solve for the wavelength and vice versa.

Problem: What is the wavelength of a radio station broadcasting at 103.9 megahertz (103.9×10^6 hertz)?

Solution: Solve for λ and enter the numbers.

$$\lambda = \frac{v}{f} = \frac{9.8 \times 10^8 \text{ feet per second}}{103.9 \times 10^6 \text{ hertz}} = 9.4 \text{ feet}$$

Problem: What is the frequency of a 10-micron (a micron is the same as a micrometer) infrared wave?

Solution: Solve for f and enter the numbers.

$$f = \frac{v}{\lambda} = \frac{3.0 \times 10^8 \text{ m/s}}{10 \times 10^{-6} \text{ m}} = 3.0 \times 10^{13} \text{ hertz}$$

That is 30 million million oscillations per second!

This same equation works for sound waves which travel about 1100 ft/s (340 m/s) at sea level when the temperature is 32°F (0°C).

contain the plasma cloud, so a blob breaks loose and escapes from the Sun into space. Billions of tons of solar material may be ejected out into space in these great outbursts called *coronal mass ejections*. See Figure 4.11 (Plate 2). During solar minimum, a coronal mass ejection occurs once every 3 to 5 days; during solar maximum they may occur 3 or 4 times a day. The fast-moving cloud takes from 2 to 4 days to reach Earth, although it could be ejected in any direction and may miss Earth entirely. If the cloud does reach Earth it produces important effects which are described below.

Solar Flares

When the energy wrapped up in the magnetic fields of an active region can no longer contain itself, it is sometimes released suddenly in a gigantic explosion called a *flare*. Flares may last from a few minutes to a few hours. Even small ones may be larger than Earth and release as much energy as a billion hydrogen bombs. A large flare will release more energy than has been consumed by all humans in all history. The frequency of occurrence of flares follows the sunspot cycle. As the sunspots increase so does the number of flares. Up to 1000 flares per month are observed at the time of sunspot maximum. However, large flares often occur after the sunspot cycle reaches a maximum and begins to decline. The very large flare in Figure 4.12 occurred nearly 4 years after the 1968 peak in sunspot index. The first flare of the current solar cycle occurred on November 4, 1997.

Some large flares break up the magnetic fields in the vicinity and trigger coronal mass ejections. Some of the particles reach extremely high speeds and arrive at Earth in less than an hour.

The light from a flare can be seen in a telescope equipped with a hydrogen-alpha filter. Once or twice in an 11-year cycle a flare is so brilliant that it can be seen against the bright photosphere without a filter. Besides visible light, bursts of x-rays, ultraviolet, and radio noise often accompany the explosion on the Sun.

Cosmic Rays

Cosmic rays are particles, not rays, traveling in all directions at extremely high speed, some of them approaching the speed

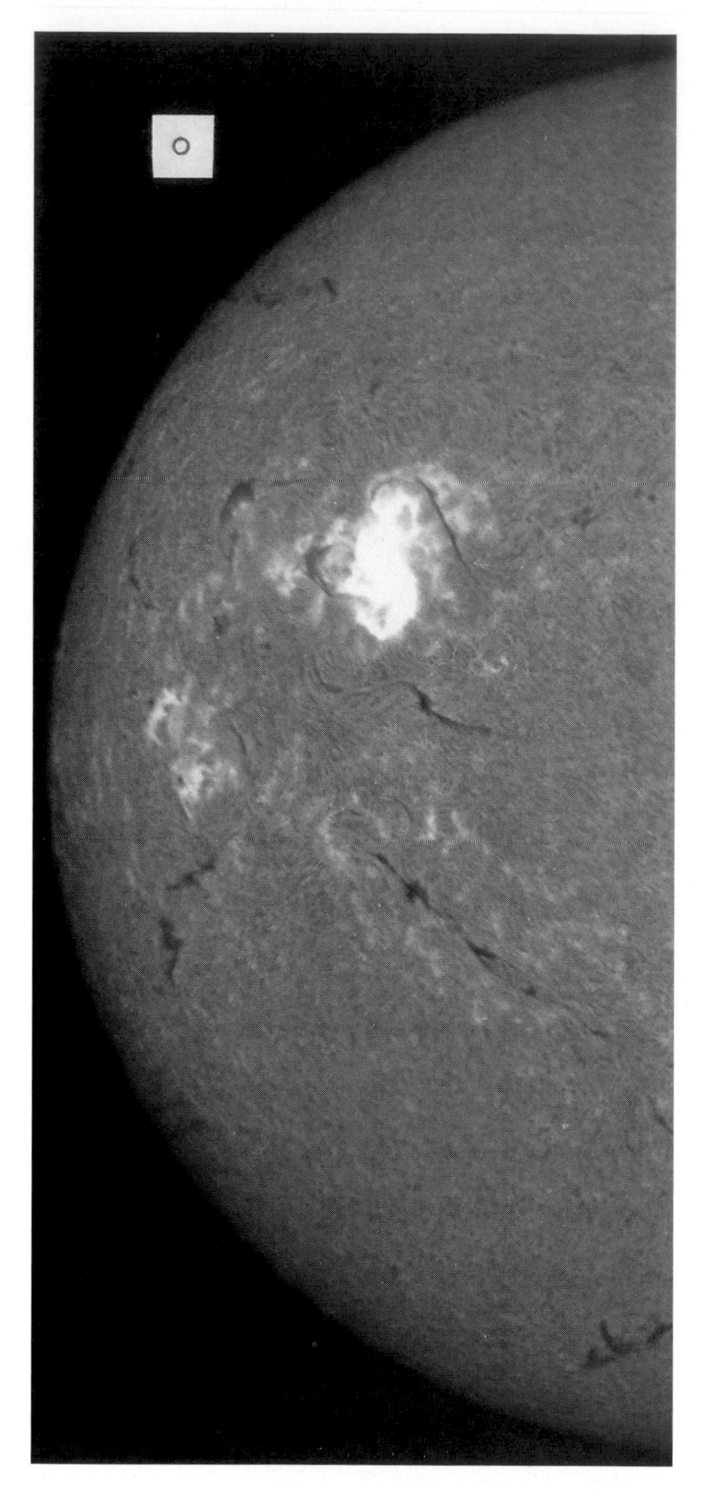

Figure 4.12 A large solar flare seen through a telescope fitted with a hydrogen-alpha filter. The small circle at the upper left shows the size of the Earth. *Courtesy of the National Solar Observatory, Association of Universities for Research in Astronomy, Inc.*

of light. Occasionally during very intense flares, protons reach speeds up to one-fourth the speed of light and fit into the classification of *solar cosmic rays*. At that speed they reach Earth in half an hour, compared to the usual 3 to 4 days for particles in the solar wind. A solar cosmic ray event, also referred to as a *proton event*, may last from several hours to a day.

A steady stream of cosmic rays also comes from outside the Solar System, and are referred to as *galactic cosmic rays*. When they were first discovered in the early part of this century they were thought to be electromagnetic radiation and were dubbed "rays." We now know they are the nuclei of atoms stripped of their electrons and therefore bearing a positive charge. They are 85–90 percent protons, 10–12 percent alpha particles, about 1 percent electrons, and 1 percent nuclei of heavy atoms such as oxygen, nitrogen, iron, and neon. Some perhaps had their origin in events during or shortly after the Big Bang that started the universe. Others come from exploding dying stars, supernovas. A few may come from other galaxies.

Sun-Earth Relationships: Space Weather

Earth intercepts only a small percentage of the great quantities of electromagnetic waves and particles that pour out of the Sun. Because of our distance from the Sun and the relatively small size of Earth in space, only about two-billionths of the total solar output falls on Earth. Yet nearly all energy consumed on Earth can be traced to the Sun. For example, plants harvest sunlight to build their "bodies" by photosynthesis, humans and animals eat plants, and humans, in turn, eat the animals. The base of the entire food chain begins with photosynthesizing plants. Fossil fuels are really stored solar energy. Oil, coal, and natural gas come from ancient living organisms which died, were buried, and were transformed into those fossil fuels. Wind is caused by uneven heating of Earth's surface by the Sun. Windmills, therefore, derive their energy from the Sun. Rain is produced by solar energy evaporating water from lakes and soil. The water vapor condenses into clouds and raindrops which fill the reservoirs from which falling water drives electric generators. Photovoltaic cells, commonly called solar cells, convert sunlight directly into electricity with no moving parts. All of these examples of useful energy on Earth originate with the energy of the Sun.

The only kinds of energy used on Earth that cannot be traced to the Sun are tidal energy, nuclear energy, and geothermal energy from the heat deep in Earth's crust. Even the tides are partially due to the Sun, though the Moon is the primary cause. The gravity of the Sun is enormous, but the Moon is a lot closer so its gravitational force is more significant in producing tidal action.

"Space weather" includes the electromagnetic waves and charged particles that emanate from the Sun and the phenomena that they produce in the vicinity of the Earth.

Electromagnetic Radiation

Life on Earth is well protected from lethal x-rays and ultraviolet radiation; the atmosphere absorbs and filters out most of the harmful rays. Even so, you should avoid too much exposure to the Sun on the beach or in the mountains. Sunburns are not only uncomfortable, but excessive exposure to ultraviolet causes skin cancer. An unprotected astronaut does not have the protection of an atmosphere and must avoid exposure to the undiluted full blast of solar ultraviolet and x-rays. Such protection is not difficult to achieve. The cabin of a spacecraft or the material of a space suit is sufficient.

X-rays and long wavelength ultraviolet are absorbed in the *ionosphere,* the ionized top part of the atmosphere above about 50 miles. The energy of these electromagnetic waves ionizes the oxygen and nitrogen atoms, that is, removes an electron from them. Therefore, the ionosphere is called a *plasma,* a mixture of neutral and ionized atoms.

Ultraviolet is also absorbed by the ozone layer some 30 miles high. Ozone is made of three atoms of oxygen bonded together to form a single molecule. Normal oxygen in the air around us consists of only two atoms bonded together. Short wave ultraviolet can break the bond of an ordinary oxygen molecule, splitting it into two individual atoms. One of those atoms then attaches itself to normal oxygen to form ozone.

Thus, the upper atmosphere absorbs hazardous x-rays and ultraviolet so they do not reach the surface of Earth to damage living organisms. On the other hand, the atmosphere is transparent to visible light, the wavelengths that the Sun emits with greatest intensity and that our eyes respond to. A good arrangement!

Particle Radiation

The electrons, protons, and alpha particles traveling from the Sun carry an electrical charge and electric charges in motion are always accompanied by a magnetic field. Earth has a permanent magnetic field, as if a giant bar magnet were buried deep in its interior. Figure 4.13 is a photograph of small bits of iron in close proximity to a bar magnet. Notice how they appear to follow curved lines which run from one end of the magnet to the other end. Compare this with Figure 4.14, a diagram of Earth's magnetic field. Remember that the solar wind blowing past Earth is composed of charged particles in motion and therefore bears a magnetic field. The interaction of the solar wind with Earth's magnetic field causes it to be distorted: compressed on the sunward side and stretched out downstream. It carves out a region of space called the *magnetosphere.* Solar wind particles follow the field lines and are deflected around the magnetosphere.

The magnetosphere acts as a shield, protecting life on Earth from deadly particle radiation. However, cosmic rays travel at such high speed that they can rip through the magnetic shield and collide with gas atoms in the upper atmosphere, smashing them into bits. These bits form secondary radiation which does reach the ground. We are all constantly exposed to this low level radiation; that is part of living. It is more intense at higher altitudes than at sea level.

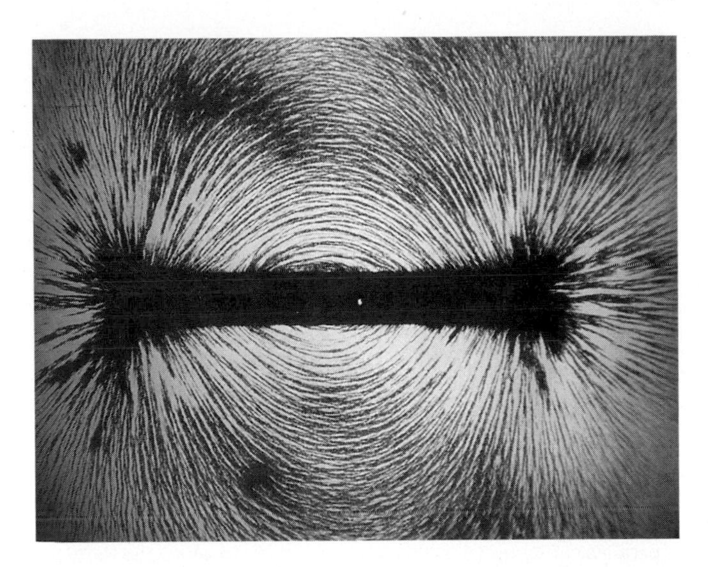

Figure 4.13 The magnetic field of a bar magnet made visible by sprinkling iron filings on a piece of glass over the magnet.

Astronauts, however, are exposed to low level radiation when traveling in the magnetosphere and to higher levels of radiation when outside the magnetosphere on the way to the Moon or other planets. Normally the solar wind does not pose a threat to life. The electrons and protons are traveling slow enough that they are not harmful for short exposure times.

Cosmic rays, both solar and galactic, are a different matter. Solar cosmic ray protons emitted during a large flare are a very serious problem. The U.S. government has set a standard for the maximum amount of radiation that workers may receive if they deal with radioactive material on a regular basis. During a proton event an astronaut could be exposed to more radiation in half an hour than is allowed a radiation worker in a year. At least two solar proton events in the past solar cycle were of sufficient intensity that astronauts exposed in the cargo bay of the Space Shuttle would have suffered from severe radiation sickness, perhaps death. For the most energetic of these events, even the cabin walls of the Space Shuttle may not provide sufficient shielding.

Passengers in supersonic aircraft flying in the polar regions at altitudes up to 60,000 feet (18,000 m) are also exposed to radiation during proton events as the particles precipitate into the atmosphere through weaknesses in the magnetosphere. Those aircraft are equipped with detectors which ring alarms when the radiation exceeds certain levels.

Galactic cosmic rays are of concern on long-duration flights. They travel at very high speeds and are very penetrating, but are relatively few in number. Many galactic cosmic rays travel at such high speed that they pass right through the spacecraft, people and all, at a rate of perhaps a dozen per square inch per hour. Buzz Aldrin and other astronauts on way to the Moon reported seeing flashes of light when they

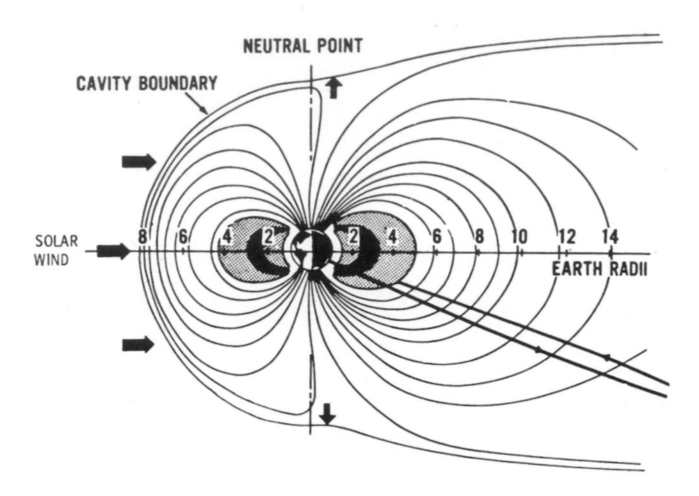

Figure 4.14 The solar wind distorts the Earth's magnetic field creating a cavity in space, the magnetosphere, in which the Earth resides. The distance scale is in "Earth radii"; the radius of the Earth is 3860 miles (6370 km). The lines extending from Earth to the lower right show the orbit of a research satellite. Most satellites fly below 4 Earth radii. *Courtesy of NASA.*

had their eyes closed, probably caused by a passing cosmic rays. For short trips such as the Apollo flights to the Moon, the dosage is not excessive. However, it has been estimated that continuous exposure to galactic cosmic rays for a year-long trip to Mars would exceed the maximum recommended dosage by some 60 times. Special shielding will be required to avoid radiation sickness.

Trapped Particles: The Van Allen Belts

Some particles do make it from the solar wind into Earth's atmosphere. There are weaknesses in the magnetosphere in the polar regions where particles can enter. In Figure 4.14 these weak places are shown by the two arrows and the broken line marked "neutral point."

Two donut-shaped regions called the Van Allen radiation belts surround the Earth inside the magnetosphere, where charged particles are trapped and held by the Earth's magnetic field. See Figure 4.15. Their existence was unknown until James Van Allen of the University of Iowa discovered them from data recorded by the first U.S. satellite, Explorer I.

The trapped particles are very energetic and high in number. The inner belt begins at 250 to 750 miles (100 to 1200 km), depending on latitude, and extends to about 6200 miles (10,000 km). It is occupied mostly by protons which have a peak count over 60,000 per square inch per second (nearly 10,000 per square centimeter) at about 2000 miles (3200 km). The outer belt contains electrons and extends from the top of the inner belt at 6200 miles (10,000 km) to 37,000 to 52,000 miles (60,000 to 85,000 km) depending on solar activity. Peak electron count is at about 9900 miles (16,000 km).

The Van Allen belts are far enough from Earth that they are no problem for space vehicles in low Earth orbits except over the South Atlantic Ocean just east of Brazil. Because of an irregularity in Earth's magnetic field in that region, they dip to their lowest altitude. In a 275 mile (440 km) orbit, a thousand-fold increase in radiation is noted by spacecraft as they pass through; a hundredfold increase is seen at 140 miles (225 km).

The Van Allen belts would be a serious threat to astronauts traversing them, but their locations in space are well mapped and they can be avoided in manned spaceflight. Spacecraft at higher altitudes such as in geosynchronous orbits at about 5.5 Earth radii over the equator need to be built with some protection against the high energy electrons. Those in highly elliptical orbits, and on escape trajectories from Earth must either avoid the belts completely or be prepared to cope with the trapped particles.

Auroras

Particles remain trapped in the Van Allen belts until an event on the Sun, such as a coronal mass ejection, disrupts the usually smooth solar wind, causing the particles to dump out along Earth's magnetic field lines and "rain" into the polar regions and cause the atmosphere to glow like a fluorescent light, the phenomenon commonly called the northern lights, but more technically named the *aurora borealis* in the northern hemisphere and *aurora australis* (southern lights) in the southern hemisphere. Oxygen atoms produce greenish colors and ultraviolet; nitrogen molecules emit a pink-violet glow. Typically auroras occur at an altitude of about 70 miles (110 km). Occasional reports from arctic explorers tell of hearing sounds like bells during an auroral display. A view of the aurora from above is shown in Figure 4.16 (Plate 2), photographed from a Space Shuttle in orbit.

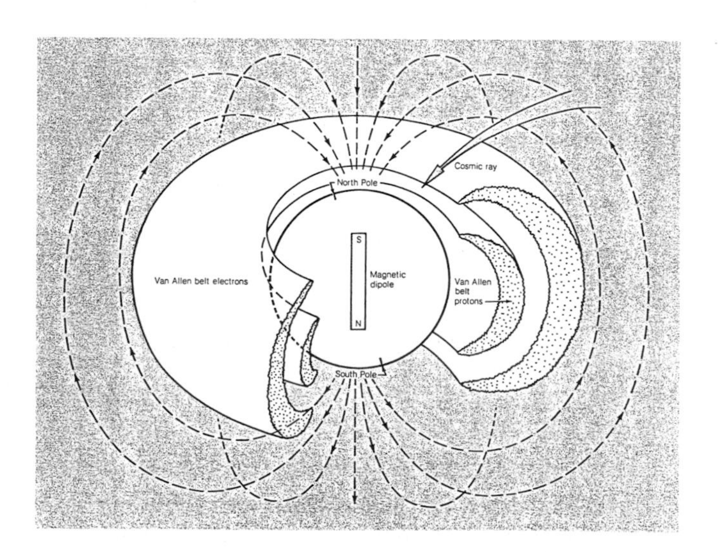

Figure 4.15 Van Allen radiation belts. Protons are trapped in the inner belt and electrons are found in the outer belt, while the dashed lines represent Earth's magnetic field. Cosmic rays can penetrate the magnetosphere especially in the polar regions. Note that Earth's north and south poles are titled with respect to the magnetic field. *Courtesy of NASA.*

Geomagnetic Storms

The plasma clouds of coronal mass ejections may intercept Earth, compress the magnetosphere, and disrupt the Earth's magnetic field. During these events, called *geomagnetic storms,* electronic equipment aboard satellites may be physically damaged, the temperature and density of the upper atmosphere increase, auroras become more intense and are seen much farther equatorward from their normal location around the north and south poles, radio communications can black out, navigation systems can give false readings, and induced currents in power lines can blow out transformers and generators.

Atmospheric Density

An important aspect of solar activity on spaceflight is its effect on the density of the air. The upper atmosphere is heated by absorbing electromagnetic energy, particularly x-rays and ultraviolet, by the auroras in polar regions, and during geomagnetic storms when the disturbed magnetic field accelerates the usually quiescent ionospheric plasma. When solar activity increases, the ultraviolet energy output of the Sun increases, auroras increase in number and intensity, and coronal mass discharges produce more frequent geomagnetic storms. All of these increase the temperature of the upper atmosphere.

You usually hear that heating causes things to expand, increasing their volume and therefore, because the same mass occupies a larger volume, the density decreases. Not so in the upper atmosphere. The atmosphere consists of 21 percent oxygen molecules, 78 percent nitrogen molecules, and 1 percent other stuff, thoroughly and uniformly mixed up to an altitude of about 60 miles (100 km). Above that, the oxygen molecules are dissociated, broken apart into atoms, by the ultraviolet radiation. The number of oxygen molecules decreases with altitude as the number of dissociated oxygen atoms increases. So we have a mixture of nitrogen molecules, oxygen molecules, and oxygen atoms. They also begin to layer out in that region of space, the heaviest nitrogen molecules at lower altitudes, then the lighter oxygen atoms, finally the little residual atmosphere above about 300 miles (500 km) consists of ionized hydrogen and helium of the solar wind.

When the atmosphere is heated by increased solar activity, it expands upward. It cannot expand downward; the Earth is in the way. This expansion carries the heavier nitrogen and oxygen molecules upward so that, at any given altitude, the density of the air increases.

The effect of a density increase on a satellite orbit can be dramatic, particularly when there is a sudden geomagnetic storm. The result, as described in Chapter 3, is a more rapid decay of the orbit and an early burn-in of the satellite. This is not all bad because it cleans up the low orbits as spent rocket tanks, dead payloads, and other space junk fall out of orbit and burn up in the atmosphere. Less than 10 percent of the large objects survive to hit the Earth, and those usually land in the oceans which cover 75 percent of the Earth's surface. At such times the orbits of satellites at low altitude may change so rapidly that those who track and control the satellites may lose track of them. Within 5 days after a storm it is possible to lose track of as many as 1300 to 1400 objects.

Effects of Space Weather on Equipment

Early in the space age, before much was known about the space environment, equipment aboard satellites sometimes failed from exposure to x-rays or cosmic rays. For example, photographic film is fogged by x-rays and cosmic rays. Computer chips and other microcircuitry are susceptible to damage from x-ray bursts or high-energy protons emitted during solar flares. Measuring instruments give false readings when high-speed protons or cosmic rays pass through, and long exposure to the contents of space invariably degrades equipment operation.

Electrons impacting a spacecraft can accumulate and build up a charge, sometimes a charge large enough to cause a spark, similar to the spark that jumps from your finger when you shuffle across the room and touch a light switch. That can seriously damage electronic equipment. As much as 20,000 volts of potential has been observed on research satellites in geosynchronous orbit.

Star sensors are used on many satellites to give the spacecraft a reference point from which it can keep itself pointed in the proper direction. It is possible to lose control of a satellite if protons entering a star sensor cause a flash of light which appears to be a star. Random flashes confuse the sensor because it cannot make a correlation with known star charts.

Scientists and engineers have learned how to properly construct and shield equipment from most of these environmental hazards. Given advance warning, controllers can shut down sensitive equipment or rotate the spacecraft's vulnerable parts away from the oncoming burst. In spite of all precautions taken to protect equipment, inexplicable events sometimes occur. On January 20, 1994, for example, operators lost attitude control of Intelsat K, a communications satellite in geosynchronous orbit on the equator over the Atlantic Ocean, and several hours later lost control of another geosynchronous communications satellite, Anik E-1. Anik E-2 suffered the same problem the following day. Operators recovered the first two by switching to backup circuits, but problems with Anik E-2 continued for months. Several other spacecraft reported problems, but none were as severe as the Anik satellites experienced.

At the same time, satellites equipped with sensors to measure particles in the magnetosphere and in geosynchronous orbit reported an unusually large flow of high energy electrons which were traceable back to the Sun. Apparently the satellites intercepted the electrons which imbedded themselves in the circuit boards. The electric fields emanating from the electrons may have caused the problem. However, such enhanced electron flow is observed from time to time by the sensors without any spacecraft malfunctions. It is difficult to understand why there should be problems with some such events but not with others.

Another problem is the nearly pure oxygen atmosphere in low Earth orbits. Although there are not many atoms at those altitudes, oxygen atoms are extremely reactive and degrade many surfaces. Shuttle astronauts report seeing the leading edge of the tail and the pods on either side of the tail glow in the dark due to the impact and reaction with atomic oxygen. Surfaces facing in the direction of motion, into the wind, intercept more atoms than other surfaces and are more likely to glow.

Space Weather Forecasting

It would be nice to know if and when solar flares, coronal mass ejections, and cosmic ray proton events are going to occur so flights are not planned at that time. Unfortunately, forecasting flares is somewhat comparable to forecasting thunderstorms, predicting solar proton events is like forecasting tornadoes, and predicting geomagnetic storms is like forecasting hurricanes. Forecasts are made routinely at a joint civilian-military forecasting center and provided to NASA. They state the probability of occurrence of a large solar flare, a proton event, and a geomagnetic storm; but a definite "yes or no" forecast giving the exact time and intensity of a significant event is beyond anyone's capability at present.

Flare Forecasting

Making flare forecasts involves evaluating the flare potential of each active region on the Sun. If there are no active regions or sunspots visible on the Sun, the forecaster can issue a near-zero probability forecast. If there are several active regions with complicated sunspot configurations and large plage areas, then the probability of a flare is high. Some of the more important criteria used in this evaluation are:

a. high magnetic field strength around sunspots and plages,

b. rapidly increasing magnetic fields,

c. magnetic complexity,

d. emergence of new magnetic fields in a stable or decaying region,

e. shearing or twisting of the field,

f. frequent small flares with increasing x-ray emissions,

g. regions moving toward each other.

There are other factors as well, but note that the most important ones are concerned with the history of the active region, how it develops and changes, particularly the magnetic field configuration.

Figure 4.17 shows three hydrogen-alpha pictures of the Sun taken at 2-day intervals. Notice how the plages and fila-

Figure 4.17 Three hydrogen-alpha photos of the Sun taken at 2-day intervals. Notice how the features appear to move across the disk from left to right. *Courtesy of the National Solar Observatory, Association of Universities for Research in Astronomy, Inc.*

ments appear to move from left to right (east to west since north is at the bottom) across the disk. This apparent motion of the features is actually the rotation of the Sun on its axis, 25 days at the equator to 32 days at the poles. If the features seen on the disk persist, they eventually disappear off the right (west) side. About 2 weeks later those active regions that rotated off the west side will come back into view on the east side. Thus their return can be anticipated. However, during the 2 weeks when they are out of view the regions may increase or decrease in complexity, or even fade away completely. Likewise, new regions are born and develop. All this must be taken into account by the space forecaster when the forecast is prepared.

The forecasts are quite reliable, especially the "no flare" forecast. Forecasters can generally tell whether a region will produce large flares, but predicting the time of occurrence varies from difficult to impossible.

Forecasting Proton Events

Proton events are even more difficult to predict. However, given that a flare has occurred, the forecaster uses several parameters to state whether a proton event is likely in near Earth space. Some of the characteristics of a proton flare are:

a. a pair of parallel ribbons,

b. coverage of the darkest portion of the sunspot,

c. the location of the event on the Sun's disk,

d. long bursts of radio waves and x-rays,

e. high intensity radio noise at certain wavelengths and less intensity at others.

If the forecaster has indications that a proton event is likely, a warning can be issued. This is similar to the way in which severe weather warnings are issued. A probability forecast is issued first, then an advisory if conditions favorable to severe weather develop, and finally a warning when the severe weather is actually sighted.

Predicting Coronal Mass Ejections/Geomagnetic Storms

Geomagnetic storms are often produced by coronal mass ejections which are only seen in ultraviolet and x-ray images of the Sun, as shown in Figure 4.11 (Plate 2). If the coronal mass ejection is observed as it happens, it would give 3 to 4 days warning time of an impending geomagnetic storm. Bright S-shaped features in x-ray images of the Sun's active regions often indicate an impending explosion. These S-shaped features are caused by intense, tightly bound, corkscrew-shaped magnetic fields in the solar corona and often precede the ejection of the plasma cloud. Several satellites have been doing research on solar x-ray emissions; in 2001 a new satellite will make x-ray images of the Sun every minute and transmit them to the ground for analysis. Computers can be taught to recognize the S-shaped feature and "ring an alarm" when one is spotted.

However, the speed of the coronal mass ejection and the direction that the will go cannot be determined until the blob is actually ejected and on its way. Several satellites measure the speed and flow of particles from the Sun and report the measurements. The forecaster also needs to know the orientation of the plasma cloud's magnetic field for forecasting the effects if it reaches Earth. When the magnetic field of the plasma cloud points south, in the opposite direction to the Earth's magnetic field, then the shielding effect of Earth's magnetosphere is weakened and energetic particles can be dumped inside.

Observing and Forecasting Network

To keep track of events on the Sun, a joint civilian-military network of solar observatories has been established in such places as Hawaii, Australia, New Mexico, and Italy. These sites were chosen for their usually clear sky for observing the Sun. They are spaced around the world so that there is always at least one watching the Sun.

Weather satellites in geosynchronous orbit are equipped with various sensors to continuously monitor the solar wind, and several research satellites report their data routinely, also. The Advanced Composition Explorer (ACE) satellite, launched in August 1997, was specifically designed for this task. From its position at the Lagrangian L1 point, about a million miles (1.6 million kilometers) from Earth toward the Sun, the satellite monitors the solar wind, the interplanetary magnetic field, and high energy particle events. (The Lagrangian points were discussed in Chapter 3.) From that vantage point, the ACE sensors detect changes in the solar wind and coronal mass discharges about an hour before they impact the Earth. A new satellite, Geostorms, similar to ACE, will be equipped with a solar sail so it can maintain a position closer to the Sun to give a longer warning time.

SOHO, the Solar and Heliospheric Observatory, also orbiting near the L1 point, monitors the Sun itself along with the Transition Region and Coronal Explorer (TRACE) and a Japanese x-ray spacecraft. They watch for precursor events on the Sun that give 2 or 3 days advance warning that a storm may be on the way.

Data from the observatories and satellites are sent immediately to the forecasters in the United States who evaluate the information and pass it on to NASA, Air Force, and other interested parties, and to the public via the Internet. (Their Internet address is www.sec.noaa.gov.) If a manned space flight is in progress, the forecasters keep close contact with mission control, updating it on any solar activity that may prove hazardous to the astronauts.

Meteoroids, Meteors, Meteorites, Asteroids

From earliest recorded history people have watched streaks of light flash across the night sky—shooting stars they call them or, more technically, *meteors*. At least a few appear every night, but there are certain dates every year when they occur in such profusion we call the event a meteor shower. These are associated with the debris left behind by passing comets. Most of the debris burns up in the atmosphere causing the streak of light. A list of some of the more intense meteor showers is given in Table 4.1.

Occasionally we hear reports of a rock or chunk of iron falling from the heavens, sometimes a whole shower of them; on rare occasions someone is hit by one. On October 9, 1992, a 27-pound rock measuring 11.5 by 6.5 by 4.5 inches (29.2 by 16.5 by 11.4 cm) struck a car sitting in a driveway in Peekskill, New York. It smashed a hole through the trunk of the Chevy Malibu and dug out a 3-inch (7-cm) deep crater in the dirt underneath the car.

In the early morning hours of March 4, 1960, near Edmonton in western Canada, a bright fireball crossed the sky and was followed a few minutes later by a thundering noise and rattling windows. The next day farmers found more than 180 pounds (80 kg) of black stones scattered over the snow in a 2 by 3 mile (3 by 5 km) area. Such objects that make it to the ground are called *meteorites.*

A mile-wide crater in northwestern Arizona was made by the impact of a large chunk of iron-nickel which fell from the sky about 50,000 years ago. Because of Arizona's desert climate Meteor Crater has not eroded very much and we see it today much as it looked when it was created.

These happenings are all related. They are caused by pieces of space debris that fall to Earth. The only difference is the size of the piece. While they are still in space we call these objects *meteoroids*. They come in all sizes, some so small you need a microscope to see them (*micrometeoroids*), some the size of small planets (*asteroids*). By far the vast majority of them have a mass less than a thousandth of a gram. (It takes 28.35 grams to make 1 ounce.) Meteoroids are classified as stony (silicates), metallic (iron and nickel), or a mixture of the two.

It is thought that meteoroids and asteroids originated at the beginning of the Solar System, that they are pieces of debris left over when the Solar System formed. They are still being swept up from space by the gravity of passing planets and moons. There is evidence of impacts on the Moon and other planets by all sizes of objects. Large craters where asteroid-sized objects impacted are obvious in photographs of the Moon, and the Apollo astronauts brought back Moon rocks that bear microscopic impact craters.

TABLE 4.1 Important Meteor Showers

Shower Name	Dates of Maximum	Typical Rate (per hour)	Parent Comet
Quadrantids	January 3–4	40–100	Unknown
Lyrids	April 21–22	15	Thatcher
Eta Aquarids	May 4–5	20	Halley
Delta Aquarids	July 29–30	20–35	Unknown
Perseids	August 11–12	40–80	Swift-Tuttle
Orionids	October 20–22	15–35	Halley
Taurids	November 1–7	5–15	Encke
Leonids	November 16–17	up to thousands	Tempel–Tuttle
Geminids	December 13–14	50–55	Ikeya (?) uncertain

We see few such craters on Earth because of our atmosphere and our weather. First, few meteoroids make it through the atmosphere to the ground. They burn up as meteors on the way down because of friction with the air. Those that do reach the ground, now called meteorites, must be quite large before their impact will leave a crater. Second, growing plants, running water, weathering, and erosion wear and tear the crater away so it is no longer recognizable after a few thousand years. Because our Moon has no atmosphere or weather, craters appear as they did when they formed, most of them 3 to 4 billion years ago.

Some meteoroids now drifting in space originated on the Moon, Mars, and perhaps other planets. An asteroid impact ejects great quantities of material. Most of it falls back immediately near the crater, some goes into orbit and settles out slowly, and a few pieces may reach escape speed and fly off into orbit around the Sun. Those may eventually find that their path intersects another planet where they finally fall to rest.

Antarctica is a favored place to find meteorites. Falling onto the ice cap, the meteorites become buried in the snow, and are carried in the glaciers as they move down toward the sea. When they reach the warmer coast, the glacial ice melts and the meteorites are left lying on the ground. Fifteen meteorites from the Moon have been identified by comparing their chemical composition with that of the lunar rocks returned by the Apollo astronauts. More than a dozen of these meteorites came from Mars. They were identified by analyzing small pockets of atmosphere trapped inside the meteorite. The chemical composition and isotope ratios match the Martian atmosphere sampled by the Viking landers. One of them contains what appears to be microfossils of once living organisms. More about this will be found in Chapter 13. Thus, planets have exchanged material through billions of years and countless collisions.

Periodic meteor showers are associated with comets' orbits. Comets have been described as dirty snowballs, composed of water ice and dry ice with various kinds of dust and debris frozen in. Some comets are periodic, orbiting the Sun in highly eccentric elliptical orbits with periods of a few years to a few centuries. Their orbits carry them out past Earth's orbit to the far reaches of the Solar System. Each time they approach the Sun they heat up to melt or sublimate some of the ices and blow off some of the dust and debris into space. It is this material that forms the comet's tail. When the comet leaves the vicinity of the Sun it once again freezes up until its next approach. The material of the tail eventually becomes distributed throughout the comet's orbit. If Earth crosses the comet's orbit, that material will fall to Earth as meteors. Comet material is very small; little of it survives the atmospheric heating to reach the ground.

Asteroids, sometimes called *planetoids,* are concentrated in the region of space between Mars and Jupiter. More than 10,000 have been catalogued. Three of them were photographed close up by interplanetary spacecraft and are shown in Figure 4.18 (Plate 3). Ceres is the largest, almost 600 miles (1000 km) in diameter; only 25 are larger than 150 miles (250 km) across. Non-cometary meteoroids are mostly the result of collisions of asteroids breaking into smaller pieces. Many of the pieces would remain clustered together due to their mutual gravitational attraction. The faster moving pieces would escape the group and go off on their own, drifting in orbits around the Sun until they collide with Earth, another planet, or a moon.

The orbits of some of the asteroids cross Earth's orbit and enter the inner Solar System. They could conceivably collide with Earth sometime in the future. Asteroid collisions have happened in the past. About 140 impact craters have been identified worldwide ranging in size from half a mile to 120 miles (1 to 200 km), the oldest being about 2 billion years old. The most famous one occurred about 65 million years ago, striking on the northern coast of the Yucatan Peninsula. It terminated the age of dinosaurs and caused the extinction of three-fourths of the Earth's animal species.

Impact craters are identified first by their appearance. They are roughly circular, the rocks along their edges are usually turned upside down as the meteoroid/asteroid drives in, some of the material from the center (ejecta) is thrown far from the center of the impact, and some of the ejecta has melted and resolidified. Its chemistry gives additional evidence—meteoroids and asteroids often contain rare elements in greater proportions than is found in Earth rock.

About 200 Earth-crossing asteroids are known, none larger than 18 miles (30 km) in diameter. However, even a small chunk of rock traveling at about 25,000 miles per hour (11 km/s) can make a big hole. The mile-wide crater in northeastern Arizona was created by a meteorite less than 100 feet (30 m) in diameter. None of the known Earth-crossing asteroids will collide with Earth in the next 200 years, but perhaps one-fourth of them will hit in the next million years. The Solar System is a large and nearly empty place and Earth is a small target.

A photographic search for other Earth-crossing asteroids continues. As viewed from Earth, all stars move at the same rate across the sky due to Earth's rotation on its axis. If you set a camera on a tripod and open the shutter for a long time exposure, the stars will appear as streaks in the photo. However, if the camera is mounted on a motor that rotates it in synchronization with the motion of the stars, they appear as dots. Asteroids do not move at the same rate or in the same direction as the stars so they appear as streaks in the photo. Astronomers use this technique to find asteroids and after photographing the same one several times over a period of a few months the orbit of the asteroid can be calculated from its location and motion among the stars.

TABLE 4.2 Meteoroid Impacts

Mass of Meteoroid	Time Between Impacts on One Square Meter Surface
One-millionth gram*	116 days
One-thousandth gram	63 years
One gram	3,200,000 years
Ten grams	320,000,000 years

*There are about 30 grams in an ounce.

Hazards to Spaceflight

An asteroid hitting a spaceship would of course be a catastrophe. Even a pea-sized meteoroid could be disastrous. But what is the probability of such a collision? Statistics on the number of meteoroids and micrometeoroids in space and their size distribution can be calculated from observing meteors in the night sky, from experience with satellites that carry impact counters, and from experiments carried on Gemini, Apollo, Skylab, and LDEF and returned to Earth for examination.

To get an idea of the meteoroid content of near Earth space, consider a square surface 1 yard (about 1 m) on a side. How frequently will meteoroids of various sizes impact that square? Table 4.2 shows a statistical estimate. However, this doesn't tell us whether or not the particle will penetrate the spacecraft. That depends on the thickness of the skin. Table 4.3 gives that information for a large aluminum space station. Credit for these calculations goes to P. M. Millman of the National Research Council of Canada.

Using a double-layered hull for the spacecraft increases the protection markedly. When a meteoroid impacts and penetrates the outer skin, it shatters into many small pieces which have a much lower probability of penetrating the inner skin.

The probability of collision with a meteoroid large enough to cause serious damage is never zero, but the hazard is relatively small compared to other hazards of spaceflight.

TABLE 4.3 Meteoroid Penetrations

Thickness of Aluminum Skin	Frequency of Penetration
0.015 inch (0.038 cm)	10 per day
0.2 inch (0.5 cm)	1 per year
2.75 inches (7.00 cm)	1 per 10,000 years

Contamination by Spacecraft

All spacecraft create a contaminated "atmosphere" which travels with them in orbit. This was first noticed in the Mercury flights. John Glenn and others reported "fireflies" following the capsule. Windows clouded over on Mercury, Gemini, and Apollo vehicles. The density of the induced atmosphere around Skylab was nearly 10,000 times greater than the normal space environment. This type of contamination has many sources:

a. Outgassing of molecules from the body of the vehicle when it is in the vacuum of space

b. Small leaks of fluids and gases from spacecraft systems

c. Exhaust gases from rocket engines and maneuvering thrusters

d. Deterioration of paints and other coatings due to ultraviolet radiation from the Sun

e. Leakage of air from the cabin

f. Water and urine dumped overboard

g. Gases from the waste management systems

This contamination follows along with the spacecraft because those molecules of gas are in the same orbit as the vehicle and, perhaps, because of the gravitational attraction between the molecules and the vehicle. As a result of this contamination, several problems arise. The surfaces of windows and optical equipment such as lenses and mirrors become fogged. Deposits on solar panels reduce their electric power output. Deposits on radiators and the outer surface of the vehicle affect their heat-absorbing and radiating properties. At certain angles, sunlight scatters off this induced atmosphere, causing a glare and making it difficult for sensitive optical instruments to work properly.

Besides gaseous contamination, an orbiter produces "light pollution." It glows as it moves through the oxygen of the thin atmosphere at 200 miles. The phenomenon is not completely understood, but it may have to do with the electrical and magnetic fields produced by the motion of the orbiter. The oxygen probably becomes ionized, then emits a glow as it recombines into atomic oxygen.

Space Debris

Objects in orbit remain in orbit forever unless some force returns them to Earth or sends them on an escape trajectory away from Earth. Some satellites and pieces of rocket have been up there for 40 years. By current estimates there are about 200,000 objects in orbit with a total mass of about 4500 tons (4000 metric tons). The Air Force Space Command is capable of tracking objects larger than 4 inches (10 cm), about the size of a baseball, in low Earth orbits and larger than 40 inches (1 m) in geosynchronous orbit. In 1999 it was tracking nearly 9000 of them. However, most of the space debris, about 100,000 to 150,000 objects, are between 0.5 inch and 4 inches in size (1 to 10 cm).

Only about 600 of the objects are working spacecraft. Half of the debris are objects discarded on the way to orbit and during payload separation from the launch vehicle, including shrouds and other protective materials that covered the satellites enroute to orbit, explosive bolts, springs, and clamps. One-fifth are pieces of rockets that went into orbit with the payload. Another one-fourth consist of dead satellites. In addition there are millions of small paint flakes and other tiny debris. As satellites age, paint comes off, thermal blankets disintegrate, and other bits and pieces are sloughed off.

The largest source of small pieces of space debris comes from satellites and rockets breaking apart or exploding either accidentally or deliberately. More than 150 have broken up or blown up since Sputnik was launched in 1957. Many of these explosions were tests of Soviet antisatellite interceptors or were satellites that fragmented during reentry. Spent rocket boosters often have a small amount of fuel left in them. When heated by the Sun, the fuel evaporates in the tank and may ignite or build up enough pressure to explode. In 1996, two years after it was launched, the upper stage of a Pegasus rocket blew into 700 trackable objects and possibly 300,000 smaller pieces.

In an unsuccessful communications experiment, the United States released 400 million 1-inch (2-cm) needles to reflect microwave radio signals over intercontinental distances. Twenty-five years later clumps of them were still in orbit.

Not enough is known about the smaller objects because they are difficult to detect with present tracking equipment. The Long Duration Exposure Facility (LDEF), shown in Figure 4.19, was a passive spacecraft that was dropped off by the Space Shuttle *Challenger* in a 284-mile (457-km) orbit in April 1984. LDEF's 12 sides and ends were covered with panels of metals, plastics, various paints, reflective coatings, and epoxy composites. The original plan was to leave it orbiting for 18 to 24 months, then retrieve it and study the panels to determine the effects of the space environment on each of the materials. One of the sides of the spacecraft was always pointed toward space and the other toward Earth. Because of the *Challenger* accident, LDEF wasn't brought back to Earth until January 1990, after it had spent 5 1/2 years in orbit. More than 15,000 impacts from both meteorites and man-made debris were counted and studied in detail. Twenty times as many impacts were on the leading side in orbit than were on the trailing side. Ten times as many microcraters were found on the end facing space than on the end facing Earth. Impacts on the end toward space were caused by micrometeorites. Figure 4.20 shows one such crater. By analyzing these pockmarks and extrapolating the data, researchers estimate that about 45 tons of micrometeorite

Figure 4.19 The LDEF spacecraft being held in place by the remote manipulator arm on the Shuttle orbiter. *Courtesy of NASA.*

material arrive at Earth annually. Besides recording impacts, LDEF also showed erosion of various materials by atomic oxygen.

With new satellites being launched every month, something will soon have to be done about the increasing amount of space debris. The most used orbits have the most debris: low Earth orbits between 250 and 600 miles (400 to 1000 km) and geosynchronous orbit at 22,400 miles (36,000 km). The smaller pieces are already a hazard, probably a greater hazard than meteoroids.

During an early Space Shuttle flight a flake of paint hit a window of the *Challenger,* causing enough damage to necessitate replacing it. A window replacement is needed after nearly every Space Shuttle flight The Space Shuttle orbiter must maneuver occasionally to avoid derelict satellites. In 1998 an orbiter had to fire thrusters to avoid colliding with a 2 square foot (0.2 m^2) object. The Russians reported that an object struck a Salyut space station window with a loud noise; they believed it to be a micrometeoroid, although it could have been a man-made object. In 1998 a spent rocket body came within 1300 feet (400 m) of the Mir space station. A French military satellite suffered severe damage to its stabilization system by a collision with a piece of an exploded rocket body in 1996. Even if they don't cause serious damage, frequent impacts by small debris gradually degrade solar panels that generate electric power for a spacecraft.

One of Air Force Space Command's tasks is to check out the planned trajectories of satellites and Shuttle orbiters and to compare them with the orbits of trackable objects to assure there will be no collision. Occasional "holds" have been necessary during Shuttle launch preparations because of passing objects. It is not possible to track all the small pieces of junk, however, and they pose the greatest hazard. A pea-sized fragment moving at high speed relative to a satellite could easily put it out of commission. During a Shuttle mission, Air Force Space Command warns the crew of known objects on a potential collision course, giving time for an orbital maneuver to avoid the object.

Many objects return to Earth naturally as atmospheric drag decays the orbit and they burn up on reentry into the atmosphere. Objects in orbit below 120 miles (200 km) will burn in within just a few days. Those at 300 miles (500 km) usually burn in within several years. However, an object in a circular orbit at 600 miles (1000 km) will remain in orbit as long as 2000 years because there the atmosphere is only about 1/300 as dense as it is at 300 miles (500 km). Geosynchronous satellites will remain in orbit essentially forever.

Geomagnetic storms help to clean up the trash in lower orbits because drag increases dramatically as the atmospheric density increases during the storm. Atmospheric drag also increases as the the solar cycle approaches a maximum. The Skylab space station fell back to Earth during the 1979–1980 solar maximum and the Salyut space station reentered in early 1991, shortly after the solar maximum in September 1989.

So how do we cope with the problem? It is not possible to eliminate the debris problem by blowing the satellite up. Instead of one object, we would then have hundreds, each in a different orbit. First, owners of spacecraft should be encouraged to burn them in as soon as possible after the job is done.

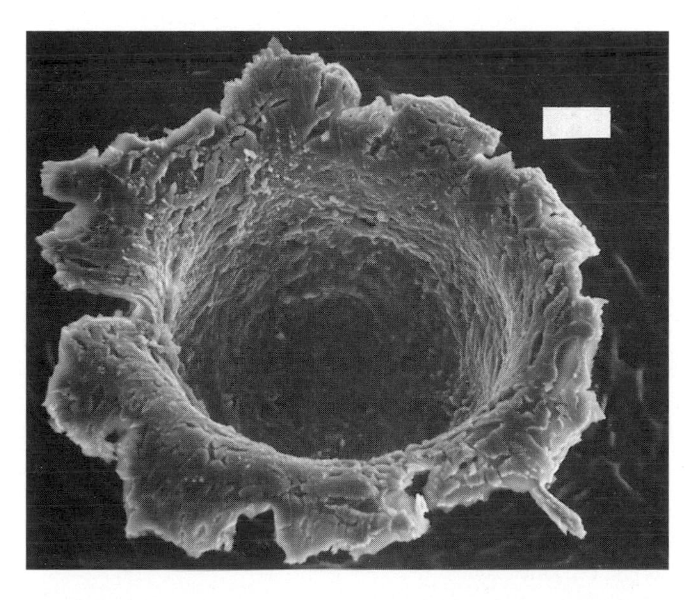

Figure 4.20 A tiny crater in one of the LDEF panels produced by impact of a microparticle. For scale, the white box at the upper right is 27 micrometers long. The hole is about 100 micrometers in diameter. *Courtesy of NASA.*

Some of the propellant used for the attitude control thrusters should be saved until after the satellite has lived its useful life. The satellite could then be rotated so the thrusters point forward and the remaining propellant burned to slow it down and cause it to burn in. Alternatively, a separate small retrorocket could be built into the satellite to deorbit it. Putting a spacecraft on a burn-in trajectory could use as much as 330 pounds (150 kg) of propellant and has some industry officials concerned about the cost, not of the fuel itself, but of carrying it into orbit specifically to burn in the spacecraft. Present cost to orbit is about $10,000 per pound!

New satellites destined for a high orbit should be parked temporarily in a low orbit first where they are checked out to be sure they are working. If they function properly, the upper stage rocket would then carry them to their final orbit. If they malfunction during checkout, they will burn in much faster from low orbit than if they were launched directly into their higher orbit and then found to be inoperable.

Because most of the Earth is covered with water and only a small part of the land area is populated, the probability of a falling satellite doing any serious damage is remote. As a general rule, anything larger than about 30 feet (10 m) will partially survive reentry. The largest object to burn in was the Skylab space station which broke up on reentry in 1979. The pieces fell into the Indian Ocean and the Australian outback. The Mir space station is over 15 years old and has become excessively expensive for the Russians to operate. At some point it will be deorbited, allowing it to break up as it enters the atmosphere so the smaller pieces burn up and the larger pieces fall into the ocean.

Many Russian spacecraft used nuclear generators for electric power. One of them reentered and fell in Canada, spreading radioactive material over a wide area. Another fell into the Indian Ocean in 1983. A recovered piece of Soviet satellite Cosmos 954 is shown in Figure 4.21. One solution is to dump the core of the reactor so the nuclear fuel burns up on reentry without reaching Earth. Another solution is to boost the satellite to a high "parking orbit" well above any remnant atmosphere where it will remain essentially forever. The Russians now take both of these actions.

There is high demand for positions in the unique 24-hour geosynchronous orbit for communications satellites, weather satellites, military surveillance satellites, and others. The atmospheric drag at geosynchronous altitude is so low, however, that satellites will remain there almost forever. To clean up geosynchronous orbit, dead satellites are boosted to a higher "disposal" orbit, at least 180 miles (300 km) above the geosynchronous altitude. This makes room for new satellites as the old ones die.

An important step in lessening the future problem is to ensure that new satellite launches do not add more unnecessary refuse. Simple things like adding devices to catch the explosive bolts, venting off or burning residual propellants in spent boosters to prevent an explosion, and adding a small retrorocket to deorbit the upper stages onto a burn-in trajectory will alleviate much of the problem. Just adjusting the attitude of the spent rocket so that the largest possible area is pointed in the direction it is moving will increase the drag and speed up the orbital decay.

Figure 4.21 A piece of the Soviet satellite Cosmos 954 which fell to Earth. It is probably part of a propellant tank for attitude control. *Courtesy of U.S. Space Command.*

Protecting spacecraft against damage from collisions with small objects is not impossible. New spacecraft can be designed with the most critical components located in the interior of the satellite where they are better protected. A meteoroid or a small piece of debris impacting a bundle of electric wiring could cause a break or a short. In one analysis, a new spacecraft under development was given only a 50–50 chance of surviving 5 years in a 500-mile (800-km) orbit. Several modifications, including simply covering the wires with multiple layers of thin insulating blankets, increased the probability of survival to 87 percent. These modifications added only 37 pounds (17 kg) to the 7000 pound (3200 kg) spacecraft.

Maneuvering the large, massive International Space Station to avoid a collision would be difficult and firing rockets would disrupt operations, particularly zero-g experiments. Debris larger than 4 inches (10 cm) can be tracked, but objects in the 0.5 to 4 inch (1 to 10 cm) range are difficult to deal with. Bumper shields may be attached to the habitats and laboratories to protect against most objects that are too small to track. The shield has three parts. Outermost is a sheet of aluminum, then one or more sheets of an insulating material called Kevlar, and finally the interior wall. The object strikes the outer aluminum and shatters. The residue is slowed or captured by the Kevlar so it cannot penetrate the interior wall. However, an object striking the shield may eject a small cloud of other smaller particles outward, thus exacerbating the problem.

Is it possible to remove the larger dead abandoned objects that are already there? The Space Shuttle has picked up several disabled satellites and returned them to Earth, but what

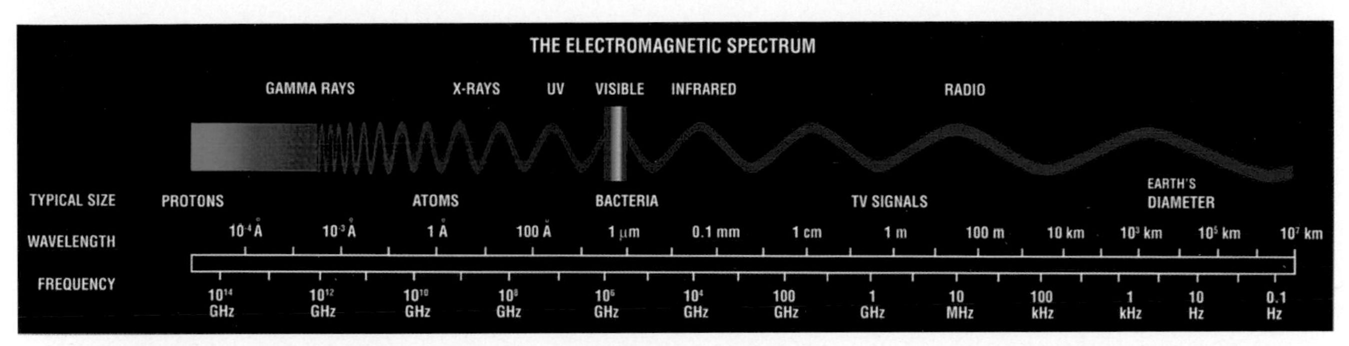

Figure 4.9 The electromagnetic spectrum. Wavelengths are given in metric units. A micrometer or micron (μm) is a thousandth of a millimeter. An angstrom (Å) is a ten-thousandth of a micrometer. The frequency is the number of oscillations of the wave per second, given in hertz (Hz) with the standard metric prefixes: k (kilo) = thousand; M (mega) = million; G (giga) = billion. *Courtesy of NASA.*

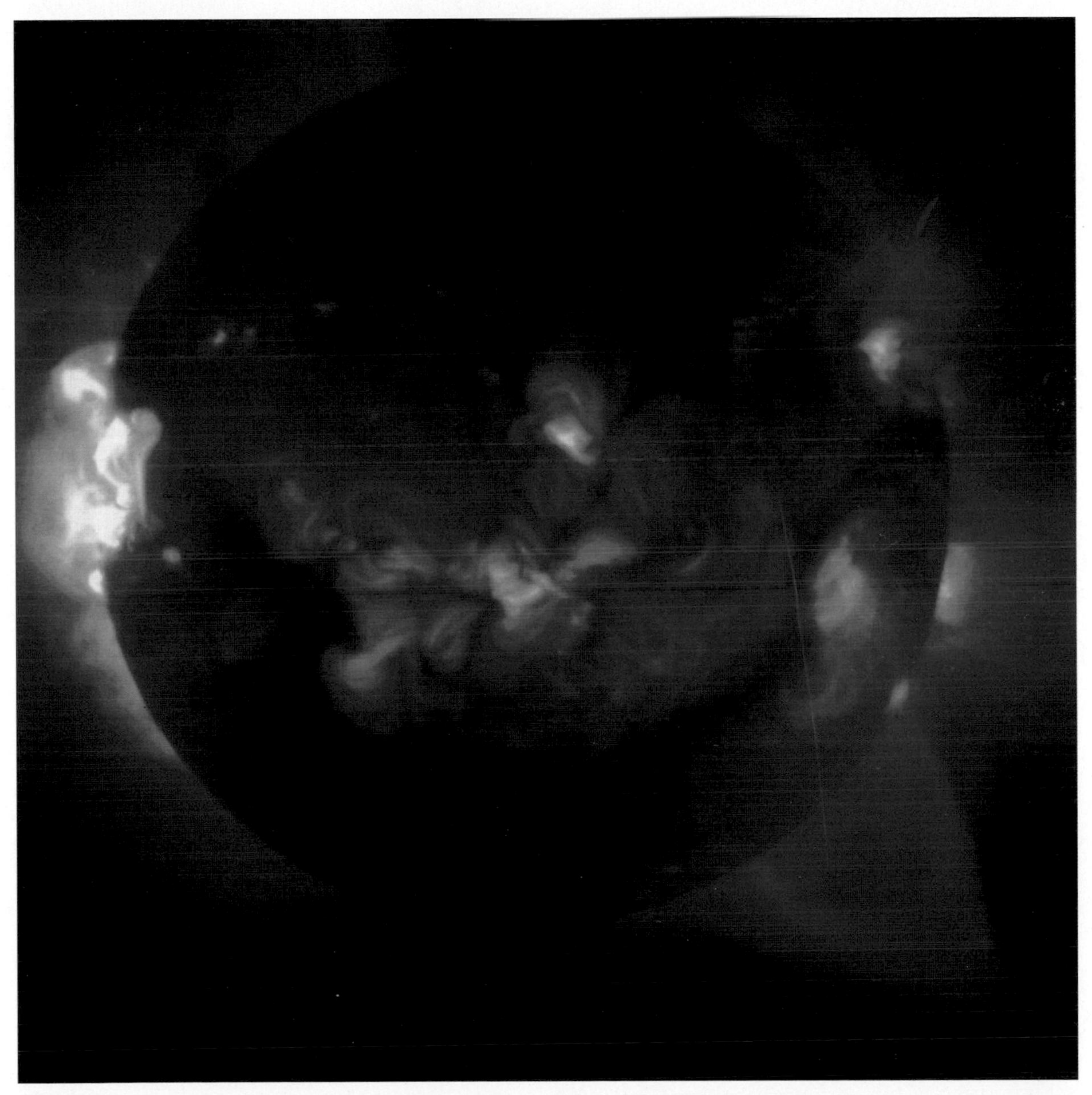

Figure 4.10 An image of the Sun's x-ray emissions. The bright areas on the left, lower right, and along the equator have strong magnetic fields that trap the hot plasma, increasing the temperatures to over 1 million degrees, causing the plasma to emit x-rays. The dark areas are coronal holes where the high speed solar wind originates and flows throughout the entire Solar System at about 400 miles (700 kilometers) per second. *Courtesy of NASA.*

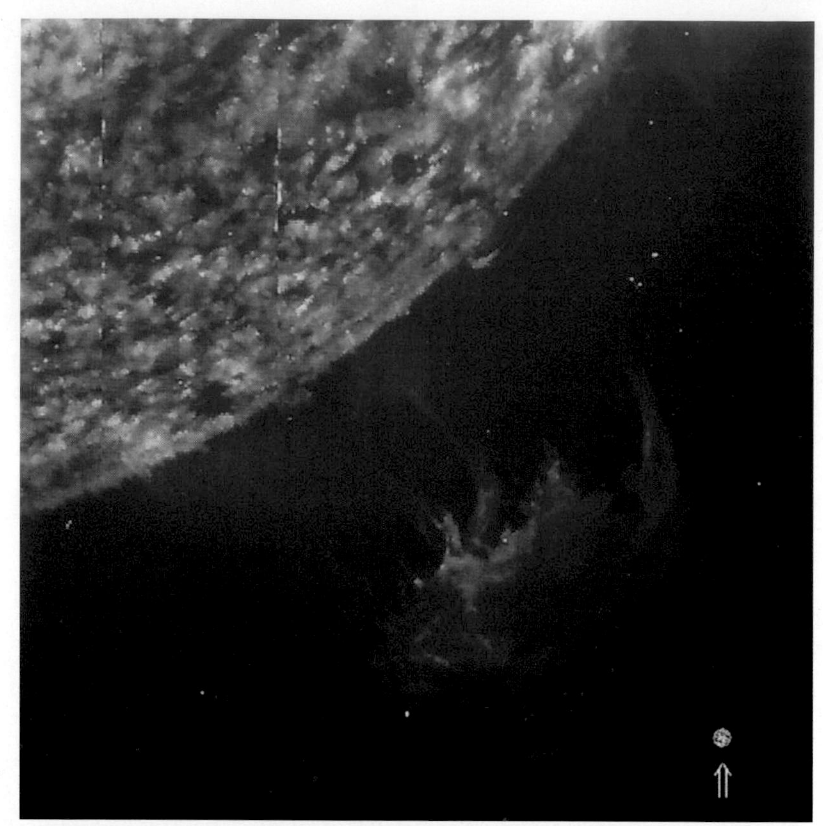

Figure 4.11 A coronal mass ejection, a huge blob of solar matter blown off into space. The arrow points to an Earth-sized circle drawn to scale. *Courtesy of NASA.*

Figure 4.16 Aurora Australis as seen from a Space Shuttle in orbit. The tail of the Shuttle is on the left. The white glow is from the maneuvering engines which had just been fired. White dots are stars. *Courtesy of NASA.*

Figure 4.18 Asteroids. Mathilde, photographed in 1997, is 39 miles (61 km) long. Gaspra, photographed in 1991, is 12 miles (19 km) long. Ida, photographed in 1993, is 32 miles (52 km) long. This composite view presents all three objects at the same scale. *Courtesy of NASA.*

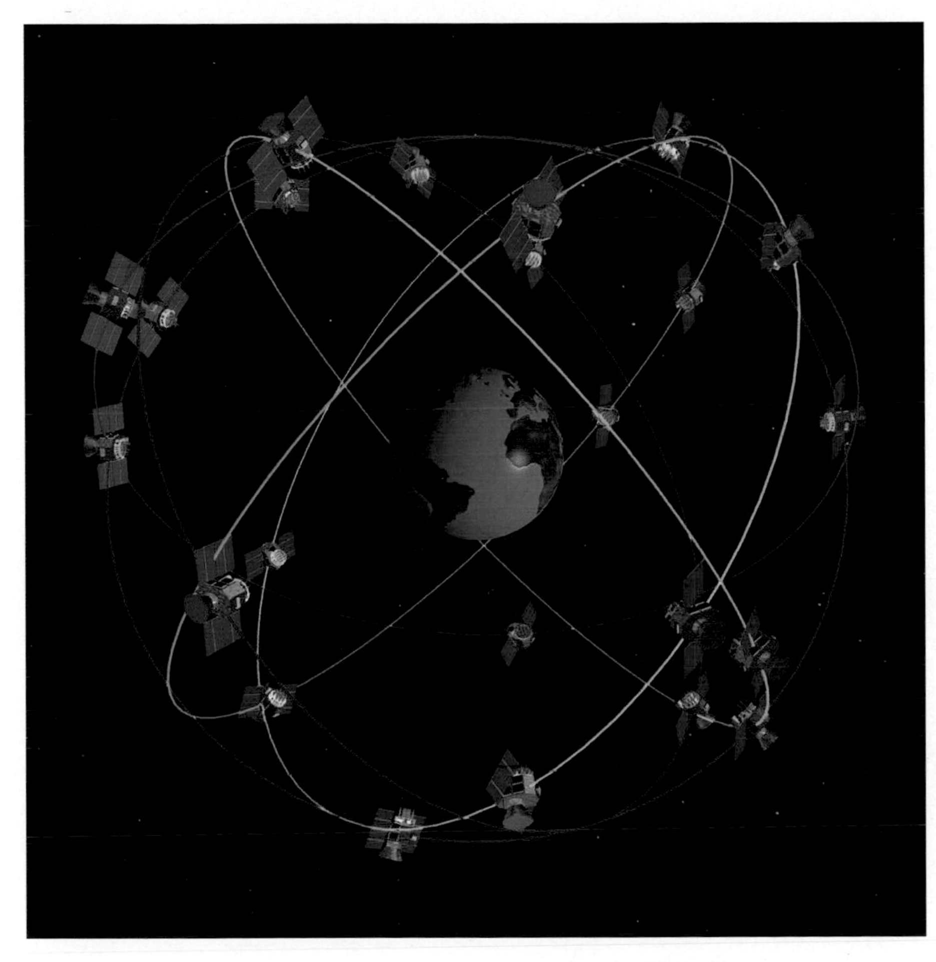

Figure 5.11 The GPS constellation of satellites. They are in orbits of 11 hours and 58 minutes, exactly one half of the geosynchronous orbit. As viewed from Earth they move slowly across the sky so at least four are visible at all times. *Courtesy of The Aerospace Corporation.*

Figure 6.10 Landsat 5 image of Copenhagen, Denmark. Bands 1, 2, and 3 were used to make this image in approximately natural color. The image was enhanced somewhat to increase the contrast between rural and urban areas. Water appears black; surf can be seen along the upper right beach. City streets, harbor docks and the airport are clearly visible. *Courtesy of EOSAT.*

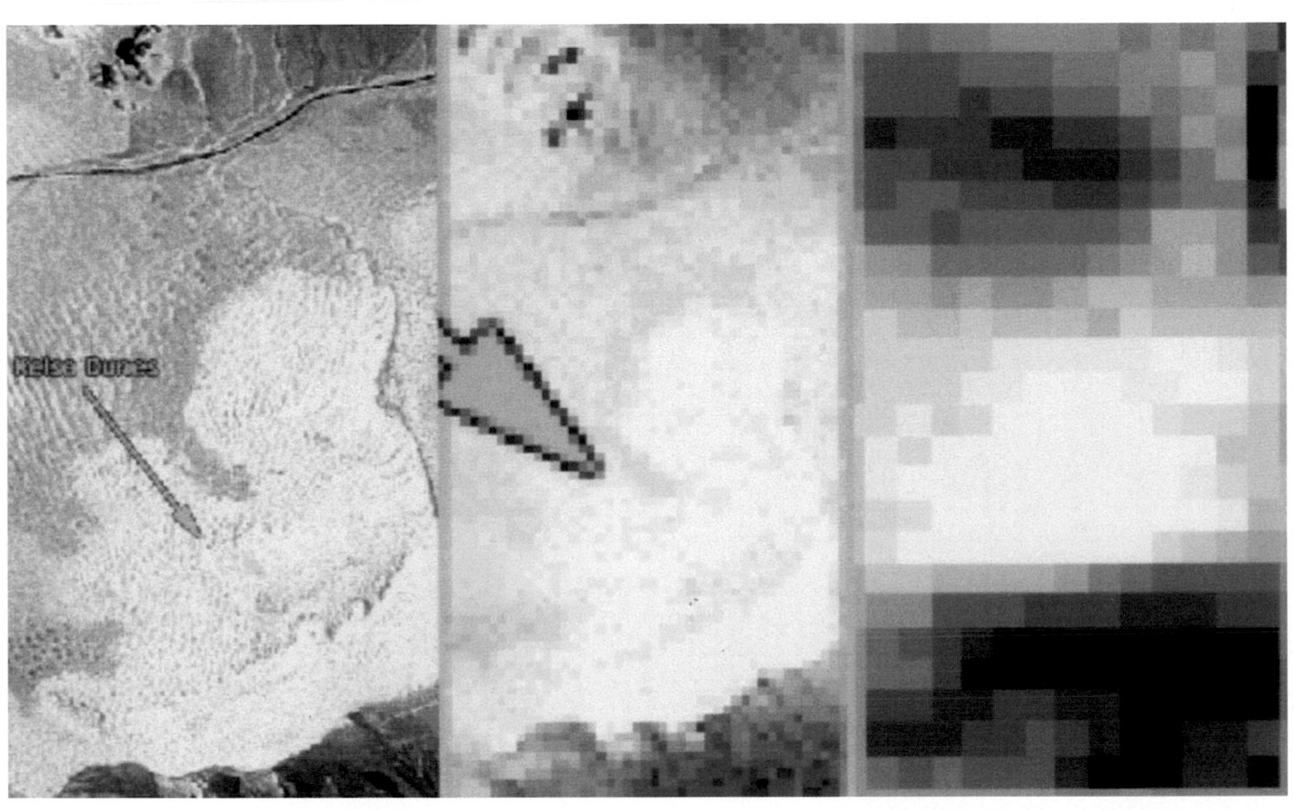

Figure 6.11 Comparison of resolution. These three images of the Kelso Dunes in southwestern United States are from three satellites with different resolutions. They are enlarged to show the same area in each picture. The right image is from a weather satellite in geosynchronous orbit with about 3280 feet (1 km) resolution. The other two are from Earth-sensing satellites. The center image has 590 feet (180 m) resolution and the left image has 100 feet (30 m) resolution. *Original images courtesy of Pat S. Chavez, Jr., U.S. Geological Survey.*

Figure 6.14 Landsat infrared image of Denver, Colorado. Vegetation is printed in various shades of red; urban and commercial areas are light blue or grey; water is black. The mountains on the left are covered with evergreen trees. To the upper right are agricultural fields, some with growing vegetation, others lying fallow. The Platte River runs through the center of the city. Streets and grassy residential areas can be clearly identified. *Courtesy of EOSAT.*

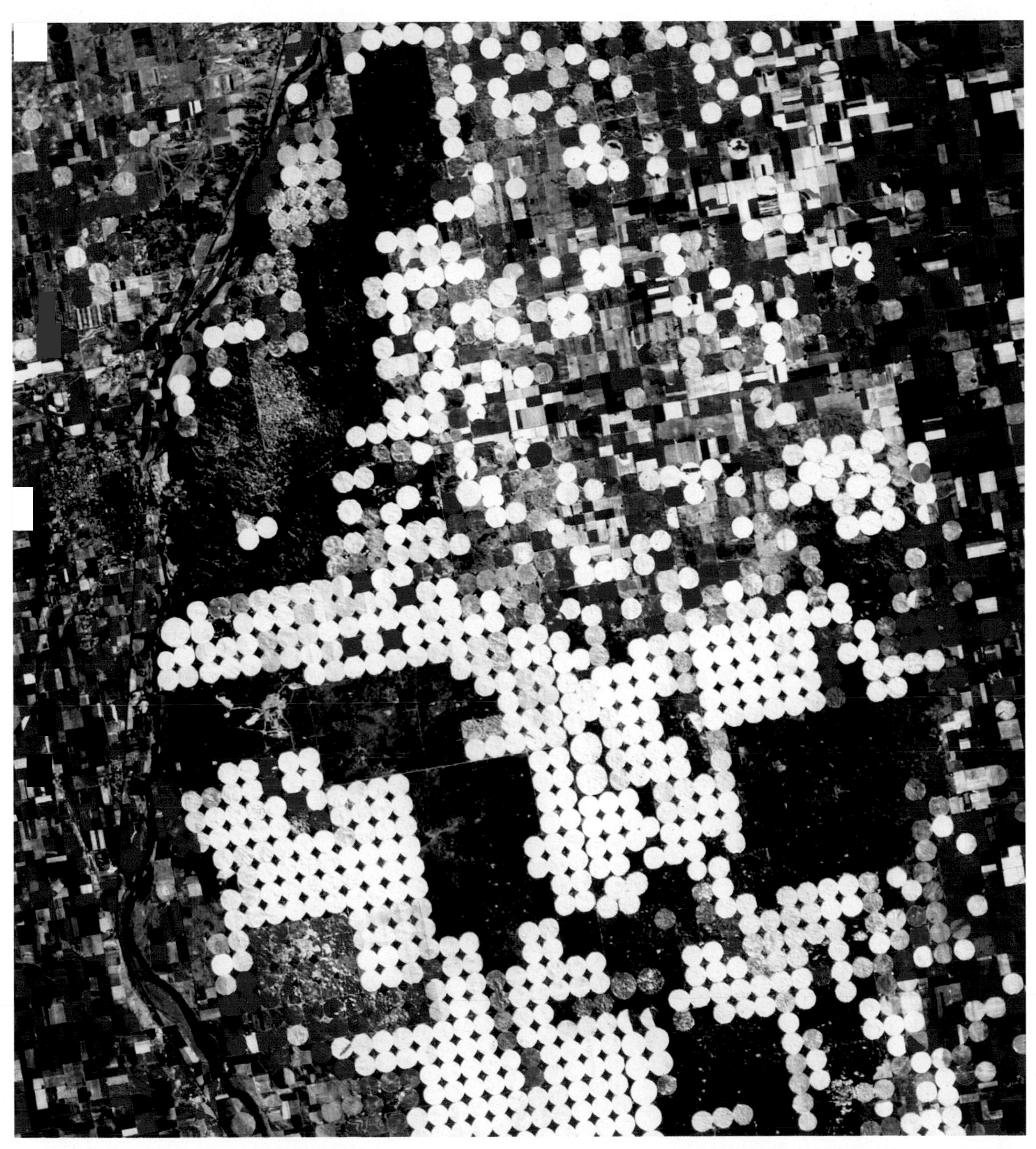

Figure 6.15 Winter wheat crops in Kansas. This infrared image was created from Landsat 5 bands 2, 3, and 4; red is used for vegetation. The circles are irrigated fields, using a sprinkler system which is pivoted at the center of the field and moves in a circle on wheels. Red fields have actively growing crops; white and grey fields lie fallow. The Arkansas River runs from the bottom to the top on the left side of the picture. Garden City is the dark-colored area left of the river. *Courtesy of EOSAT.*

Figure 6.20 A radar image of the Great Wall of China about 430 miles (700 km) west of Beijing taken from a Space Shuttle. The orange band running from top to bottom is the wall constructed about 600 years ago. Just to the right is another thinner orange band, an older wall built about 1500 years ago. Archaeologists knew of the older wall, but much of it is covered with sand and only small sections are visible from the ground. Rectangular patterns are agricultural fields, primarily wheat. At the lower center is a dry lake bed from which salt is being extracted. *Courtesy of NASA.*

Figure 6.21 Earthrise over the Moon with the Apollo 11 lunar module in the foreground. Views like this emphasize the isolation of our small planet with its closed ecosystem. *Courtesy of NASA.*

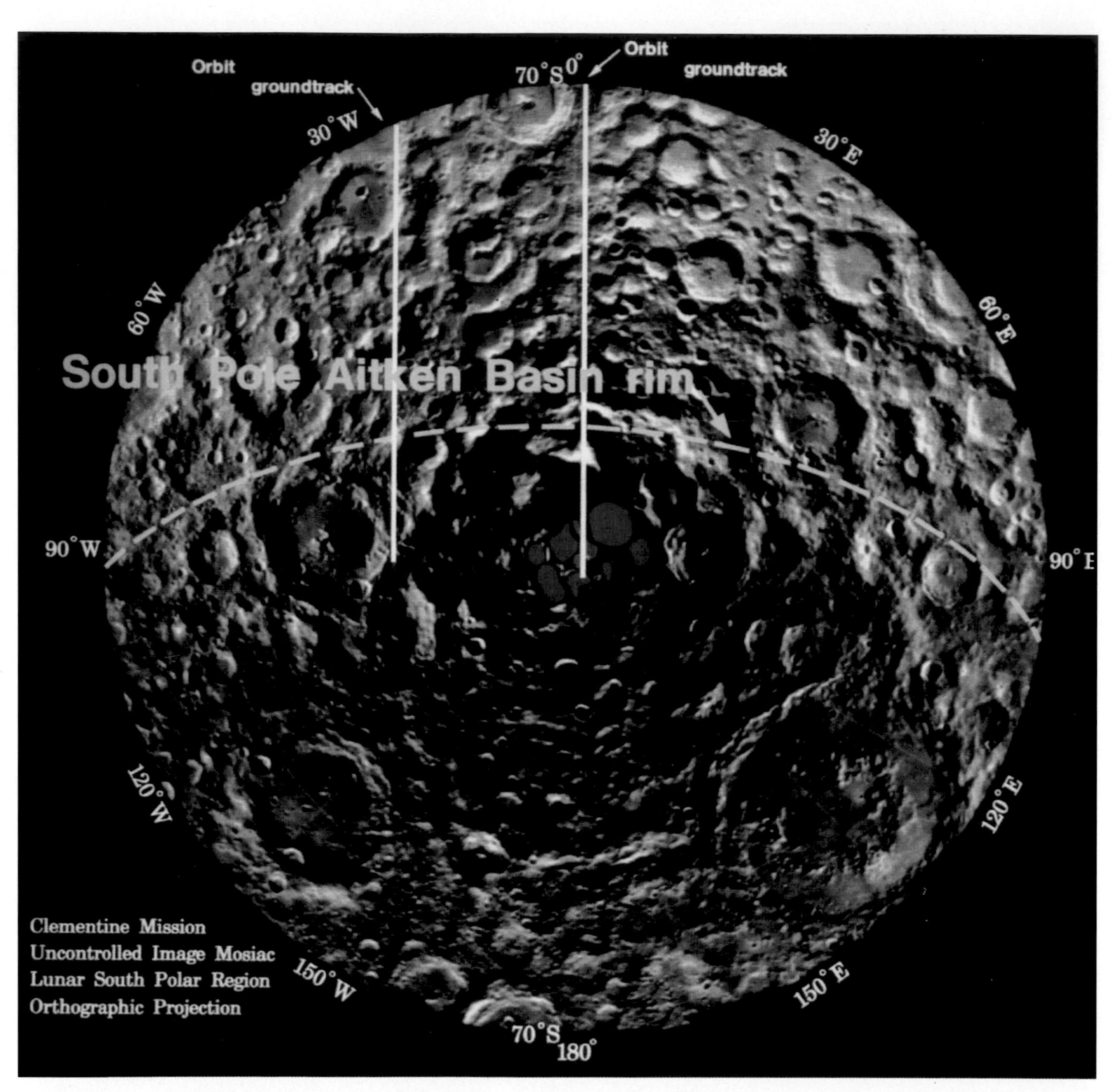

Figure 7.4 Clementine's image of the Moon's south pole. Computer-colored blue spots mark the locations of water ice. The circular image covers from the pole, 90° south at the center, to 70° south latitude at the edge. Numbers around the edge give the longitude. Segments of Clementine's orbital passes are shown by the white lines. *Courtesy of U.S. Naval Research Laboratory.*

Figure 7.9 Sojourner pushes its spectrometer against the rock called Yogi. Everything is covered with fine-grained red dust. The dark areas in the dirt are wheel tracks made by Sojourner as it did a "wheelie" to get positioned correctly against Yogi. The rock is about 1 yard (1 m) high. *Courtesy of NASA / JPL / CalTech.*

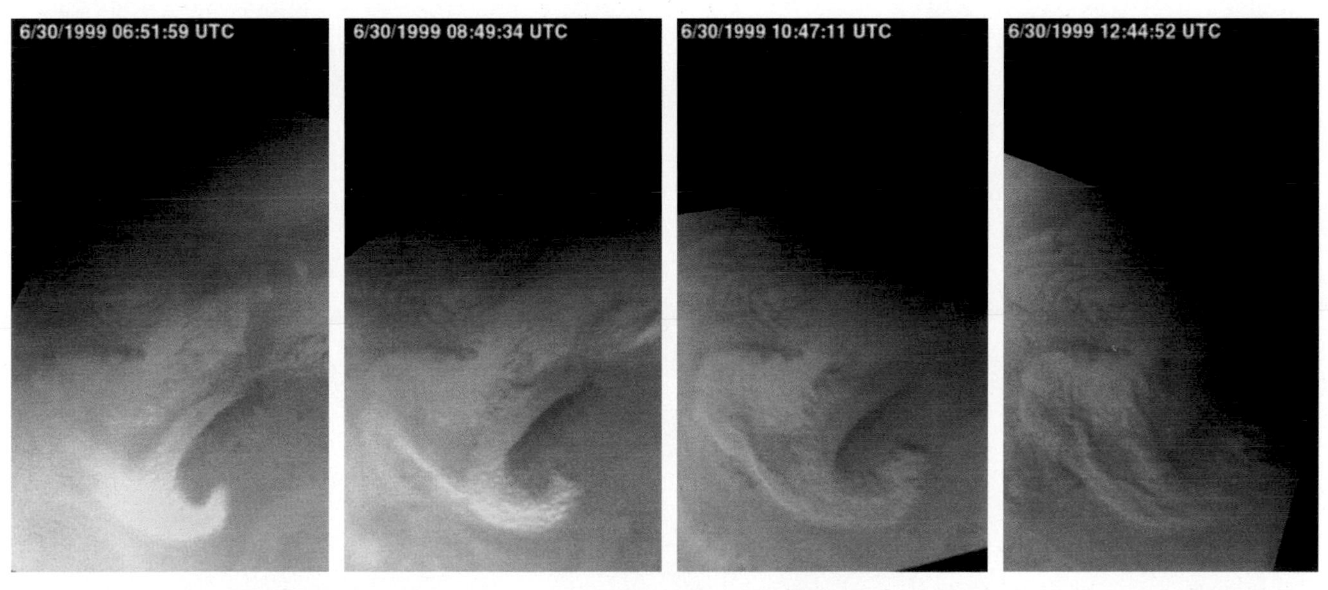

Figure 7.10 Storm clouds over the north polar cap of Mars look similar to weather patterns on Earth. The pictures were taken at 2-hour intervals. White areas are ice clouds, reddish clouds contain dust. *Courtesy of NASA / JPL / MSSS.*

Figure 7.16 Jupiter with its Galilean moons. The planet has great bands of colorful clouds and a giant red spot. Upper left is Io, center is Europa, lower center is Ganymede, and the lower right corner is Callisto. None of these Voyager pictures is to scale. See Table 7.1 for actual sizes. *Courtesy of NASA.*

Figure 7.17 Saturn with its rings and its major moons. Dione is in the front. Tethys and Mimas are to the right, Enceladus and Rhea are to the upper left, and Titan is at the top. *Courtesy of NASA.*

Figure 7.18 Uranus, a blue featureless planet, and its five major moons. From the largest to the smallest in this picture (not their real relative sizes) are Ariel, Miranda, Titania, Oberon, and Umbriel, named after characters in Shakespeare's plays. *Courtesy of NASA.*

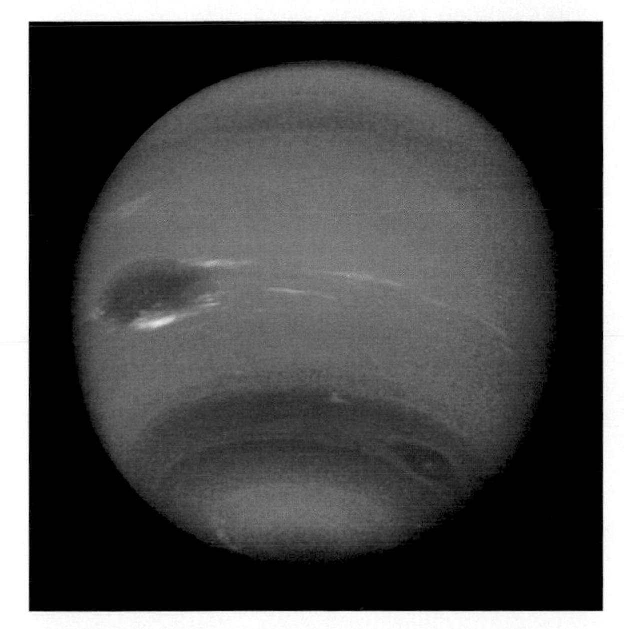

Figure 7.19 Neptune. A great dark spot is attended by wispy white clouds and a smaller dark spot is seen near the bottom. Between them is a white feature nicknamed Scooter because it moved faster than the other features. *Courtesy of NASA.*

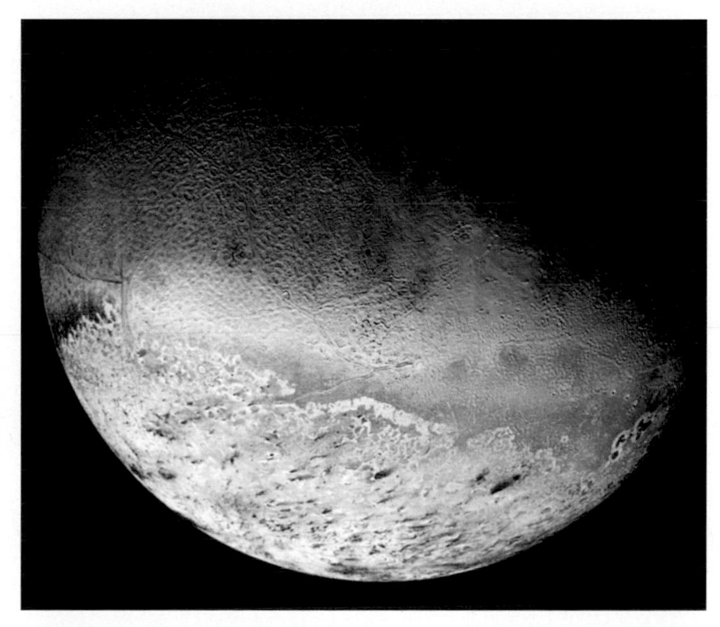

Figure 7.20 Triton, Neptune's largest moon, shows an incredible variety of terrain. The south polar region at the bottom of the picture is probably frozen nitrogen. The "cantaloupe terrain" at the center is unexplained. *Courtesy of NASA.*

Figure 7.22 Galileo spacecraft passing Jupiter. *Courtesy of Harris Corporation; artist J. Harrelson.*

Figure 7.26 The four Galilean moons of Jupiter. Left to right, in order of distance from Jupiter, they are Io, Europa, Ganymede, and Callisto. *Courtesy of NASA.*

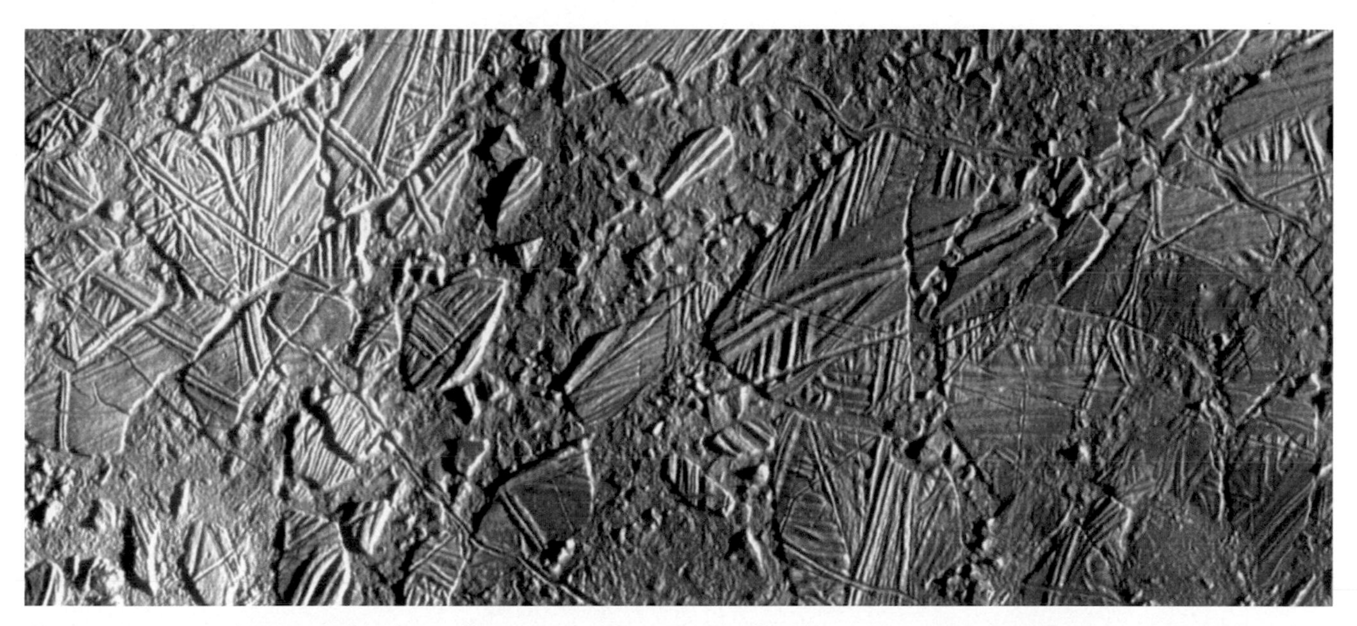

Figure 7.27 Europa is covered with a layer of ice and perhaps has an ocean of liquid water beneath the ice. This picture covers an area about 45 by 20 miles (70 by 30 km). The disrupted appearance was caused by an impact crater some 620 miles (1000 km) to the south which blew large blocks of ice and fine ice particles into this area. The brownish color comes from minerals in the water; the white and blue areas are covered with ice particles. *Courtesy of NASA.*

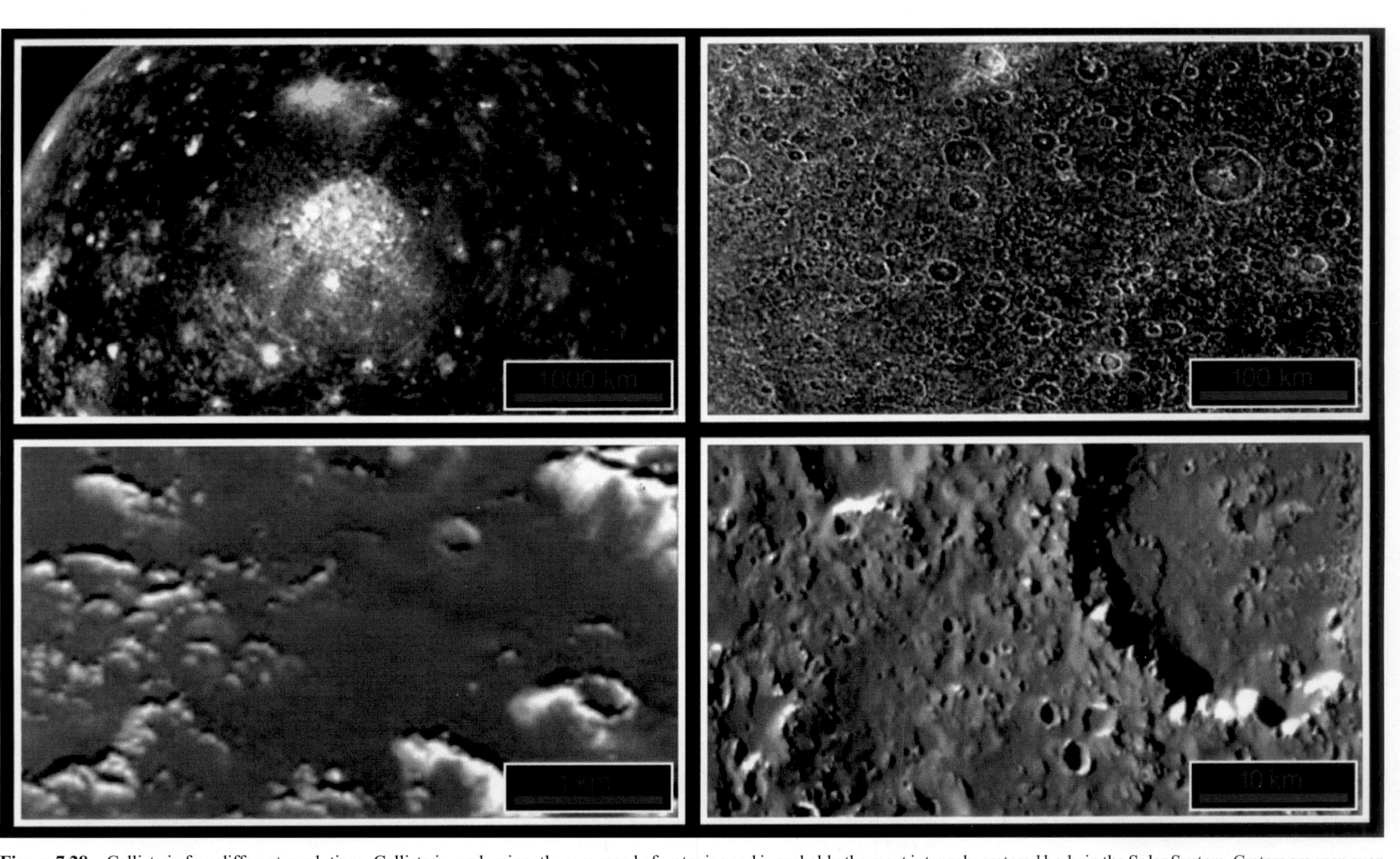

Figure 7.28 Callisto in four different resolutions. Callisto is predominantly composed of water ice and is probably the most intensely cratered body in the Solar System. Craters are seen even at the highest resolution. *Courtesy of NASA.*

are those few among so many? Remember that each piece of debris is in its own orbit and the retrieving vehicle must rendezvous with it in order to capture it. Rendezvous requires time and propellant. The real problem of the millions of smaller pieces remains. Using the Shuttle is expensive, hazardous to the astronauts, and limited to altitudes below 360 miles (600 km). In addition, the Shuttle has better things to do. This is not a practical solution.

A number of schemes have been proposed for cleaning up the orbital trash. NASA, concerned about objects in the size range of 0.5 to 4 inches (1 to 10 cm), studied the possibility of using a laser operated from the ground. High-power lasers now under development as weapons may some day be able to completely disintegrate an orbiting object down to submillimeter size. Current lasers are powerful enough to evaporate some of the surface material, not enough to destroy the object, but enough to change its orbit so it will burn in. To strike a small target, a laser requires a radar detection and tracking system to steer the laser; a radar with that capability now operates in Massachusetts. Two optical tracking systems in Hawaii and New Mexico could also detect the debris but cannot see during bad weather. A far-out possibility would use a radio receiver that is capable of receiving microwave transmissions from communications satellites that reflect off the debris. By one calculation it would be possible to eliminate all debris 1 to 10 centimeters in diameter up to 1500 kilometers in attitude over a 3 year period for a cost of about $160 million.

Several types of space sweepers have been proposed. One kind would employ a cone-shaped device to capture objects up to 0.8 inch (2 cm) in diameter by spraying them with swirling droplets of water or a gas and then depositing them in a receptacle which is ejected on a reentry course. An on-board tracking system would find the objects. Another type would use a constellation of satellites with a large surface area, perhaps several square miles. A hard surface would capture the debris or deflect it into a reentry orbit. A soft or spongy surface would capture it or slow it down so it burns in.

The problem is intercepting the debris. Recall the discussion of rendezvous in Chapter 3. If both the sweeper and the debris are in precisely the same orbit, they must be traveling at precisely the same speed and will not meet. If the sweeper and the debris are in circular orbits at different speeds, then they must be at different altitudes and again they will not meet. If the sweeper's orbit is elliptical or at an inclination different from the debris, the interception is possible because their paths cross. Another problem is the closing speed. If the closing speed is high, then the collision could produce more debris particles than are collected. A large number of such collisions could eventually destroy the sweeper itself. The sweeper would have to carry a propulsion system and fuel so it can change orbit to intercept debris or to avoid a disastrous collision with an active satellite. It is an expensive method; several hundred sweeper satellites would be needed to do the job, costing perhaps $1.5 billion plus another $1 billion to launch them.

A tether system has been proposed for capturing larger objects such as dead satellites and rocket bodies. A small satellite approaches the object and spins up a rotating disk with attached tethers (strings) like a yo-yo. One end of each tether is attached to the disk and weights are attached to the free ends. When the weights are released they pull the tethers out and, because they are rotating, wind themselves like a bola around the debris. The capturing satellite then fires a retrorocket to deorbit, dragging its captive piece of space junk with it. One satellite would be needed to dispose of each large piece of debris. At $5 million each for some 1200 large objects, plus launch costs, this system gets quite expensive.

There is no easy solution to the problem of space debris. But it will not go away; it will only get worse as more satellites are launched.

DISCUSSION QUESTIONS

1. Why doesn't the solar wind blow satellites out of orbit? (Look at the diagram of the magnetosphere, Figure 4.14).
2. The Sun is composed mostly of hydrogen. Why, then, does the solar wind consist mostly of protons and electrons?
3. Auroras are frequently seen in the polar regions but seldom seen in mid-latitudes and equatorial regions. Why? (Look at the Earth's magnetic field lines in Figures 4.14 and 4.15.)
4. If the Earth's atmosphere suddenly became transparent to all electromagnetic radiation, what changes, if any, would you expect to take place on Earth?
5. If the Earth's atmosphere had always been transparent to all electromagnetic radiation, how might Earth be different?
6. If you had an important space mission to fly but there was a high probability of a solar proton storm, would you go anyway? What factors would go into your decision?
7. When was the last sunspot maximum? When was the last minimum? What was the Sunspot Index at those times? Search for the answers on the Internet.
8. Give some additional suggestions as to how to reduce the amount of space debris in orbit around Earth.

ADDITIONAL READING

Heath, Gloria W., ed. *Space Safety and Rescue 1997*. Volume 96, Science and Technology Series, American Astronautical Society, 1999. Symposium proceedings.

Hunton, Donald E. "Shuttle Glow." *Scientific American*, November 1989. Technical but readable.

Johnson, Nicholas. "Monitoring and Controlling Debris in Space." *Scientific American,* August 1998. Summary of the problem and possible solutions.

Johnson, Nicholas, and Joseph P. Loftus, Jr. "Reducing Orbital Debris: Standards and Practices." *Launchspace,* March/April 1999. Easy to understand article about the international concerns.

Johnson, Nicholas, and Darren McKnight. *Artificial Space Debris.* Updated edition, Krieger Publishing Company, 1991. Detailed analysis of debris problems.

Lang, Kenneth R. "SOHO Reveals the Secrets of the Sun." *Scientific American,* March 1997. About the Solar and Heliospheric Observatory spacecraft.

Millman, Peter M. "The Meteoritic Hazard of Interplanetary Travel." *American Scientist,* November–December 1971. Technical, but readable.

NASA. *Our Prodigal Sun.* NASA Facts, KSC 116–81. Government Printing Office, 1982. Easy reading pamphlet.

Nesme-Ribes, et al. "The Stellar Dynamo." *Scientific American,* August 1996. Summary of current understanding of how the Sun and stars work.

Smith, Edward J., and Richard G. Marsden. "The Ulysses Mission." *Scientific American,* January 1998. First spacecraft to fly over the poles of the Sun.

Stafford, Edward P. *Sun, Earth and Man.* NASA EP-172. Government Printing Office, 1982. Easy reading booklet.

Tascione, Thomas F. *Introduction to the Space Environment.* Second edition, Krieger Publishing Co., 1994. Thorough summary.

Van Allen, James. "Radiation Belts Around the Earth." *Scientific American,* March 1959. Discovery of the Van Allen belts.

White, Oran R. *The Solar Output and Its Variation.* Colorado Associated University Press, 1977. Technical treatise.

Zimmerman, Robert. "Sun Up." *Star Date,* July/August 1999. Easy reading article about the approaching peak in solar activity.

NOTES

Chapter 5

Satellites

Earth has one natural satellite: the Moon. Sputnik, the first artificial satellite of Earth, was launched by the Soviet Union on October 4, 1957. It carried a small transmitter that beeped a signal to ground stations following its trek across the sky. That is all it did. By the end of 1999, over 9000 man-made objects smaller than 4 inches (10 cm) in diameter were being tracked in their orbits around Earth, only 600 of which were useful satellites. The rest were pieces of burned-out boosters, satellites whose transmitter or sensors had gone dead, and a few small objects such as bolts and an astronaut's glove. In addition, at least 150,000 objects too small to track are in orbit.

What are they all doing up there? Some of the practical uses for satellites include communications relay, weather observing, military surveillance, global positioning, monitoring ocean wind and waves, geodetic mapping, and monitoring of Earth resources. Satellites have been used for search and rescue, locating fishing grounds, observing icebergs, and for scientific research in astronomy, space physics, and planetary science. After years of investing tax dollars in the space program, taxpayers are now getting personal payoffs with direct satellite TV broadcasts, low-cost reliable long-distance telephone service, planet-wide cell phones, satellite location finding and navigation, high resolution images of Earth, and more innovations to come. Government, through NASA, continues to do the basic space research necessary to make these things happen, but commercial enterprises are now developing many of these systems and selling their services. In 1999, for the first time, there were more commercial launches than government-sponsored (NASA and military) launches. This is what free enterprise is all about: government began and developed the capabilities to leave Earth and do practical things in space; now industry can do profitable things in the commercial marketplace.

This chapter will deal with satellites for communications, navigation, energy production, and defense. Satellites that view Earth will be covered in Chapter 6. Spacecraft for astronomical research and exploration will be discussed in Chapter 7.

Communications Satellites

Radio waves propagate in straight lines and cannot travel around the curvature of the Earth. Thus, a radio or television receiver cannot pick up broadcasts from a transmitter that is beyond the horizon; see Figure 5.1. The *ionosphere,* a naturally occurring ionized layer of the atmosphere from about 60 miles to 200 miles (100 to 320 km) altitude, reflects certain frequencies so they can be heard over the horizon. Long-distance shortwave broadcasts may be received by ionospheric reflection, but the ionosphere is irregular, disrupted by solar activity, and somewhat undependable. If an artificial "mirror" could be placed in orbit, communications could be reliably established between any two points on Earth.

The first U.S. experiment in communicating via satellite was Echo 1, a 100-foot (30-m) aluminized mylar balloon launched into space folded in a small package which opened and inflated on arriving in orbit. How do you fold a 100-foot (30-m) balloon into a 28-inch (70-cm) diameter canister? One of the engineers was inspired by his wife's plastic rain cap, which folded into a long narrow strip and unfolded into a perfect hemisphere that fit over her head. It worked.

Because it carried no receiver or transmitter, an Echo balloon could be used only passively as a reflecting surface to bounce radio signals beyond the transmitter's horizon (Figure 5.2). The spherical-shaped Echo balloons, dubbed "satelloons," scattered the radio waves in many directions, so that only a small amount arrived back at the ground. High-powered transmitters, 10 kilowatts or more, were necessary to assure that a detectable amount of radio signal reached the receiver. Large, moving dish antennas on Earth tracked them across the sky.

The balloons were launched into elliptical orbits between 600 and 1100 miles (1000 to 1800 km) initially but, because of their size, air drag had a major effect on them and the orbits varied substantially from week to week. Echo 1 was launched August 12, 1960, and remained in orbit for nearly 8 years before burning in. Echo 2, 135 feet (40 m) in diameter, was launched in 1964 and lasted about 5½ years.

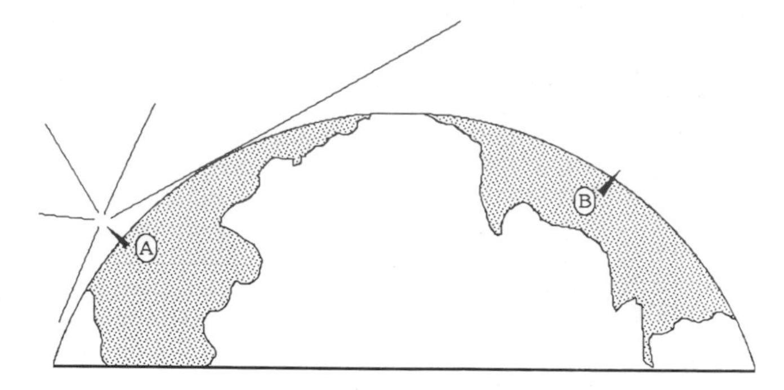

Figure 5.1 Radio signals travel in straight lines; they do not curve around Earth. A radio receiver at B cannot receive the broadcast from the transmitter at A.

Later, active relay satellites replaced the passive balloons. Active relay communications satellites carry receivers and transmitters. The signal is sent to the satellite, picked up by its receiver, and retransmitted on another frequency to a ground station.

Placing relay stations in space was a natural follow-on to microwave relay stations on Earth. During the 1940s and 1950s, based on technology developed in World War II, communications companies had set up a network of microwave relay towers across the country to replace the long wires and to extend telephone and television services to remote areas. At each relay station the signal was received and retransmitted to the next station in line. Because radio waves do not follow the curvature of the Earth, the microwave towers had to be placed within line of sight of each other. On flat terrain the towers were built about 50 miles (80 km) apart. This communications technology was well developed, and when satellites came along there was the opportunity to place the relay tower in space where broadcasts could be transmitted to vast areas from a single transmitter in orbit.

The first active repeater satellite was Courier, a Department of Defense experiment that operated from October 4 to October 21, 1960. Spherical in shape, it was covered with solar cells, the first satellite to use solar cells instead of batteries for power.

The first commercial active repeater satellites were Bell Telephone's Telstar 1 and RCA's Relay 1, both launched in 1962. Telstar 1 weighed 170 pounds (77 kg) and was capable of handling one television channel or 600 telephone circuits. By comparison, later communications satellites weigh nearly 8000 pounds (3600 kg) and can handle more than 250 television channels or 50,000 voice channels simultaneously.

All these early communications satelllites were in relatively low orbits which meant that they rose above the horizon, ambled across the sky, and set below the horizon. Of course, they were only useful when they were above the horizon. Relay 1's orbit had an apogee at about 4600 miles (7400 km), perigee near 800 miles (1300 km), and a period of 185

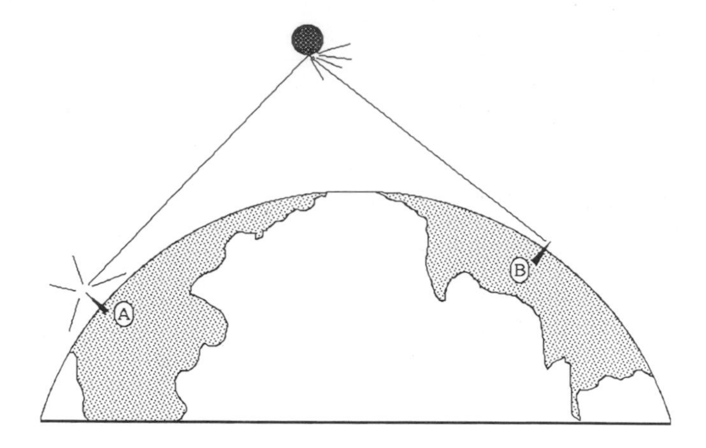

Figure 5.2 The Echo balloon in orbit acted as a reflector so signals from the transmitter at A bounced to the receiver at B. Later, relay satellites received the signal from the transmitter on the ground and re-transmitted it to receivers beyond the horizon.

minutes. This allowed it to "hang" at apogee in position to provide communications between two ground terminals for more than 2 hours. To provide continuous communications between any two points would require a "constellation" of satellites so that at least one was visible from the ground at all times.

Geosynchronous Satellites

The favored orbit for communications satellites is the geosynchronous orbit, a circular orbit with an inclination of zero degrees, 22,300 miles (36,000 km) above the equator. (Refer back to Figures 3.17 and 3.18.) At that altitude the orbital period is 23 hours and 56 minutes, the time it takes the Earth to make one rotation on its axis. Therefore, the satellite's angular speed is the same as the Earth's, 360 degrees per day. To a ground observer, the satellite appears to remain stationary in the same spot in the sky. This has three advantages for communications satellites. First, the satellite can be used continuously; it never goes below the horizon. Second, the large dish antennas do not have to track a moving satellite. Antenna construction is much simpler and cheaper. Third, from a geosynchronous altitude, a satellite can "see" almost one-third of the Earth's surface. Theoretically, only three satellites are needed to cover the entire Earth except for the high latitude regions around the north and south pole. Practically, four satellites are needed with some overlapping antenna patterns to assure reliable coverage. Figure 5.3 shows the view from geosynchronous orbit at three points on the equator, 120 degrees apart. Notice how their viewing areas overlap at the equator up to midlatitudes.

These ideas originated with science and science fiction writer Arthur C. Clarke, who coauthored the movie *2001: A Space Odyssey* with Stanley Kubrick. In the 1940s and 1950s, Clarke was active with the British Interplanetary Society, a far-sighted group of scientists and engineers that

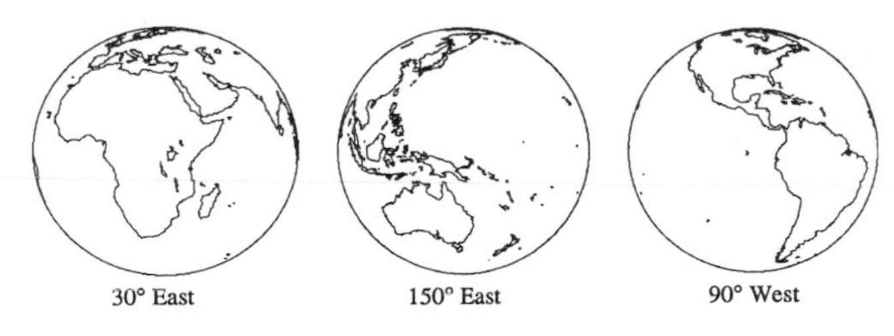

Figure 5.3 View from geosynchronous orbit over spots on the equator at 30° E, 150° E, and 90° W longitude.

studied and developed many futuristic ideas about space. In the October 1945 issue of *Wireless World,* Clarke published a paper showing the technical feasibility of communications satellites in geosynchronous orbit. Actually, the Russian schoolteacher, Constantin Tsiolkovski, had recognized the existence of the geostationary orbit around Earth decades earlier, but the idea of using it for communications satellites originated with Clarke. Therefore, the geosynchronous orbit is often called the Clarke orbit in western countries.

The first geosynchronous satellite was Syncom II, which achieved orbit on July 26, 1963. On August 19, 1964, Syncom III was locked into a stationary location near the international date line. It carried the first trans-Pacific television broadcast, the Tokyo Olympic Games. By 1995 there were nearly 150 communications satellites in geosynchronous orbit.

To prevent interference between satellites, they are each assigned specific operating frequencies and are spaced at least 4 degrees apart as viewed from Earth, about 2000 miles (3200 km) in orbit. This limits the number of satellites that can operate on any given frequency in geosynchronous orbit. The International Telecommunications Union, an organization under the United Nations, is responsible for assigning frequencies and orbit positions so there is no mutual interference.

Most communications satellites operate in the C-band, 3700 to 4200 MHz (3.7 to 4.2 GHz). Recently satellites have been moving to higher frequencies in the Ku-band, 11.7 to 12.2 GHz. This allows the transmitter to carry more channels or conversations on a single channel. At these frequencies, absorption by large, shower-type raindrops is greater than by drizzle or small cloud droplets. A climatological study of the receiver site gives a clue as to the percent of time that the downlink may be wiped out due to heavy precipitation. The engineers' solution to the problem has been to increase the power of the satellite transmitters. However, as higher frequencies are called into use, the problem becomes more acute.

The antennas on the satellites may be designed for maximum area coverage like the whip antenna on an automobile, or they may be narrow beams using large dish reflectors, beaming their signals to relatively small regions on the Earth. A NASA Tracking and Data Relay Satellite, its antenna open for inspection is shown in Figure 5.4. It is folded into a compact package for launch (Figure 5.5). Figure 5.6 shows it deployed in space.

Direct-to-Home Satellites

In the 1970s it was necessary to have a 30 foot (9 m) dish antenna to receive signals reliably and clearly, so communications satellites were used directly only by telephone companies, television networks, and other businesses who could afford the expensive installations. By the 1980s an 8-foot (2.5 m) dish would do the job and many people installed them in their backyards to listen in on television broadcasts. Practically all television programming, except for local programs, is relayed around the world by satellite. A dozen or more satellites, each carrying half a dozen or more programs, can be seen from a single location. Even television broadcasts via Molniyas can be picked up from the northern United States, although the color and sound are not compatible with a standard American television set.

In the 1990s a new generation of direct broadcast satellites (DBS) began transmitting superb quality television broadcasts direct to homes. Three satellites are parked in geosyn-

Figure 5.4 A NASA Tracking and Data Relay Satellite (TDRS) antenna open for ground testing. *Courtesy of NASA.*

Figure 5.5 The TDRS satellite being lifted from a Space Shuttle orbiter for dropping off in space. The spacecraft folds into a compact package to fit into the cargo bay. A folded umbrella-like antenna and a dish antenna can be seen inside the folded solar panels. A rocket attached to the bottom will boost it from the Shuttle's orbit to geosynchronous orbit. *Courtesy of NASA.*

chronous orbit near 101 degrees west longitude. Dish antennas only about 18 inches (0.5 m) in diameter are able to receive the broadcasts because the satellites transmit with higher power at higher frequencies, 12 and 17 GHz. However, large raindrops absorb these higher frequencies and the signal may drop out during a rainstorm. The system carries hundreds of TV channels and is capable of digitizing and compressing the image for high definition TV (HDTV), the newest innovation in home entertainment. The service operates similar to a cable system; that is, users pay a monthly fee and must purchase receivers and decoders to unscramble only those channels they subscribe to.

Molniya Satellites

Geosynchronous satellites, positioned over the equator, as they must be, are difficult to use in high latitude regions because they appear so low on the horizon. At a low angle, the radio waves must traverse a longer path through the atmosphere, making them more subject to atmospheric fluctuations and disturbances. The Soviet Union had a unique solution to this problem its their Molniya communications satellites, which it began to launch in 1965. The Molniyas are placed in 63 degree, 12 hour orbits of high eccentricity with perigee in the southern hemisphere at 330 miles (530 km) and apogee in the northern hemisphere at 24,400 miles (39,500 km) as described in Chapter 3. The Russians continue to use the Molniya orbits. ("Molniya" is Russian for "lightning.") Besides dwelling for many hours over the high latitudes, less energy is required to launch into a 64 degree orbit than into a geosynchronous orbit from their northern launch site at Plesetsk. Three of them equally spaced around the orbit can provide continuous communications links; as one goes down below the horizon, another is rising. By the year 2000, more than 135 Molniya satellites had been put into orbit, many of which have since burned up on reentry back into the atmosphere.

Direct-to-People Satellites

As the state of the art has advanced, the cellular phone concept has gone into orbit. Electronic components have become smaller, more reliable, and cheaper. It is now possible to make satellites smaller, lighter weight, smarter, cheaper to build assembly line fashion, and cheaper to send into orbit on small rocket launchers. Improvements in power supplies and advances in digital technology have made it possible to communicate directly to and from satellites with handheld radiotelephones. By keeping the satellites in low orbits, the transmitters on both ends of the conversation need less power and simpler, smaller antennas than are needed for satellites in geosynchronous orbit. Also, low orbiting satellites do not have the time delay problem associated with geosynchronous satellites, a minimum of ¼ second for the round trip, a frustrating delay when trying to carry on a conversation. The main problem with this concept was mentioned before: low orbiting satellites spend much of their time out of sight below the horizon. The solution to this problem is to place large constellations of small satellites in several orbital planes to assure that several are always in view.

Two main contenders have dominated the cell-phone-in-space business: Iridium and Globalstar. Iridium originally planned to put 77 satellites (plus spares) in 480 mile (780 km) orbits. Its name comes from element number 77 in the periodic table of the elements. In the final configuration it was downsized to a constellation of 66 working satellites, 11 plus a spare in each of 6 polar orbits. (Element 66 is dysprosium, not quite as easy to say as iridium.) They operated in the 1.5 to 30 GHz range. The higher frequencies were used for satellite-to-satellite links so they could hand-off phone calls from one to another so a user in one location could communicate with someone at any other point on Earth. Each satellite had 2800 voice channels capable of receiving signals from handset transmitters with only a few watts of power. In March 2000 the company went bankrupt and plans were made to deorbit the satellites. The anticipated number of customers did not materialize because the cost of the handset, the monthly fee, and the per-minute charge were all greater than individuals or companies were willing to pay.

The Globalstar constellation has 48 satellites plus on-orbit spares in 870 mile (1400 km) orbits inclined at 52 degrees, operating in the 1.5 to 6 GHz range. The satellites were launched four at a time on a U.S. Delta 2 or a Russian-French

Figure 5.6 Artist's concept of a NASA Tracking and Data Relay Satellite (TDRS) in orbit. Six antennas (can you find them?) cover four frequency bands. The two largest ones, over 16 feet (5 m) in diameter, are steerable. Receivers and transmitters are in the hexagonal box in the center. Arrays of solar cells on the two paddles provide power. *Courtesy of NASA.*

Soyuz-Ikar booster rocket. As designed, the completed system is capable of handling 7.5 million users. Globalstar connects with existing ground-based wire and cellular phone systems where they are available. It has a better chance of success because operating costs are lower and customer fees are less than Iridium.

We are seeing the beginning of a new world order in personal communications. By their very nature, these satellite telephone systems are worldwide. Half the people in the world have never used a telephone, but because of geosynchronous satellites there are now more TV sets in the world than telephones. Huge areas of the world have never been wired and do not have cellular systems. Several other satellite systems for voice and data channels are under development in the United States and abroad. High-speed data links via satellite for graphics, imagery, video, and Internet connections at home and in businesses are likely to become a reality in the near future. The number of potential users is enormous if the price is right. As with any new technology, the price is high for the first customers. But prices are sure to drop as more customers sign on.

The next step in personal satellite service will be channels for providing broadband Internet connections. A voice channel requires only a few kilohertz bandwidth, but a data channel to handle worldwide web photographs and graphics requires a few megahertz. You need half a dozen voice channels to send one data channel. Satellite communications systems will soon be available for direct Internet connections.

Military Communications Satellites

The defense communications relay satellites are similar in operation to their civilian counterparts. To prevent outsiders from listening in, the messages are generally encoded electronically before transmission to the satellite and automatically decoded at the receiving station. Narrow beam antennas reduce the possibility of jamming.

The United States is probably more dependent on satellites for military communications than other countries. American forces are spread around the globe on all continents; positive command and control are essential worldwide. The Milstar satellite system is the newest, most highly capable communications system designed specifically for military use. The original plan called for four satellites in geosynchronous orbits and three satellites in highly elliptical (Molniya-type) orbits to provide worldwide command, control, and communications for land, air, sea, and undersea forces. A number of spare satellites may be placed in 100,000 mile (160,000 km) orbits. It is easier and quicker to move a satellite from a high orbit to geosynchronous altitude than it is to prepare and launch a replacement from the ground, especially during wartime.

Navigation Satellites

Using satellites it is now possible to find your precise location at any time and place on Earth with an accuracy of about 30 feet (9 m). How is it done? Look first at an example of how you determine distances traveled in your car.

Let's take a trip from Colorado Springs to Denver. We leave at 2:30 on the dot and we arrive in Denver at exactly 3:45. Assume we traveled at a constant speed of 60 miles per hour (96 km per hour). How far is it from Colorado Springs to Denver? Time elapsed from 2:30 to 3:45 is 1 hour and 15 minutes or 1.25 hours. Multiply that times the speed and we get 75 miles (120 km). The accuracy of our answer depends on whether our watch gained or lost time during the trip and how precisely we held to a constant speed.

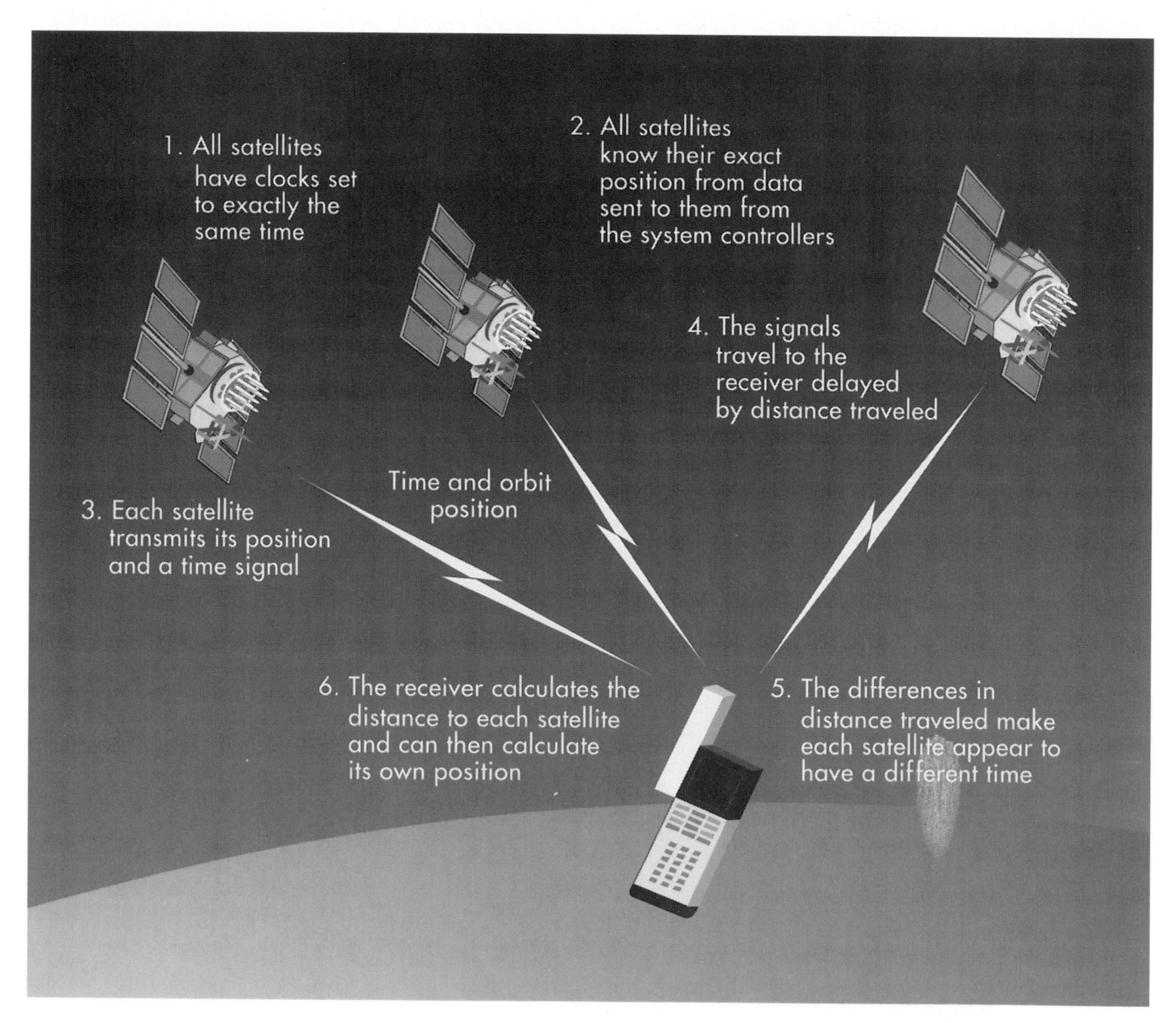

Figure 5.7 How the Global Positioning System (GPS) works. Courtesy of The Aerospace Corporation.

Using navigation satellites to locate our position on Earth is based on the same principles. Figure 5.7 shows schematically how it works. A satellite transmits a radio pulse at a given time. The clock in the receiver tells what time the pulse was received. Assume that the radio pulse travels at a constant speed, the speed of light, and assume that the clocks are synchronized together so we can precisely determine the elapsed time. We can then calculate the range, the distance to the satellite.

The range by itself is not sufficient information; it is also necessary to know the location of the satellite. The latest ephemeris (orbital data) is also transmitted from the satellite so we can calculate its precise location. Now that we have the satellite's location and its range we know that we are somewhere on a sphere with a radius equal to the range and with the satellite at the center. See Figure 5.8.

Next we obtain the same information from a second satellite to locate ourselves on a second similar sphere. If we are on both spheres at the same time, we must be at the place where the two spheres intersect. See Figure 5.9 (it looks like a snowman). Notice that two spheres intersect in a circle (the waist of the snowman) so we know we must be somewhere on that circle. Getting warmer! Reading out a third satellite gives us another circle which intersects at two points: one will be way out in space or deep in the Earth, but the other is our precise location (X on Figure 5.10), our latitude, longitude and altitude, *provided our clock is accurate.*

If the user is in a moving vehicle, its speed and direction can be calculated by getting two positions in a short time and finding the change in location during that time. Velocity can also be calculated by measuring the Doppler shift in the frequency of the signal coming from the satellite. A Doppler shift is an apparent change in frequency of a wave when the source of the wave or the receiver or both is moving toward or away from each other. You have perhaps noticed the change in pitch of an ambulance siren as it moves past you. When it is moving toward you, the siren sounds higher in pitch than when it is moving away. The faster the vehicle moves toward you, the higher the pitch. Similarly, when it moves away the pitch sounds lower, and the faster it moves away from you, the lower the pitch. In this example we are referring to sound waves, but the same Doppler effect applies to electromagnetic waves. The GPS satellite transmitter operates on a very stable frequency. If the receiver can precisely measure the received frequency and detect a shift in that frequency, then the speed of separation or approach between the satellite and the observer can be calculated. Using the orbital elements of the satellite included in the transmitted signal, the receiver can calculate the velocity of the satellite and subtract it from the Doppler shift leaving only the observer's velocity with respect to that satellite. Then, by making similar measurements of velocity with respect to four satellites, the absolute velocity of the observer is calculated.

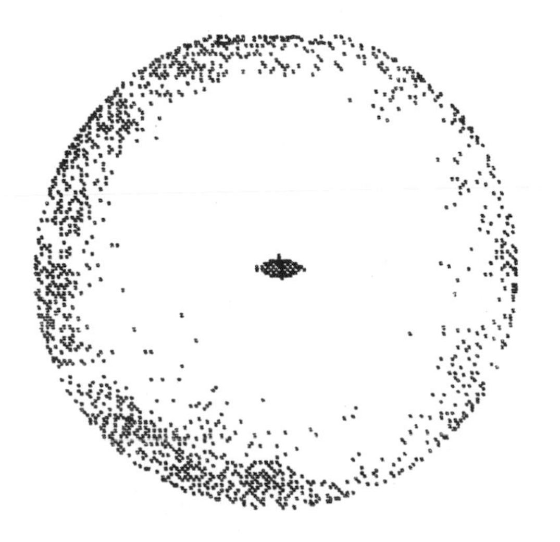

Figure 5.8 A sphere with a satellite at the center. The radius of the sphere is the distance (range) from the observer to the satellite.

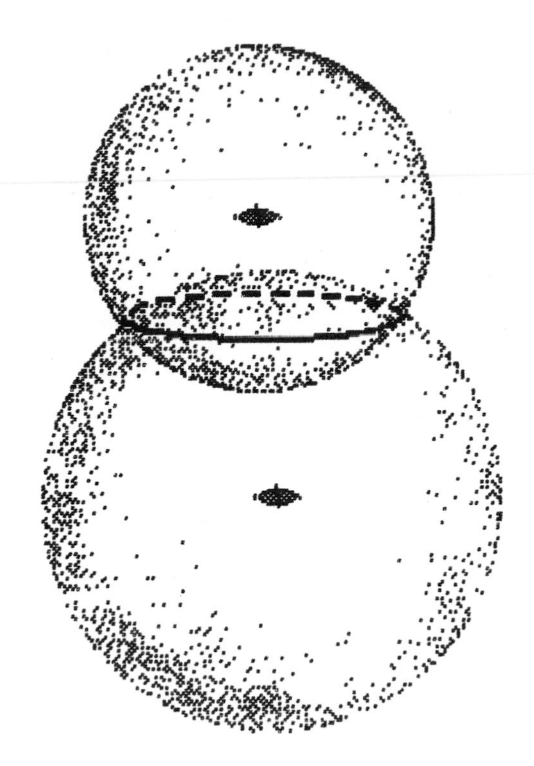

Figure 5.9 Two overlapping spheres, with radii equal to the ranges of the two satellites. They intersect in a circle.

Errors in Position Finding

There are three difficulties with putting this relatively simple idea into practice.

First, in order for this to work, the clocks in the satellite's transmitter and the user's receiver have to be exactly synchronized. Each satellite carries four atomic clocks which are updated and synchronized daily using atomic clocks on the ground which are accurate within 1 second in 300,000 years. However, atomic clocks are not portable enough and the

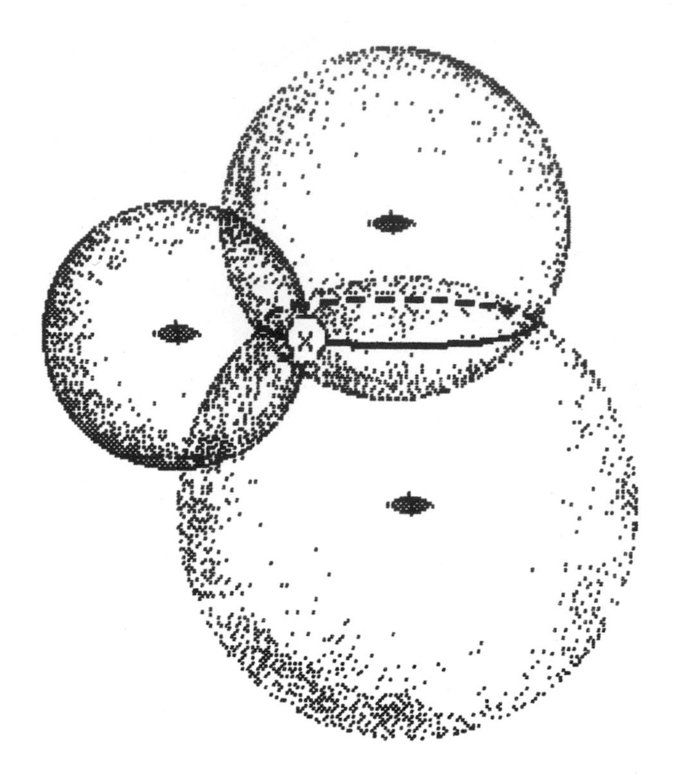

Figure 5.10 Three overlapping spheres with radii equal to the ranges of the three satellites. The position of the observer is marked "x".

clock in a mobile receiver cannot be made accurate enough to precisely measure the travel time of the pulse. A tiny error in measuring the travel time produces a large error in the range calculation. MATHBOX 5.1 shows some sample calculations. An error of only 50 billionths of a second in time produces a range error of 50 feet (15 m). If the receiver's clock is running fast, the spheres are too big; if the receiver's clock is slow, the spheres are too small. In either case the spheres do not intersect at a point.

The solution to this problem is to read out a fourth satellite and treat the receiver's clock error as a fourth unknown. Then adjust the clock error up and down until a unique solution is found, the exact position of the receiver in three-dimensional space. Finding this solution is a mathematical problem of no small magnitude. It involves solving four equations in four unknowns: latitude, longitude, altitude, and clock error. MATHBOX 5.2 shows the algebra. Receivers specially designed for use with the navigation satellites incorporate a microcomputer chip preprogrammed to handle the calculations. In actual practice, the user receives transmissions from at least four satellites and the computer in the receiver does the rest.

A second problem comes from the assumption that the radio transmission from the satellite travels at 186,282 miles per second (299,792 km/s), the speed of light in a vacuum.

MATHBOX 5.1

Range Error

Radio waves travel at the speed of light, about 186,000 miles per second (300,000 km per second). At a satellite range of 12,000 miles (approximately 19,320 km), travel time is

$$t = d\backslash v = 12{,}000\ \text{miles}\backslash 186{,}000\ \text{miles per second} = 0.0645\ \text{second.}$$

In metric units,

$$t = d\backslash v = 19{,}320\ \text{km}\backslash 300{,}000\ \text{km/s} = 0.0645\ \text{second.}$$

An error of 1% in measuring the time would produce an error of 1% in computing the distance. In our example, a 1% timing error is 0.0006 second; that would produce a range error of 120 miles (193 km)! A navigator can guess his position better than that!

Suppose we want to reduce the range error to 50 feet. The distance to the satellite is

$$12{,}000\ \text{miles} \times 5280\ \text{feet per mile} = 63{,}360{,}000\ \text{feet.}$$

What percentage of this distance is 50 feet?

$$(50/63{,}360{,}000) \times 100\% = 0.00008\%$$

We must reduce the timing error to

$$0.00008\%\ \text{of}\ 0.0645\ \text{second} = 0.00000005\ \text{second}$$

That is 50 billionths of a second! (Try this in metric units. The answer will come out the same.)

MATHBOX 5.2

Navstar GPS Equations

Before we examine the equations used by the Navstar Global Positioning System (GPS) in three dimensional space, let us consider a more manageable problem—determining the position of an object, *P*, on a two dimensional surface with respect to another object, *A*.

Look at Figure 5.2.1. The point *A* is at position $A_x = 4$, $A_y = 3$. This is written (4, 3). Similarly, *P* is at $P_x = 1$, $P_y = 1.5$, written (1, 1.5). Notice the right triangle formed by a horizontal line from *P*, a vertical line from *A* and the line *r* connecting *A* to *P*. Our question is, how far apart are *A* and *P*? Because *r* is the hypotenuse of a right triangle,

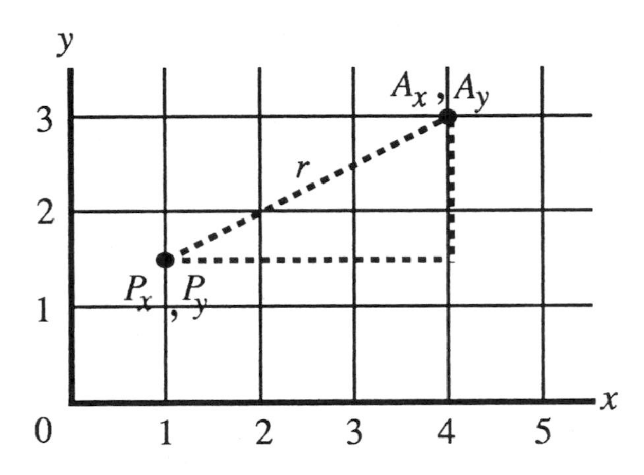

Figure 5.2.1

$$(A_x - P_x)^2 + (A_y - P_y)^2 = r^2.$$

In this case, $r^2 = (4 - 1)^2 + (3 - 1.5)^2 = 9 + 2.25 = 11.25$. So $r = \sqrt{11.25} = 3.4$.

When we deal in three dimensional space, we must include a third term, the *z* term:

$$(A_x - P_x)^2 + (A_y - P_y)^2 + (A_z - P_z)^2 = r^2$$

Let us apply this to GPS. The distance to satellite *A* is r_A; the distance to the satellite is also the speed of light times the time of travel: $r_A = vt_A$. But we cannot be sure our clock is correct, so we must add a correction factor, *e*, to the time; $r_A = v\,(t_A + e) = vt_A + ve$.

Now we are ready to read out four satellites (*A, B, C,* and *D*) and write four equations, one for each of them.

$$(A_x - P_x)^2 + (A_y - P_y)^2 + (A_z - P_z)^2 = (vt_A + ve)^2$$
$$(B_x - P_x)^2 + (B_y - P_y)^2 + (B_z - P_z)^2 = (vt_B + ve)^2$$
$$(C_x - P_x)^2 + (C_y - P_y)^2 + (C_z - P_z)^2 = (vt_C + ve)^2$$
$$(D_x - P_x)^2 + (D_y - P_y)^2 + (D_z - P_z)^2 = (vt_D + ve)^2$$

The four unknowns in these equations are our position (P_x, P_y, P_z) and the clock error, *e*. The positions of the four satellites are given in their transmissions, the travel time of the wave is measured by our receiver and the speed of light is a constant. These are knowns. Because we have four equations we can solve for the four unknowns. (That is, we can let the computer do it).

But here the pulse does not travel in a vacuum. It travels through the Earth's atmosphere during part of its trip to the ground, and its speed depends on atmospheric conditions, particularly conditions in the ionosphere. The ionosphere is the upper part of the atmosphere extending from about 50 miles to 250 miles (80 to 400 km). In this region the intense solar x-ray and ultraviolet radiation ionize the air, and the free electrons cause an apparent slowing of the radio pulse. The density of free electrons determines how much the actual speed of the pulse deviates from the speed of light in a vacuum. Electron density in the ionosphere varies in a somewhat predictable fashion through the day, through the seasons of the year, and with the 11-year solar cycle. However, auroras and geomagnetic storms produced by coronal mass discharges from the Sun cause large and unpredictable variations in the electron content of the ionosphere which in turn cause errors in the GPS range calculations.

The lower atmosphere, the troposphere, contributes to an error in the assumed speed. However, this error is smaller and less variable. Uncorrected, these ionospheric and tropospheric effects on pulse speed cause range errors amounting to nearly 20 feet.

The problem of pulse speed can be handled in either of two ways, one more accurate than the other. The amount of error depends on the frequency of the transmitted pulse: the higher the frequency, the lower the error. If a satellite carries two transmitters operating on different frequencies, it is possible to reduce the ionospheric error to less than a foot by receiving the signals on both frequencies and comparing the range calculations. However, a less accurate correction can be made using information about the ionosphere based on the average daily and seasonal variations in the ionosphere. For this method of correction the user needs to listen to only one frequency. While the correction is not exact, it is accurate enough for many purposes.

Another source of error comes from Einstein's theory of relativity; both gravity and time dilation affect the rate at which time passes. Time dilation means that time passes more slowly for an object in motion. Thus, the clocks in the GPS satellites, which travel at orbital speeds of 8600 miles per hour (14,000 km per hour), run slow compared to clocks on Earth. Gravity affects time also—if gravity is greater, time runs slower. Thus, a clock at a higher altitude where the force of gravity is less than on Earth will run faster than clocks on the ground. The gravity effect is greater than the time dilation so the net result is that the GPS clocks run fast. There is an easy correction to this error—the manufacturer of the clocks sets them to run slower by just the right amount to compensate so they keep the same time as Earth clocks. To the GPS receiver on the ground, the satellite clock appears to be running at the same rate.

The Navstar Global Positioning System

The Navstar Global Positioning System (GPS) was designed, launched, and funded by the Department of Defense primarily for military use. However, it is available to anyone with a commercially available GPS receiver. The system consists of 21 satellites and 3 spares in six circular 12-hour orbits at 12,500 miles (20,000 km) altitude, with an inclination of about 56 degrees. Figure 5.11 (Plate 3) shows the constella-

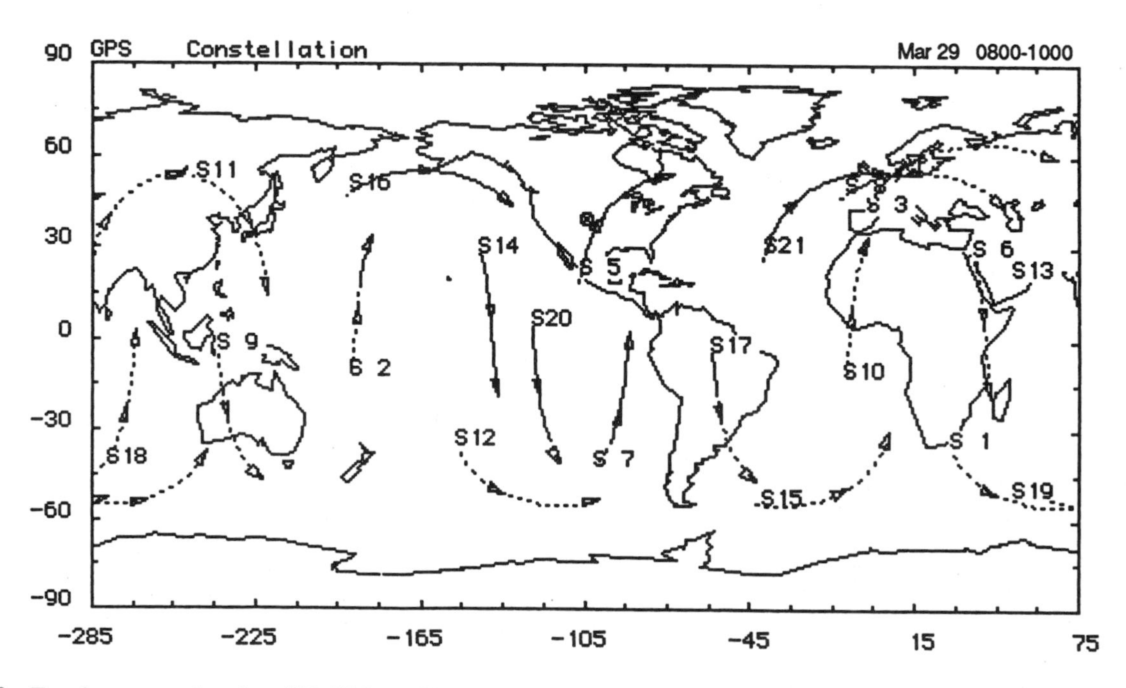

Figure 5.12 Two-hour ground tracks of 21 GPS satellites. The arrows on the traces indicate the 9 a.m. and 10 a.m. locations. Notice that they are evenly spaced over the entire Earth. Notice also that the satellites move due east at about 56° N and 56° S latitude, the inclination of the orbits. The bullseye marks Colorado Springs.

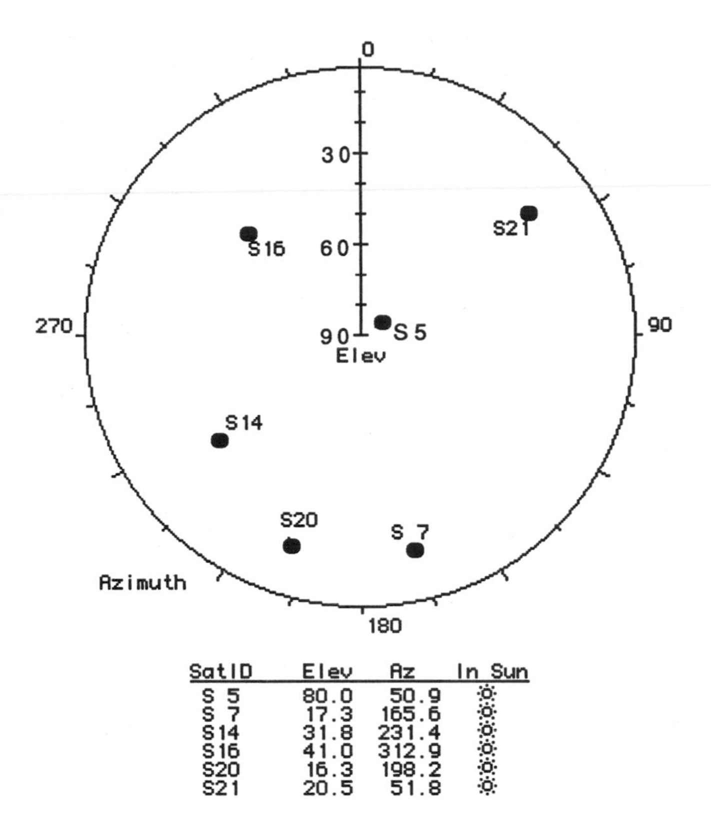

SatID	Elev	Az	In Sun
S 5	80.0	50.9	☼
S 7	17.3	165.6	☼
S14	31.8	231.4	☼
S16	41.0	312.9	☼
S20	16.3	198.2	☼
S21	20.5	51.8	☼

Figure 5.13 The GPS satellites visible from Colorado Springs at 9 a.m., the time of the traces shown in Figure 5.12. The outer circle is the horizon marked 0° at north, 90° at east, 180° at south, and 270° at west. The center of the circle is the point overhead, the zenith. The line from north to the zenith is marked off in degrees above the horizon. Only satellites higher than 5° are shown.

tion, Figure 5.12 shows their ground traces for a 2-hour period, and Figure 5.13 shows those visible from the ground at a specific time. All transmit on two frequencies, 1227.60 MHz and 1575.2 MHz. The satellites are powered by two large panels of solar cells which are kept oriented toward the Sun while the transmitting antennas remain pointed toward Earth. Military controllers at a ground station near Colorado Springs send messages to the satellites several times daily to update their clocks and the orbital parameters.

GPS gives an unlimited number of users the capability of determining their position within 30 feet (9 m), their velocity within 0.07 mile per hour (0.11 km per hour), and the time within a millionth of a second. The Department of Defense (DOD) can insert deliberate errors in the clock or in the orbital information, which are supposed to prevent a military adversary from getting precision data from the system.

It didn't take civil users long to find a way around both the natural and deliberate errors in the system, however. A GPS receiver is located at a position whose *exact* coordinates are known. That "base" receiver continuously reads out the GPS satellites and notes the error in the calculated position. The error is then converted to a clock error that is sent to other GPS receivers in the area, often by radio, so they can insert the correction into their receivers. This scheme is called *differential GPS* and is available from many sources worldwide. Using it, accurate locations can be found within a few inches, an accuracy never expected by the designers of the system.

In April 2000 the DOD turned off the deliberate errors and users everywhere now have access to the most precise data.

GLONASS

The Soviet Union simultaneously developed a similar system called GLONASS for both military and civilian use. Its theory of operation is identical to the Navstar GPS. GLONASS has about 15 satellites in three orbits at inclinations of 64.8 degrees. The broadcast frequencies are near 1600 and 1250 MHz. Only a few hundred GLONASS receivers were built in the Soviet Union. Manufacturers in both countries have developed equipment that is able to receive and process both GPS and GLONASS signals.

A European Union satellite navigation system called Galileo is in the planning stages. It may consist of 18 to 22 satellites in medium orbits and two or three in geosynchronous orbit, but other configurations are under study. The proposed system will be compatible with GPS and GLONASS, but will be designed and operated primarily for civilian rather than military users.

GPS Users

The obvious military use of GPS is for ground troops, aircraft, ships, and tanks to know where they are and where they are going. In the Persian Gulf War troops used GPS receivers to find their way day or night across trackless desert terrain without visible landmarks. Some GPS receivers allow the users to enter the coordinates of their destination. The receiver then finds their current location and calculates the direction and distance to their destination. Soldiers who had to make a trip away from home base would enter the location of their camp before leaving. GPS then continually updated the distance and direction back to camp.

At the beginning of the Persian Gulf action several thousand military GPS receivers were available. Many were mounted permanently in tanks and other vehicles and were capable of receiving and decoding GPS signals on both frequencies. The number was insufficient to meet the demand. When the U.S. forces realized how valuable the small handheld units could be, several thousand more were purchased off-the-shelf from companies in the United States and shipped to the Gulf area. Some units were available at aircraft and marine supply stores in the United States and numerous soldiers wrote or called back home to ask relatives to send them one. Some soldiers simply called suppliers and bought them on their credit cards. When thousands of commercial GPS receivers were in use in the area, DOD, assuming that enemy soldiers did not have access to U.S. supplies and credit cards, turned off the clock error.

The International Civil Aviation Organization, an agency of the United Nations, has endorsed the use of satellite navigation systems for commercial aviation, recognizing that GPS and GLONASS provide redundancy for each other in case one becomes inoperable. A single channel receiver receives only one satellite at a time; it must switch to four satellites in succession requiring several minutes to get an accurate position. That may be adequate for some applications, but because of their speed, aircraft must have multichannel receivers which can read out six or more satellites simultaneously.

Private and business airplanes have been using GPS for navigating for many years. The Federal Aviation Administration (FAA) soon will require all airports and commercial airliners to use GPS for navigation and landing. The goal is to use GPS as the sole means of navigation for commercial aviation. Differential GPS correction information is now available to aircraft near major airports. FAA is developing a system to calculate the corrections at 25 ground stations and broadcast them nationwide. The air traffic control system which now uses radar tracking and radio communications between aircraft and controller will be overhauled. Reliance on differential GPS will make the system more accurate and less costly to operate. FAA and others have asked that any new GPS satellites be equipped to transmit two additional unencrypted and undegraded frequencies so civilian users can have the same accuracy as military users.

With GPS, merchant ships can follow the most favorable routes more precisely, taking advantage of ocean currents. For example, the Maritime Administration estimates that a large tanker using GPS could save $17,000 in time and fuel in crossing the Atlantic. Sea and air collisions can be reduced by each craft knowing its exact position. A GPS combined with a walkie-talkie radio would be extremely useful to rescue downed aircrews who become separated from their aircraft. Differential GPS corrections are available in the U.S. Coast Guard radio transmissions for ships entering major harbors. It seems strange that two agencies of the federal government, the FAA and the Coast Guard, are willing to pay millions of dollars to undo the errors introduced deliberately by DOD, another government agency.

GPS works in space as well. Eventually, the Space Shuttle, International Space Station, and all new NASA satellites will be equipped with GPS receivers so we know precisely where they are. That information is useful for orbit calculation, attitude determination, and the location where it makes images of Earth. Satellites will be able to maintain their attitude and orbital position within about 15 feet (5 m) autonomously, without control from ground stations.

Some automobile manufacturers are now installing GPS equipment in their new cars. The software converts latitude and longitude to a location on a map display and voice commands direct the driver to the destination. Map sets for different areas are stored on a CD-ROM.

Iridium satellite telephones contained a GPS receiver that constantly updated the subscriber's location so the network knew which satellite must be used to send calls. Cellular phones do the same thing: they routinely send a short message so the network knows which cell the subscriber is in and can route incoming calls correctly. Connecting GPS with cell phones, either ground-based or satellite-based, gives trucking companies a method of keeping track of their vehicles. The GPS receiver regularly updates the truck's location and sends it to the company's control center. When an order comes in, the dispatcher can direct the nearest truck for a pickup.

Buses, taxis, trains, or any other mode of transportation may find that capability useful in their operations. After a robber threatened a driver at knifepoint, a city in Sweden installed GPS equipment in local taxis. If a driver is in trouble he presses two hidden buttons and holds them for 2 seconds. That keys the transmitter and the dispatcher knows immediately where the taxi is located. In the same city 600 buses, trams, and trains were equipped with GPS receivers that feed into a computerized vehicles allowing the controllers to continuously monitor their movements. Although the individual entrepreneurial taxi drivers resisted having "big brother watching," the others were not so concerned.

Besides the obvious uses, there have been many innovative applications. For example, geologists searching for oil under the seas may locate a likely spot but then have difficulty finding it again later. With precise position information from GPS, it is easy to return to the exact spot. When archaeologists locate prehistoric and historical artifacts they make precise maps of the sites using compass, topographic map, transit, and measuring tapes. Now, with GPS receivers, an area can be mapped in far less time, with greater accuracy, and sites that become covered or otherwise lost from view by vegetation, fire, flood, or drifting sand can be easily found again at a later date.

Because electrons in the ionosphere and water vapor in the atmosphere cause errors in the travel time of the GPS signal, meteorologists are able to "work backward" to extract ionospheric electron content and water vapor information from the error data. This information is useful to predict the state of the ionosphere and to aid in weather forecasting.

GPS has become a government-operated public utility. Receivers have been miniaturized down to pocket size and prices have come down to an affordable level for most people. Carry a GPS receiver with you and you will never be lost!

Energy Satellites

Proposals have been made to use satellites to collect solar energy and beam it to our energy-hungry Earth. The prime source of energy consumed by humans now comes from the fossil fuels: coal, oil, and natural gas. But they are not inexhaustible; they will eventually run out. In addition, most of Earth's air pollution problems are traceable to the combustion of fossil fuels. Even nuclear energy is not an unlimited source of power. Nuclear electric generators run on uranium fuel but uranium is not inexhaustible either. An additional factor is the strong public concern over how to dispose of the radioactive waste from nuclear power plants.

One practically endless source of nonpolluting energy exists: the Sun, which will continue in its present state of energy production for another 5.5 billion years. The problem is how to tap into it efficiently and economically.

There has been much popular interest in solar energy in recent years. We see solar collectors on rooftops and sunspaces attached to houses. In larger installations they are designed to heat the building interior; smaller collectors just heat water for household use. In principle, a solar collector is simply a glass-covered box which gets hot when the Sun shines into it, like the interior of an automobile sitting in the Sun. Interior objects absorb the sunlight, raising their temperature. The heat produced does not readily escape through the glass cover. Water is pumped through pipes in the box or air is blown through ducts. The heated water or air is then sent to the building or to a water storage tank to do its job. In the early 1980s, interest was so high in solar energy that the federal government and some state governments gave tax breaks to people who installed solar collectors or sunspaces in homes or businesses.

Photovoltaic Cells

Another way to collect and use solar energy involves *solar cells*. More properly called *photovoltaic cells,* they generate electricity when light shines on them. They are semiconductor devices made of thin wafers of pure silicon, like transistors and computer chips, but they are much larger in diameter. Figure 5.14 shows a typical cell. There are no moving parts and no fuel is needed to produce clean, quiet, nonpolluting, free electricity. As a bonus, another country cannot cut off the source of supply.

Figure 5.14 Typical photovoltaic cell. *Courtesy of Solar Energy Research Institute.*

You could install an array of solar cells on the roof of your house or in your yard and generate electricity to run your house. A 20 by 30 foot (6 by 9 m) panel would yield a peak of 5000 watts at midday and an average of more than 1000 watts over the year. There are a few problems, however. Most obvious, what do you do after sunset and on rainy days? One solution is to charge up storage batteries during the daytime and run the house from them at night. Car batteries are common, reliable, and have the lowest cost of appropriate storage batteries, but even they would be quite expensive. A car battery costs perhaps $50.00 with a 4-year warranty. To assure that you have enough electricity for overnight plus periods of heavy cloudiness when the cells generate less electricity you would need perhaps 40 batteries. If they needed to be replaced every 4 years, the annual cost of your "free" electricity is $500.00 per year. In the late 1990s, the total operating cost of a photovoltaic system with backup batteries, amortized over the life of the system, is about 30 cents per kilowatt-hour. By comparison, electricity purchased from utility companies varies from 6 to 20 cents per kilowatt-hour in different parts of the United States.

Photovoltaic cells have been used for low power requirements in rural areas and in isolated locations on Earth where it may be too expensive to run an electric power line (Figure 5.15). A mountain cabin, an isolated boat or fishing dock, park campgrounds, country billboard lighting are all situations where photovoltaic cells have proved to be most useful. A rule of thumb is that if power requirements are less than 500 watts and you have to extend the electric company's line by more than one pole then a photovoltaic system is probably the best choice. Photovolatic cells are also used to power pocket calculators and exposure meters in cameras.

Solar cells are quite inefficient. Cells made of gallium arsenide cells can convert about 18 percent of the light falling on them into electricity while the older silicon cells convert only 12 percent at best. Experimental cells have been constructed that have an efficiency of 32.3 percent. Also, sunlight is lost on the way down through the atmosphere by scattering, clouds, and pollution. One solution to such inefficiency is simply to build a larger array.

The cost has been high because present-day photovoltaic cells are manufactured from single crystals of silicon that require much handwork, similar to the processes involved in making semiconductors a few years ago. Techniques for mass production have been under development for several years and there is some hope that prices will come down dramatically just as they have for other mass-produced semiconductor devices. A promising approach is to use a triple layer of cells, each converting a different wavelength of sun-

Figure 5.15 A traffic warning sign in the mountains of Colorado powered by photovoltaic cells. A storage battery is on the ground next to the post. *Photo by the author.*

Figure 5.16 Power for the Dynamics Explorer is provided by solar cells covering the main body of the satellite. *Courtesy of NASA.*

light into electricity. Some light is converted to electricity in the top layer of silicon while the two lower layers of silicon-germanium alloys produce electricity from wavelengths that passed through the top layer. Although each layer is inefficient in itself, the combination of the three converts about 10 percent of the sunlight into electricity. While this is not as efficient as pure silicon or gallium arsenide, the layered cells can be mass-produced inexpensively by depositing hot gases containing silicon and germanium onto a substrate such as stainless steel. This new development could bring the cost down to 12 to 16 cents per kilowatt-hour, competitive with utility companies in high cost or isolated areas.

Solar Cells for Spacecraft

Photovoltaic cells are commonly used as power sources for spacecraft. In fact, the first practical application of solar cells was for powering satellites. You can see them covering the bodies of spacecraft as in Figure 5.16 or installed on large winglike panels attached to the vehicle like TDRS (Figure 5.6). Skylab, the first U.S. space station, shown in Figure 1.14, was powered by an array that produced 10 kilowatts of electricity.

Most solar cells used in space applications have been similar to the one shown in Figure 5.14. They are heavy and expensive to produce, but highly reliable. However, the new methods of depositing thin-film photovoltaic cells on lightweight substrates described above appear to be practical. These cells may not be as efficient as the silicon crystals, but they are lightweight and can be folded into compact packages.

An experiment in deploying a large photovoltaic array was carried on the Space Shuttle *Discovery*. See Figures 5.17 and 5.18. The accordion-pleated array, 13.5 feet wide and 105 feet tall, was extended and retracted 17 times successfully. When folded into its box in *Discovery's* cargo bay, it was only 4 inches (10 cm) thick! For the experiment, only one of the panels had active solar cells. If completely covered with photovoltaic cells it would have produced 12.5 kilowatts of electrical energy.

Power Stations in Space

From favorable experiences with photovoltaic power supplies came the idea to convert solar energy to electricity in space for use on Earth. The concept is to erect very large arrays of solar cells in geosynchronous orbit to produce electricity, convert the electricity to microwaves, beam the microwaves to Earth, and reconvert them to electricity. The

Figure 5.17 The solar cell array during preflight tests prior to installation in *Discovery*. *Courtesy of Lockheed.*

electricity thus produced would be fed to customers via the commercial power grid. The night problem and atmospheric loss problem are absent in space. A satellite in geosynchronous orbit is in almost continuous sunlight and can harvest solar energy continuously, 24 hours a day.

A study in the 1970s showed the feasibility of solar power satellites. The photovoltaic array would be very large, 20 square miles (52 km^2) in area, covered with millions of solar cells. On Earth it would weigh about 75 million pounds (34 million kg). The electricity would be converted to microwaves and transmitted to Earth in a narrow beam using a large dish antenna half a mile (0.8 km) in diameter (Figure 5.19). The Earth stations receiving the beams would be located in remote areas where there would be no hazard from the microwaves (Figure 5.20). Receiving antennas would cover perhaps a square mile (2.6 km^2). Because the satellite power station would be in geosynchronous orbit it would appear fixed in the sky and the antennas on both ends would point in fixed directions.

Although the microwave power density in the center of the beam would be more than double the safe limit, the edge would be at very low power. A restricted area would have to be designated to keep airplanes from flying into the beam. (There are many such restricted areas in the world where aircraft are prohibited.) An automatic system on the satellite would shut the beam off if it strayed from its designated path. One concern is the effect of the microwave beam on the ionosphere as it passes through. Microwave beams affect the ionosphere, but the extent and consequences are not well known.

At the ground station the microwaves would be converted back to electricity and fed into the established national electric power grid just like power from a coal-fueled generator or a hydroelectric dam generator. One such satellite would produce 5 gigawatts (5 million kW) of electricity. MATHBOX 5.3 shows how to calculate this. Two of them would meet the maximum requirements of Manhattan or Chicago or Houston. Projections showed that 10 percent of the 1980 United States energy needs could be met by 300 power satellites.

There are, of course, a few problems. Building such a huge power station in space would be extremely expensive in time and money. The cost of lifting the required mass into orbit is enormous. There would be three possible approaches. One, carry the pieces into low Earth orbit, assemble the structure there, and boost it to geosynchronous altitude. This approach requires carrying boosters and fuel as well as construction materials and workers to LEO. Also, the structure would need added bracing to withstand the stresses of acceleration. Another approach is to build the power station in geosynchronous orbit. The problem here is that the Space Shuttle cannot reach geosynchronous altitude. New heavy lift launch vehicles would have to be designed to carry workers and materials to that altitude. A third approach is to manufacture the parts on the Moon and transport them to Earth orbit. Launching from the Moon to geosynchronous orbit requires less than 8 percent of the energy required to launch from Earth to geosynchronous orbit. A colony on the Moon is a prerequisite, which we will discuss in Chapter 12.

Recent research by S. D. Potter sponsored by the Space Studies Institute suggests a different approach to the con-

Figure 5.18 Artist's concept of the collapible solar array carried to orbit on the maiden voyage of the Space Shuttle *Discovery*. *Courtesy of Lockheed; artist Joe Boyer.*

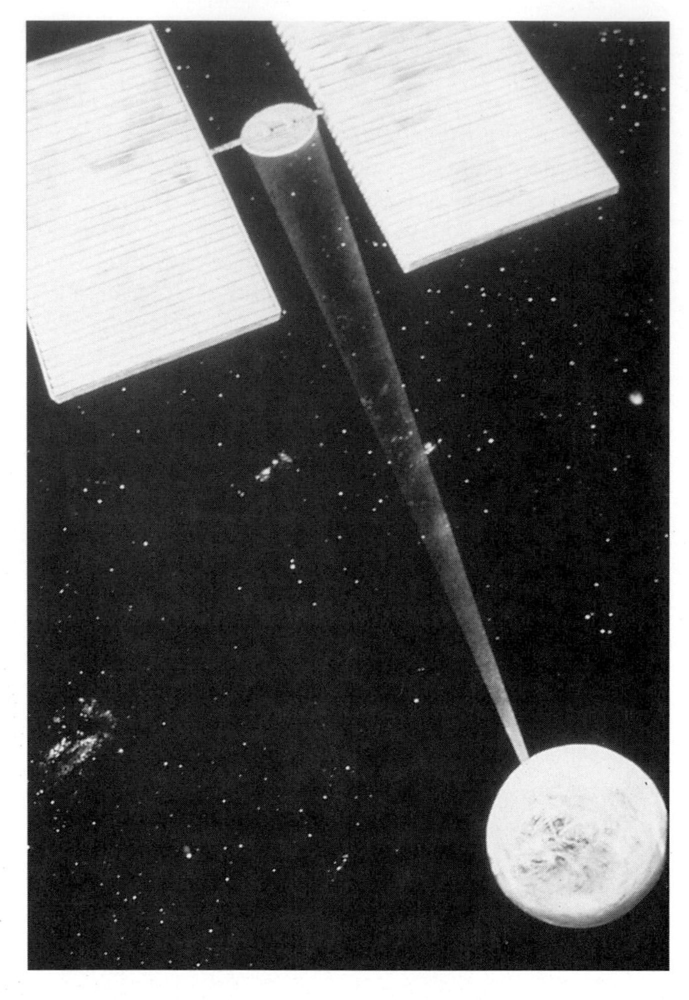

Figure 5.19 Artist's concept of a solar power satellite, transmitting microwave energy to a receiving station on the ground. The photovoltaic array may have an area of 20 square miles (52 km^2). *Courtesy of Solar Energy Research Institute.*

struction of solar power satellites that may be more practical. It reduces their mass to less than 10 percent of the earlier concept. Two recent technological developments contribute to this mass reduction: (1) thin film deposition of photovoltaic cells on a thin plastic material such as Kapton and (2) miniaturization of solid-state microwave transmitters which can be deposited among the solar cells on the same substrate. Thin film solar cells have been made only experimentally and only on heavy substrates such as stainless steel. Deposition on lightweight substrates is still in the research stage. However, results to date look very promising.

In the earlier power satellite concept, high power bus wires carried the current from the entire array to the microwave transmitter and large dish antenna. The transmitter used klystron tubes. In Potter's approach, small solid-state microwave transmitters are deposited over the entire surface of the substrate among the photovoltaic cells. Integrated "wiring" runs to the transmitters only from nearby solar cells. Each miniature transmitter has its own small antenna; thus

Figure 5.20 Antenna farm for receiving microwaves transmitted from a solar power satellite. The antenna elements are arranged in a circle because the beam coming from the satellite has a circular cross section. *Courtesy of Solar Energy Research Institute.*

the transmitting antenna is an array of small antennas as large as the entire photovoltaic array. The larger the antenna, the more concentrated is the beam reaching Earth. This fact limits the size of the array in order not to exceed the safe intensity level of the microwave beam.

One possible design is in the shape of a bicycle wheel. The support structure is shown in Figure 5.21. For clarity, the size of the hub is grossly exaggerated in the diagram. The wheel is about 1¼ mile (2 km) in diameter while the hub is about 6 feet (2 m) in diameter and 4 inches (10 cm) thick. Eight spokes on each side provide rigidity. The spokes are made of a very thin but strong silicon carbide fiber. The rim is made of the same material, four times as thick. The structure is then covered by the plastic material with the solar cells and microwave transmitters deposited on it.

The logical location for power satellites is, of course, geosynchronous orbit so they always appear to be in the same spot in the sky. As the bicycle wheel power station revolves around the Earth, though, it is impossible for it to always face both the Sun and the Earth at the same time. At midnight it is on the side of Earth directly opposite the Sun, sunlight falls fully on it, and its illuminated side faces the Earth like the full Moon. At noon, however, it passes between Earth and the Sun, like the new Moon, so the side facing the Earth is not illuminated and not producing electricity. There are several possible ways to overcome this problem. One is to build a large mirror in orbit near the power satellite oriented so that the sunlight is reflected onto the dark side. Such a mirror could be made out of aluminized plastic film and would weigh about as much as the power satellite itself. Another solution is to cover the dark side with microwave transmitters deposited on a lightweight substrate and supply them with power from the sunlit side by connections through the interior space between the two substrates.

MATHBOX 5.3

Solar Energy

The intensity of solar radiation decreases as the square of the distance from the Sun. At the Earth, 93 million miles (150 million km) from the Sun, solar energy above the atmosphere is about 128 watts per square foot (1377 W/m^2). Consider an array of solar cells covering 20 square miles (51.8 million m^2). In common units, 20 square miles $\times$ (5280)^2 square feet per square mile = 558 million square feet

The array would intercept

$$558 \text{ million square feet} \times 128 \text{ watts per square foot} = 71 \text{ billion watts} = 71 \text{ gigawatts of solar energy.}$$

In metric units the result is, of course, the same.

$$51.8 \text{ million square meters} \times 1377 \text{ watts per square meter} = 71 \text{ gigawatts}$$

Solar cells at best are only about 14 percent efficient. The array would produce

$$71 \text{ gigawatts} \times 0.14 = 10 \text{ gigawatts of electricity.}$$

Converting the electricity to microwaves and transmitting them to the ground is about 50 percent efficient, so only about 5 gigawatts would reach the receiving site on Earth.

The earlier design produced 5000 megawatts from one satellite weighing about 75 million pounds (34 million kg). It would take 46 bicycle wheels at 109 megawatts each to equal that. However, their total weight would be only 5 million pounds (2.3 million kg), one-fifteenth as much. Recall that the cost of delivering material to orbit is the limiting factor in space exploration and exploitation. This new concept for building power satellites could make them economically feasible.

Solar radiation pressure would have to be counterbalanced by some means to keep the power station in the proper orbit with the proper orientation. Micrometeoroid impacts and the intense sunlight above Earth's protective atmosphere will slowly degrade the array of photovoltaic cells so the electrical output will gradually decrease. Even so, it should be possible to build a power station that will last 30 years. They are expensive to build, but cheap to operate.

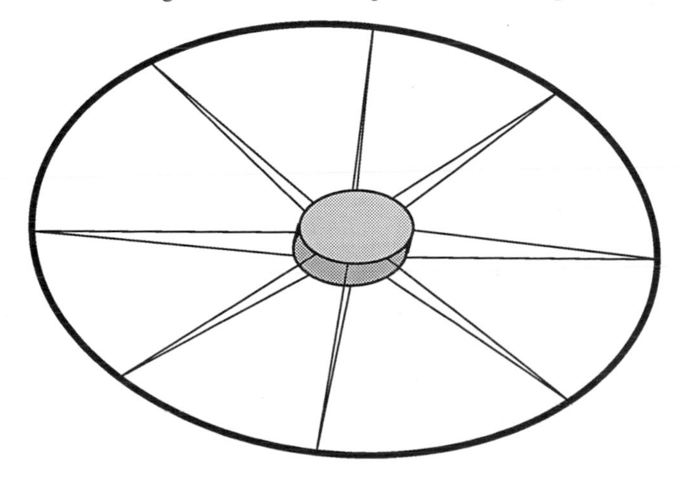

Figure 5.21 Bicycle wheel solar power satellite concept. For clarity, the central hub is shown oversized.

Space Defense

Space defense operations have been conducted since the beginning of the space age. We often hear slogans opposing the militarization of space; but it is naive to think that space is not already militarized. Nuclear-tipped intercontinental ballistic missiles were operational before the first satellite went into orbit. In fact, it was the development of those heavy-lift missiles in both the United States and the Soviet Union that made it possible to launch satellites. When the Russians successfully orbited their first Sputnik in 1957, President Eisenhower announced that we had a satellite in preparation, but our highest priority was the continued development of the ICBM, and other space activities had a secondary role.

In addition to ballistic missiles, both countries immediately recognized the advantages of satellites for many purposes. By far the greatest number of satellites in orbit today are for defense. Indeed, much of the spacecraft technology now in use was developed for defense purposes and had its origin in military satellites. Communications relay and navigation satellites were discussed earlier in this chapter. Other defense satellites such as photo reconnaissance, electronic intelligence gathering, weather observing, and warning of missile launches are in the category of remote sensing which will be covered in Chapter 6.

The defense activities in space listed above are obvious, contribute substantially to national security, and are accepted by all adversaries. International agreements on the uses of space do not preclude any of these defense satellites. All sides take advantage of their capabilities. Some things have been prohibited by international agreement, however. Nuclear weapons in orbit are prohibited but nuclear power generators are not. Establishing military bases on the Moon or other celestial bodies is prohibited, but military personnel may be used for exploration and scientific pursuits.

Defense Against Ballistic Missiles

The 1972 antiballistic missile (ABM) treaty allowed for the deployment of a single system to protect an ICBM missile complex and one to protect its seat of government. While the Soviets elected to deploy an ABM system around Moscow, the United States began construction of a system to protect some of its missile silos. It used large northward pointing radars which could see warheads (and decoys) rising over the horizon. Their detection information was passed to terminal guidance radars that then tracked the objects. Spartan antimissile missiles would be fired to intercept the incoming warheads above the atmosphere using x-rays from a nuclear explosion to disable as many of the incoming objects as possible. The lethal range for x-rays from a nuclear detonation in space is a few kilometers. As the remaining objects entered the atmosphere, the decoys would burn up while the heat-shielded warheads continued on toward their targets. Sprint antimissile missiles would then be fired and steered by radar to intercept and destroy the warheads before they reached their targets. By the time the real warheads were identified, the Sprints would have less than 30 seconds to do their job.

One major problem that apparently was never overcome is that x-rays produced in a nuclear weapon explosion above the atmosphere will cause increased ionization in the ionosphere which the ground-based radar waves could not penetrate, blacking them out. Another problem is that unless the destroyed warheads vaporize completely, their radioactive residue would scatter over the Earth beneath.

Strategic Defense Initiative

President Reagan announced in 1984 that it appeared to be technologically possible to destroy ICBMs during their launch phase and that the United States would undertake a research program called the Strategic Defense Initiative (SDI to investigate the feasibility of developing a defense against nuclear missiles. The program came to be better known as Star Wars.

Even though the ICBM threat seems to have ended, shorter range missiles are still of concern and new scientific research and technological developments continue under SDI. An ABM system stationed in space might be able to destroy a missile over enemy territory while it is still in the boost phase with engines running. Figure 5.22 shows the various stages

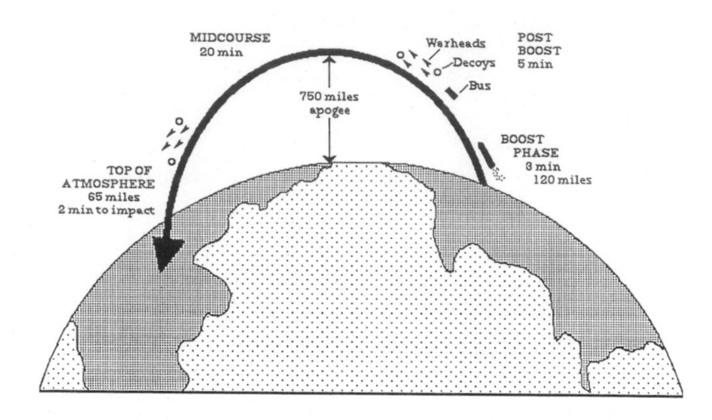

Figure 5.22 Typical ballistic missile trajectory. For about 10 minutes after booster burnout, the bus uses thrusters to aim and release warheads and decoys. On reentry into the atmosphere, shielded warheads glow with heat; lighter decoys slow and fall behind. *After Office of Technology Assessment, A. Carter.*

of an ICBM missile flight launched from a land site. The boost phase lasts 2 minutes, warheads separate 2 minutes later and coast for 7 minutes, reentry phase lasts only a minute. A submarine-launched ballistic missile or a short-range missile like the Scud has a similar but much shorter flight.

There are numerous advantages to attacking the missile while it is still under power. First, it is easier to detect and track using infrared sensors. Second, the missile and its bus carrying the payload is only one target; later, up to 10 warheads and 100 decoys may separate from the bus, becoming a multitude of targets. Third, a rising missile under power and full of fuel is particularly vulnerable.

A space-based early warning detection system now in operation detects missile launches and gives some information on their trajectories, but it does not track the missiles with the precision required to aim and fire a weapon against them. The original task of SDI was to detect, track, and destroy hundreds of ICBMs—each carrying several warheads and decoys, launched within a few minutes of each other, and each having a slightly different trajectory from its launch site to its target a third of the way around the globe, that is, a mass raid on the United States.

A mass raid would take place too rapidly for human decision making. A computer program would have to be written which would completely and automatically manage such a battle. It could never be tested under actual battle conditions; it would be written for a one-time use and would have to be 100 percent reliable. Some computer scientists doubt that a 100 percent reliable computer program of such complexity could ever be written. It must be noted, however, that computers are now doing things reliably that were undreamed of only 20 years ago. For example, extremely complex programs handle our charge cards and our banking; we can withdraw money from our bank account via a computer-driven machine anywhere in the country. Industrial computer programs control machines that build everything from automobiles to computer chips. The military computers have never had to operate under real battle conditions; yet simulations are routinely done which give credibility to their ability to operate properly should a mass raid occur. In addition, rapid advances are being made in artificial intelligence techniques for programming computers and the construction of reliable programs is a major area of research in computer science.

So how do you shoot down an ICBM? Two types of antimissile weapons were being considered for SDI: *directed energy weapons and kinetic energy weapons.* Directed energy weapons send a beam of electromagnetic waves to heat and burn through the outer skin of the missile, while kinetic energy weapons fire projectiles into the path of the missile, destroying it by collision.

Laser Weapons

One type of directed energy weapon is a powerful laser that produces a narrow intense beam of light with a single wavelength and focuses it on the target. Lasers are now used for microsurgery, spot welding, cutting metal, and other applications. Most of them involve heating the area or shocking it with a short, high intensity pulse. To attack a missile, the laser beam must be focused on the surface long enough to burn through or at least to weaken the skin and cause it to rupture. A laser weapon has a very large advantage: the destructive beam travels at the speed of light. The nuclear weapon aboard the missile would not be triggered, but would be destroyed in the explosion.

Lasers with sufficient power to destroy a missile pour out tremendous quantities of power, tens of millions of watts, and require very large power supplies. At least two concepts were being considered. First, the laser and its power supply could be stationed in orbit where it would shoot at rising missiles. An orbiting laser weapon would be about 80 feet (24 m) long and weigh 50 tons (45,000 kg) at launch with its large power supply. Second, the laser could remain on the ground with mirrors in orbit to reflect and focus the laser beam. This concept saves placing the power supply in orbit, but requires several orbiting mirrors. One mirror would be positioned at geosynchronous altitude, stationary like a communications satellite, and other mirrors would be in low orbit to redirect the laser beam to the targets. The laser could be located on U.S. territory and the constellation of mirrors could reflect the energy to any spot from which an attacking missile might rise.

Particle Beams

Another type of directed energy weapon uses a particle beam accelerator, similar to the ion engine described in Chapter 2, but many times more powerful. A beam consisting of charged particles would be deflected from its intended path by Earth's magnetic field. Therefore, the particles must be neutralized as is done with an ion engine.

The beam diverges and loses its intensity if the particles collide with other atoms such as atmospheric gases; therefore it is effective only when the target missile is well above the atmosphere. When a high-speed hydrogen atom hits the skin of the missile it splits into a proton and an electron; the proton penetrates deeply into the metal and does the damage. Electronics hit by the beam would be disrupted by only a small dose; very heavy doses would be needed to melt the aluminum skin.

Kinetic Energy Weapons

Exotic laser weapons get a lot of attention, but as with any new technology, they would require a great deal of research and development before becoming practical. Kinetic energy weapons, on the other hand, are "old-fashioned rifle bullets" that damage a missile by hitting it. It appears that developers consider this approach more feasible than the others mentioned above. There are two general types of kinetic energy weapons: those that accelerate the interceptor "bullets" with rockets and those that use electromagnetic devices to fire the projectiles.

The first type operates rather like a retrorocket by taking the projectile out of a higher storage orbit onto a collision course with the rising missile or the warheads on their ballistic path. The interceptor carries sensors so it can home in on its target and may be able to differentiate between warheads and decoy. It does not carry any explosive; it simply collides with the target (see Figure 5.23). Rockets have been miniaturized so an orbiting platform could carry a large number of such interceptors.

One type of electromagnetic accelerator is the rail gun. It has two parallel rails carrying a high electric current. When a piece of metal shorts across the rails, the high current flow-

Figure 5.23 Artist's concept of a kinetic energy weapon in space. A number of interceptor rockets are carried in the platform at the center. The spacecraft at the right is a radar which tracks the rising missile and directs the interceptor on a collision course. *Courtesy of Lockheed; artist Louis De La Torre.*

ing through it causes it to suddenly evaporate to a high-pressure gas which accelerates a projectile out the barrel. Velocities of over 50,000 miles per hour (80,000 km/h) may be achievable in space. A barrage of small plastic projectiles fired into the path of a missile could destroy it by collision. Rail guns are a difficult technology. Problems with handling the high current or damaging the rails and gun barrel during rapid repeated firing would have to be overcome.

Countermeasures

Whenever a new military system is developed, opponents immediately try to find ways to counter it. Several countermeasures against a directed energy weapon have been suggested. The attacking beam of energy must dwell on the missile outer shell for a period of time to weaken the metal sufficiently to cause it to rupture. Spinning the missile as it rises would spread the energy of the weapon over the missile so it would not dwell on one spot. Since the beam width is similar in size to the missile itself, spinning is probably ineffective as a countermeasure.

A shiny or glossy coating on the missile may reflect rather than absorb the energy. However, a shiny surface which we usually think of as reflecting light may not reflect x-rays or infrared delivered by a laser weapon. Perhaps a coating could be found which does.

A skirt over the rocket engine's nozzle to hide the heat of the flame from the infrared detectors may prevent the tracking of the rising missile. However, the hot exhaust plume is larger than it appears to the eye and the added weight of a large skirt would reduce the effective payload of the missile.

Because the missile is best attacked when it is in the boost phase before the warheads and decoys separate from the bus, rapidly burning boosters would be an effective countermeasure. They would be more difficult to attack if they burned out in, say, 90 seconds, before they left the dense part of the atmosphere. However, replacing an entire fleet of missiles with fast burn boosters would be extremely expensive.

Antisatellite Weapons

With the high value placed on observation, detection, and communications systems in space, it is not surprising that both superpowers developed means of disabling an opponent's spacecraft. Compared to missile defense, it is relatively easy to "shoot down" a satellite in low orbit. Once its orbit is determined, its exact position in space can be predicted with great accuracy, unless it is maneuvered into another orbit.

For many years the Soviet Union had the capability of disabling or destroying a satellite in orbit. A rocket interceptor was designed to rendezvous with the target spacecraft and, using a high explosive warhead, blast it with pellets. If rendezvous is accomplished during the interceptor's first orbit, the target spacecraft does not have much warning time to make an evasive maneuver. The Soviets also had a powerful ground-based laser which may have been capable of damaging optical sensors and electronics on orbiting satellites.

The United States developed and successfully tested a satellite interceptor. A two-stage rocket was carried to high altitude by an F-15 airplane and launched to intercept with the target spacecraft. It carried a homing device and destroyed the satellite by collision. Congress forbade any further tests, however.

Conclusion

The problems in developing a space-based antimissile system are enormous, as with any other new undertaking. The objective of SDI has been to complete enough research and development to determine the feasibility and possible architecture of a space defense system.

Cost estimates of such a system run into the trillions of dollars, depending on whom you talk to and whether you are talking about development or deployment, space-based or ground-based, conventional or nuclear, directed energy or kinetic energy weapons. Even if each of the various components proves to be technologically feasible, many notable scientists doubt that a reliable working system could be built at any price. A system that would guarantee protection of the entire United States against all attacking missiles, they argue, would be entirely too complex and could not be realistically tested to see if it worked. And, they add, anything less than complete protection would not be justifiable. However, scientists are not invariably correct. Such criticisms have frequently been voiced against just about every new state-of-the-art undertaking. As late as the 1920s one prominent scientist said that a rocket ship would not work in space because there was no air for the exhaust to push against. A few years ago, a friend of Dr. Edward Teller, father of the hydrogen bomb, said, "When Edward says something can be done, it can be done. When he says something cannot be done, he is frequently wrong."

The dissolution of the Soviet Union reduced the danger of a massive exchange of nuclear-tipped intercontinental ballistic missiles. Deployment of an SDI type of system has become a moot question as the two major world powers dismantle their stockpile of those weapons. That does not mean that a defense against such missiles will never be needed. Several other countries are developing both nuclear weapons and short-to-medium range rockets. Iraq's Scud missiles fired at Israel and Saudi Arabia during the Gulf War are a case in point. Their launches were detected by early warning satellites. In fact, in 1992, Russia and the United States, former adversaries, jointly announced a concept for a global missile defense including a joint early warning center. If a danger to the world arises in the future, the research and development done by SDI will be useful in building a defense against those missiles.

DISCUSSION QUESTIONS

1. What has been the social impact of international television broadcasts in less developed countries and countries with closed borders?
2. Name possible users of a precision positioning system other than those mentioned in this chapter.
3. If you could build your own satellite, what kind would you build and what would you use it for?
4. Should the United States continue to develop a defense against nuclear missiles?
5. What factors might be considered in determining whether or not to deploy an antimissile weapons system in orbit?
6. Many cable TV companies are now replacing old wire systems with fiber-optic cables that are capable of high-speed data handling such as Internet connections. Compare the advantages and disadvantages of fiber optics versus satellite links.

ADDITIONAL READING

Caprara, Giovanni. *The Complete Encyclopedia of Space Satellites*. Portland House, 1986. Describes over 1000 civil and military satellites of all nations launched or planned as of 1986.

Garwin, Richard L., et al. "Antisatellite Weapons." *Scientific American,* June 1984. Argues against them.

Hansen, James R. "The Big Balloon." *Air & Space, Smithsonian,* April/May 1994. History of Project Echo.

Herring, Thomas A. "The Global Positioning System." *Scientific American,* February 1996. How the system works. Technical, understandable.

Hudson, Heather E. *Communications Satellites, Their Development and Impact*. The Free Press, 1990. Political, economic, military, cultural, and social impact of communications satellites.

Kaplan, Marshall H. *Modern Spacecraft Dynamics and Control*. John Wiley and Sons, 1976. Calculus level math.

King-Hele, Desmond. *Observing Earth Satellites*. Van Nostrand Reinhold, 1983. Do-it-yourself visual observing of satellites.

Logsdon, Tom. *The Navstar Global Positioning System*. Van Nostrand Reinhold, 1992. Comprehensive description of GPS and its applications for both novices and professionals.

Patel, C., and Nicolaas Bloembergen. "Strategic Defense and Directed Energy Weapons." *Scientific American,* September 1987. Findings from American Physical Society study.

Potter, Seth D. *Low Mass Solar Power Satellites Built from Lunar or Terrestrial Materials: Final Report*. Space Studies Institute, 1994. Research report on use of thin-film materials for photovoltaic solar power satellites.

Strategic Defense Initiative Organization. *Report to the Congress on the Strategic Defense Initiative*. Government Printing Office, April 1987. Summary of SDI concepts and research.

U.S. Congress, Office of Technology Assessment. *Ballistic Missile Defense Technologies*. Government Printing Office, September 1985. Nonjudgmental survey of SDI.

Williamson, Mark. *Dictionary of Space Technology*. Adam Hilger, 1990. Comprehensive.

Wilson, Andrew, ed. *Interavia Space Directory 1994–95*. Jane's Information Group Inc., 1994. Excellent reference on space programs and aerospace industry of all nations, and more; updated regularly.

PERIODICALS

Air & Space/Smithsonian. P.O. Box 53261, Boulder, CO 80322-3261. Bimonthly magazine with two or three articles about space in each issue.

Aerospace America. American Institute of Aeronautics and Astronautics, 370 L'Enfant Promenade, S.W., Washington, DC 20024. The monthly popular journal of the AIAA.

Aviation Week & Space Technology. P.O. Box 1505, Neptune, NJ 07754-1505. Weekly magazine, current events.

GPS World. P.O. Box 7677, Riverton, NJ 08077-9177. Monthly magazine covering news and applications of GPS.

Lauchspace. 1201 Seven Locks Rd., Ste. 300, Potomac, MD 20854. The magazine of the space industry, excellent semitechnical reading.

Space News. 6883 Commercial Drive, Springfield, VA 22158-5803. Weekly newspaper covering space business, policy, and some technology.

SSI Update. Space Studies Institute, P.O. Box 82, Princeton, NJ 08542. Bimonthly newsletter devoted mostly to space manufacturing, lunar mining, power satellites, and space colonization.

NOTES

Chapter 6

Remote Sensing

When you look across the room at an object you are "sensing remotely." Your eyes are sensors, sensitive to light waves and your brain interprets their message. When you use a camera to make a photograph you are also sensing remotely, using a machine to sense the object and make a permanent record on the light sensitive film. The camera becomes an extension of your eyes.

Hearing is also remote sensing. Ears and their associated anatomy are sensitive to sound waves. Microphones, amplifiers, and speakers are machines that extend the capabilities of the human ear.

Even human touch sensors can operate remotely. Consider that you can feel a flame without actually touching it. The touch sensors are sensitive to the infrared heat waves emitted by the hot object.

Remote sensing, then, is learning something about an object at a distance, without coming into physical contact with it.

Many spacecraft are remote sensors, acting as extensions of human senses. They look up into the universe and down to Earth. One of the earliest and most practical applications of remote sensing has been for weather observing using both visible and infrared waves. Mapping and cataloging the resources of Earth are other applications of remote sensing with widespread benefits. Military early warning satellites detect launches of missiles and spacecraft by means of infrared sensors that respond to the heat of the rocket exhaust. Spy-in-the-sky satellites looking and listening from their orbits keep each side informed in remarkable detail what the other side is up to. The news media are using satellites to cover news events from space. A satellite may soon carry a sensor designed expressly for the media, capable of seeing things as small as a truck, in full color, available within hours, with global coverage.

The science of astronomy is based on remote sensing, first using earthbound telescopes, now using spacecraft carrying sensors above the atmosphere for a better view. They observe the skies not only with sensors of visible light but also radio waves, infrared, ultraviolet, x-rays, and gamma rays. Many of these emanations from stars and galaxies are not visible at earthbound observatories because they are absorbed in the atmosphere before they reach the ground.

This chapter will cover the principles of remote sensing and their applications to Earth observing. The next chapter will deal with applications of remote sensing to astronomy.

Electromagnetic Waves

Radio, infrared, visible light, ultraviolet, x-rays and gamma rays are all electromagnetic waves, differing only in their wavelengths. The electromagnetic spectrum was described in Chapter 4 and Figure 4.9 (Plate 1). The length of a light wave, measured from crest to crest or from trough to trough, is 0.4 to 0.7 micrometer. A micrometer, also called a micron, is one-millionth of a meter. Therefore, visible light waves are very short electromagnetic waves. Shorter still are the ultraviolet, x-rays, and gamma rays, ranging down to a trillionth of a micrometer. On the other hand, infrared, microwaves, television, and radio waves are all electromagnetic waves that are longer than visible light, ranging up to many miles for AM radio.

The Sun is the primary source of electromagnetic waves by which we see things on Earth. Three things can happen to sunlight falling on an object. It may be absorbed by the object, reflected off its surface, or transmitted through the object. For example, a mirror reflects all of the light falling on it while a totally black object absorbs all of it. A green leaf absorbs red and blue light but reflects green light. Some of the light striking a piece of glass reflects from it but most of it passes through the glass.

Besides reflecting sunlight, objects emit their own electromagnetic waves. Anything which has a temperature above absolute zero emits electromagnetic energy. An object of high temperature radiates shorter waves; an object at a lower temperature emits radiation of longer wavelengths. The photosphere of the Sun at 6000°C (about 11,000°F) emits most of its radiant energy in the visible light portion of the spectrum. On the other hand, a human body at 98.6°F (37°C)

MATHBOX 6.1

Wavelength of Maximum Emission

In the late 1800s Wilhelm Wien found a simple relationship between the temperature of a body and the wavelength at which it emits the maximum amount of electromagnetic radiation:

$$L = \frac{2900}{T}$$

where T is the temperature in kelvins (degrees above absolute zero) and L is the wavelength in micrometers. (To find the temperature in kelvins, just add 273 to the Celsius temperature.)

Wien's law says that the higher the temperature of an object, the shorter the wavelength of maximum emission. Very hot objects emit ultraviolet; cold objects emit radio waves.

For example, the Sun's surface temperature is about 5700 K. So

$$L = \frac{2900}{5700} = 0.51 \text{ micrometer}$$

in the middle of the visible light part of the spectrum.

A human body is at 98.6°F which is 37°C or 310 K. It emits its peak or maximum radiation at

$$L = \frac{2900}{310} = 9.4 \text{ micrometers}$$

in the near infrared. An IR (heat) detector reveals that people glow at that wavelength.

emits longer infrared waves. Since everything emits some amount of infrared, it is a favored region for remote sensing. The infrared portion of the electromagnetic spectrum is sandwiched between visible light and radio waves, with wavelengths from 0.8 micrometer to about 1 millimeter (1/25 inch). MATHBOX 6.1 shows how to calculate the wavelength at which an object radiates the maximum amount of energy when you know its temperature.

We are accustomed to viewing the world around us by the light reflected from the objects surrounding us. See Figure 6.1. These principles also apply to other wavelengths of the electromagnetic spectrum—ultraviolet and infrared, for example—but human eyes are sensitive only to visible light so we are not generally aware of what is happening to the other electromagnetic waves surrounding us. However, our skin can detect infrared waves in the form of heat. When you turn on the burner of an electric stove, you can sense the infrared before you see the glow. As the burner heats up, it becomes red hot and emits visible light as well as infrared. Pit viper snakes can detect infrared and deduce the direction it is coming from; they can "see" warm objects in the dark when there is no visible light. Insects can see ultraviolet and the world looks quite different to them.

Sensors can be made to detect any wavelength of the electromagnetic spectrum. Imaging devices which are sensitive to other wavelengths give quite a different view of the world. For example, doctors use x-rays striking a piece of film to create an image of the inside of the body. X-rays pass through skin but are absorbed variably by flesh and bone on the way through. When they strike a piece of film they create a shadowgram of the interior. Figure 6.2 shows an x-ray image of a broken leg.

With proper equipment it is possible to create an image of an object solely by the infrared heat waves that it gives off. Photographic film sensitive to infrared is available at camera stores. Photographs made on infrared film look quite different from those made on regular film. For cooler objects which emit longer infrared waves, a curved mirror is used instead of a lens to form an image because the longer infrared waves are absorbed by glass lenses and do not reach the film.

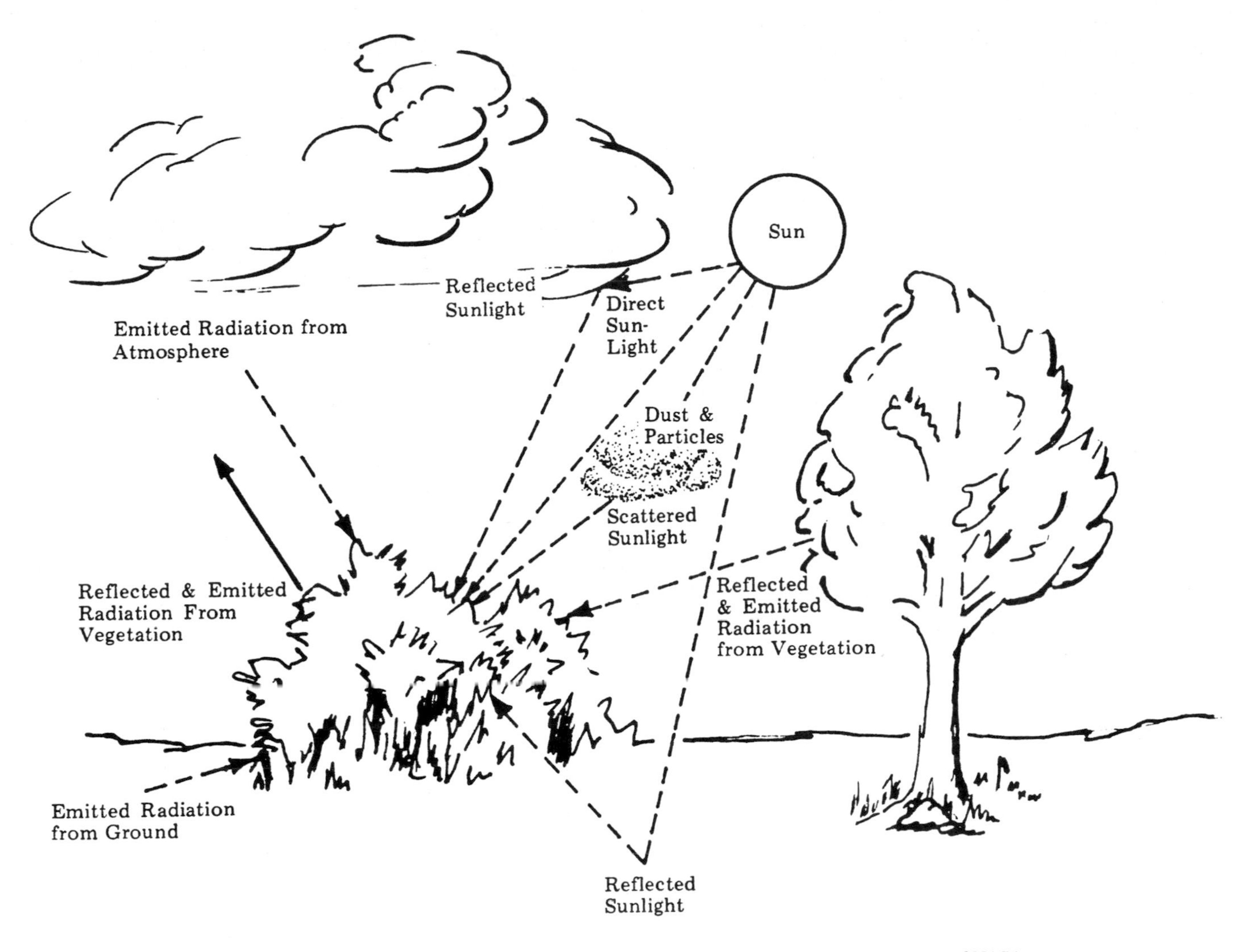

Figure 6.1 Emitted and reflected electromagnetic radiation surrounds us. *Courtesy of NASA.*

Remote Sensing from Space

The exploration of space has led to an unprecedented and remarkable new exploration and understanding of Earth. Landsat satellites have been recording infrared and visible light images of the Earth's surface for 30 years. The first one was launched in 1972. Landsat 5 has been in orbit since 1984. It was designed for a 3-year lifetime, but was still working 15 years later. Landsat 6 was launched in 1993, but failed to reach a stable orbit when the apogee kick motor failed and the satellite fell back to Earth. Landsat 7 was launched in April 1999. It is shown in Figure 6.3.

A typical Landsat orbit is shown in Figure 6.4. It is *Sun-synchronous* which means that it reaches the equator at the same local time on every pass. For Landsat 7, this is between 10:00 a.m. and 10:15 a.m. It makes images only on the north-to-south part of the orbit. The south-to-north part on the other side of the Earth is in darkness. At that time of day shadows in the picture help to identify the objects in the image. Also, it is easier to determine changes taking place in an area by comparing a new image with past images if they are taken under the same lighting conditions. It is also important that the orbit be perfectly circular so that all images are taken from the same distance from the Earth and therefore have the same scale.

Similar to Landsat, the French Space Agency's Satellite Pour l'Observation de la Terre, Spot for short, has been operating since 1986. The current spacecraft, Spot 4, was launched in 1993, into a Sun-synchronous orbit with an inclination of 98 degrees at an altitude of 511 miles (822 km), crossing the equator from north to south at about 10:30 a.m. local time. Segments of orbits of Spot 3 and Landsat 5 are shown in Figure 6.5; they were almost identical, but cross the equator at different times with respect to the Sun.

Landsat's rotating mirror produces east-west scan lines making a continuous strip picture 115 miles (185 km) wide as shown in Figure 6.6. Figure 6.7 shows the strips covered by three consecutive orbits. Because of the rotation of the Earth under the satellite, different areas are covered on each

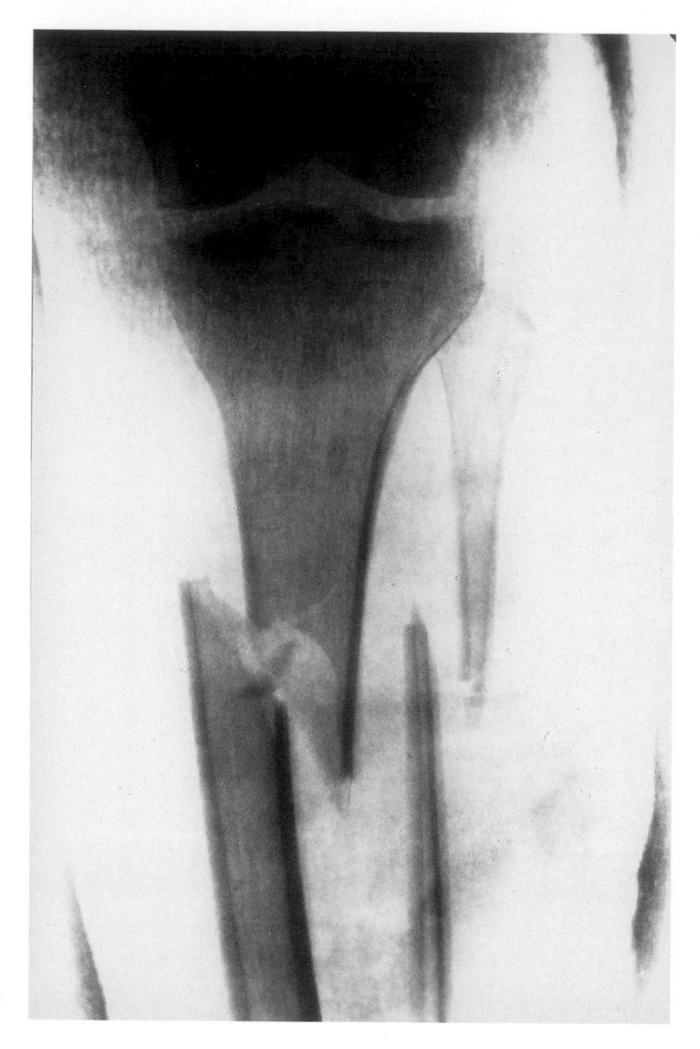

Figure 6.2 X-ray photo of a broken leg. Both the large bone (tibia) and the small bone (fibula) are broken through. The knee joint can be seen at the top. *Courtesy of C. Hudson.*

Figure 6.3 Artist's concept of Landsat 7 in orbit. There are no actual photographs of the spacecraft with the solar panels open because they will not support their own weight on the ground. They are unfolded only when the spacecraft is weightless in orbit. *Courtesy of NASA.*

orbit with gaps between them. On following days the strips will fall in between until after 16 days the entire area has been covered and the pattern begins to repeat itself. See Figure 6.8. Notice in Figure 6.8 that on any pass, the strip covers an area just west of the strip produced one week earlier, and the next strip west will be covered one week later. Except for small regions around the poles, the entire Earth is completely covered every 16 days.

Spot scans a swath 73 miles (118 meters) wide so it covers the entire Earth in 26 days as compared to Landsat's 16 days. However, an adjustable mirror allows the satellite to scan areas up to 27 degrees either side of the point directly beneath it. This permits it to cover a selected area on 7 successive passes in equatorial regions or on 11 successive passes at midlatitudes. This also allows any point on Earth above 40 degrees latitude to be scanned any day. Only about 5 percent of the area at the equator cannot be covered every day; in that area, images of a specific point can be made every other day. In addition, stereo pairs of pictures can be made by using both Spot satellites.

Spectral Bands

A remote sensing satellite records electromagnetic waves that strike its sensors. When looking at the Earth, those waves must pass through the atmosphere to reach the satellite. Some wavelengths are scattered or absorbed by gases and particles in the atmosphere, particularly by water vapor and carbon dioxide. However, there are "windows" in the atmosphere that let certain wavelengths pass through with little loss. See Figure 6.9. It is those wavelengths that are most useful for remote sensing by satellites.

On Landsat, electromagnetic waves reflected from the scanning mirror pass through a device which divides them into different *spectral wavelength bands*. These beams then pass into the detectors which measure their intensities. Table 6.1 lists the spectral ranges for the eight sensors on Landsat 7. Earlier Landsat spectral bands were nearly the same, but they had no *panchromatic* (grey scale) sensor (band 8).

Spot carries four sensors for multispectral imaging: the green band from 0.50 to 0.59 micrometer; the red band from 0.61 to 0.68 micrometer; a near infrared band from 0.79 to 0.89 micrometer; and a longer wavelength infrared band from 1.58 to 1.75 micrometers. Notice that these are the about the same as Landsat bands 1 to 5 except that Spot covers portions of bands 1 and 2 with a single green band sensor. Instead of a separate sensor for panchromatic black and white pictures, Spot uses its 0.61 to 0.68 micrometer (red) band.

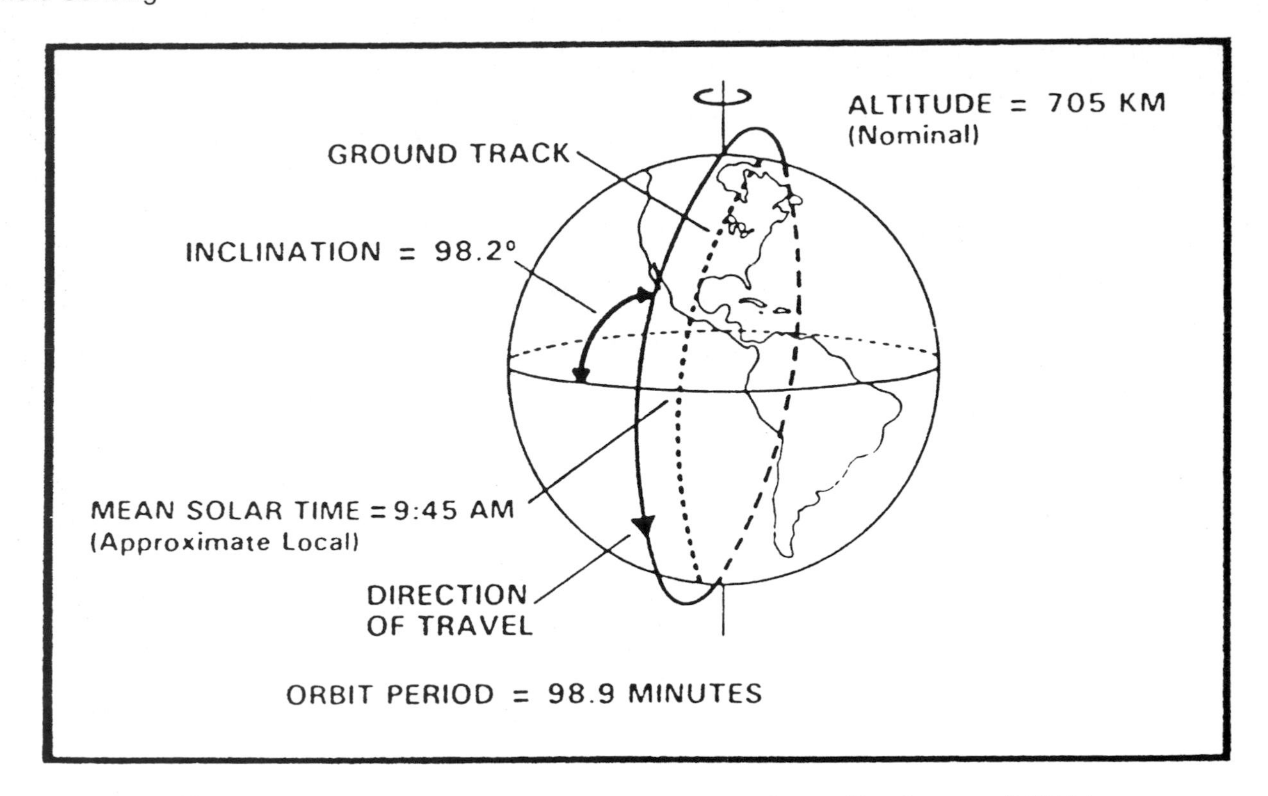

Figure 6.4 Typical Sun-synchronous orbit of a remote sensing satellite. *Courtesy of EOSAT.*

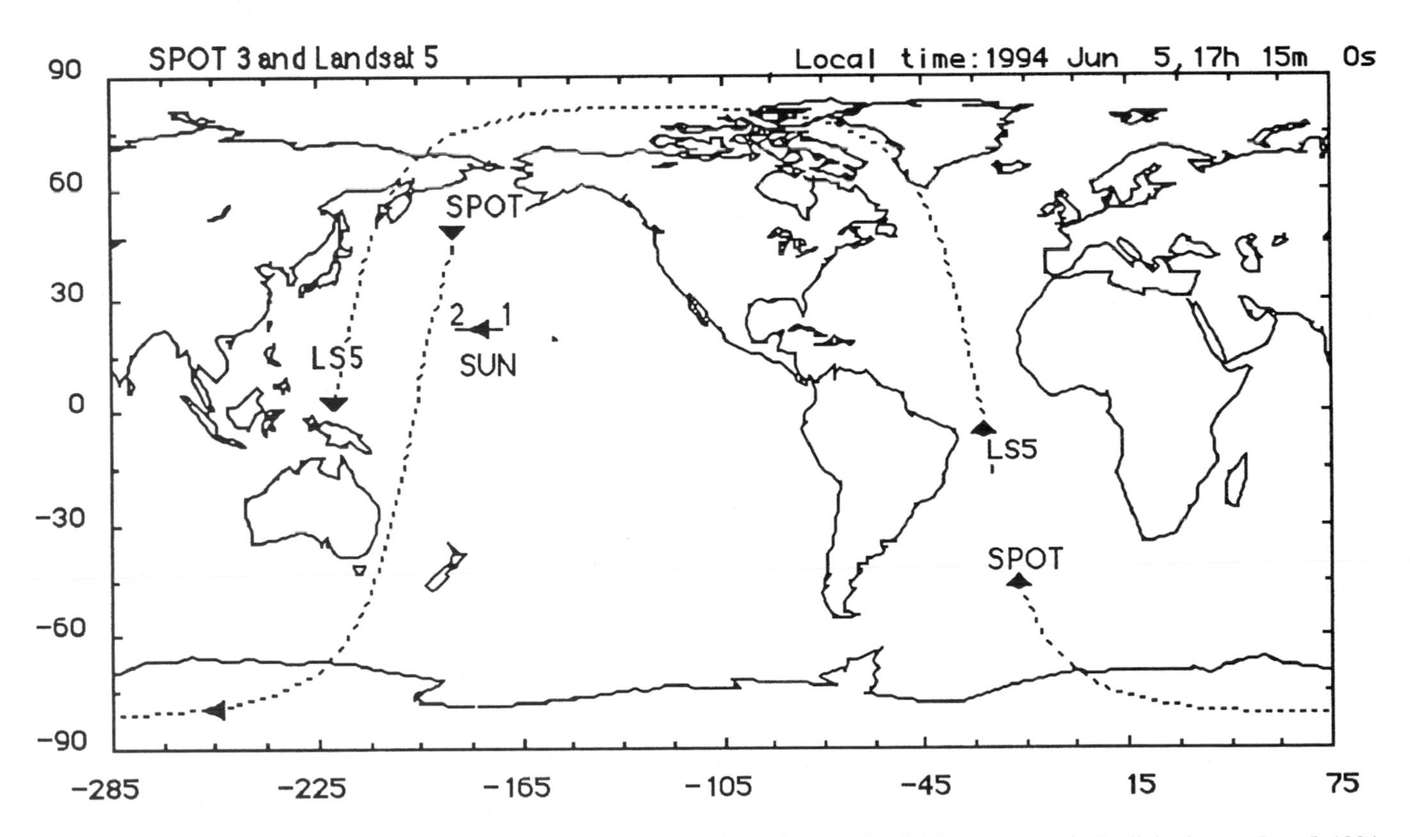

Figure 6.5 Ground traces for 55 minutes of travel for Spot 3 and Landsat 5 from 4:20 to 5:15 p.m. mountain daylight time on June 5, 1994. At the start time Spot is over the Bering Sea, Landsat is off the east coast of Brazil, and the Sun is over the Pacific at the position marked 1. Spot crosses the equator in the Pacific at about 10:30 a.m. before the Sun is overhead at noon. Landsat reaches the Pacific equator later when the Sun is at position 2, but at 9:37 a.m., farther ahead of the Sun than Spot. At the end of the time period, Spot is approaching the point in the Atlantic where Landsat began. Spot's next orbit will trace almost precisely over Landsat's previous one.

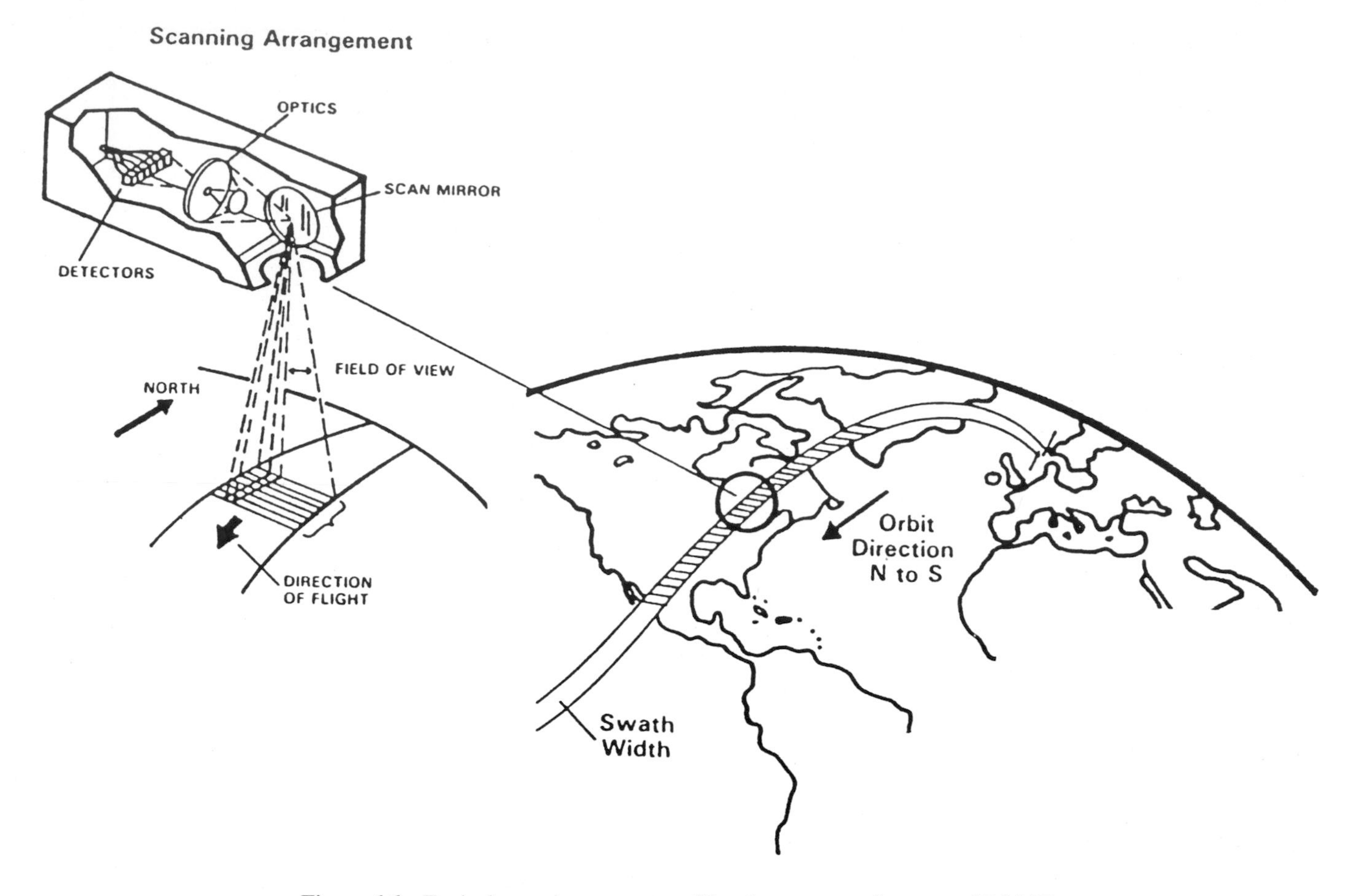

Figure 6.6 Typical scanning geometry of Landsat sensors. *Courtesy of EOSAT.*

Resolution

Scan lines are broken into short bits called picture elements or *pixels* which make up the image. To be detected, an object must be at least one pixel large. Landsat has 100 foot (30 meter) resolution. See Figure 6.10 (Plate 4). Its pixels are 100 feet (30 meters) square at the surface of the Earth, about one-third the area of a football playing field. Spot pictures have higher resolution than Landsat. A pixel is 33 feet (10 meters) square for grey-scale panchromatic images and 66 feet (20 meters) square for multispectral images. Other satellites may have smaller or larger pixels depending on their application. Figure 6.11 (Plate 5) compares the resolution of three satellites observing the same size area on the ground. Typical weather satellites have a resolution of about 0.6 mile (1 km); that is sufficient to see details in cloud and water vapor patterns. Photoreconnaissance satellites have much higher resolution, perhaps a few inches, so very small details are visible.

Of course, higher resolution means a greater number of bits of data. There must be a compromise between the desired resolution and the quantity of data that the satellite must collect and transmit. An image of a 1 kilometer square with a resolution of 1 kilometer requires 1 pixel. At a resolution of 1 meter, that 1 kilometer square contains 1 million pixels.

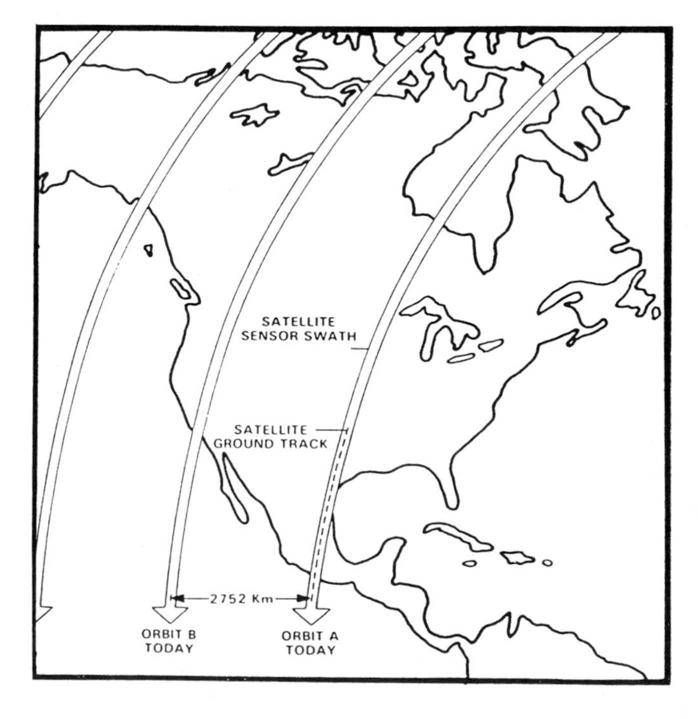

Figure 6.7 Landsat coverage on each orbital pass. *Courtesy of EOSAT.*

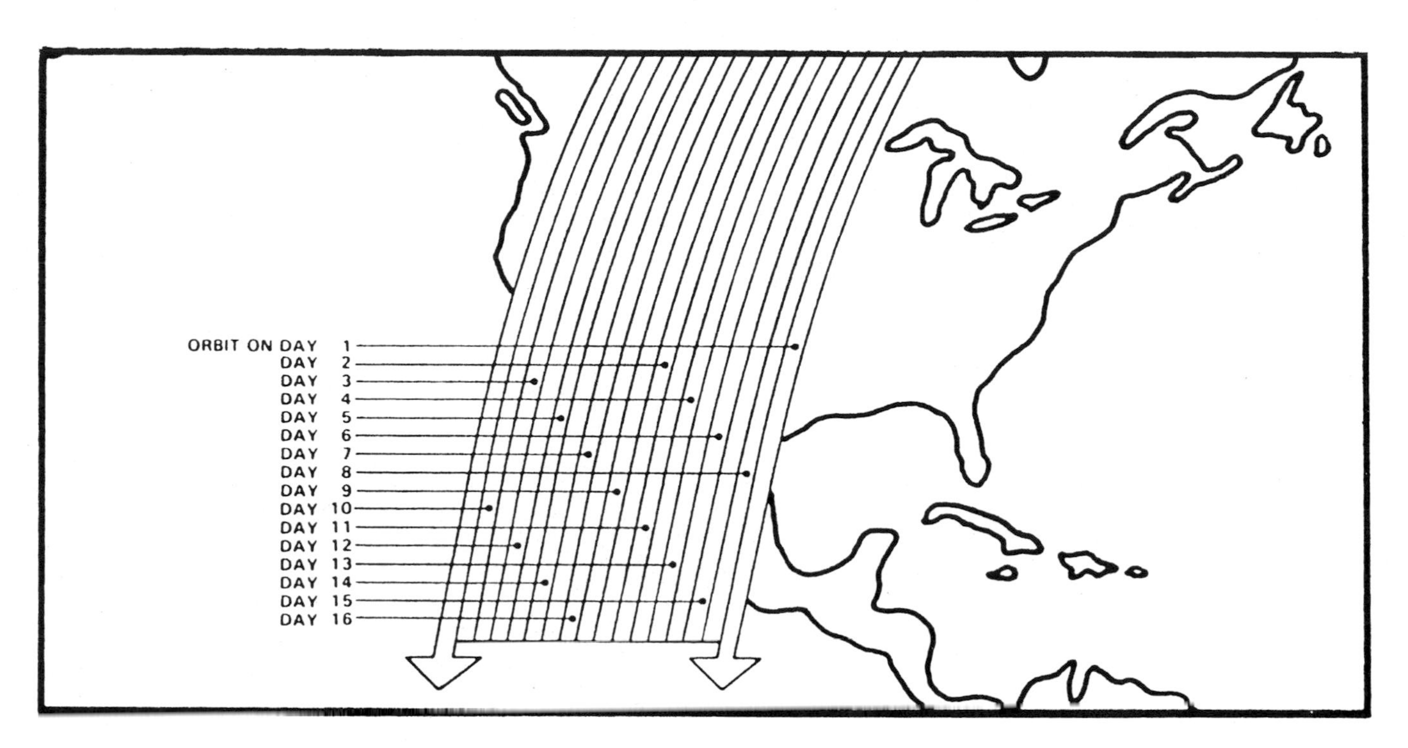

Figure 6.8 Typical Landsat coverage pattern. A different swath is covered each day until the entire region is covered in 16 days. Then the pattern repeats itself. *Courtesy of EOSAT.*

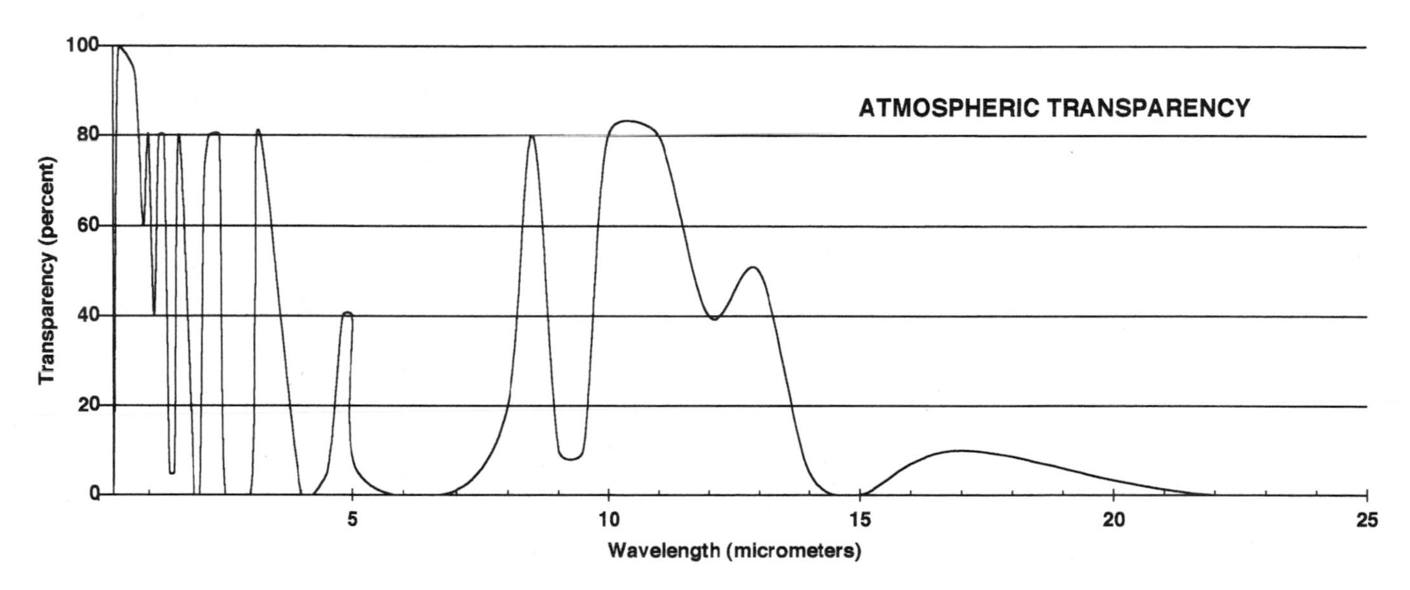

Figure 6.9 Atmospheric transparency in the infrared wavelengths. On the far left side of the graph is visible light where the atmosphere is nearly 100 percent transparent. On the far right, the atmosphere is opaque to wavelengths greater than about 22 micrometers. Between these extremes are a number of narrow windows where the atmosphere is partially transparent. These are the wavelengths that are most useable for remote sensing from space.

TABLE 6.1 Spectral Bands of Landsat 7 Sensors

Band	Spectral Range	Color	Principle Applications
1	0.45–0.52 μm	Blue-Green	Coastal water mapping. Soil/vegetation differentiation. Deciduous/conifer tree differentiation.
2	0.53–0.61 μm	Green-Yellow	Reflected by healthy vegetation. Soil organic content.
3	0.63–0.69 μm	Red	Chlorophyll absorption for plant species differentiation.
4	0.78–0.90 μm	IR	Biomass surveys. Water body delineation. Sewage dumps.
5	1.55–1.75 μm	IR	Vegetation moisture measurement. Snow/cloud differentiation.
6	10.4–12.5 μm	IR	Plant heat stress measurement. Thermal mapping.
7	2.09–2.35 μm	IR	Mineral discrimination. Hydrothermal mapping.
8	0.52–0.90 μm	Vis/IR	Panchromatic imaging

Note: μm means micrometer, one-millionth of a meter, also called a micron.

Figure 6.12 A 1-meter resolution image from the Ikonos spacecraft showing part of lower Manhattan at the approaches to the Brooklyn Bridge, October 19, 1999. Individual cars are visible. *Courtesy of Space Imaging.*

Each pixel is viewed by each sensor which measures the intensity of the electromagnetic waves received from the pixel, represented by a number which tell the intensity of the waves in each of the spectral bands. The satellite then transmits the stream of numbers to a ground station below.

Space India has launched a series of six remote sensing satellites dating back to 1988 and several more are in the planning stages. Similar in concept to Landsat and Spot, they were designed primarily to assist the country to better manage its water, agricultural, and mining resources. At the time it was launched in 1996, the Indian Remote Sensing Satellite (IRS-P3) had better panchromatic resolution than either Landsat or Spot: 16 feet (5 m). Because of increasing demand for high resolution imagery which was not available from other spacecraft, India made the IRS data available to commercial users worldwide.

Three commercial companies in the United States, recognizing the increasing demand for Earth images, are now operating remote sensing satellites for profit. The highest resolution space images currently available come from Ikonos, launched and operated by Space Imaging, Inc. Spectacular 1-meter resolution panchromatic images became available shortly after the spacecraft was launched in September 1999. Figures 6.12 and 6.13 are samples of Ikonos imagery. Bear in mind that these pictures were made from orbit, about 450 miles (725 km) from the ground.

Image Processing

The streams of numbers transmitted to Earth from both Landsat and Spot are processed by a computer into grey shades or colored images. A 100-mile square (161 km^3) image at a resolution of 33 feet (10 m) contains nearly 260 mil-

Figure 6.13 A 1-meter resolution image from the Ikonos spacecraft showing San Francisco's Pier 39. About 200 boats are parked at the docks inside the breakwater. *Courtesy of Space Imaging.*

lion pixels. Each pixel has a set of numbers associated with it, giving the intensity of the waves for each spectral band. Each intensity number can be associated with a shade of grey for printing or displaying on a monitor. Of course, it is easier to view and analyze an image with colors than it is with grey shades. Visible light pictures can be printed in their natural colors: green ink for the green band, red ink for the red band, and so on. A full color picture is a composite of superimposed red, blue, and green images, just as a color picture on a TV or a monitor is composed of red, blue, and green dots. Figure 6.10 (Plate 4) is just such a composite of the three Landsat visible bands.

An infrared image is different. The various IR bands can be interpreted as different "colors" of infrared but, because our eyes are not sensitive to infrared, the "color" of IR waves has no meaning for us. Therefore, the picture is printed with a different visible color assigned to each of the bands. Often the green band is printed in blue, the red band is printed in green, and the IR is printed in red. Figure 6.14 (Plate 6) is a false color image of Denver; compare it with Figure 6.10 (Plate 4). Such false color images have become popular, not only for infrared images, but for images made in other wavelength bands as well. The data can be manipulated in a number of different ways. Colors can be adjusted to emphasize any particular land use: urban, cropland, vegetation, wetland, or sediment, for example. Cropland in Kansas is shown in Figure 6.15 (Plate 7). A 1-meter panchromatic image can be merged with a 4-meter multispectral scanner image to produce color pictures that appear to have higher resolution than either one by itself.

Unique patterns of light intensity across the visible and IR bands can be used to identify features on the Earth which may not be apparent to the human eye. For example, a field of soybeans, once its spectral intensity pattern has been identified, can be located on a satellite image. See Figure 6.16. The light intensity levels of the various spectral bands of the soybean field are as unique as a person's handwriting. Other areas on the same image that have the same *signature* can be interpreted as being the same type of soybean fields. Finding the right combination of spectral intensities for a soybean field in an image is done by computer evaluation of the pixel data.

Fine details can be recognized in these spectral signatures, even to determining the health of crops. Healthy vegetation looks green because the chlorophyll absorbs blue and red wavelengths (Landsat bands 1 and 3) and strongly reflects green (band 2). Water reflects very little visible light and absorbs infrared, so Landsat bands 5 and 7 are indicative of water content. Crops suffering from a drought have less water content so they do not absorb as much infrared and they reflect more blue and red. Therefore their signatures are different from normal healthy crops. Using this information, estimates of crop production can be quickly assessed. In some areas, farmers are using multispectral images along with GPS positioning to irrigate or apply fertilizer to very specific small sections that require attention rather than waste water or fertilizer over a large area where it is not needed.

Trees and lawns in a residential area, asphalt and cement areas, different kinds of exposed rocks, meadows, and fields—all can be recognized. Sediment and pollution in rivers give a different signature from clear water. Mineral deposits can be located precisely. New uses for satellite remote sensing are being applied to many different problems.

To accomplish this kind of mapping of Earth resources by aerial photography would require 100-photographs and months to assemble and analyze a 100 mile square scene. A satellite can cover the same area in a single picture and the computer analysis can be done in a few hours. The timeliness of the images makes them especially useful in agricultural and environmental studies.

Would it be possible to correctly identify the object in each and every pixel using more spectral bands? If six spectral band are good, are 100 spectral bands better? Imagine the signatures in Figure 6.16 with 100 data points instead of six. Such sensors could potentially identify specific crops, minerals, pollutants, and other things in far more detail than is

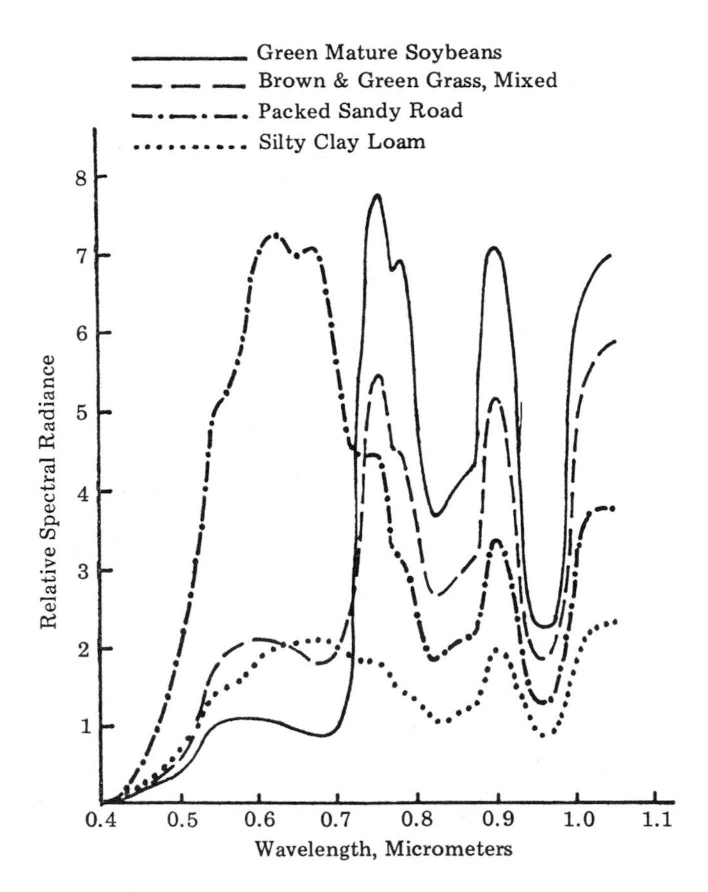

Figure 6.16 Signatures of four different materials. The packed sandy road is brightest in visible light (0.4–0.7 micrometer) and loam is dark. Soybeans and grass have similar signature patterns but the intensities differ. *Courtesy of EOSAT.*

possible with present sensors. Department of Defense research has been done in *hyperspectral* imagery to see if more detailed identification of pixels is possible. Reportedly hyperspectral sensors can detect tanks under camouflage nets and differentiate between vehicles and decoys because they each have unique signatures. Because of this, restrictions have been placed on distribution of hyperspectral data with better than 20-meter resolution.

Weather Satellites

Tiros, the first true weather satellite, was launched on April 1, 1960, and carried television cameras that viewed Earth's cloud cover by visible and infrared light from a 450-mile orbit. A total of 10 Tiros satellites were launched over a 5-year period. Tiros 8 carried automatic picture transmission (APT) equipment so that anyone with a simple receiver and a facsimile recorder could receive the pictures. Nimbus was a second-generation weather satellite with improved cameras, an APT system, and an infrared sensor that allowed pictures to be made at night. Seven Nimbus satellites were put into orbit.

GOES weather satellites were placed in geosynchronous orbit beginning in the 1970s to transmit images almost continuously. Two of them are constantly on duty. One is located at about 75° west longitude covering eastern North America and the Atlantic Ocean and the other is at about 135° west longitude watching western North America and the eastern Pacific Ocean. Figures 6.17 and 6.18 show parts of images from these two satellites. Earthbound meteorologists have a near real-time view of the broad weather patterns as they emerge and change. The satellites can also provide an enlarged image of a small area in which hurricanes, tornadoes, or severe thunderstorms may be developing. A series of several dozen pictures taken over a period of several hours can be assembled into an animated movie showing speeded-up motion of the clouds, a display often seen on television weather programs.

For many years the U.S. military had its own Defense Meteorological Satellite Program (DMSP) flying in polar orbits at lower altitudes for greater resolution to meet military needs. Figures 6.19 is an interesting nighttime image from such a military satellite. The polar orbits enable them to observe the weather anywhere on Earth where geostationary satellites can provide useful cloud information only to about 60° north and 60° south latitudes. Recently the DMSP satel-

Figure 6.17 Weather over the United States at midday on December 31, 1999. This is the view the GOES 8 satellite has from 22,400 miles in geosynchronous orbit. Besides the clouds, some surface features such as the mountains of Mexico can be seen. *Courtesy of NASA/Goddard Spaceflight Center.*

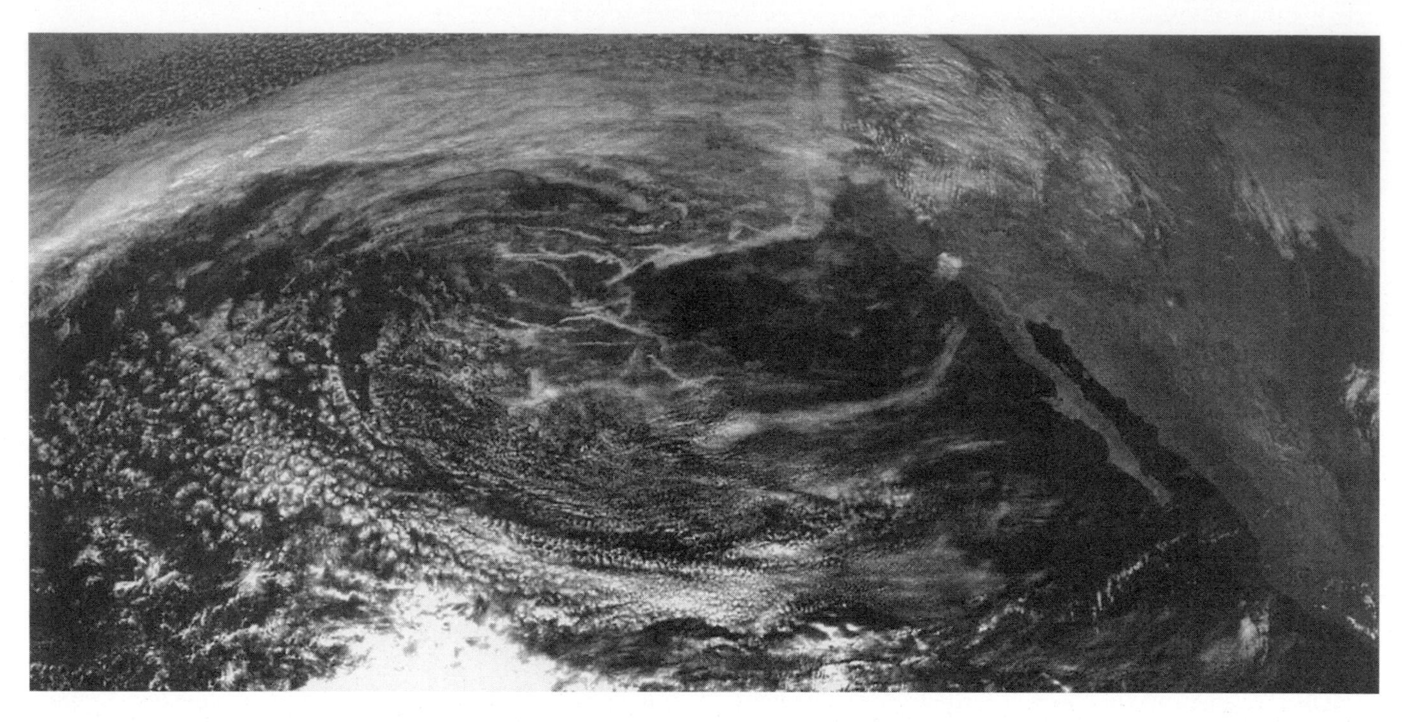

Figure 6.18 Weather over the western United States and eastern Pacific from GOES 10. Clear skies appear over Mexico and Baja California to the right. Jet stream clouds curve across the top. *Courtesy of NASA/Goddard Spaceflight Center.*

lites and the civilian National Weather Service polar orbiting satellites were merged into one national system.

At least three bands of the visible and infrared spectrum are aboard most of the weather satellites: (1) the visible band from 0.4 to 0.7 micrometer, (2) a band in the range from 5 to 7.5 micrometers which is sensitive to water vapor, and (3) a thermal infrared band in the 8 to 15 micrometer range which indicates temperature. Visual pictures of clouds, storms, and weather fronts are interesting and useful, but the infrared bands show the water vapor in the air and the temperatures of clouds. From this, their altitudes can be estimated—most important information for the weather forecaster.

In addition to detecting visible and infrared waves, the polar-orbiting satellites carry sensors that record four bands of microwaves to measure the temperature and humidity at several altitudes in clear air between clouds. These microwave emissions have wavelengths of 1.6 to 6 millimeters (0.6 to 2.3 in). The satellites can also measure ozone in the high atmosphere which is of intense interest to the inhabitants of planet Earth because it seems to be decreasing. Ozone in the stratosphere is important to humans, plants, and animals because it absorbs dangerous solar ultraviolet, preventing it from reaching the ground. The decrease may be related to chlorofluorocarbons released by industrial manufacturing processes as well as by refrigerators and air conditioners. For this reason, international restrictions have been placed on the use of chlorofluorocarbons.

It must be emphasized that the satellites are observing tools only. They do not make forecasts, but they give the human forecaster a much broader, more complete, and more timely view of weather systems. Newly developing storms can be observed without delay and their motion monitored. Before the advent of satellites, meteorologists had to rely on ground-based observations from weather stations and occasional reports from ships at sea and pilots of commerical aircraft. New storms developed and old storms intensified or weakened, often between ground stations where the changing pattern was not observed. Also, information concerning the patterns of wind flow in the higher atmosphere, in particular the jet stream, came only from weather balloons sent aloft by a widely scattered network of observing sites. Data from mountainous areas, polar regions, and oceans were almost nonexistent. Weather satellites have changed all that. Now, forecasters, armed with more complete and timely information about the state of the atmosphere, are able to make significantly more accurate forecasts.

Radar Mapping

Radar is another method of remote sensing. The remote sensing satellites discussed above are passive receivers of reflected sunlight and infrared emitted by objects on Earth. A radar, however, transmits microwaves which can penetrate clouds, and receives those waves reflected back to it from the ground. For this reason radar is known as an active form of remote sensing. Consequently, unlike passive visible and infrared sensors, radar can operate at night and through clouds.

Radar waves reflect differently from various surfaces. Interfaces between dissimilar kinds of rock, diverse types of

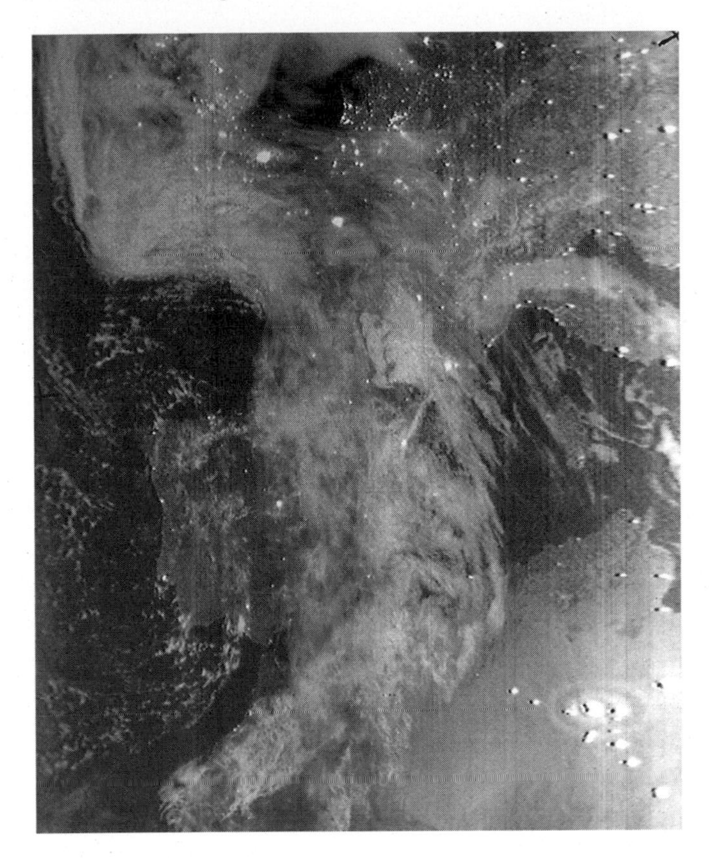

Figure 6.19 Nighttime DMSP weather satellite picture of western Europe and North Africa. Clouds are illuminated by moonlight. Bright spots are city lights. Eastern Spain is covered by clouds, western Spain is clear. Madrid is at the edge of the clouds. The two brightest spots at the upper center are London and Paris. The "cities" in the desert of north Africa are oil well fires (bottom right). *Courtesy of U.S. Space Command.*

vegetation, sea ice, and paved surfaces, for example, all produce different radar returns. Because the antenna points at an angle to the ground, vertical objects have a three-dimensional look in the picture. Also, a radar measures the travel time of the signal to and from the ground so the distance to the object and its altitude can be calculated. From this a three dimensional map can be constructed.

One spectacular result is the discovery of dry river channels buried under the sands of the Sahara Desert. Radar pulses can penetrate dry soil, but reflect from water droplets in wet ground; in effect, they can "see" through the dry sand. In New Mexico, ground penetrating radar detected the fossilized remains of dinosaurs were detected under the dry desert. Paleontologists later excavated a seismosaur at the site. Archaeologists have located traces of ancient cities and roads which appear on multispectral images but cannot be seen from the ground. Figure 6.20 (Plate 8) is a radar image of the Great Wall of China. Sections of an earlier wall were known to archaeologists but the picture shows it to be continuous for many miles. These examples and many other applications have generated high interest in satellite radar images.

Images created by a radar operating on only one frequency are like photographs on black and white film. In April 1994, Space Shuttle *Endeavour* carried imaging radar which operated on two frequencies. Recall our discussion above about signatures obtainable from the multiband Landsat or Spot data. Combining the two radar frequencies into one image gives more information than either frequency by itself.

The Space Shuttle has carried imaging radars on several subsequent flights. Topex/Poseidon is a satellite which carries a radar altimeter to measure the height of ocean waves and sea roughness. Wind speed can be inferred from the strength of the reflected radar wave.

Canada launched Radarsat-1 in 1996 into a 480-mile (800-km) orbit. It was the first continuously operating radar satellite with a resolution of 8 meters (26 ft). One of its primary uses is to map iceburg locations and movement for ships plying the Arctic. It can differentiate between ice and snow and measure the movement of glaciers. It can also track oil tankers and watch for spills. Radarsat-2 with 3-meter (10-ft) resolution is under development.

The Magellan spacecraft completely mapped Venus by radar while orbiting the planet. With its perpetual overcast of heavy clouds, the surface features are not visible by any other means. The details are in Chapter 7.

Reconnaissance

Spy satellites operate somewhat like Landsat, Spot, and Ikonos but with much higher resolution. The capabilities of intelligence-gathering satellites are, of course, highly classified. It is a little frightening to realize what intelligence agencies must have if 1 meter images with the clarity of Figures 6.12 and 6.13 can be purchased on the Internet. Rumors circulate about spy satellites with sufficient resolution to read license plates and identify people from a 100-mile (160-km) orbit. Current U.S. photoreconnaissance spacecraft reportedly have a resolution of about 4 inches (10 cm).

The camera on one early U.S. satellite, called Big Bird, photographed the area of interest, the film was processed like a Polaroid snapshot, and the picture was transmitted to the ground. If still higher resolution was needed, Big Bird could eject a capsule containing the film itself. The capsule deorbited like a miniature spacecraft, and an airplane flew out to the landing point to catch its parachute in midair before it reached the water. Using a different approach, the Russians recovered the entire satellite and its load of film when a mission was completed.

Film returned to Earth yields higher resolution images than pictures transmitted to the ground by radio. The difficulty in transmitting very high resolution images is the quantity of data that must be sent. A picture covering 1 square mile (1.6 km^2) with 1-foot (30-cm) resolution requires nearly 30 million pixels (and a very high pointing accuracy to photograph the area desired). Saving images from a large area or many smaller areas for subsequent transmission to a ground station perhaps halfway around the globe may strain the storage capability of the spacecraft. With greater storage capacity in computer memory chips, newer reconnaissance satellites will undoubtedly overcome these problems and will be able to transmit images as fast as they are taken, with higher resolution, in several spectral bands, and perhaps with radar to make images through clouds.

For best resolution it is necessary, of course, to operate as low as possible, down to 90 miles (about 150 km), which means rapid decay of the orbit due to atmospheric drag. Therefore, spy satellites were equipped with thrusters and fuel to periodically boost them back up to higher altitudes. Big Bird had a lifetime of 6 to 7 months; Russian spy satellites operated at lower altitudes and remained in orbit for about 2 weeks. That explains why the United States launched about three per year during the Cold War, while the Soviets usually launched more than 30. Newer Russian photoreconnaissance satellites are highly maneuverable, changing orbit frequently, dropping to a low perigee over a place of interest, taking their photos, then returning to a higher orbit before atmospheric friction brings them down. They have also been seen to change apogee which changes the local time they fly over the target for better lighting for their photography.

Electronic intelligence (ELINT) satellites have listening devices which tune in on telephone conversations, radio messages, radar signals, and data transmissions during missile launches. The United States keeps several such satellites in low orbit and at least two in geosynchronous orbit. The Russians use a constellation of satellites in six different orbital planes, about 400 miles high (about 650 km), at about 82 degrees inclination.

Ocean surveillance is another major effort by intelligence agencies. Watching the activities of naval surface fleets is relatively easy by radar and ELINT satellites. Keeping track of nuclear submarines with their load of ballistic missiles is more difficult, but several factors make it possible. Although the submarines remain underwater for weeks at a time, when they are under power they leave a trail of hot water which may be followed by infrared sensors. They also leave a small wake which may be detectable by radar or by the glint of sunlight from the surface.

Cosmonauts in space stations probably have had intelligence-gathering duties, also. Salyuts carried telescope-cameras with 33-foot (10-m) focal length lenses. While solar observing was their announced purpose, such a camera in low orbit is better adapted for Earth observing.

Both the U.S. and Russian governments recently began releasing images that were made by reconnaissance satellites. The once super-secret grey-scale images have a resolution of just 6 to 16 feet (2 to 5 m), not as good as the images now available to the public from commercial satellites. Stereo images were also made at a resolution of 33 feet (10 m). The U.S. image archive covers 1960 to 1972, while the Russian images date from the early 1980s.

Whatever their capabilities, governments have become highly dependent on the information satellites gather. Even though the Cold War has ended, the need for intelligence information is as important as ever. Just knowing that others are probably monitoring your every move is a stabilizing influence. It is unlikely that any nation could mount large-scale military activities without detection. The fact that other sides are watching and listening is a strong deterrent against any actions that may appear offensive.

Early Warning

Several U.S. early warning satellites located in geosynchronous orbit were designed to watch the Soviet ICBM launch complexes, their space launch cosmodromes, and the Atlantic and Pacific Ocean areas from which submarines could launch missiles. These early warning satellites carry infrared sensors which respond instantaneously to the enormous quantity of heat emitted when a rocket engine is burning. Within a few seconds the detection is reported to the North American Aerospace Defense Command (NORAD) inside its Cheyenne Mountain Complex near Colorado Springs. Computer analysis and human interpretation of the intensity, duration, and location of the infrared signal make it possible to determine what kind of launch is taking place.

These satellites may seem to be of little use now that Russia and the United States are disarming and disposing of their nuclear weapons arsenals. However, they have proven their value in smaller regional conflicts such as the war with Iraq. The Iraqi Scud missiles were observed igniting and were tracked long enough to give a short advance warning of their probable targets. The satellites operated near the limit of their capabilities because they were designed to detect much larger rocket boosters.

The infrared sensors on early warning satellites also respond to the heat from nuclear explosions worldwide for the purpose of monitoring treaty compliance and for damage assessment.

Mission to Planet Earth

The views of Earth returned by the Apollo astronauts caught the attention of people everywhere and placed a new perspective on our home planet. See Figure 6.21 (Plate 9). People began to view the Earth as a limited, closed particle, isolated from the rest of the universe by vast nothingness. The Moon, our nearest neighbor, appeared to be a desolate, forbidding place. Earth seemed to be all we had, and we had better take good care of it. The exploitation of natural resources, decline and extinction of plant and animal species, potential for global warming, depletion of ozone in the stratosphere, and pollution of water resources have increased the concern over human guardianship of the Earth. We suddenly realized that we have the power to dramatically change the environment, not always for the better, and perhaps irreversibly. We also realized that we are woefully ignorant of how the Earth's systems work and interact: the biosphere, the atmosphere, the oceans, and the land surfaces. Especially significant, we do not understand the changes produced by human activities.

Studies of the Earth's environment from the ground have been severely limited because only small areas could be examined for short periods of time, making it difficult to see the big picture. What is needed is comprehensive and continuous observation of everything we can measure about the Earth and its environment. Remote sensing from space provides a unique capability to examine the Earth systems individually and as a whole, to study and to better understand the Earth as a single, unified living system.

Mission to Planet Earth is the name given to an international program to measure just about everything there is to measure about Earth's environment during a 15-year period. From these global observations scientists try to understand the links between the oceans and the atmosphere, the ozone layer and the troposphere, the rain forests and global climate, greenhouse gases and global warming, El Niño in the Pacific Ocean and weather elsewhere, and many other complex interactions among the Earth's systems.

Measurements of ocean currents, waves, and topography are particularly important in understanding climate. Heat from the Sun is most intense in the tropics where it is absorbed mostly by ocean water and then transported toward the poles by ocean currents. The warm Gulf Stream flowing from south to north in the Atlantic Ocean along the coast of North America is a prime example. Perhaps surprisingly, some regions of the ocean surface are higher in altitude than other regions. The difference is not much, about 7 feet (2 m) from peaks to valleys, but it is significant in determining ocean circulation patterns. Although the circulation of ocean water is a most important factor in determining weather and climate, it is not well understood and is being studied in detail during the Mission to Planet Earth.

Chemical reactions between atmospheric gases and pollutants are not well understood, either. For example, we are just beginning to realize that chlorofluorocarbons from man-made sources on the ground react with ozone in the stratosphere to deplete the ozone concentration. Ozone in the stratosphere is essential to life on Earth because it absorbs ultraviolet rays from the Sun and prevents them from reaching the ground. Earth as a habitat for humans could be jeopardized if the ozone depletion continues.

Periodically, the waters of the eastern Pacific near South America become abnormally warm. Air moving over warm water becomes warmer and more humid causing dramatic changes in the weather, including torrential rains along the coast of South America. This anomaly is named El Niño for the Christ Child because the effects are most pronounced near Christmas time. At other times that same water becomes colder than normal, called La Niña or El Viejo.

El Niño causes abnormal weather conditions in other parts of the Earth also. In the southeast United States, heavier than normal winter rains and cooler winter temperatures usually occur, while warmer than normal winter temperatures occur in the north central United States. Meanwhile, on the other side of the Pacific, Indonesia experiences drought. These conditions are somewhat reversed during La Niña. The Pacific Ocean oscillates back and forth between El Niño and La Niña. The Pacific Ocean oscillates back and forth between El Niño and La Niña irregularly every few years. Before satellite measurements of ocean temperatures were available, meteorologists did not realize how widespread the effects of El Niño and La Niña are. Now, with a better understanding of the impacts of these phenomena, meteorologists can make reliable forecasts for some regions a few months in advance.

International Participation

Remote sensing satellites like Spot and Landsat have been making their rounds for many years along with the weather satellites of the United States, Russia, and the European Space Agency. Now, exploiting this experience, space agencies of 20 countries are cooperating in a coordinated and integrated data-gathering program to study all aspects of Earth's crust, oceans, atmosphere, and biosphere in an attempt to understand their interactions.

The United States and Russia continue to operate their existing constellations of remote sensing satellites as well as develop new more capable sensors and spacecraft. One such instrument built for the U.S. Earth Observing Satellite (EOS) carries 36 sensors observing in 36 different wavelength bands from 0.4 to 14.4 micrometers, as compared to Landsat 7's eight bands. They are designed to measure land, atmosphere, cloud, and ocean temperatures, as well as ocean color, atmospheric water vapor, cloud tops, and ozone.

The European Remote Sensing satellite (ERS-1) was launched to orbit in July 1991. In addition to infrared sensors, it carries a radar imaging instrument which can see the ground through clouds, day or night. It monitors the ocean temperatures, waves, currents, drifting ice floes, and even oil spills. Follow-on satellites called Envisat and Metop carry advanced radar and atmosphere-monitoring equipment. Topex/Poseidon is a joint U.S.-French spacecraft for measuring ocean currents and surface topography. Canada's Radarsat will gather global data on crops, forests, ocean ice conditions, and geological structures.

Japan built two new Earth-observing satellites, ADEOS and JERS. ADEOS was launched into a Landsat type orbit in August 1996 carrying ocean wind and wave measuring instruments and a variety of atmospheric sensors. Unfortunately, the breakdown of a soldered part at the base of the solar panels caused structural damage and it ceased operation in July 1997. JERS has both radar and optical sensors. India also operates a series of remote sensing satellites observing in four bands similar to the Landsat bands with much higher resolution as discussed above. China and Brazil are cooperating in building a Spot type of Earth resources satellite.

The vast quantity of data anticipated from all these satellites would be useless without some means of acquiring them, archiving them, and making them easily accessible to all researchers. Formerly, prime investigators on a project had exclusive use of data from their experiments for a year or more. Not so with Mission to Planet Earth. A data center is expected to make raw data available to users within 48 hours of collection and processed data within 96 hours. This speed of handling is unprecedented in the research community.

Not only does Mission to Planet Earth offer some hope of discovering how the Earth works as a closed system, it also is a rare opportunity for international cooperation on an extraordinary scale for a common purpose.

DISCUSSION QUESTIONS

1. What is the value of remote sensing of the Earth's resources?
2. Could very high resolution photo reconnaissance satellites of the future be considered an invasion of your privacy?
3. Do you see any objection to the use of high resolution satellite photographs by the news media?
4. What uses could be made of satellites other than those described so far in this book?

ADDITIONAL READING

Belokon, William F. *Multispectral Imagery Reference Guide*. Logicon Geodynamics, Inc., 1997. Fundamentals of remote sensing and spectral imagery, with real problem case studies. Comprehensive.

Brugioni, Dino A. "The Art and Science of Photoreconnaissance." *Scientific American,* March 1996. The work of photointerpreters.

Coyle, Harold P., Irwin I. Shapiro, and Sharon M. Stroud, editors. *Project Image*. Kendall/Hunt Publishing Co., 1998. A middle school–high school level manual of image-based activities.

Drury, S. A. *A Guide to Remote Sensing, Interpreting Images of the Earth*. Oxford University Press, 1990. A thorough introductory book with case studies.

El-Baz, Farouk. "Space Age Archaeology." *Scientific American,* August 1997.

Evans, Diane L., et al. "Earth from Sky." *Scientific American,* December 1994. Space Shuttle radar images of Earth.

Hafemeister, David, et al. "The Verification of Compliance with Arms-Control Agreements." *Scientific American,* March 1985. Includes satellite photography.

Launchspace. "Remote Sensing." *Launchspace*. September 1998. Special issue of the magazine.

Massonnet, Didier. "Satellite Radar Interferometry." *Scientific American,* February 1997. Some new concepts not discussed in this book.

NASA. *Landsat Tutorial Workbook, Basics of Satellite Remote Sensing*. Government Printing Office, 1982. Comprehensive reference book.

NASA. *Planet Earth Through the Eyes of Landsat 4*. NASA Facts NF-138, Government Printing Office, March 1983. Pamphlet, select example images in color.

Vasyutin, Vladimir V., and Artur A. Tishchenko. "Space Colorists." *Scientific American,* July 1989. A study of colors seen from orbit.

Weaver, Kenneth F. "Remote Sensing: New Eyes to See the World." *National Geographic,* January 1969. Somewhat dated, but excellent fundamentals.

NOTES

Chapter 7

Astronomy from Space

Astronomers learn about the universe by using remote sensing of the remotest kind—at first with naked eyes, then with telescopes, and now with spacecraft. The people of bygone centuries probably knew more about the location and arrangement of the stars in the sky and the motions of the Sun, Moon, and planets than their modern descendants. Without electric lights and TV, they spent much time outside at night just looking and wondering what it was all about. We have come a long way since the ancient astrologers watched the motions of the celestial bodies with the unaided eye and tried to relate heavenly events to human affairs. At the beginning of the 21st century and the third millennium, astronomers have incredible new tools in their hands, powerful new telescopes on Earth and on spacecraft above Earth. Not since Galileo pointed his primitive telescope at the sky has there been such a transformation in the way astronomers go about their business. Spaceflight has aided astronomers in two ways: first, by transporting sensors and men to other celestial bodies, moving up close for a better look, and second, by carrying the instruments above the "muddy" and turbulent atmosphere of Earth for a clear view in all wavelengths of the electromagnetic spectrum.

Exploring the Solar System

Distances in the Solar System are not so great as to preclude sending people and robotic spacecraft to investigate its planets, moons, asteroids, comets, and smaller objects. During the past 40 years we have been doing just that. Pluto is the only major body that has not yet been visited. Granted, the journeys may take several years and the problems of life support for human exploration are many, but technology is moving forward at high speed and we are making progress toward the goal of sending people to Mars in a decade or two.

Missions to the Moon

The greatest space spectacular of all human experience was undoubtedly the Apollo Moon landings. For the first time, men left Earth, landed on another celestial body, and returned home. Astronauts walked on the Moon, photographed its surface, installed instruments for measuring moonquakes, and brought back hundreds of pounds of lunar rocks. (See Figure 7.1.) From their work we have a clearer understanding of the origin of the Moon and the Solar System. The Apollo program was summarized in Chapter 1.

We have learned far more about our Solar System by sending robotic spacecraft to distant places than we have by sending people simply because we have sent many more robots than people. To survive in space, people require oxygen, food, water, and a secure environment, which add weight and space, as well as stricter safety standards. Robots have needs, too, but not as demanding.

Pre-Apollo Lunar Exploration

Thirteen unmanned flights preceded Apollo to the Moon. Three Rangers transmitted closeup pictures of the lunar surface as they approached and crash-landed. Five Surveyors went to the Moon, three soft-landed and two not-so-soft. They tested the composition of the soil, and transmitted more

Figure 7.1 An Apollo scientist-astronaut scoops up rock and soil from the surface of the Moon. *Courtesy of NASA.*

Figure 7.2 Apollo 12 Astronaut Charles Conrad examines the TV camera on Surveyor 3 which had been on the Moon more than 2 1/2 years before his arrival in the lunar lander seen in the background. *Courtesy of NASA.*

than 17,000 pictures from the lunar surface. Apollo 12 landed so close to Surveyor 3 that the astronauts could just walk to it, look it over, and see how it had survived the trip and the lunar environment for 2½ years. See Figure 7.2. Finally, five lunar orbiters made photographs of 99 percent of the lunar surface to aid in selecting Apollo landing sites. When the orbiters finished their job, they fired retrorockets and crash-landed on the Moon so they would not interfere with future Apollo activities.

Clementine

When the Apollo program ended, lunar exploration was in hiatus for many years while much discussion and debate ensued as to the value of returning to the Moon. (Chapter 12 will examine this question in detail.) Finally, on January 25, 1994, a spacecraft named Clementine was launched toward the Moon, the first lunar mission in almost 22 years. It is pictured in Figure 7.3.

Clementine was injected into a highly elliptical polar orbit around the Moon with a near point, perilune, at about 270 miles (435 km) and a far point, apolune, at 1840 miles (2960 km). These points are sometimes called *periselene* and *aposelene* after Selene, the Greek goddess of the Moon, comparable to the Roman goddess Luna. Clementine carried two cameras, each with CCD imaging devices similar to those found in camcorders. A strip of data was taken from the south pole to the north pole on the perilune side of the orbit when the spacecraft was closest to the surface. Then Clementine turned toward Earth as it approached apolune and the data were transmitted. Because of the elliptical orbit, pixel size varied from 20 to 30 feet (6 to 10 m). Data were taken at each pixel through 11 filters to make images of the Moon in 11 different "colors" in visible light and infrared wavelengths. The spacecraft also carried a Laser Imaging Detection and Ranging (LIDAR) system which is similar to a radar except that it operates with laser light instead of microwaves and yields much higher resolution measurements of the heights of lunar surface features.

Clementine sent back 25,000 pictures per day during more than 2 months in lunar orbit, a total of more than 1.5 million images covering the entire 15 million square mile (39 million km^2) lunar surface. Like Landsat, Spot, and other remote sensing satellites, these multispectral images can be analyzed by the signature at each pixel. Data through the 11 filters at each pixel are combined to determine rock, mineral, and soil types and to produce geological maps. (Landsat 5 sensors cover seven wavelength bands and Spot views in four. See Chapter 6.) For the first time we have a complete high-

Figure 7.3 Clementine. Four solar panels provide power. Instrument package is on the top face. The orbital maneuvering rocket is at the right and a dish antenna is on the left. *Courtesy of NASA*

resolution digitized map of the Moon showing surface features and topography.

One of the most interesting findings was Clementine's discovery of water ice in the polar regions. Scientists had long speculated that ice could exist in the deep craters near the north and south poles where the Sun never shines. When the Sun is shining at the Moon's equator, the surface temperature reaches 250°F (120°C). Water anywhere on the Moon would rapidly sublimate or evaporate and would quickly disappear into space because of the Moon's low gravity. Water could exist only in permanently shadowed places. In the deep polar craters, with no sunshine or atmosphere to keep them warm, the temperature never rises above −280°F (about −170°C). Previous spacecraft had not passed over the polar regions but Clementine used radar to detect the presence of ice in the south pole craters. See Figure 7.4 (Plate 10). The spacecraft transmitted a signal toward the area in the S-band, the 2 to 4 GHz range. The signal reflected off the Moon to a sensitive receiver on Earth with a 230-foot (70-m) diameter dish antenna. Radar waves reflecting off ice have different characteristics from waves reflecting off rock. Those from this experiment showed characteristics of an ice reflection.

After completing its lunar mapping mission, Clementine was to head for the asteroid belt to study surface features of asteroid Geographos. However, a software bug showed up shortly afterward that caused the spacecraft to fire thrusters until fuel for the attitude control system was used up, spinning it out of control. Controllers finally gave up on restoring it to proper operation.

Clementine, also called the Deep Space Program Science Experiment, is a unique mission for two reasons. First, it was designed and built by the Ballistic Missile Defense Organization and the U.S. Naval Research Laboratory to test the durability of new military sensors and electronics in the radiation environment of space and the ability of a spacecraft to operate autonomously. With the end of the Cold War, much defense research has been more open. The Navy engineers got together with space scientists, decided they could combine missions, and sent Clementine to the Moon. The scientists got their long-sought-after Moon map and the military tested its equipment. Second, it is a small spacecraft built with new lightweight components: star tracker, reaction wheels for attitude control, nickel-hydrogen battery, and solar panel. After the failure of several very large, and expensive spacecraft in recent years, NASA has been stressing the "faster, better, cheaper" approach. That is, send more smaller, simpler spacecraft instead of a few large expensive ones. The loss of a small craft may be a disaster but it is not necessarily a catastrophe.

Space technology has advanced by leaps and bounds. Miniaturizing spacecraft components means that electric power requirements are reduced. Consequently the power sources can be made smaller. Clementine's sensors weighed less than 20 pounds (8 kg) and consumed less electricity than a 75-watt lightbulb. In addition, making smaller spacecraft from lighter weight composite materials instead of heavy metals means that they can be launched on smaller rockets. The 300-pound (135-kg) Clementine was built in 2 years for $55 million and was launched on a $20 million Titan 2 rocket. Compare that with the 25,000-pound (11,000-kg), $1.5 billion Hubble Space Telescope which took nearly 20 years to design, build, and place in orbit. Such a comparison may be unfair because the Space Telescope is a far more capable and versatile piece of equipment. Nonetheless, 20 Clementine-class spacecraft could be built for the price of one Hubble.

Lunar Prospector

Lunar Prospector, another "faster, better, cheaper" spacecraft was launched to an orbit around the Moon in January 1998. It was a small 4.6 feet by 3.9 feet (1.4 m by 1.2 m) drum-shaped vehicle covered with solar cells. See Figure 7.5. Besides providing electric power in the sunlight, the solar cells kept the batteries charged for use up to 47 minutes while the craft was in the Moon's shadow.

Five experiments were carried to the Moon. Three masts extended out about 8 feet (2.5 m) to hold the instruments away from the main body of the spacecraft. A spectrometer measured gamma rays, very short wavelength electromagnetic waves, coming from the Moon's surface. Refer back to Figure 4.9 (Plate 1). There are two sources of gamma rays: (1) the spontaneous decay of radioactive elements such as uranium and (2) elements which are bombarded by cosmic rays. Each element has a characteristic set of wavelengths that it emits. When a gamma ray enters the spectrometer, it passes through a crystal and causes the atoms in the crystal to emit a short flash of light. The brightness of the flash depends on the wavelength of the gamma ray; the wavelength, in turn, depends on what element released it. The instrument was calibrated to record the characteristic gamma ray wavelengths from 10 elements: thorium, uranium, potassium, aluminum, calcium, iron, silicon, oxygen, magnesium, and titanium. A gamma ray spectrometer must operate continuously for an extended period of time, accumulating and summing the flashes. One orbit is not enough because a great number of flashes at each wavelength are needed to separate out the random flashes from those produced by the 10 elements.

The neutron spectrometer on Lunar Prospector was designed to search for ice by detecting hydrogen. It does this by looking for low-energy (low-speed) neutrons which come from the surface of the Moon. When cosmic rays collide with matter they can dislodge neutrons, gamma rays, and other subatomic particles. Some of the neutrons released in this way fly off into space at high speed. Some move into the lunar rocks and soil, ricochet off the large mineral atoms like a ball off a brick wall, and slow down only moderately. However, if they strike something their same mass, like a cue ball hitting a bil-

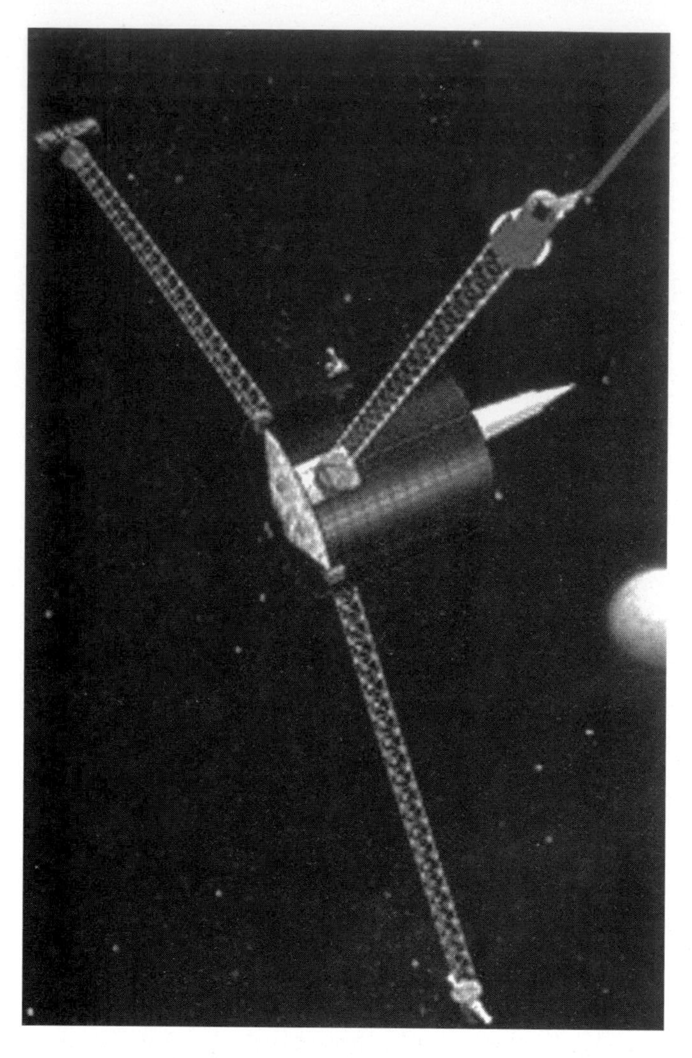

Figure 7.5 Lunar Prospector. Three masts hold the instruments away from the drum-shaped body which is covered with solar cells. *Courtesy of NASA.*

liard ball they may slow down to a crawl. Hydrogen atoms are the only atoms that have the same approximate mass as neutrons, so finding slow neutrons coming from the Moon implies the presence of hydrogen, most likely in water molecules. This instrument on the Lunar Prospector confirmed Clementine's finding of water ice in both polar regions.

To search for gases venting out from the Moon's interior, Lunar Prospector carried an alpha particle spectrometer. Alpha particles are the nuclei of helium atoms, consisting of two protons and two neutrons bonded together by nuclear forces. They are emitted by radioactive elements and their presence usually indicates that radioactive material is in the area. An alpha particle's kinetic energy (speed) indicates what kind of atom it came from. Gases venting from the Moon's interior include radioactive elements so finding alpha particles is a way of locating the source of the gas. Two other instruments, an electron reflectometer and a magnetometer, measured the magnetic fields surrounding the Moon.

When Lunar Prospector's mission ended on July 31, 1999, the little spacecraft gave its all to further search for water. Controllers directed it to crash into one of the craters at the south pole (87.7°S latitude and 42.1°E longitude) to see if the impact would melt some of the ice and spray it out into space. Instruments on Earth and on satellites in LEO watched carefully, but no water was detected.

Chapter 12 will discuss what we know about the Moon and will consider the possibility of establishing a colony there.

Missions to Mars

Two Viking spacecraft began their yearlong flight to Mars in 1975. Both had orbiters and landers that examined the red planet, studied its geology, and searched for living organisms. They sent back spectacular and surprising information. Mars is no longer just a point of light in the sky; it is a real place, a world that is both very much like Earth and very different from Earth. More recent Mars missions have included some disastrous failures as well as several outstanding successes.

The Mars Observer was the next U.S. mission to the red planet, departing Earth in September 1992 and arriving at Mars in August 1993. Unfortunately, controllers lost contact with the spacecraft just 3 days before it was to enter orbit around Mars. Commands had been sent to pressurize the propellant tanks in preparation for firing the thrusters which would insert the spacecraft into a long elliptical orbit. The transmitters were shut down during that time to avoid damage to their electronics when explosive devices were fired to open the propellant valves. Ten minutes later, when the tank pressurization was completed, the transmitters were supposed to turn back on. Mars Observer was never heard from again. There have been a number of suggestions as to what happened to this complex and expensive spacecraft, but the real cause of its demise may never be known.

In November 1996 a Russian craft, Mars 96, carrying an orbiter and landers was destroyed when the launch vehicle failed. Nozomi, a Japanese Mars orbiter, was launched to Mars on a circuitous route in July 1998. It has a propulsion problem and will arrive at the red planet in December 2003.

The U.S. Mars Climate Orbiter was launched in December 1998 and arrived at Mars in September 1999. A final command was sent to fire the engines and direct the vehicle to skim into the upper Martian atmosphere for aerobraking into an elliptical orbit. The command was sent to the controllers in English units, but the spacecraft interpreted it to be metric units, resulting in an error factor of nearly 5 in the engine burn. (One pound is 4.45 newtons.) With an insufficient delta-v the spacecraft crashed into the planet.

Pathfinder and Sojourner

On the Fourth of July 1997, after a 7-month interplanetary cruise, the Mars Pathfinder spacecraft landed on a rock-strewn ancient flood plain in the Ares Vallis in the northern hemisphere. It was the first spacecraft to set down on Mars since Viking over 20 years before. Following the "faster, better, cheaper" principle, this spacecraft was designed, built, and launched in only 3½ years with a cost between $200 million and $300 million, only one-tenth the cost of previous large-scale interplanetary flights.

The most difficult part of the trip was the entry into the Martian atmosphere and landing on the surface. Figure 7.6 shows this rapid sequence of events. Half an hour before touchdown, the lander separated from the cruise stage. An aeroshell heatshield protected the lander from the heat of reentry as it slowed from 13,600 miles (22,000 km) per hour to about 900 miles (1500 km) per hour. Then, during the final 2 minutes, a parachute deployed, a radar altimeter began measuring the distance to the ground, airbag cushions inflated, and retrorockets fired to further slow the descent to a touchdown speed of about 30 miles (50 km) per hour. The craft bounced at least 15 times before rolling to a stop. Then the airbags deflated and the craft unfolded its petals.

The Pathfinder lander, shown in Figure 7.7, carried a multispectral imaging camera, a weather station, a transmitter for communicating with Earth, solar panels for electric power, and a small rover. Accelerometers measured the deceleration as it fell through the atmosphere and hit the ground. That information was used to calculate atmospheric density.

Sojourner, the rover shown in Figure 7.8, measured 2 feet (65 cm) long, 1.5 feet (48 cm) wide, 1 foot (30 cm) high, and weighed 8.8 pounds (4 kg). A solar panel covering the top provided electric power for the motor, the radio transmitter for communicating with the lander, and an alpha proton x-ray spectrometer. Cleats on the six wheels gave good traction for driving in the dusty soil and around rocks. Two cameras on the front made stereo images.

Sojourner would back up to a rock, push the spectrometer against it, and bombard it with alpha particles. The varied wavelengths of the return emissions from the rock identified the chemical elements present in quantities of at least 0.1 percent. It is shown in Figure 7.9 (Plate 11). A full study of a rock took 10 hours.

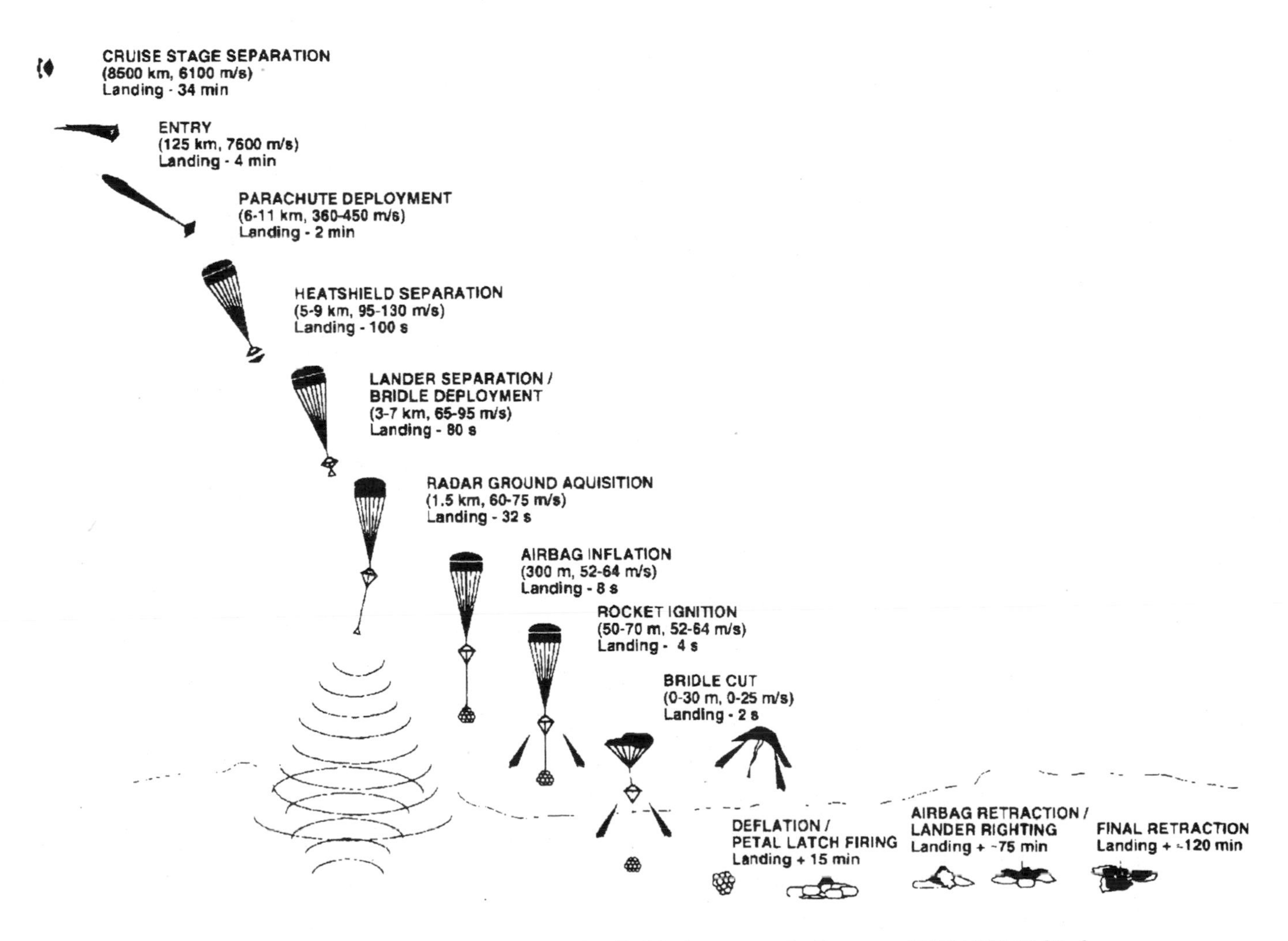

Figure 7.6 The landing sequence of the Mars Pathfinder spacecraft. *Courtesy of NASA/JPL/CalTech.*

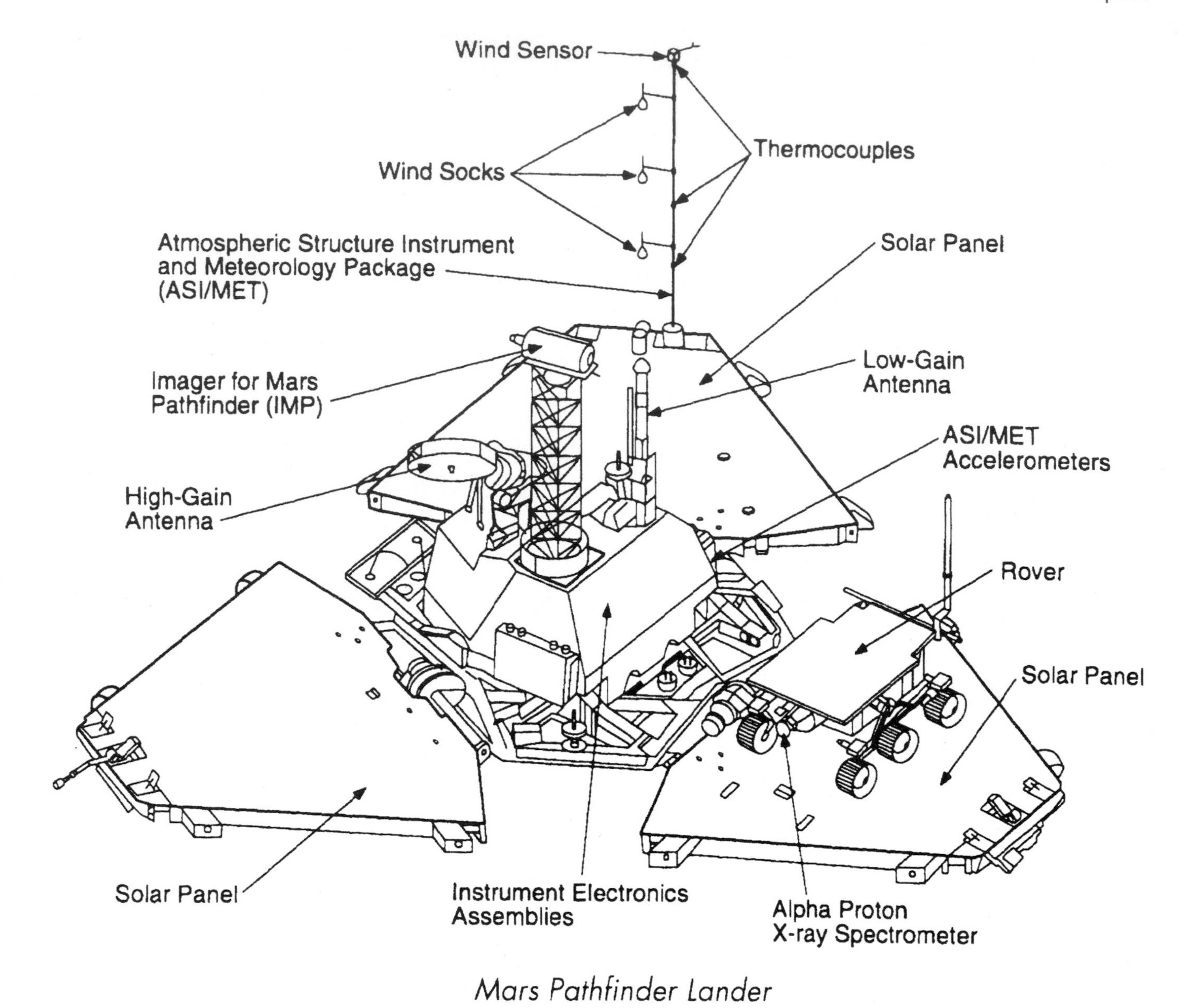

Figure 7.7 Diagram of the Mars Pathfinder lander. *Courtesy of NASA/JPL/CalTech.*

Mars Polar Lander

The Mars Polar Lander with Deep Space 2 microprobes arrived at the red planet in December 1999, after an 11-month cruise. It was to use similar techniques for descent and landing as Pathfinder but it would use a retrorocket for touchdown instead of air bags. The two small probes were to deploy as the spacecraft entered the atmosphere, crash land on the surface, penetrate about a meter, take some measurements, and transmit their findings. The spacecraft appeared to enter the atmosphere on a proper trajectory, but all further attempts to communicate with the lander or the probes failed.

Global Surveyor

Mars Global Surveyor, a U.S. spacecraft, was launched November 7, 1996, and arrived at Mars on September 12, 1997, two months behind the Pathfinder spacecraft. To reduce the weight of the craft, it did not carry enough fuel to circularize its orbit around Mars. Instead, the spacecraft was injected into a long elliptical orbit and then *aerobraked* to circularize it. At periapsis, the closest point to the planet, it would skim the top of the Martian atmosphere and the drag would slow the vehicle slightly so it did not go as far out at apoapsis. Apoapsis was lowered a little at a time and after hundreds of such passes the orbit would be nearly circular. (This was dis-

Figure 7.8 Front view of the rover, Sojourner, sitting on the surface of Mars. Solar cells cover the top. A radio antenna for communicating with the lander extends vertically on the corner. Stereo cameras are on the front just under the solar cells. The Alpha Proton X-ray Spectrometer is not visible from this viewpoint. *Courtesy of NASA/JPL/CalTech.*

cussed in Chapter 3.) Then the engine would be fired at apoapsis to raise periapsis up out of the atmosphere so the craft would not burn in.

Because of their large area, the atmospheric drag would be greatest against the solar panels, so they would endure most of the stress of aerobraking. However, one of them had not unfolded completely when commanded to do so after launch. Engineers decided that a piece of the mechanism had broken off and was stuck in the hinge. They felt that it would be able to handle the stress by turning it around so the "wind" blew against the face instead of the back as originally planned. After aerobraking began, however, the panel went through some unexpected motions and aerobraking was stopped for several weeks. A circular orbit is essential to a successful mapping mission so that all pictures of the planet are made from the same distance. Engineers tested an identical panel on the ground and their analysis indicated that the panel would be able to survive the stresses. Therefore, aerobraking was continued and the orbit was successfully circularized.

The Mars Global Surveyor began mapping the ice caps, volcanoes, plains, and deep valleys at higher resolution than had been obtained during any previous mission. An altimeter on the spacecraft permitted mapping in three dimensions. The ice caps were found to be nearly 1.5 miles (2.3 km) thick, while areas in the northern hemisphere are the flattest plains yet found in the Solar System, varying in altitude by only a few feet. Mars has a dynamic environment. The Global Surveyor has tracked cloud formations and weather storms, as

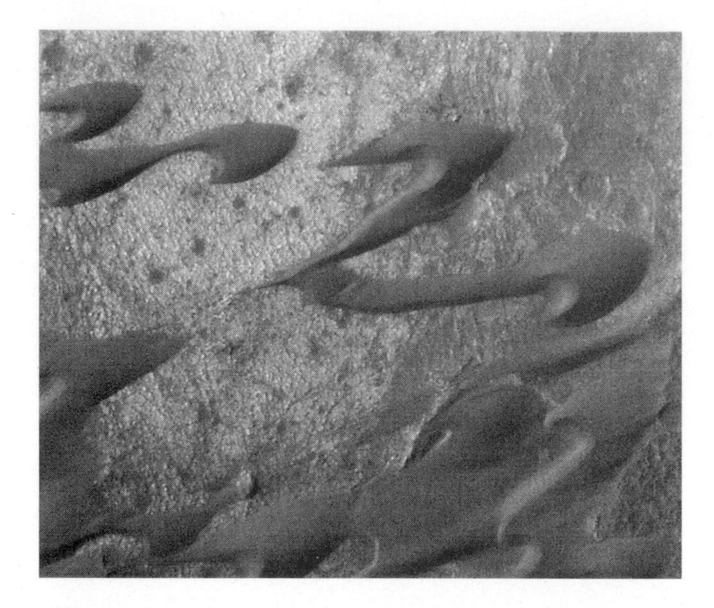

Figure 7.11 Crescent-shaped sand dunes in Syrtis Major are similar in appearance to barchan dunes on Earth. The dunes are migrating from the upper right to the lower left. Wind blows particles of sand up the gently sloping windward side of the dune and they fall down the steep side on the inside of the curve of the crescent. Thus the "horns" of the crescent point downwind. *Courtesy of NASA/JPL/MSSS.*

seen in Figure 7.10 (Plate 11), the advancing southern spring thaw and northern winter frost, the motions of dust devils and sand dunes (Figure 7.11), and many other dynamic features.

The findings of these recent missions to Mars as well as the earlier Viking results will be discussed further in Chapters 12 and 13 in the context of establishing a human presence and searching for life on the red planet. More probes will be sent approximately every 26 months for the next 10 years, each time the planets are properly aligned for Hohmann transfers. They include searching further for possible water environments; drilling deep into the subsurface material; and collecting rock and soil samples for return to Earth. They will land in different types of terrain carrying rovers and balloons, penetrators and drills, and ascent vehicles to carry the samples to Mars orbit. Later a separate vehicle will arrive from Earth to retrieve the samples from Mars orbit and return them to Earth. At some point a communications relay satellite will be placed in orbit around Mars to enhance communicating with Earth. United States, France, Italy, and other countries will cooperate in this extensive robotic exploration.

Voyager

Perhaps the most successful planetary probe has been the pair of Voyager spacecraft that made closeup inspections of the outer planets, Jupiter, Saturn, Uranus, Neptune, their moons, and their rings between 1979 and 1989. They took more than 100,000 images and millions of measurements of spectra, magnetic fields, and particles in the interplanetary solar wind.

Twelve Mariner and Pioneer spacecraft had flown past Mercury, Venus, Mars, Jupiter and Saturn from 1962 to 1978. They were the forerunners of the more sophisticated and technically capable Voyager spacecraft. Figure 7.12 is a diagram of Voyager.

Once every 175 years the outer planets line up on the same side of the Sun, making it possible to send one spacecraft to all of them on one flight. This NASA-funded "Grand Tour" was conceived of and carried out by the Jet Propulsion Laboratory. The plan was that as Voyager fell toward Jupiter, the gravitational force would increase the spacecraft's speed sufficiently and curve its trajectory around the giant planet just properly so it would go on to Saturn, the next planet in line. Similarly, at Saturn Voyager would pick up enough speed and change course for Uranus, and so on to Neptune. These trajectories are shown in Figure 7.13. The Voyager 1 trajectory past Jupiter and its moons is shown in Figure 7.14. Voyager 2's trek through the Saturnian system is shown in Figure 7.15.

Two Voyager spacecraft were sent on their way in 1977. Voyager 2 was launched first on August 20; Voyager 1 was next launched into a faster trajectory on September 5 and reached Jupiter first on March 5, 1979. Voyager 2 arrived 4 months later on July 9.

Voyager at Jupiter

What the Voyagers found was awesome. Jupiter and its four largest moons are shown in Figure 7.16 (Plate 12). These moons are called the Galilean moons because they were discovered by Galileo in the early 17th century. None of these Voyager pictures are to scale. Table 7.1 lists the sizes of the four planets and their major moons.

The giant red spot in Jupiter's atmosphere had been observed for three hundred years. It was found to rotate counterclockwise, like a hurricane on Earth, once in 6 days at the outer edge. Lightning bolts and aurora were seen and a magnetic field and magnetosphere were discovered. In addition, Jupiter was found to have a thin ring which could be seen only by looking back into the glare of the Sun from the far side.

Planetary scientists had expected the moons of Jupiter to be cold, rocky bodies riddled with craters. The surprises were many. Other than Earth, Io is the most geologically active body yet found in the Solar System with at least eight active volcanoes and a surface strewn with red-yellow-orange sulfur compounds from the volcanic eruptions. Europa appeared to be as smooth as a billiard ball covered with water ice and dark linear markings. Standing on Europa, you would see Jupiter in the sky 20 times larger than our Moon. You

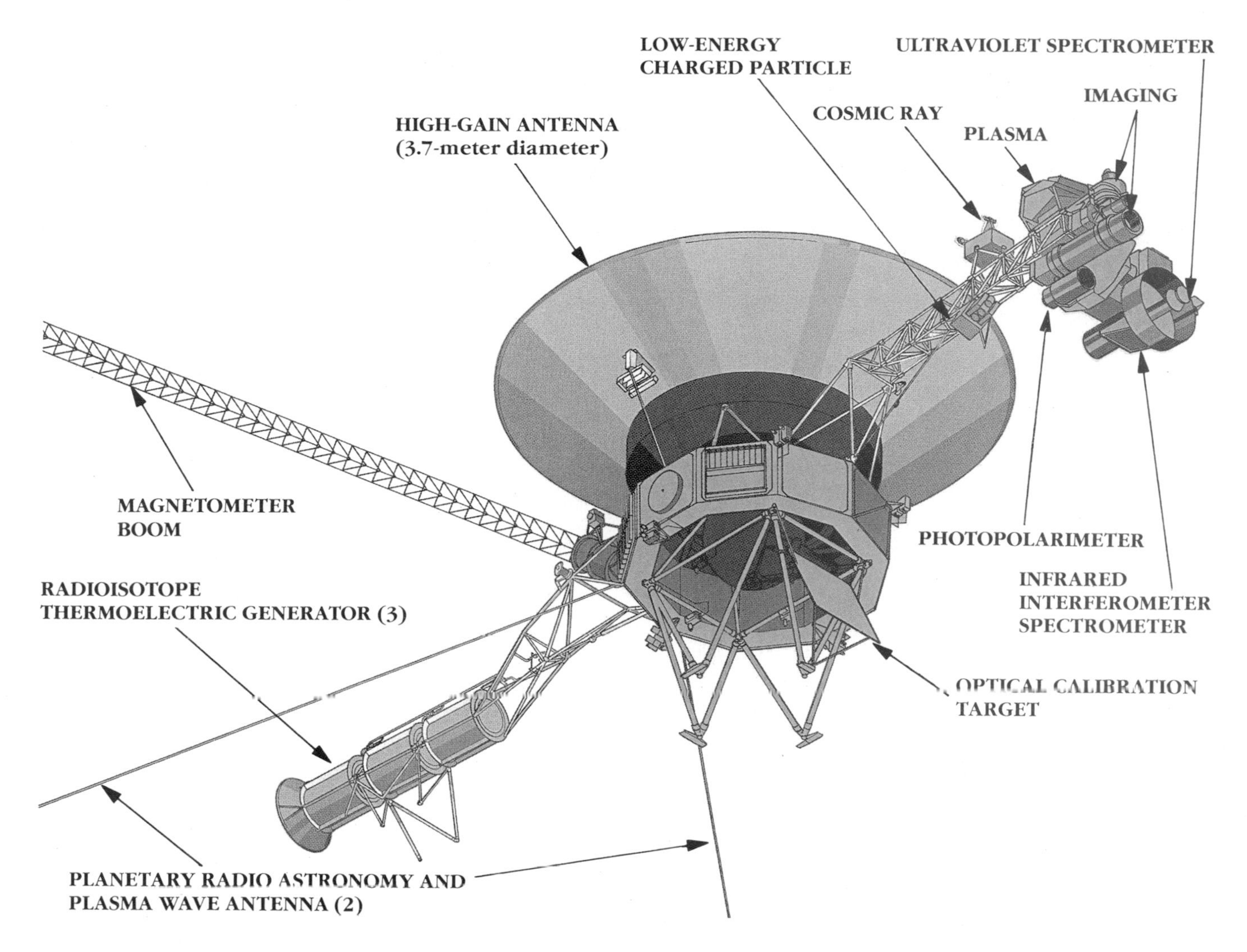

Figure 7.12 Voyager spacecraft. The spacecraft was powered by nuclear electric generators because the intensity of solar energy is so low at the outer fringes of the Solar System that solar cells cannot be used. The communication antenna is over 12 feet (3.6 m) in diameter. Other components are sensors which measure particles and electromagnetic radiation in space and in the vicinity of the target planets. The imaging devices created spectacular photos of the planets and their moons. *Courtesy of NASA.*

could watch the giant planet rotate once every 10 hours and see it go through phases like our Moon every 3.5 days. Ganymede is the largest moon in the Solar System and is covered with both craters and grooved ice terrain. Callisto has wall-to-wall craters, probably the most intensely cratered body in the Solar System. Other lesser moons were photographed as well and three new ones were discovered.

Voyager at Saturn

Voyager 1 arrived at Saturn on November 12, 1980, after a 39 month flight, only 12 miles (20 km) off course; Voyager 2 followed on August 25, 1981, 2.7 seconds early and 30 miles (48 km) from the aim point. The Voyager 2 trajectory through the Saturnian system is shown in Figure 7.15. A montage of images is shown in Figure 7.17 (Plate 12).

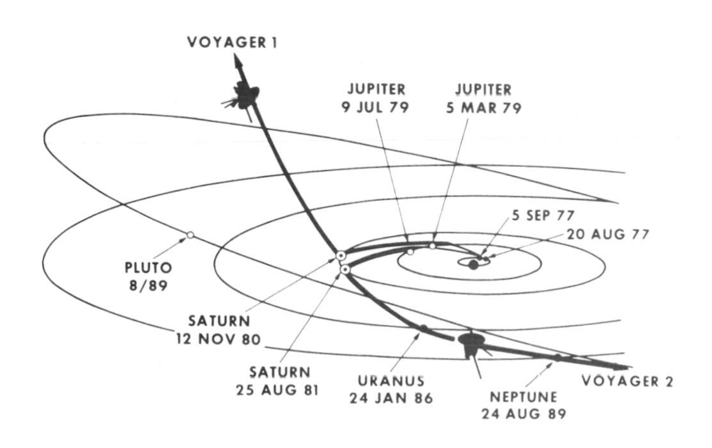

Figure 7.13 Voyager spacecraft "Grand Tour" of the outer Solar System. *Courtesy of NASA.*

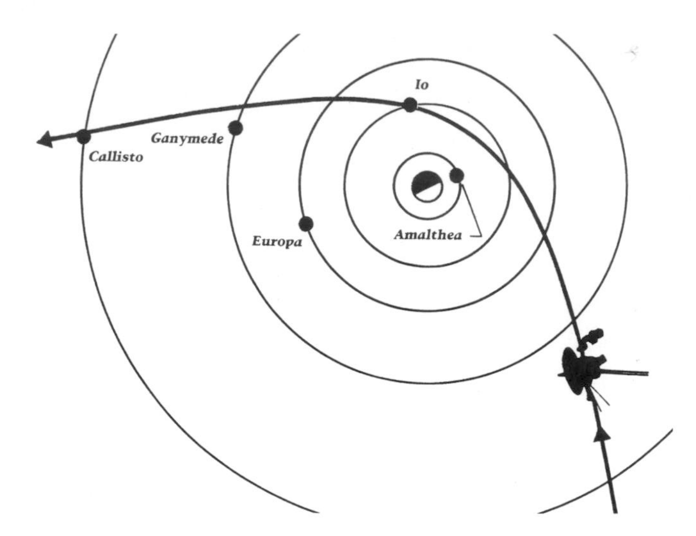

Figure 7.14 Voyager 1 track past Jupiter and its moons. The curvature of the trajectory is due to the gravitational force as Voyager approached Jupiter, which deviated the spacecraft from its elliptical orbit around the Sun. *Courtesy of NASA.*

Again the discoveries were many. Saturn's density is less than water. Spoke-like features were seen in the rings. Small moons were found on the edge of some of the rings; they act to keep the ring particles in place and the edge of the ring sharply defined and consequently are called "shepherd moons." Cloud features are similar to Jupiter but not as pronounced. Wind at the equator was measured at 1100 miles (1770 km) per hour. Saturn was also found to have a magnetic field and thus a magnetosphere.

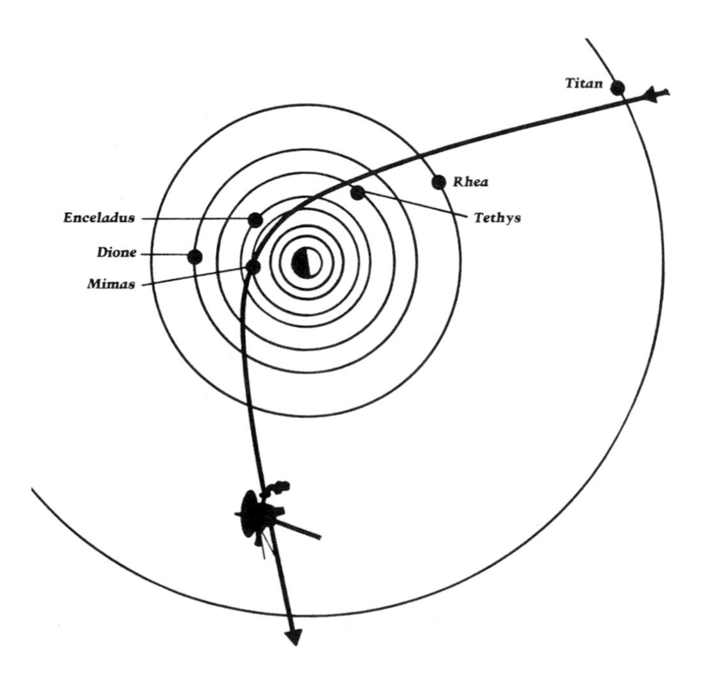

Figure 7.15 Voyager 2 trajectory through the Saturnian system.

TABLE 7.1 Major Moons of the Giant Gas Planets

Planet/Moon	Distance from Parent		Diameter	
	Thousands of miles	Thousands of km	miles	km
JUPITER				
Io	262	422	2255	3630
Europa	417	671	1949	3138
Ganymede	665	1070	3268	5262
Callisto	1170	1883	2981	4800
SATURN				
Mimas	116	186	243	392
Enceladus	148	238	311	501
Tethys	183	295	655	1055
Dione	234	376	696	1120
Rhea	327	526	950	1530
Titan	758	1220	3199	5150
URANUS				
Miranda	81	130	298	480
Ariel	118	190	714	1150
Umbriel	165	266	733	1180
Titania	270	435	994	1600
Oberon	362	583	950	1530
NEPTUNE				
Triton	220	355	1677	2700

Titan, the largest moon, has a dense nitrogen atmosphere, with haze layers formed from organic compounds such as methane, propane, and acetylene. The surface temperature on Titan is −288°F (−178°C). At that temperature methane could play much the same role as water does on Earth. The innermost moons are composed mostly of water ice.

Voyager at Uranus

Voyager 1 was steered under Saturn and allowed to accelerate to escape velocity from the Solar System. It headed upward out of the plane of the ecliptic where the orbits of the planets lie. Voyager 2 took a trajectory around Saturn and on to Uranus, arriving there on January 24, 1986. Uranus and its largest moon are shown in Figure 7.18 (Plate 13). Clouds on Uranus were not as outstanding as the clouds of Jupiter and Saturn; Uranus showed a nearly blank bluish colored ball. Strangely, Uranus lies on its side as compared to the other planets, that is, its axis of rotation lies close to the plane of its orbit. Consequently, as the planet moves in its orbit, the north pole is pointed toward the Sun for about 20 years, then the south pole is aimed toward the Sun for the same time. The equatorial regions receive less sunshine than the poles. Even so, winds were clocked at 375 miles (600 km) per hour and the cloudtop temperatures are quite uniform over the entire planet.

The surprise of the encounter was the moon Miranda. It has geological features and patterns never before seen in planetary exploration. One cliff is 16 miles (26 km) high; considering Miranda's gravity, if you dropped a stone off the cliff it would take 9 minutes to reach the bottom! A possible explanation for the weird structure is that it collided with another moon-sized object, broke apart, and the pieces rejoined a bit out of order.

Voyager at Neptune

The final planet visited by Voyager 2 was Neptune on August 25, 1989. It is the smallest of the four giant gas planets, about 31,000, miles (50,000 km) in diameter, almost four times the diameter of Earth. Voyager was then nearly 3 billion miles (nearly 5 billion km) from the Sun, 30 times farther than the Earth. At first Neptune, too, appeared as blank as Uranus, but with some processing and enhancing of the images, clouds and a giant blue spot can be seen. See Figure 7.19 (Plate 13). These features all move at different speeds. The highest wind at cloud tops was over 700 miles (1100 km) per hour.

Triton, the largest of Neptune's moons, is seen in Figure 7.20 (Plate 13). It shows an incredible variety of terrain. Its surface temperature is down to −400°F (−240°C) and it has a thin nitrogen atmosphere with some methane mixed in. It is thought that Triton may have been captured into orbit after Neptune formed because its orbit is retrograde, that is, it goes around Neptune in the opposite direction from the other moons.

One of the significant discoveries of Voyager is that, although they may look superficially alike, each body in the Solar System has unique features all its own, different from every other body.

Pioneer and Voyager Beyond the Solar System

At the beginning of the new millennium four spacecraft were beyond the orbit of Pluto. Their trajectories are shown in Figure 7.21. On January 1, 2000, Pioneer 10 was 6.9 billion miles (11 billion km) from the Sun, 74 times farther from the Sun than Earth is and its 8-watt transmitter was still sending data. A radio message travelling at the speed of light takes

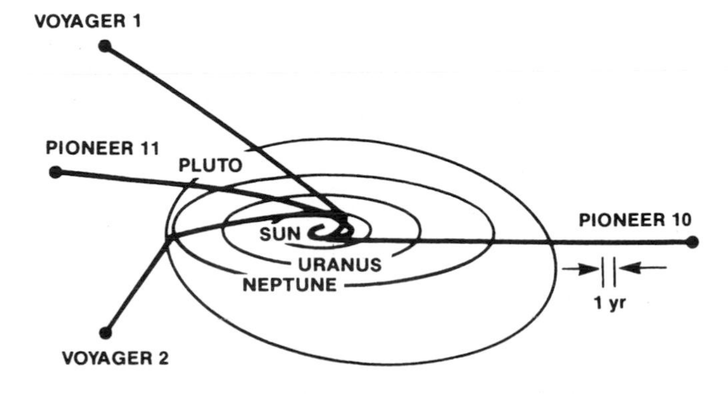

Figure 7.21 The trajectories of four spacecraft leaving the Solar System for interstellar space. *Courtesy of NASA.*

more than 10 hours to reach Earth at that distance. Pioneer 10's speed, about 7.5 miles (12 km) per second, is greater than escape speed from the Solar System so that, even after its power supply dies, it will travel on forever in interstellar space, unless it collides with something. Launched in 1972, it was the first spacecraft to fly through the asteroid belt and explore the outer Solar System. Pioneer 11, launched the following year, is going in the opposite direction from Pioneer 10. The last message from Pioneer 11 was received in November 1995. Its power supply is exhausted but it continues on its journey toward the constellation Aquila; it will arrive there in about 4 million years.

The two Voyager spacecraft are also going where no one has gone before. Voyager 1 passed by Saturn and, with a gravity assist under Titan, is headed up and out of the Solar System at an angle of 35° to Earth's orbital plane (the ecliptic). It is travelling faster than Pioneer 10, about 10.7 miles per second (17.3 km/s). In 1998 it surpassed Pioneer 10's distance but in a different direction. At the turn of the century it was the most distant manmade object, 7.1 billion miles (11.4 billion km) from the Sun. By 2015 it will reach 12.4 billion miles (19.9 billion km). Voyager 2 is moving slower and is headed downward at an angle of 25° below Earth's orbit. "Up" and "down" are roughly the north and south directions from Earth's point of view.

Even at their great distance from the Sun, the spacecraft still detect disturbances in the solar wind which can be traced back to solar flares and coronal mass ejections.For example, in February 1999 Pioneer 10 reported a fluctuation that corresponded to the large solar flare experienced at Earth in April 1998.

Just as Earth is located within a magnetosphere created by the interaction between the Earth's magnetic field and the solar wind, the Solar System is located within the *heliosphere* created by the interaction between the solar wind expanding outward against the interstellar gas. Inside the heliosphere the solar wind dominates; outside the heliosphere lies the interstellar gas. The boundary between the solar wind and the interstellar gas is called the *heliopause*. These spacecraft are looking for that edge of the Solar System. Theoretically, before they reach the heliopause they will encounter a place called the *termination shock* where the solar wind speed abruptly drops from supersonic to subsonic as it approaches the heliopause. Voyagers and Pioneer 10 should be able to detect the termination shock and the heliopause and let us know just where they are.

How long the Voyagers will be able to communicate with Earth depends on several factors. They are powered by electricity produced by heat from the decay of plutonium and do not depend on the Sun for power. The radioactive material eventually will be used up as will the hydrazine fuel used by the thrusters to keep the large antennas pointed toward Earth. Actually, the Voyagers' sensors cannot detect the Earth but they can detect the Sun and use it as a reference to point the antennas correctly. Some of the instruments and communications equipment have failed and other equipment may stop functioning also. Unless a catastrophic failure occurs, we will continue to hear from the Voyagers until perhaps 2020. Where will they be then? Barely out of the Solar System. Alpha Centauri, the nearest star after the Sun is about 4 light years away. (A light year is a measure of disance, not time. It is the distance light travels in a year, about 6 trillion miles.) At Voyager speed, about 19,000 years to travel 1 light year, it would take about 76,000 years to travel that distance. However, neither Voyager is headed in that direction,

Unless they collide with something, the spacecraft themselves will remain intact essentially forever. Perhaps some alien species will find them someday. Science fiction has it that V—-ger will be picked up by the crew of the starship *Enterprise* sometime in the 23rd century. Messages from Earth to whomever may find them are attached to the Voyager spacecraft. They are sounds and pictures recorded on a gold-plated record carried in an aluminum case. Instructions on how to play the record are included. It is assumed that anyone smart enough to intercept Voyager will be smart enough to decode the instructions. We will have more to say about this in Chapter 13.

Galileo to Jupiter

Voyager results at Jupiter whet the appetite of planetary scientists for more. Everything discovered by Voyager was so completely beyond expectations that the mission raised many new questions. Galileo is a spacecraft designed to take a more thorough look at the Jovian system, visiting Venus, two asteroids, and the Earth-Moon system twice, before heading for Jupiter. It is shown in an artist's rendering in Figure 7.22 (Plate 14).

Taking the long way around to Jupiter by way of Venus was necessary because of a change in the booster rocket that would boost Galileo out of Earth orbit onto an escape trajectory. Galileo was to be launched in the cargo bay of the Space Shuttle but following the *Challenger* accident, NASA decided it was not safe to carry a liquid-fueled Centaur rocket as an upper stage for the interplanetary spacecraft. The solid propellant Inertial Upper Stage (IUS) was to be used instead, although the IUS does not have the power of a Centaur and cannot send a large spacecraft like Galileo on a direct path to the outer planets. Therefore, it was necessary to devise a flight plan which used Venus and Earth for gravity boosts. The trajectory was called VEEGA for Venus-Earth-Earth Gravity Assists. The planets had to be located in the correct positions at launch so each would be at the correct spot in space when Galileo arrived. At each encounter with Venus and Earth, the spacecraft fell toward the planet to increase its speed and extend its apogee until it could reach Jupiter. Figure 7.23 shows this complex trajectory. Follow the path and notice the change in aphelion after each planetary encounter.

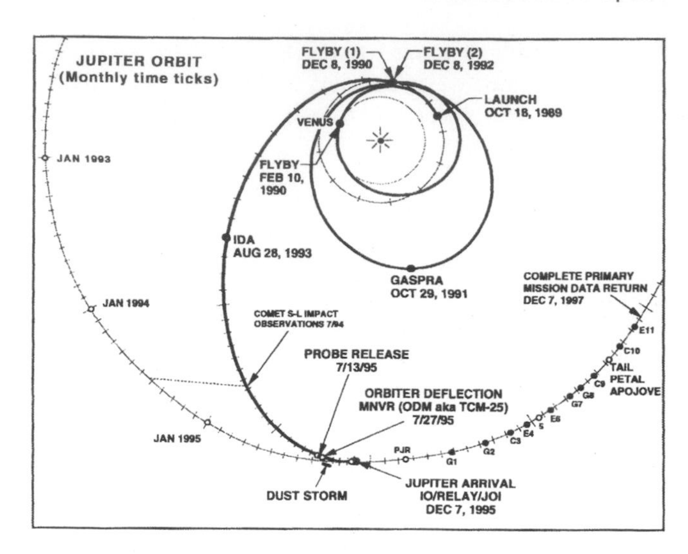

Figure 7.23 Galileo's complex trajectory to Jupiter via Venus, Earth twice, and the asteroid belt. Jupiter's orbit is shown with monthly tick marks. After arrival the marks on the orbit indicate Galileo's encounters with the moons: G for Ganymede, E for Europa, and C for Callisto. *Courtesy of NASA Jet Propulsion Laboratory.*

Unfortunately, the 16-foot (5-m) antenna used for communicating with Earth, which was designed to open like an umbrella, did not open all the way. This severely limited the capacity for speedy transmission of the data collected by the sensors. Using the smaller antennas, data transmissions were very slow: 10 bits per second instead of the 131,000 bits per second possible if the balky antenna were fully open. At this leisurely rate it took days to send a single picture. Two things were done to improve the situation. First, the spacecraft's computer was reprogrammed to compress the data before sending it. Second, the sensitivity of the ground receivers was improved. These actions increased the data rate to 100 bits per second.

While enroute to Jupiter, Galileo observed and recorded many events of scientific interest. At Venus, it used its infrared sensors to map the turbulent middle level cloud layer about 30 miles (48 km) above the surface where winds blow more than 150 miles (240 km) per hour. It also searched for lightning in the clouds of Venus and measured the atmospheric composition. Closeup photos of asteroids Gaspra and Ida were captured. Scientists were surprised to find that Ida has a moon. See Figure 7.24. The possibility had been suggested previously by amateur astronomers who observed stars "blink out" momentarily when an asteroid moved in front of them. On some occasions the star would blink off and on more than once, suggesting that there was more than a single asteroid in the vicinity. The picture of Ida was the first confirmation that asteroids may indeed have their own moons. In addition, Galileo captured images of the string of pieces of the Shoemaker-Levy comet as they crashed into Jupiter and recorded the most intense dust storm ever encountered by a spacecraft.

Figure 7.24 Asteroid Ida and its moon. Ida is about 35 miles long and 15 miles wide. By preliminary estimates its moon is about a mile in diameter and roughly 60 miles away from the asteroid. Both Ida and its moon appear to be made of silicate rock. *Courtesy of NASA*..7

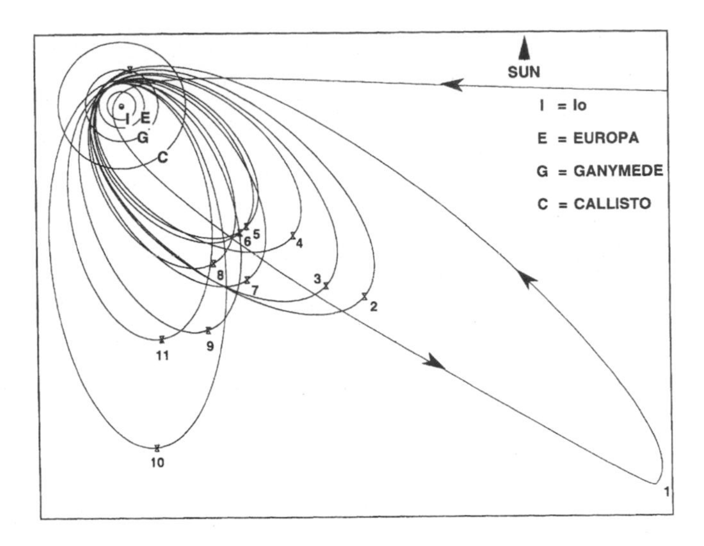

Figure 7.25 Galileo's first 11 orbits around Jupiter which carried it to close encounters with the Galilean moons. The first orbit past Io had the highest apoapsis (apojove) with a period of about 7 months. Other orbital periods were from 1 month to 2 1/2 months. *Courtesy of NASA/Jet Propulsion Laboratory.*

Both Earth flybys were made in December during the southern hemisphere summer. Earth was photographed in 1000 frames to make a movie of its 24-hour rotation. South America, Australia, Africa, and Antarctica are featured in the images. From Galileo's vantage point, about 1.2 million miles away, Earth appears as a distant planet with white wispy clouds, blue oceans, brown land, and white ice cap. These features, which are important to life, distinguish it from all the other planets and moons we have seen.

A probe to investigate Jupiter's atmosphere was released from Galileo in July 1995, months before it arrived at the planet. The probe entered the atmosphere on December 7 at the same time that Galileo put on the brakes to enter an orbit around Jupiter. The probe's heat shield protected it from destruction while it slowed from 105,000 miles per hour (170,000 km/hr) to parachute speed. As it dropped into Jupiter's atmosphere it measured temperature, pressure, sunlight intensity, wind speed, lightning, and composition of the atmosphere. The data were transmitted for nearly an hour to the Galileo spacecraft orbiting above until the pressure reached 24 times Earth's atmospheric pressure and the temperature reached 306°F (152°C) about 100 miles (160 km) below the top of the atmosphere. The probe continued to fall, melted, and then evaporated into molecules which are now part of Jupiter's atmosphere.

For the next 2 years Galilio made side trips to many of the Jovian moons for closer looks than were made by the Voyagers in their quick flybys. The orbits are shown in Figure 7.25. The first trajectory past Io and Jupiter provided the gravity boost to take it into orbit 1, an elongated 7-month orbit. Each subsequent orbit depended on a gravity assist from Jupiter and one of the moons. The spacecraft was steered by adjusting its trajectory slightly so as it passed by the planet and a moon it got just the proper abount of gravity assist to shift into the desired orbit for an encounter with another moon. The spacecraft could have accomplished its mission by firing its rocket for each orbital change, but using the gravity assists required only 1/60 as much fuel. Thus Galileo visited all the moons in 11 orbital changes in less than 2 years, some of them several times, with fuel left over to continue for several more years. Enroute to each moon the spacecraft measured the particles and magnetic fields in the Jovian environment.

A sampling of images from Galileo are shown in Figures 7.26 to 7.29 (Plates 15, 16, 17) and Figure 7.30. The spacecraft's cameras can make infrared as well as visible pictures with much higher resolution than Voyager. Small details in the surfaces of the moons can be seen.

Even though it is running low on fuel and has been bombarded by energetic particles in the severe radiation environment near Jupiter, Galileo still has a promising future. In 2000 it has further encounters with Ganymede and Europa, and in February it flies within 124 miles (200 km) of Io to make an extremely close examination of the volcanoes. This is particularly hazardous because of the severe radiation. Not only can the energetic particles do physical damage to the sensors and electronics, but they also produce noise in the data which shows up as speckles in the images.

On December 30, 2000, the Cassini spacecraft will come past Jupiter for a gravity assist on its way to Saturn. If Galileo survives the close encounter with Io and still has fuel left, it will "meet" Cassini and the two will take simultaneous measurements of the particles and fields environment around Jupiter. That will probably be Galileo's final effort although it may continue in orbit around Jupiter for thousands of years.

Figure 7.30 The surface of Ganymede, showing a chain of 13 craters probably formed by pieces of a comet that broke up in the intense gravity of Jupiter. The area covered is about 133 by 135 miles (214 by 217 km). *Courtesy of NASA.*

Cassini to Saturn

Cassini is on its way to investigate Saturn, its surrounding environment, and its 22 moons. The objective is to develop an understanding of the entire Saturnian system as a whole: the interactions among the planet, moons, ring system, particle environment, and magnetic fields. Also, scientists are very interested in comparing the atmospheres of the giant gas planets, Saturn and Jupiter, to discover their similarities and differences. Sensors on Cassini will examine the atmosphere of Saturn in great detail so the data can be compared with the data on Jupiter gathered by Galileo.

Of particular interest is Saturn's largest moon, Titan, because, like Earth, it has an atmosphere composed mostly of nitrogen, methane, and complex carbon based molecules. Even though the temperatures on Titan are much colder than on Earth, *photochemical* processes are going on which manufacture hydrocarbon compounds. The reaction is probably similar to that which occurred in Earth's atmosphere billions of years ago before oxygen and living organisms existed here. It is likely that there are lakes or oceans of liquid nitrogen and hydrocarbons. Carl Sagan suggests that Titan is a planet just waiting for a global warming so biological reactions can develop from the chemical reactions that have been going on for millennia.

When Galileo first viewed Saturn through his telescope and notice bumps on both sides of the planet, he referred to Saturn as the planet with ears; his simple telescope could not resolve the bumps into rings. Nearly 50 years later, Huygens, with an improved telescope, was able to discern the rings as a disk, and correctly surmised that the rings change appearance from time to time because we see them from different points of view as Earth and Saturn revolve around the Sun. Huygens also discovered Titan, the largest moon of Saturn. In the 1670s Cassini, the first director of the Paris Observatory, saw that the ring was not solid but there was a dark region that looked like a blank space dividing it in two. The dark region is now called the Cassini division. Cassini also discovered several of Saturn's moons and was the first to report the giant red spot on Jupiter.

Cassini is a very large, complex, and expensive spacecraft. Price tag for the mission is over one billion dollars. It carries a dozen experiments including a radar to map Titan, spectrometers to measure various bands of electromagnetic radiation, and a magnetometer to measure magnetic fields. It stands about 30 feet (9 m) tall and has a large antenna for communicating with Earth. See Figure 7.31. Many of the experiments are attached to two movable rotating platforms on long booms so they may be pointed at objects of interest. The magnetometer is also mounted on a long boom to avoid interference from magnetic fields in and on the spacecraft body. Cassini will release a probe built by the European Space Agency named Huygens to investigate the atmosphere of Titan and to land on the surface.

Cassini was launched on a Titan 4 rocket on October 12, 1997. Like Galileo, to get where it is going it needs four gravity boosts. To accomplish this it flew by Venus in April 1998 and again in June 1999, then past Earth in August 1999, and finally by Jupiter in 2001. These gravity boosts save about 2 years travel time by putting the spacecraft on a faster trajec-

Figure 7.31 The Cassini spacecraft. At the top is the communications antenna. Instrumentation is attached to brackets and booms sticking out in all directions. The disk on the left is the European Space Agency's Huygens probe which will be ejected and descend to Saturn's moon, Titan. *Courtesy of NASA.*

tory. It will reach Saturn June 25, 2004, fly very close to the rings, and pass them before the rockets are retrofired to inject it into a highly elliptical orbit around the planet. Periapsis will lie close to the planet and apoapsis will be out beyond the rings and moons.

Again like Galileo, Cassini will make use of gravity assists to steer it through the Saturnian system. Each time it passes close to Titan its orbit changes inclination, rotates, or changes plane. Some orbits pass near other moons for closeup study; some move up out of the equatorial plane for a different perspective of Saturn's rings and magnetic fields. It is expected to maneuver through 40 orbits in 48 months. During each pass by Titan, the radar makes an image of the strip of the surface beneath the spacecraft at a resolution of several miles (kilometers). Some selected areas of the moon can be mapped with a 1000 foot (300 m) resolution. The Cassini radar will be able to see through the clouds of Titan.

On the first flyby of Titan, Cassini is steered on a collision course with the moon so the Huygens probe can be released from the spacecraft. This is shown in Figures 7.32 and 7.33 (Plates 18, 19). Once the probe is released, the Cassini orbiter monitors Huygens's motion for several hours, then Cassini fires a thruster so it doesn't hit Titan too! A shield protects the probe from overheating and slows its fall. A parachute opens and slowly carries it on a 2 to 3 hour descent through the atmosphere, striking the surface at about 1 mile (1.6 km) per hour. It will float if it lands in a pool of liquid nitrogen.

The Huygens probe contains instruments to measure the atmospheric pressure, temperature, density, and composition; a collector to study aerosol particles; a camera that looks down to take pictures of the surface, then up to measure the light transmission through the atmosphere; a lightning and radio wave detector; and surface-measuring instruments. A small transmitter on the probe sends its data to the orbiter which then relays the information to Earth. That way the probe doesn't need a high power-transmitter, power supply, or large antenna. The Huygens probe will last only last a minute or so on the surface before the battery runs out. The next pass of the orbiter is 120 days later and it would add too much weight to supply batteries to last that long.

Magellan to Venus

Venus has a very dense carbon dioxide atmosphere and is completely shrouded in cloud cover. Images through telescopes show only a bright white disk. Venus became the target of intense effort by the Soviet Union's planetary space program. At least 25 spacecraft were launched with the intention of studying that planet. Although many failed, many were phenomenally successful. Venera 1, launched February 12, 1961, was the first spacecraft to escape Earth orbit and reach another planet as it flew by Venus in May of that year. On February 27, 1966, Venera 3 was the first man-made object to land on another planet. In total, the Soviet spacecraft made seven successful soft landings on Venus between 1965 and 1985. They found the surface temperature to be about 900°F (480°C) and the atmospheric pressure to be 90 times that at the Earth's surface. At least two of the Venera spacecraft survived the heat long enough to send pictures of the dry, rocky surface.

The Venusian cloud cover is so dense that it is impossible to photograph its surface in the usual way with CCD type cameras. In 1978 a U.S. Pioneer spacecraft arrived to begin mapping 92 percent of the planet by radar with a resolution of 60 miles (100 km). Like Earth, Venus has high continents and low basins. The basins, of course, contain no water. Water cannot exist in liquid form because of the extreme heat. Pioneer Venus finally ran out of gas for the thrusters and fell out of orbit into the dense Venusian atmosphere in October 1992 after sending its messages for 14 years. A better radar map of Venus came from Venera 15 and Venera 16 which arrived at Venus in 1983. Their radar had a resolution of 6500 feet (2 km) and covered about 30 percent of the planet.

Magellan, launched from the Space Shuttle *Atlantis* on May 4, 1989, carried an even higher resolution imaging radar to map Venus through the clouds. It could distinguish objects as small as 350 feet (106 m) across and measured altitudes with an accuracy of 100 feet (30 m). From its highly eccentric orbit with apoapsis of 5255 miles (8460 km), periapsis of 180 miles (290 km), and period of 3.26 hours, the radar recorded an image of strips of the ground 12 to 15 miles (19 to 24 km) wide from pole to pole as it flew close to the planet on the periapsis side of its orbit. Then, as it slowed down approaching apoapsis, it transmitted its findings to Earth. Mapping data were taken for 37 minutes of each orbit and played back to Earth in 114 minutes, leaving about 45 minutes of each orbit for turning, scanning the stars for navigating, and spacecraft "housekeeping." As Venus rotated on its axis, Magellan scanned a different strip of the surface on each orbit. It took one Venusian day, 243 Earth days, to map almost the entire planet, missing only a small area near the south pole. After three complete mapping cycles from mid-September 1990 to mid September 1992, 98 percent of the planet had been covered.

The radar was not aimed straight down, but looked sideways at angles from 14 degrees to 52 degrees which gives a more pronounced perspective to the images. Combining the radar data with the altimeter data produces spectacular three-dimensional pictures. As with every other spacecraft that has visited another planet, Magellan found some surprising terrain on Venus. Three of the many interesting images are shown here. Figure 7.34 shows some highly unusual circular dome-shaped features that are thought to be lava flows. Three large impact craters are shown in Figure 7.35. Figure 7.36 (Plate 19) shows a Venusian mountain, similar in structure to the Hawaiian shield volcano, Mauna Loa. It rises 5 miles (8 km) above the surface and the lava flows extend in the foreground for more than 100 miles (160 km).

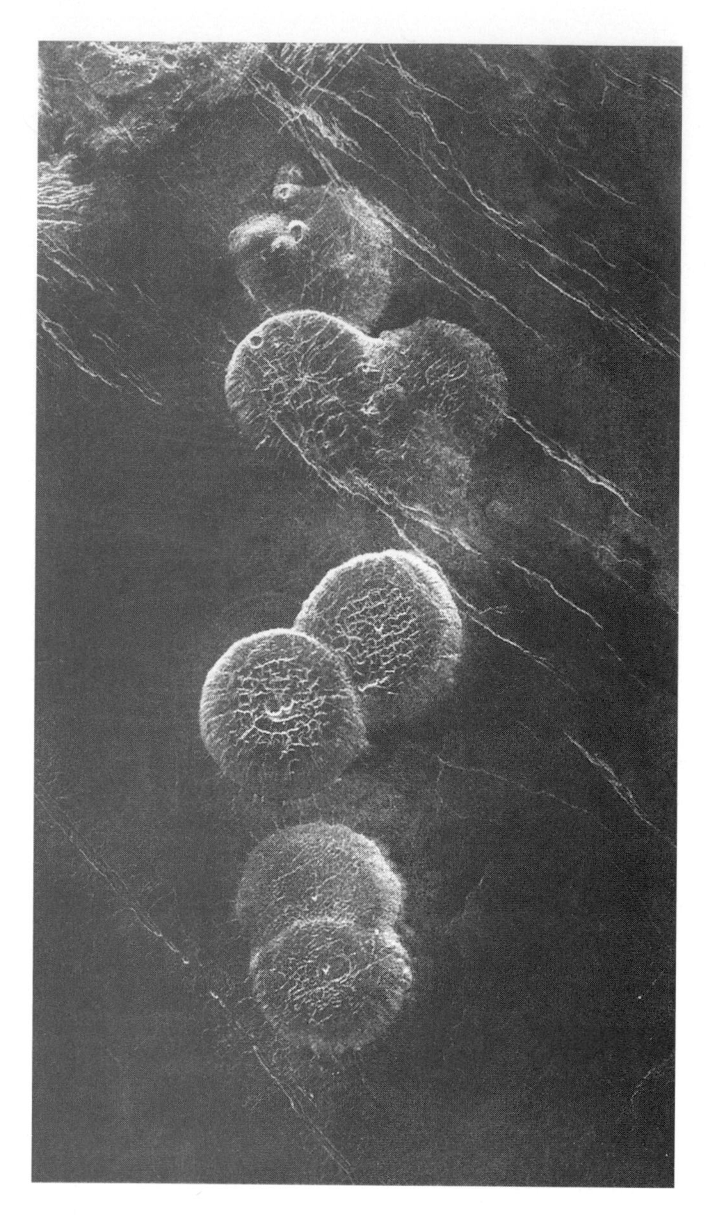

Figure 7.34 These unusual dome-shaped features apparently formed when very thick lava flowed over flat terrain on the hot surface of Venus like pancake batter on a hot griddle. They are about 15 miles in diameter and 2,500 feet high. *Courtesy of NASA.*

Ulysses to the Sun

The Solar System explorers discussed so far have all traveled in or near the *ecliptic,* the plane of the Earth's orbit projected to the celestial sphere. At the end of their missions, two Pioneer spacecraft and the two Voyagers were steered out of the ecliptic plane and are leaving the Solar System. (Refer to Figure 7.21.) No spacecraft had been launched out of the ecliptic plane toward the Sun until Ulysses, a spacecraft designed to fly over the poles of the Sun, uninhabited regions of space never before explored.

To leave the plane of Earth's orbit requires more energy than a booster rocket can provide. Therefore, Ulysses was sent in the direction of Jupiter, steered over Jupiter's northern hemisphere, and with a gravity assist from the giant planet, headed south below the ecliptic plane toward the south pole of the Sun. It was launched on October 6, 1990, reached Jupiter in February 1992, and passed beneath the Sun in mid-1994. In 1995 the spacecraft moved back up through the ecliptic plane and over the Sun's north pole.

Ulysses carries detectors to measure solar x-rays, gamma rays, solar wind, high and low energy particles, solar magnetic fields, interstellar gas and dust, and radio waves. Contrary to expectations, Ulysses found the intensity of the magnetic field at the solar poles to be similar to that at the solar equator; it did not find magnetic north and south poles. However, the spacecraft didn't come any closer than about twice the Earth-Sun distance so there may be a diffuse magnetic pole that does not show up at that distance. It found that in the polar regions above 30 degrees latitude the solar wind blows about 2 million miles per hour (3.2 million km/h), twice as fast as in the equatorial regions.

By 1998 it was back near Jupiter's orbit ready for another trip over both poles during the high part of the solar cycle.

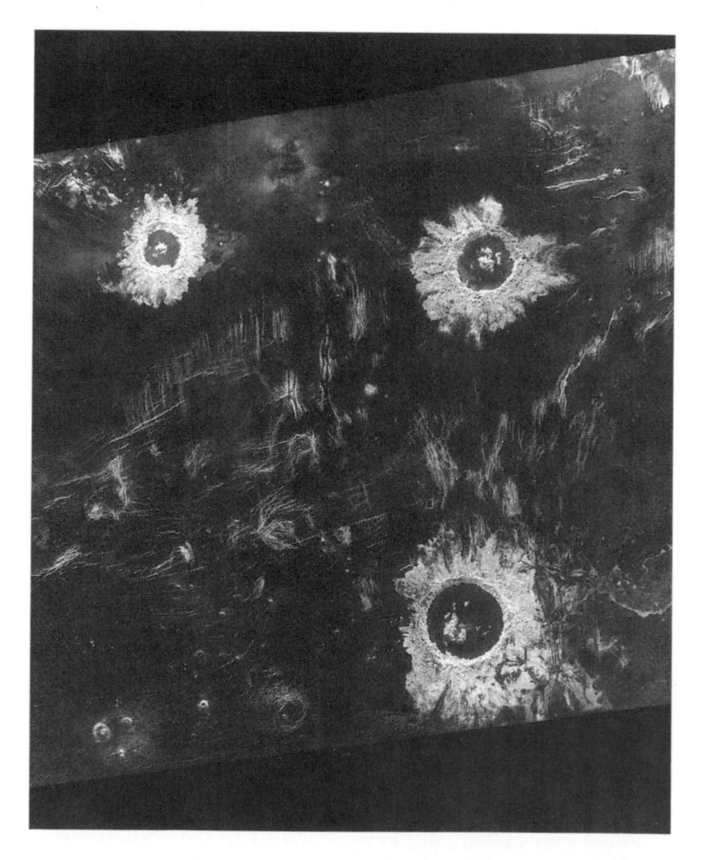

Figure 7.35 Three large impact craters on Venus. The smallest is 23 miles across, the largest is 31. Because of Venus's deep, dense atmosphere, the meteoroids that hit the surface must have been quite large or they would have burned up before reaching the ground. Several volcano vents are located in the lower left corner. *Courtesy of NASA/JPL.*

Ulysses was built and is operated by the European Space Agency. It was launched by Space Shuttle *Discovery*.

Pluto-Kuiper Express

The only planet left unvisited is Pluto and its moon Charon. Pluto's 248-year orbit is the most eccentric of any of the planets. Its distance from the Sun varies between 2.8 billion miles (4.5 billion km) at perihelion and 4.6 billion miles (7.4 billion km) at aphelion. In the 1990s it was near perihelion, closest to the Sun. The orbit has the greatest eccentricity of any of the planets, so great that at perihelion Pluto is closer to the Sun than Neptune is. It crossed inside Neptune's orbit in 1979, passed perihelion more than 60 million miles (100 million km) inside Neptune's orbit in 1989, and moved back out in March 1999. With an orbital period of 248 years, Pluto's next perihelion passage will occur in the year 2237. So the present time, when it is closest to the Sun and the voyage would be shortest, is the best time to start a spacecraft on its way to Pluto.

Planetary scientists believe that Pluto is probably very much like Triton. They are about the same size and density, and Pluto's atmosphere, as observed from Earth, is similar to Triton's. Being so far out on the edge of the Solar System, Pluto is a very cold place. Even at perihelion, the energy from the Sun is only about one-thousandth the intensity of solar energy at Earth. The atmosphere, composed of nitrogen and methane, is in gaseous state for only about 50 years when it is nearest perihelion. It will freeze up and fall to the ground as it moves away from perihelion. The 1990s would have been a good time to send a spacecraft.

Pluto has one icy moon, Charon, just half the size of the planet, and very close to it. Its circular orbit is about 12,000 miles (20,000 km) in radius, only about three times the radius of the Earth! Because of their comparable size and proximity, Pluto and Charon can be considered a "double-planet system." Charon orbits the planet in the same time that Pluto rotates on its axis, 6.4 days. Thus its orbit is "geosynchronous"; it hangs continuously in the same spot in Pluto's sky. Viewed from Pluto it would appear three times bigger than our Moon does from Earth and it would go through phases, new to crescent to full to crescent to new, every 6.4 days.

A mission to Pluto and Charon, named Pluto-Kuiper Express, is being planned for launch in December 2004, arriving at Pluto in December 2012 with a gravity assist from Jupiter in 2006. Still in its planning stages, the spacecraft will be very small and lightweight using miniaturized electronics. An artist's conception is shown in Figure 7.37. The dish antenna, the largest part of the vehicle, is 5 feet (1.5 m) in diameter. The entire craft with fuel for the attitude control thrusters may weigh as little as 45 pounds (20 kg). By launching this tiny craft on a high-powered booster it will have a higher initial speed and can make it to Pluto in about 8 years instead of a 30-year minimum energy Hohmann transfer orbit.

Figure 7.37 The tiny Pluto Fast Flyby spacecraft passing Pluto and its moon, Charon. The 1.5-meter dish antenna points toward Earth while the sensors point toward the planet. *Courtesy of NASA/JPL.*

The flyby past Pluto will last only a couple of hours. Data will be stored in memory and sent back to Earth at very slow speed during the following 6 months so that only a small transmitter and antenna need to be carried along. "Minimal weight" and "faster, better, cheaper" are the key phrases for planning this mission. To minimize the chance of failure, project planners recommend sending two separate spacecraft on two separate rockets.

If the encounter with Pluto is successful the mission could be extended to reach the Kuiper belt, a region of asteroid-sized objects orbiting the Sun beyond Pluto.

Beyond the Solar System

This is a truly remarkable time in history. With telescopes in orbit, astronomers are literally discovering the universe in the same sense as explorers in sailing ships discovered the Earth in the past few hundred years.

One of the problems that has always plagued astronomers is having to look through Earth's atmosphere. Light coming from the universe is refracted, that is, bent by the air. Refraction causes stars to twinkle; above the atmosphere, they shine with a steady light. Air currents and bubbles of warmer or cooler air cause the light waves to refract as they pass through. Because the air is in constant motion, the refraction varies from moment to moment and the image in the telescope appears to shimmer and dance around.

In addition, atmospheric gases and particulate matter cause light to scatter away from a straight line. Sunlight is a mixture of all colors. As it penetrates the air, the longer wavelengths pass straight through while the shorter wavelength blue light scatters in all directions. That is why the sky is blue. The blue part of the sunlight comes at us from all directions. We cannot see the stars during the daytime because the sky is too bright from the scattered sunlight. Scattering is also the reason that the night sky is often rather bright. Starlight, moonlight, and city lights scatter and reflect from the gas and dust in the air, giving the sky a background glow which interferes with astronomical observations.

Another problem is that the atmosphere selectively absorbs many wavelengths of electromagnetic energy so they never reach the ground. This was pointed out in Chapter 4 and is shown in Figure 7.38. Unfortunately for astronomers, but fortunately for the rest of us, ultraviolet, gamma rays, and x-rays are absorbed by the ionosphere and ozone layer. Certain infrared wavelengths are absorbed by molecules of water vapor and carbon dioxide. Some radio waves are absorbed by the ionosphere or reflected off its top. Some few infrared and ultraviolet observations are possible from the tops of mountains or from balloons above the densest part of the atmosphere. Until the coming of the space age, astronomers had to be content with looking at the sky primarily by visible light, radio waves, and certain wavelengths of infrared, "windows" in which the atmosphere is partially transparent.

A final problem: everything emits electromagnetic waves and the atmosphere is no exception. It emits infrared. This fact is a boon to meteorologists whose satellites can detect variations in the temperature of the ground and atmosphere from the infrared emissions. But this same fact is the bane of astronomers who would like to see the universe, not the atmosphere, with infrared sensors on the ground.

Before the days of spaceflight, astronomers used rockets and balloons to send their instruments above the atmosphere for short looks. Putting their observing equipment above the atmosphere allows astronomers to examine all wavelengths of electromagnetic energy coming from the universe. Many astronomy satellites were launched into orbit during the 1960s and 1970s. Uhuru discovered many x-ray sources in its 4-year lifetime. Explorer 53 discovered, among other things, stars that produce enormous bursts of x-rays every few hours. Two Orbiting Astronomical Observatories and the International Ultraviolet Explorer searched the universe for ultraviolet sources. Three High Energy Astronomy Observatories further studied x-ray sources. One of them searched out cosmic ray particles and gamma ray sources. Advanced spacecraft for detecting everything from gamma rays to microwaves are operating or are planned.

To emphasize the significance of observing the sky in several wavelengths, Figure 7.39 (Plate 20) shows the visible light, x-rays, radio waves, and infrared emissions from a single object. Notice the similarities and differences in the sources of these emissions. For this reason, astronomers are intent on examining objects in the sky at several different wavelengths. Each wavelength represents some particular physical phenomenon occuring in the object.

X-ray Telescopes

X-ray bursts are often observed coming from the Sun during a large solar flare. However, some early rocket experiments carried sensors above the atmosphere and found x-rays coming not only from the Sun, but from all directions in the sky. Earth seemed to be bathing in an x-ray bath. Further experiments on rockets, balloons, and finally on spacecraft showed that the sources include remnants of supernovas, clusters of stars, neutron stars, pulsars, quasars, superheated gas be-

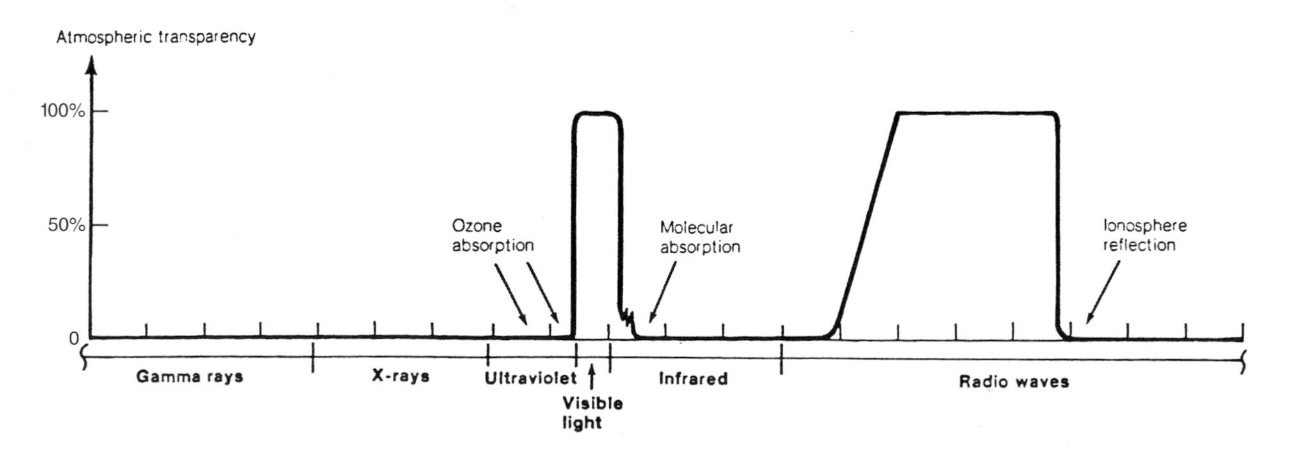

Figure 7.38 Transparency of Earth's atmosphere. *Courtesy of NASA.*

tween galaxies, and black holes. Detecting and studying x-rays can give a better insight into the workings of these objects.

Because x-rays penetrate most substances, the normal telescope design cannot be used for x-rays. They will, however, reflect from a surface if they strike it at a very shallow angle, a glancing blow, so to speak. See Figure 7.40. To accommodate this phenomenon, the reflecting surfaces are not flat, but are curved concentric rings made of fused quartz and coated with a thin layer of nickel.

Several x-ray telescopes have been launched into orbit. The most recent one is the Chandra X-ray Observatory which was carried by a Space Shuttle in July 1999 into a highly elliptical 64-hour orbit with apogee about a third of the way to the Moon. The x-ray picture of CasA, Figure 7.39b (Plate 20), one of its first images, shows the remnant of a supernova, an expanding shell of hot gases which may have a neutron star or black hole in the center.

The Chandra X-ray Observatory is one of three large astronomical spacecraft in Earth orbit, 45.3 feet (13.8 m) long and weighing 10,560 pounds (4784 kg) plus a ton of fuel for its attitude control thrusters. Its mirrors are the smoothest ever made. If the state of Colorado were as smooth, Pikes Peak would be less than an inch tall. The telescope resolves objects 10 times smaller and 100 times fainter than its predecessor x-ray telescopes. It could resolve the letters on a stop sign 12 miles away.

The European Space Agency's X-ray Multi-Mirror spacecraft (XMM) is also a high resolution x-ray satellite launched December 1999. Its orbit has a perigee at 4300 miles (7000 km) and apogee at 70,000 miles (114,000 km) with an inclination of 40 degrees. In this highly elongated 48-hour orbit it spends 40 hours above the Van Allen radiation belts to get clean data without interference from the radiation.

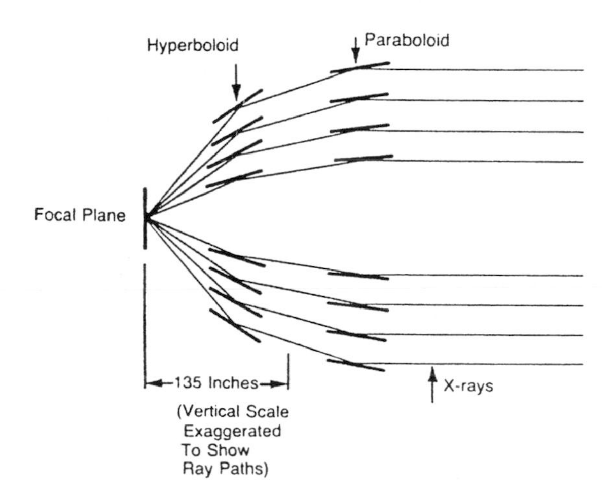

Figure 7.40 Diagram of an x-ray telescope. X-rays enter the telescope from the right, skip off two reflecting surfaces, and converge at the focal plane. "Paraboloid" and "hyperboloid" refer to the shape of the surface. The reflectors are actually rings; we see them edgewise in the diagram. The telescope is about 2 feet (0.6 m) in diameter at the entrance. *Courtesy of NASA.*

Gamma Ray Observatory

The Compton Gamma Ray Observatory is another of the large astronomical observatories in orbit. Named for a Nobel Prize winner, physicist Arthur Compton, it was carried to orbit by the Space Shuttle *Atlantis* in 1991. It transmits a continuous real-time data stream to astronomers listening on the ground. Gamma rays are emitted from supernovas, quasars, pulsars, and black holes, so the telescope detects some of the most violent events in the universe. Its discoveries are many. It found that gamma ray bursts are evenly distributed over the entire sky which indicates that they originate outside our Milky Way galaxy. Solar flares emit short bursts of low energy gamma rays and longer bursts of high energy gamma rays. Mutual destruction of matter and antimatter which releases gamma rays is concentrated in the center of the Milky Way. Gamma rays are given off by interactions of molecules as well as by the decay of radioactive atoms.

After one of its gyros failed, the Gamma Ray Observatory was deliberately deorbited on June 4, 2000; it landed in the Pacific Ocean southeast of Hawaii.

Infrared Astronomy from Orbit

One of the most successful of the astronomical satellites, both operationally and scientifically, was the Infrared Astronomical Satellite (IRAS). It was a joint undertaking with U.S., British, and Dutch scientists participating.

A diagram of the satellite is shown in Figure 7.41. It is 12 feet (3.7 m) long, 7 feet (2.1 m) in diameter and weighed 2365 pounds on Earth (a mass of 1075 kg). It uses a 22-inch (56-cm) diameter concave mirror, shaped like a shaving or makeup mirror, to focus the infrared waves. The mirror is made of beryllium, a very lightweight but strong metal. A secondary mirror reflects the waves to the focal plane where the 62 silicon arsenide and germanium gallium detectors are positioned. The detectors are sensitive to four wavelength bands from 119 micrometers to 8.5 micrometers, to observe objects as cold as 27 Fahrenheit degrees above absolute zero (15 K) to about 230°F (110°C). The short wavelength sensors detect hot sources such as stars, while the longer wavelength sensors observe cooler or extended objects such as dust clouds.

As we have pointed out several times before, everything at a temperature above absolute zero emits infrared, even the telescope itself. A cold object emits less infrared at a longer wavelength than a warm object. Therefore, to keep interference to a minimum, the telescope was kept cold with liquid helium. This cryogenic system kept the optics, the barrel of the telescope, and the sensors at a temperature just a few degrees above absolute zero. All the liquid helium needed for the lifetime of the spacecraft had to be put aboard before launch. The

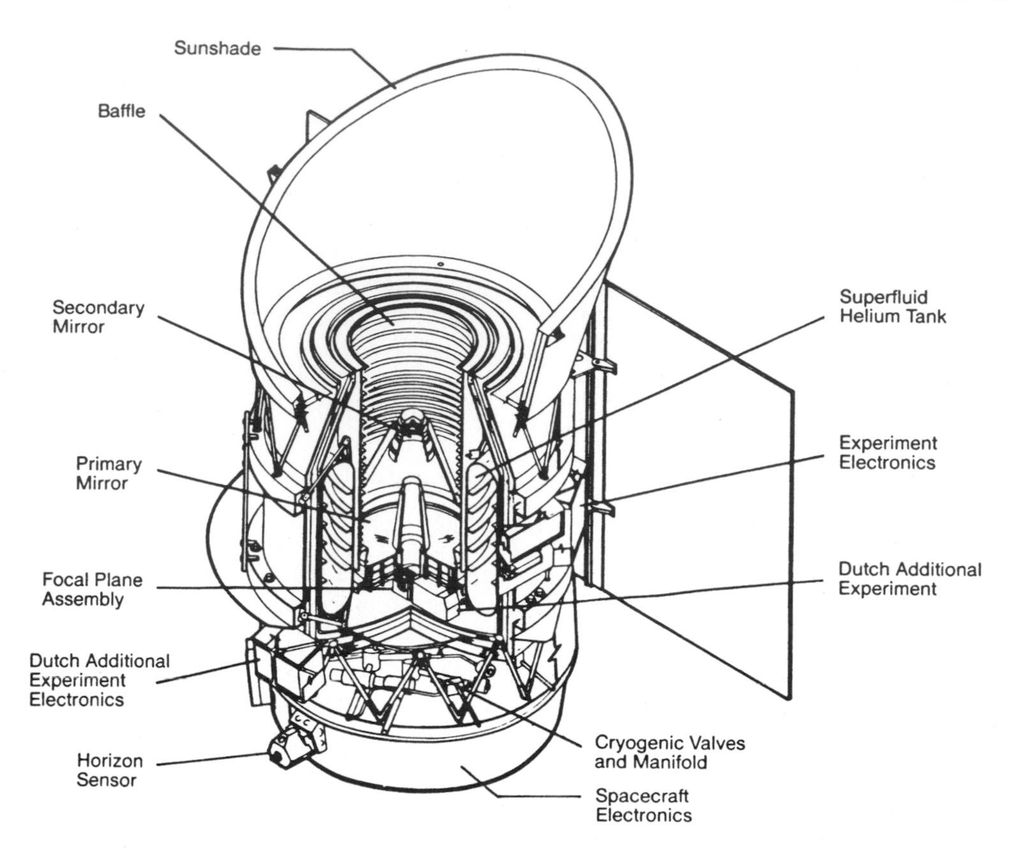

Figure 7.41 Anatomy of the Infrared Astronomical Satellite (IRAS). *Courtesy of NASA.*

125 gallons (473 l) carried in a "thermos jug" was expected to be a 6 months supply, but it evaporated more slowly than expected and the satellite continued to operate for 10 months.

IRAS was launched on January 25, 1983, from Vandenberg Air Force Base California, into a 560-mile (900-km) polar orbit with a period of 100 minutes. That altitude is well above the atmosphere, but below the Van Allen belts except in the South Atlantic Anomaly region. There the data become contaminated by many false readings from protons in the radiation belts.

The Sun-synchronous orbit was oriented over the *terminator,* the sunrise-sunset line between daylight and darkness on Earth's surface. The orbit rotated about 1 degree per day so that as the Earth moved around the Sun the satellite remained over the terminator. It was important to keep the orbit oriented in that direction because the spacecraft had to point more than 60 degrees away from the Sun and more than 88 degrees from the Earth's horizon so the sensors would not pick up the Sun or Earth. A sunshade and baffle aided in keeping stray infrared from reaching the sensors.

The spacecraft scanned a 0.5 degree wide strip on each orbit which overlapped the previous strip by 0.25 of a degree. It did a complete survey of 95 percent of the sky. IRAS catalogued 250,000 individual sources within the limits of sensitivity of its detectors. Most of them are stars in our Milky Way galaxy, some surrounded by cooler material. The nucleus of the Milky Way is readily discernible in the data, giving further proof that we live in a spiral galaxy. At least 10,000 galaxies outside the Milky Way were also observed. A few sources were seen to move and therefore had to be nearby. Some turned out to be asteroids; five were comets. A ring of dust particles was discovered around the star Vega, the first evidence of cool solid matter around a star other than the Sun. Some newly born stars were also seen, hot bodies that had not turned on as stars yet.

IRAS did a complete all-sky survey for infrared sources. See Figure 7.42 (Plate 21). The European Space Agency's Infrared Space Observatory (ISO) operated from November 1995 to May 1998. It was designed with high resolution to study individual objects. It was capable of detecting infrared sources as faint as an ice cube 60 miles (100 km) away, and was able to lock onto objects the size of a person as seen from 60 miles (100 km). With more than 550 gallons (2080 l) of liquid helium, ISO was expected to have a 20-month lifetime. By careful control of consumption it operated for 28 months. To avoid interference to the detectors from trapped protons and electrons in the Van Allen radiation belts, the spacecraft was inserted into a highly elliptical 24-hour orbit with perigee at 62 miles (1000 km) and apogee at nearly 44,000 miles (70,500 km). It spent about 17 hours a day near apogee, outside the radiation belts where the detectors could operate at their highest sensitivity.

Radio Wave Astronomy from Orbit

The Cosmic Background Explorer (COBE) searched for microwave noise left over from the Big Bang. The existence of such radiation was predicted more than 40 years ago and was first detected on the Earth as a hiss of radio noise in 1965. The COBE spacecraft carried a radio receiver to listen for noise from space in a broad range of wavelengths. According to theory, the noise should be most intense at a wavelength of about 2 mm and be less intense at both longer and shorter wavelengths. COBE found exactly that. In addition, COBE made maps of the microwaves over the entire sky. After known sources of microwaves, such as the Sun and Earth, were removed, the maps were extremely smooth. No evidence of early galaxies was found and the origin of galaxies is still uncertain.

The Submillimeter Wave Astronomy Satellite (SWAS) records data in the wavelength region between infrared and microwaves. These emissions come from gaseous clouds where new stars are forming, especially from molecules of water, oxygen, carbon monoxide, and carbon dioxide. SWAS was launched in December 1998 and was still operating at the beginning of the new millennium.

Hubble Space Telescope (HST)

This Space Telescope is named after Edwin Hubble, an American astronomer whose research on galaxies early in the 20th century led to the realization that the universe is expanding. The telescope, shown in Figure 7.43, is 43 feet (13 m) long and on Earth weighs about 25,000 pounds (mass of about 11,400 kg). It can observe for longer periods of time unhindered by clouds and scattered sunlight, up to 4500

Figure 7.43 Artist's conception of the Hubble Space Telescope in orbit. *Courtesy of NASA.*

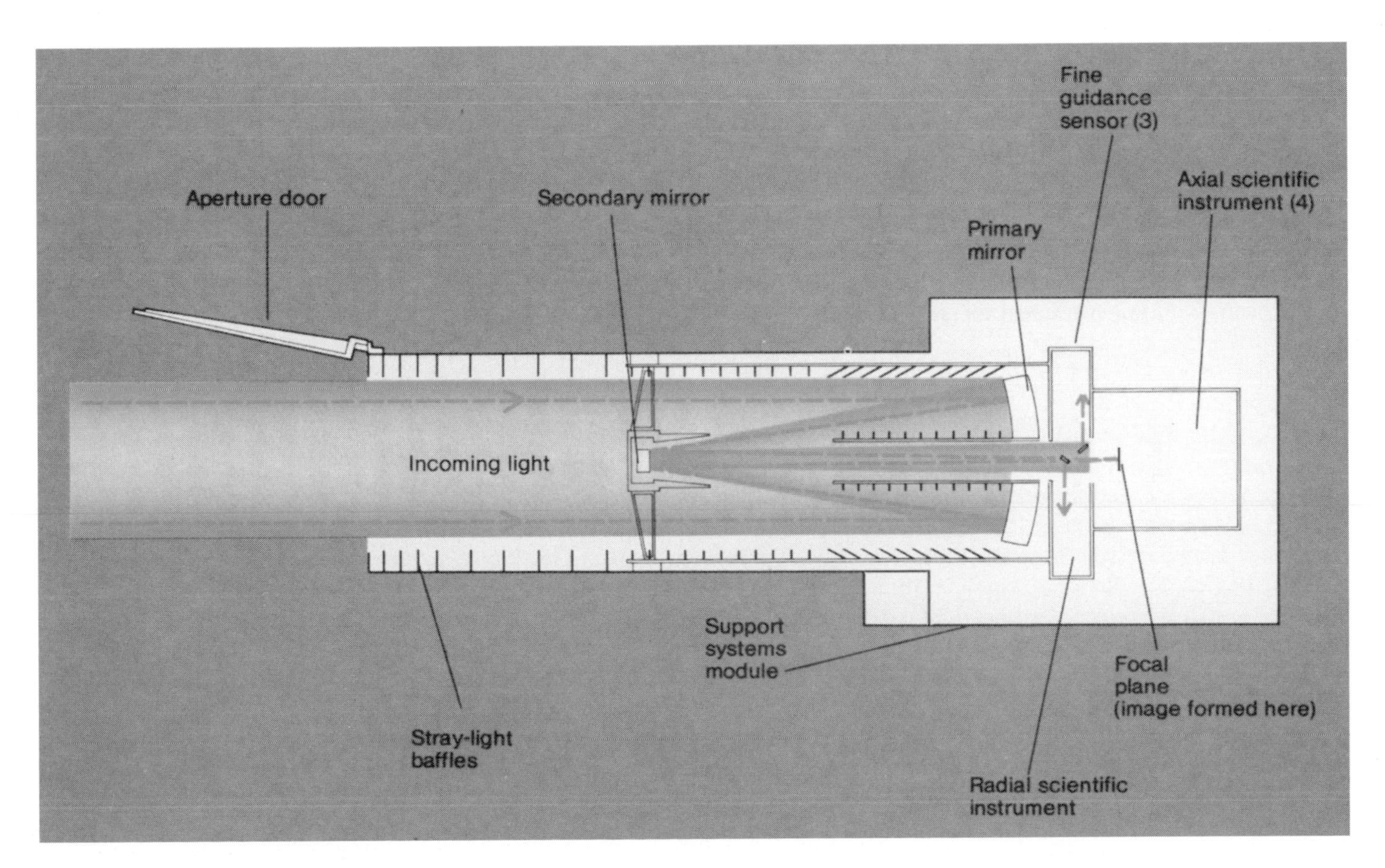

Figure 7.44 Light path through the Hubble Space Telescope. *Courtesy of NASA.*

hours per year compared with perhaps 2000 hours per year at the best ground observatories.

The working end of the telescope is an 8-foot (2.5-m) diameter mirror which brings electromagnetic waves from space into focus. The size of this mirror makes it one of the largest instruments ever built. A cutaway view in Figure 7.44 shows the light path through the instrument. The telescope is a Cassegrain design, a common design for Earth-based telescopes. The aperture door is opened and light from the object under study enters. A system of baffles keeps stray light out. The light rays reflect from the large primary mirror to a 12-inch (30-cm) secondary mirror mounted in the center of the tube, 16 feet (5 m) from the primary. The light beam is reflected back from secondary mirror and it passes through a 24-inch (30-cm) hole in the center of primary mirror to the instruments enclosed in the aft shroud.

Star trackers and fine guidance sensors keep the telescope pointed with unprecedented precision. It could remain locked onto a dime at 450 miles (725 km). Five instruments for analysis of the light are mounted behind the primary mirror. Two spectrographs break up white light into its rainbow of colors, two cameras are for wide angle and planetary photography and for faint object photography, and a photometer measures light intensity. The resolution of the HST is ten times better than ground-based instruments.

The instrument can detect objects only one-fiftieth as bright as can be seen from the ground. Thus it is able to see objects seven times farther away than ground-based telescopes. (The square root of 50 is about 7. See MATHBOX 7.1.) Ground-based telescopes can detect objects 2 billion light years away; HST can observe objects as far away as 14 billion light years. Since the universe is thought to be about 15 billion years old, the objects seen by Space Telescope are observed as they were at the beginning of time, shortly after the Big Bang. This expands the observable volume of space to 350 times that which is now observable. (Seven cubed is about 350. See MATHBOX 7.2.)

The Hubble Telescope is powered by six nickel-cadmium (Nicad) batteries kept charged by two arrays of solar cells, each providing up to 2000 watts of power when the instrument is in the sunlight. This operates much like an automobile electrical system where the battery is kept charged by an

MATHBOX 7.1

Light Intensity

Light waves expand outward from their source spherically in all directions. The area of a sphere is $a = 4\pi R^2$ where R is the sphere's radius.

Refer to Figure 7.1.1. Sphere A has a radius of 1 mile and an area of 4π. Sphere B has a radius of 3 miles and an area of 36π.

Although the radius of sphere B is three times as large, its area is nine times larger: $3^2 = 9$.

As it expands from sphere A to sphere B, light energy from a source at the center would spread out over nine times the area and would therefore be only one-ninth as intense by the time it reached sphere B.

Thus, light intensity decreases as the square of the distance from the source. If a light source is 7 times farther away it is only 1/49 as bright. If two lights are actually of the same intensity but one appears 49 times fainter, then it must be 7 times farther away.

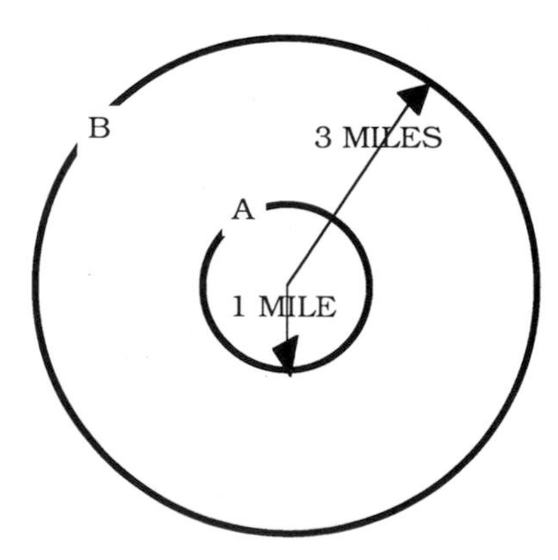

Figure 7.1.1 Spherical expansion.

MATHBOX 7.2

Volumes of Spheres

The volume of a sphere is $V = 4\pi R^3/3$ where R is the radius of the sphere.

Refer to Figure 7.2.1.

Sphere A has a radius of 1 mile and a volume of $4\pi/3$. Sphere B has a radius of 2 miles and a volume of $32\pi/3$.

Although the radius of sphere B is twice as large, its volume is eight times larger: $2^3 = 8$.

If the Space Telescope can see objects 49 times fainter than are now visible from Earth, then it can see seven times farther and can examine a volume of space $7^3 = 343$ times greater.

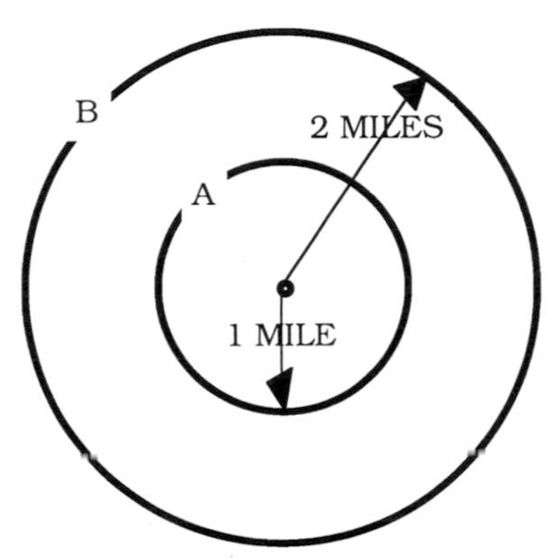

Figure 7.2.1 Spherical volume.

alternator attached to the engine. The batteries charge when the telescope is in sunlight and provide power to the equipment when it is on the dark side of Earth. This charge-discharge cycle recurs 16 times each day as the telescope orbits Earth every 90 minutes.

HST is remotely controlled from the ground. Earthbound astronomers at the Space Telescope Science Institute at The Johns Hopkins University in Baltimore evaluate proposed research projects from astronomers worldwide and set the schedule for observing time.

In April 1990, the Hubble Space Telescope was carried by the Space Shuttle to a 320-mile (515-km) circular orbit with an inclination of 28.5 degrees. As with any other satellite, atmospheric drag gradually lowers the orbit. When it gets down below 300 miles (about 480 km) the pointing accuracy is degraded by the air drag and it has to be boosted to a higher orbit again every so often, especially during the peak of a solar cycle when the atmosphere is at its greatest expansion.

Almost immediately after launch, operators discovered that the primary mirror had been ground to the wrong specifications. For a light ray to come to a sharp focus, the mirror must be curved in the shape of a paraboloid. If it is curved like a sphere, the light rays reflecting from the mirror do not all converge to the same point and the image is fuzzy. This defect, known to every amateur astronomer who has tried to grind a mirror, is called spherical aberration. The HST mirror had been ground to a spherical shape. For an 8-foot (2.5-m) mirror, there is not much difference between a spherical surface and a parabolic one. The edge was too flat by less than the width of a human hair.

What could be done? Flawed optical devices can generally be fixed by adding another optical device, just as human eyes are corrected with glasses or contact lenses. A very small mirror ground to correct the spherical aberration could be placed in the light path of the HST. The telescope had been designed to be serviced and repaired in orbit by astronauts. In December 1993, the Shuttle *Endeavour* came within 30 feet (9 m) of the orbiting telescope and hauled it into the cargo bay. In five space walks, astronauts inserted the corrective optics, replaced jittering solar panels, added some computer memory, replaced two pairs of faulty gyros, and replaced the magnetometers with improved ones. *Endeavour* then raised it to a higher orbit and released it. The corrective optics did the trick. See Figure 7.45 (Plate 21) for a before-after comparison.

Some of the many impressive pictures sent down to astronomers on Earth are shown in Figures 7.46, 7.47, and 7.48 (Plates 22, 23).

DISCUSSION QUESTIONS

1. Referring to Figure 7.13, what gravitational forces are acting on Voyager at each point along the trajectory?
2. Would it be better to build a few very expensive, very capable astronomy spacecraft or many smaller less capable ones? What are the arguments for each?
3. Why is there so much interest in astronomy? There doesn't seem to be much practical use for the information.
4. Why was IRAS oriented carefully so it was never pointed toward Earth or the Sun?
5. What limits the useful life of an infrared space telescope?

ADDITIONAL READING

Aichele, Jean H., ed. *Galileo, The Tour Guide*. Jet Propulsion Laboratory publication JPL D-13554, June 1996. Thorough, well-illustrated, readable summary of the mission.

Bowyer, Stuart. "Extreme Ultraviolet Astronomy." *Scientific American,* August 1994. Results from the EUV Explorer spacecraft.

Burrows, William E. *Exploring Space, Voyages in the Solar System and Beyond*. Random House, 1990. The human side as well as the technical side of space exploration.

Davies, John K. *Satellite Astronomy*. Ellis Horwood Ltd., 1988. Technical but readable.

DiChristina, Mariette. "Telescope Tune-Up." *Popular Science,* September 1999. Easy reading article about the third Hubble Space Telescope repair and maintenance mission.

Dunne, James A. *The Voyage of Mariner 10*. NASA SP-424. Government Printing Office, 1978. Detailed results of mission to Venus and Mercury.

Fimmel, Richard O., et al. *Pioneer Venus*. NASA SP-461, Government Printing Office, 1983. Very complete document of results

Giacconi, Riccardo. "The Einstein X-Ray Observatory." *Scientific American,* February 1980. Results from HEAO-2 spacecraft.

Golombek, Matthew P. "The Mars Pathfinder Mission." *Scientific American,* July 1998. Results of the lander and Sojourner rover miss

Habing, Harm J., and Gerry Neugebauer. "The Infrared Sky." *Scientific American,* November 1984. IRAS results.

Hart, Douglas. *The Encyclopedia of Soviet Spacecraft*. Exeter Books, 1987. Comprehensive and well illustrated.

Johnson, Torrence V. "The Galileo Mission." *Scientific American,* December 1995. Description of the mission, written before it reached Jupiter.

Kohlhase, Charles. *The Voyager Neptune Travel Guide*. JPL Publication 89–24, Government Printing Office, 1989. Detailed description of Voyager spacecraft and their mission in anticipation of Voyager 2's encounter with Neptune.

Laeser, Richard P., et al, "Engineering Voyager 2's Encounter with Uranus." *Scientific American,* November 1986. How spacecraft problems were solved.

Luhmann, Janet G., James B. Pollack, and Lawrence Dolin. "The Pioneer Mission to Venus." *Scientific American,* April 1994. The 14-year history of the spacecraft.

Saxena, S. K., ed. *Chemistry and Physics of Terrestrial Planets*. Springer-Verlag, 1986. Prior understanding of chemistry, physics and geology is necessary.

Yenne, Bill, et al. *Interplanetary Spacecraft*. Exeter Books, 1988. Descriptions of U.S. spacecraft, components and subsystems, trajectories and communications; well illustrated and readable.

PERIODICALS

Aerospace America. A monthly magazine of the American Institute of Aeronautics and Astronautics. Technical, but readable.

Astronomy. A popular monthly magazine with frequent articles about astronomy from space.

The Planetary Report. Bimonthly popular magazine published by the Planetary Society for its members.

Sky and Telescope. Monthly magazine for amateur astronomers with frequent articles on spacecraft.

Space News. A weekly newspaper devoted to the business, politics, and technology of space activities.

NOTES

Chapter 8

Space Shuttle

The Space Shuttle is one of the most complex machines ever built. It is the first reusable vehicle for carrying people and cargo into orbit and returning them to Earth. As part of the Space Transportation System, the Shuttle takes off vertically like a rocket, orbits Earth as a spacecraft, and lands like an airplane. It is the outgrowth and merging of two technologies: the Mercury-Gemini-Apollo "man-in-the-can" approach to spaceflight combined with the experimental rocket-powered aircraft of the 1950s. Instead of a capsule sitting atop a heavy-lift booster rocket, the STS approach has a rocket-powered airplane, the orbiter, attached to a large external fuel tank with two solid propellant booster rockets attached to it, as shown in Figure 8.1. All except the external tank are recovered, reconditioned, and reused. The orbiter was originally designed for 100 missions; the solid rocket boosters were designed for 20 flights.

The Space Shuttle is the backbone and workhorse of the U.S. manned space program. It was expected to be a low-cost bus-truck service to space for 7 to 14 day missions. It can carry satellites into orbit and drop them off; pick up satellites and repair them on site or return them to Earth; conduct scientific, technological, and industrial research; bus people to or from a space station; or act as a platform for construction of large objects in space such as a space station . All of these tasks have been done in more than 100 flights. They are cataloged in the Appendix.

The Shuttle carries up to seven astronauts. It was designed to open spaceflight to men and women who do not have the test pilot background and training that previous astronauts had. Any qualified person in reasonably good health with the need to go into space for scientific or commercial ventures could do so.

While maximum altitude for a Shuttle orbit is nearly 700 miles (1100 km), most missions are conducted at lower altitudes. If a satellite must be delivered to a higher orbit, it is equipped with an upper stage booster rocket. When the Shuttle reaches orbit, the satellite and its upper stage are released from the cargo bay, checked out, and left behind. At the proper time a command from the ground ignites the upper stage and the satellite is carried to its higher orbit, often to geosynchronous altitude.

A robotic mechanical arm, designed and built in Canada, is used to move things in and out of the cargo bay and to assist in other work. Astronauts in space suits leave the cabin

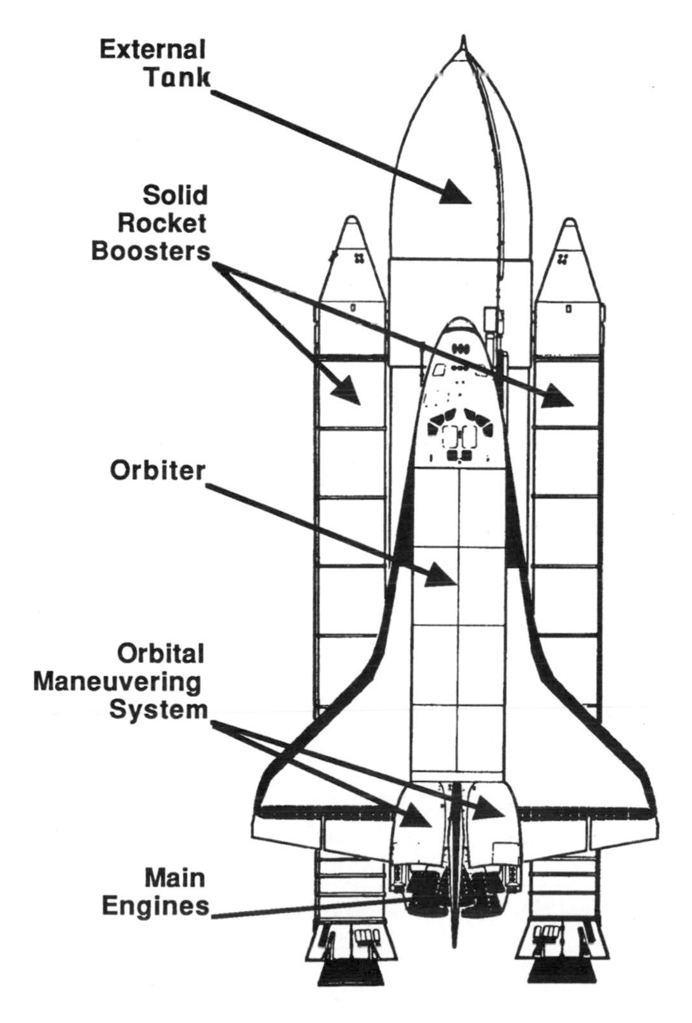

Figure 8.1 The Space Shuttle. The orbiter is attached to the external tank, flanked by two solid propellant rocket boosters. *Modified NASA artwork.*

through an airlock to work outside. A rocket-powered backpack called the manned maneuvering unit (MMU) is used to fly away from the Shuttle to repair satellites or other jobs. Spacelab is a laboratory which is carried in the cargo bay to provide a shirtsleeve environment for scientific research. All of this and more will be discussed in detail in Chapter 10, Working in Space. This chapter is concerned with the structure and flight profile of the Shuttle.

The Space Shuttle has three major components: the *orbiter,* a pair of *solid rocket boosters* (SRBs), and the *external tank* (ET) as shown in Figure 8.1.

Solid Rocket Boosters

The two solid propellant booster rockets burn for about 2 minutes, providing most of the initial thrust, along with three engines in the orbiter, to lift the 4.5 million pound (20 million newton) vehicle from the launch pad and boost it to an altitude of about 28 miles (45 km). While resting on the launch pad, the entire weight of the vehicle is supported by the bottom skirt of the two booster rockets.

The solid rocket boosters (SRBs) of the Space Shuttle are the largest ever flown, the first designed for reuse, and the first to power a manned spacecraft. Figures 8.2 and 8.3 show the SRB structure and Table 8.1 lists SRB statistics. Overall each booster is nearly 150 feet (46 m) tall and over 12 feet (3.7 m) in diameter, constructed of half-inch thick steel. Because of their size, it is not possible to manufacture the motors and ship them to Kennedy Space Center in one piece. They are made in eleven pieces at a factory in Utah, then partially assembled, filled with propellant, and shipped in four segments on special flatbed railway cars to Florida for final assembly. Figure 2.17 shows how the four segments are fitted together.

The propellant is composed of aluminum powder fuel, ammonium perchlorate oxidizer, a polymer binder, an epoxy curing agent, and iron oxide catalyst. When hardened into a grain it looks and feels like a hard rubber typewriter eraser. The perforation, the hollow core that runs up the center of the grain, is an 11-point star in cross section. (Refer back to Chapter 2 for a discussion of perforations.) To ignite the propellant, a small rocket motor is fixed at the upper end of the perforation. Its flames ignite the entire exposed inner surface of the perforation and the booster comes up to full thrust in less than half a second.

Each booster contains 1.1 million pounds (500,000 kg) of propellant and develops a thrust at liftoff of 2.9 million pounds (12.9 million N). As the fuel burns, the weight of the booster decreases. As it rises into less dense atmosphere, the drag decreases, and as it leaves the surface of the Earth the gravitational force decreases. All these contribute to an increase in acceleration. The Shuttle is designed to withstand accelerations of up to 3 g. Therefore, grains are cast in such a way as to produce a regressive burn 55 seconds into the flight to keep the acceleration from exceeding the 3 g design limit. This also keeps stress on the astronauts and the payload within tolerable limits. Figure 8.3 shows the grain structure.

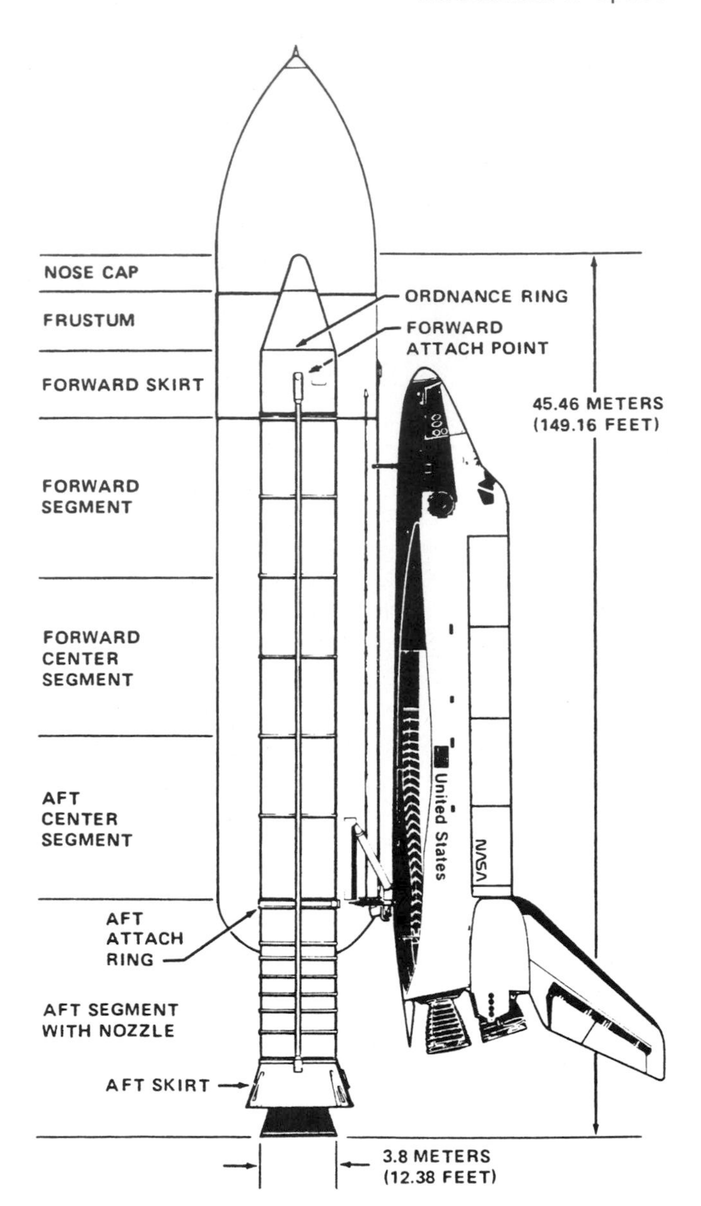

Figure 8.2 Details of the solid rocket booster. *Courtesy of NASA.*

The exhaust nozzle can be swiveled up to 8 degrees from the central axis of the booster to change the direction of thrust and help steer the Shuttle during ascent. Each booster has its own auxiliary power units that operate the hydraulic pumps to steer the nozzle. Guidance system sensors inform the control computer of the speed and direction of motion of the vehicle. The computer signals the booster when a change in direction of the nozzle is required. Electric power is supplied to the booster through a cable from the orbiter's electric system.

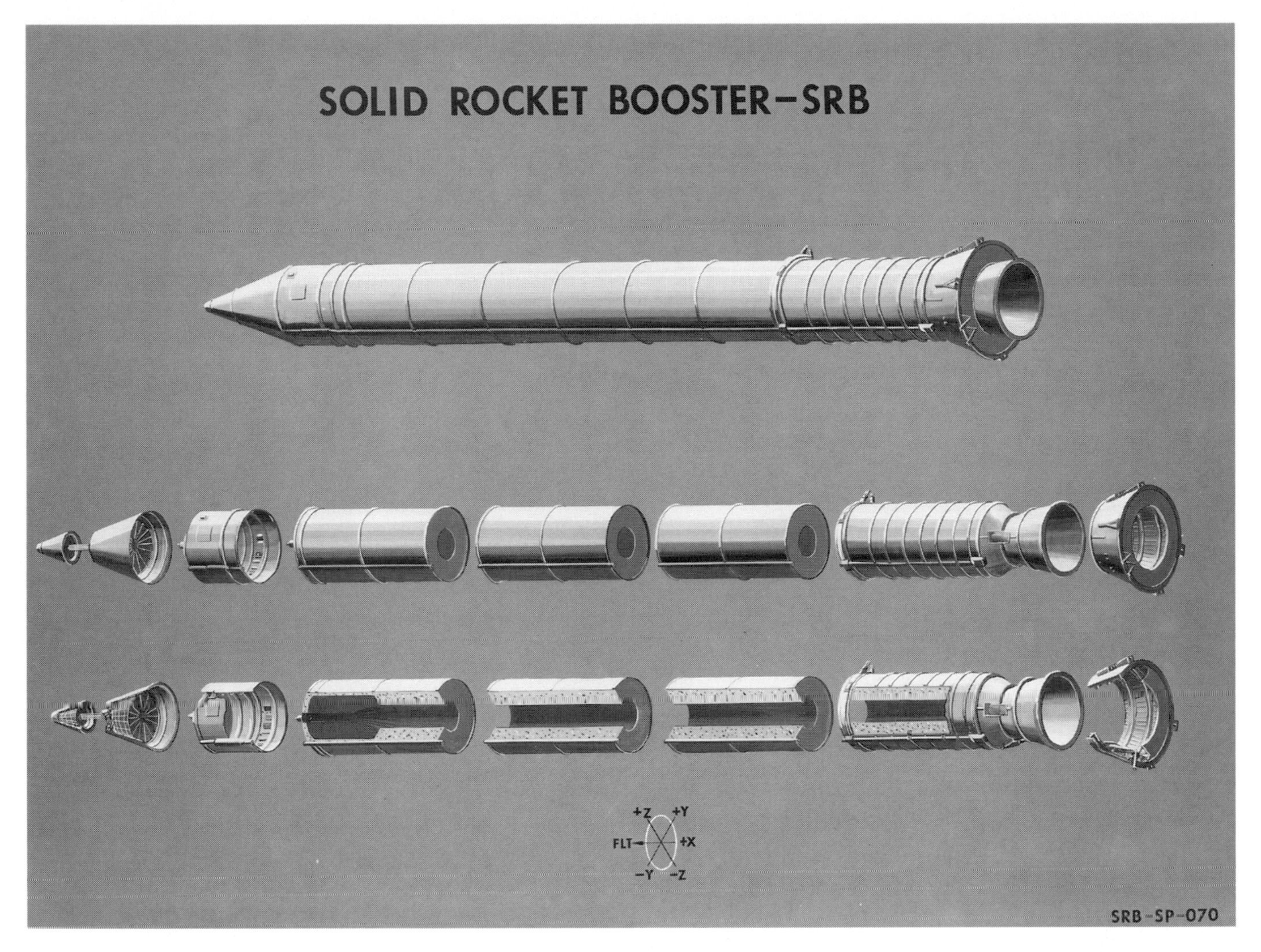

Figure 8.3 Grain design of the solid rocket booster. Four segments contain propellant. The bottom (right) propellant grain is slightly tapered and flared out at the end with the nozzle attached; exhaust gases from the entire engine must escape through this section. The two center segments have cylindrical grains which produce increasing thrust with time, as the burning surface enlarges. The upper (left) half of the topmost section contains less propellant and has slots like a star-shaped grain. This segment burns very rapidly, contributing to the high thrust needed at liftoff but burns out quickly so the maximum desired thrust level is not exceeded. The thrust-time graph is also shown. *Courtesy of NASA.*

TABLE 8.1 Solid Rocket Booster Statistics

Overall dimensions	
Height	149 feet (45.5 m)
Diameter	12.2 feet (3.72 m)
Propellant composition	
Aluminum powder (fuel)	16.0 percent
Ammonium perchlorate (oxydizer)	69.6 percent
Iron oxide (catalyst)	0.4 percent
Polymers (binder)	14.0 percent
Weight (mass) each booster	
Empty	188,400 pounds (85,600 kg)
Propellant	1,112,000 pounds (505,000 kg)
Total	1,300,000 pounds (591,000 kg)
Thrust at liftoff	
Each booster	2,900,000 pounds (12,900,000 N)

The Orbiter

The *orbiter* is the airplane-rocket ship shown in Figure 8.4 and its dimensions are given in Table 8.2. It is about the size of a DC-9 jet airliner. The main structure of the orbiter is constructed of aluminum, similar to the way an airliner is built. There are three main sections: the forward fuselage with the crew compartment; the midsection which includes the payload bay and its clamshell doors; and the aft fuselage which includes the engines, pods, and vertical tail. Empty it weighs about 175,000 pounds (mass of 80,000 kg).

Five orbiters have been built. *Columbia* was the first, delivered to NASA in March 1979. It was used for the first four orbital test flights and has been modified and modernized several times as new technology developed. *Discovery* and *Atlantis* were delivered in 1983 and 1985, respectively. The

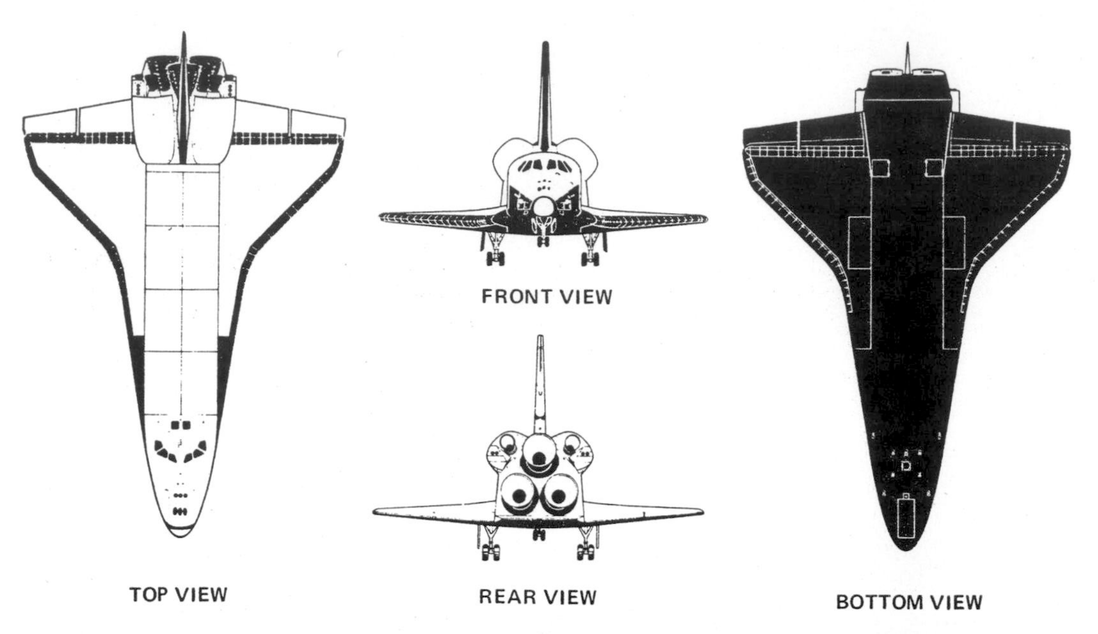

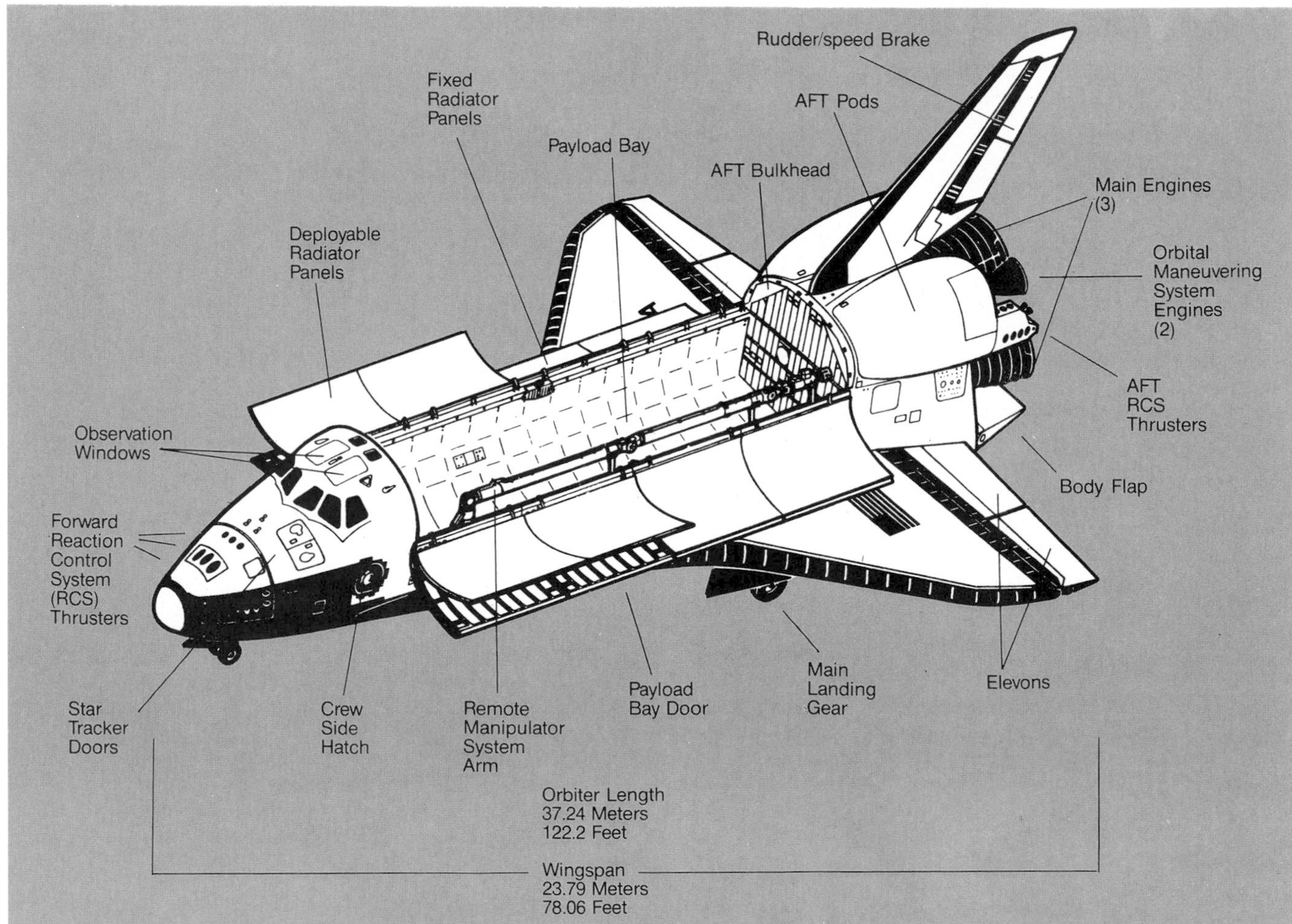

Figure 8.4 The Space Shuttle orbiter. *Courtesy of NASA.*

TABLE 8.2 Orbiter Dimensions

Total length	122.2 feet (37.2 m)
Height	56.6 feet (17.3 m)
Rudder height	26.3 feet (8.0 m)
Wing	
Span	78.1 feet (23.8 m)
Maximum thickness	5.0 feet (1.5 m)
Body flap	
Width	20.0 feet (6.1 m)
Area	135.8 square feet (12.6 m^2)
Aft fuselage	
Length	18.0 feet (5.5 m)
Width	22.0 feet (6.7 m)
Mid fuselage	
Length	60.0 feet (18.3 m)
Width	17.0 feet (5.2 m)
Main wheel span	22.7 feet (6.9 m)
Crew cabin	2525 cubic feet (71.5 m^3)
Payload bay	
Length	60.0 feet (18.3 m)
Diameter	15.0 feet (4.6 m)

newest orbiter is *Endeavour,* built as a replacement for *Challenger* which was destroyed in an accident in January 1986. Each orbiter is thoroughly inspected, refurbished, and modified about every 3 years. Modifications have included a docking port and airlock changes to allow docking with the Russian Mir space station and the new International Space Station (ISS).

Payloads are carried in the huge cargo bay. The maximum payload weight depends on the direction of launch. Eastward launches take place from Kennedy Space Center in Florida. Polar launches were planned from Vandenberg Air Force Base, California, but no launches ever took place from there and after the *Challenger* accident in 1986, the facility was closed. Polar launches do not take place from Florida because the spacecraft would overfly land areas while under power. Because of the rotation of the Earth, Cape Canaveral is moving eastward with respect to space at a speed of 914 miles per hour (1472 km/h). Given that initial speed while still on the launch platform, the Shuttle can carry a payload of 65,000 pounds (about 30,000 kg) into low Earth orbit on an eastward launch from Kennedy Space Center. A maximum of 39,700 pounds (18,000 kg) could be carried into polar orbit because the initial eastward velocity of Earth's rotation does not help in a polar launch. Up to 32,000 pounds (14,500 kg) of cargo can be returned from orbit to Earth.

The cargo bay is not pressurized. Its doors open like a clamshell to expose the orbiter's cooling radiators (Figure 8.4). The front radiators tilt so that heat may escape from both sides. It is absolutely essential that the doors be opened immediately on achieving orbit and that the radiators be exposed to space. Otherwise the orbiter would overheat within a few orbits and the mission would have to be aborted. On the other hand, the doors must be closed for reentry or air friction would quickly destroy them and probably the entire cargo bay. If the automatic door opener fails, a manual opener can be used.

Engines

The orbiter is powered by 49 engines used in various combinations to launch to orbit, maneuver while in orbit, and return to Earth. Its three main engines are the most advanced liquid-fueled rockets ever built. Figure 8.5 is an exploded view of the main engine. Propellants are liquid hydrogen and liquid oxygen carried in the external tank. Each develops 393,800 pounds (1,750,000 N) of thrust at liftoff. They can be throttled from 65 percent to 109 percent of rated thrust by varying the flow of the fuel from the tanks, just as you operate the accelerator in your car to change the engine speed.

Rockwell International, who built the main engines, presents the following incredible facts. The fuel pumps are not much larger than an automobile engine, yet they generate as much horsepower as 28 locomotives, 100 horsepower for each pound of weight. By comparison, an automobile engine generates about 1/2 horsepower for each pound of its weight. Output pressure from the pumps could send a column of liquid hydrogen 180,000 feet (55,000 m) in the air. The fuel pump runs at 37,000 revolutions per minute; a typical automobile engine runs at 2500 revolutions per minute when going 60 miles an hour (100 km/h). Liquid hydrogen at −423°F (−253°C) is the second coldest liquid on Earth (liquid helium is colder). When hydrogen burns in the combustion chamber, the temperature reaches 6000°F (3300°C), higher than the boiling point of iron. The three engines empty the half-million gallon (2 million liter) external tank in 8.5 minutes.

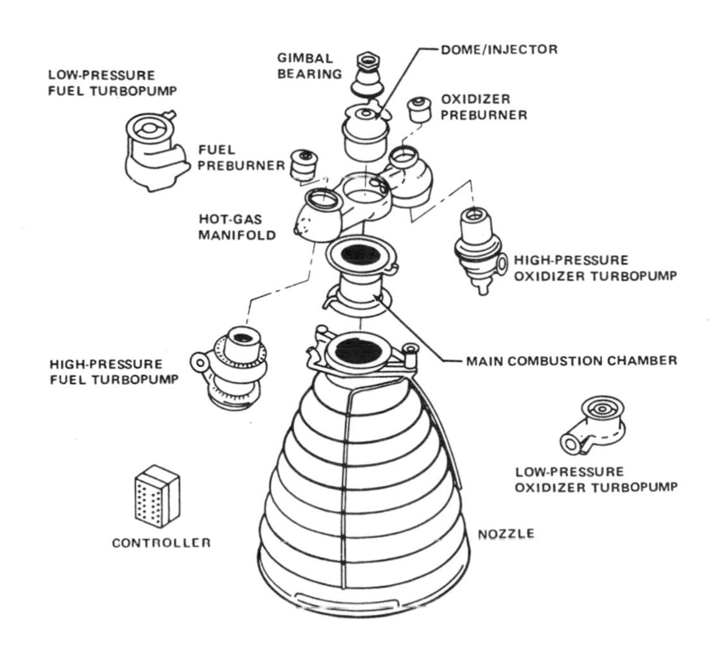

Figure 8.5 Component parts of the Space Shuttle main engine. *Courtesy of NASA.*

The orbiter engines are mounted in such a way that they can be gimbaled, moved to change the direction of the thrust. In conjunction with the solid rocket boosters, this provides the means for steering the Shuttle during powered flight.

Orbiter Thermal Protection

When returning to Earth, the orbiter must reduce its speed from 18,000 miles per hour (29,000 km/h) in orbit to a 200 mile per hour (320 km/h) landing speed in about half an hour. Its kinetic energy of motion must be dissipated in some way. When an automobile is brought to a stop by applying the brakes, its kinetic energy is converted to heat in the brake lining. Similarly, the Shuttle reenters with the large, flat bottom surface leading the way so that the atmospheric frictional drag acts like a brake. During reentry into the atmosphere, the frictional heat produces temperatures on the orbiter varying from 600°F (315°C) to 2750°F (1510°C). Without a protective covering of insulation, the aluminum would melt. Ablative heat shields were used on the Mercury, Gemini, and Apollo spacecraft, but none of those vehicles was intended for reuse. The heat shield simply charred, flaked off, and ablated away, carrying the heat with it. But the Space Shuttle orbiter was designed as a 100-mission vehicle. It would be too expensive to replace the heat shield after each flight, so a reusable one had to be developed.

Figure 8.6 shows how the various areas of the orbiter are covered by different types of insulation. The nose tip and leading edges of the wings are subject to the greatest heat

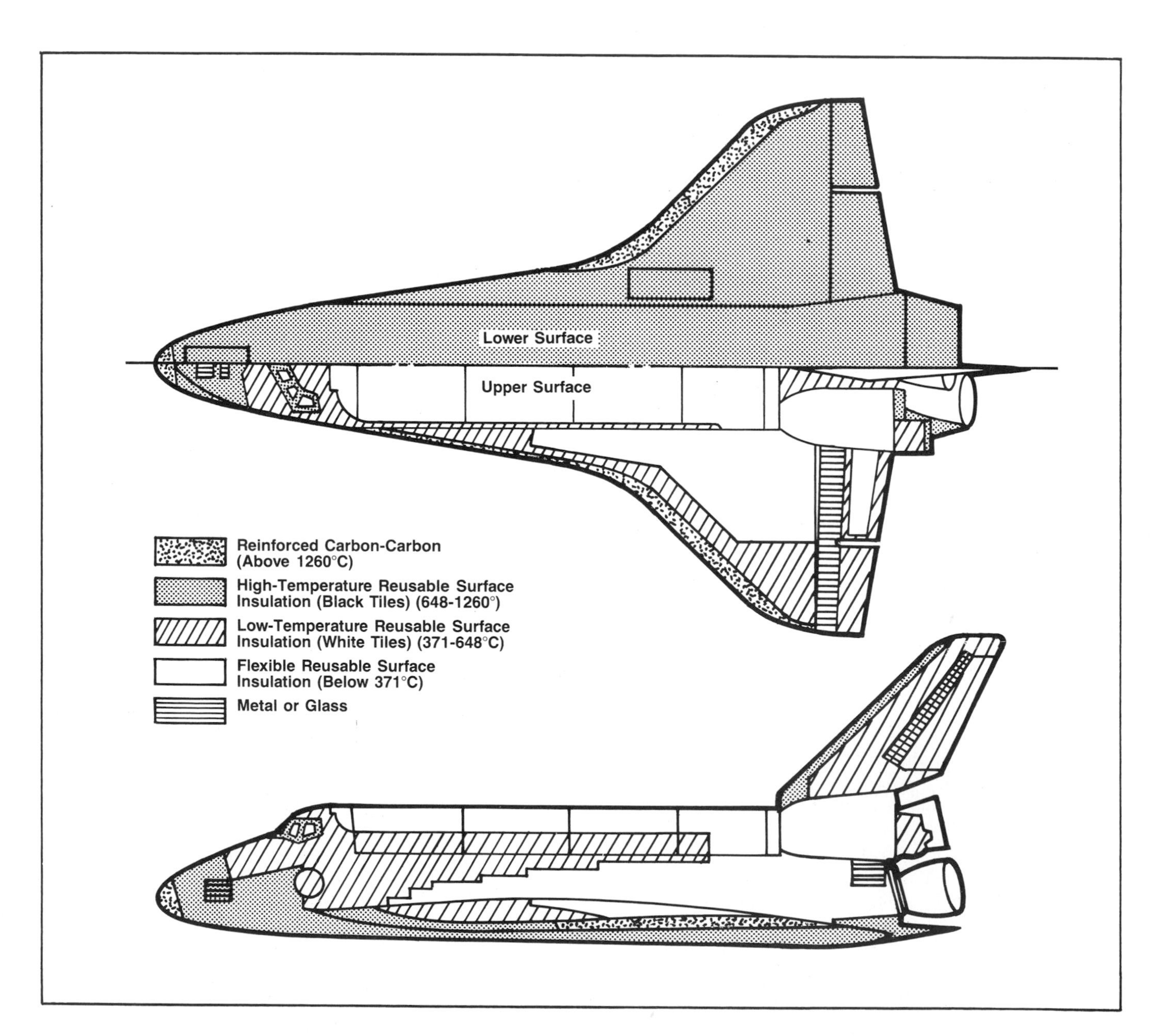

Figure 8.6 Four different kinds of insulation cover the surfaces of the orbiter. *Courtesy of NASA.*

stress at a temperature of 2750°F (1510°C) and are protected by a carbon composite consisting of layers of graphite cloth in a carbon matrix. The outer layers are chemically converted to silicon carbide, the same material used to make grindstones. Not shown in Figure 8.6, reinforced carbon-carbon has been added between the nose tip and the nose wheel door to provide better insulation of that area.

Perhaps the most remarkable parts of the thermal protection system are the silica fiber tiles, labeled High-Temperature and Low-Temperature Reusable Surface Insulation in the diagram. This material absorbs great quantities of heat but transfers it very slowly through its interior. Therefore, the atmospheric friction during reentry heats the tiles to as much as 2300°F (1260°C), but the aluminum skin of the spacecraft never exceeds 350°F (177°C). A spectacular demonstration of the thermal properties of the material is shown in Figure 8.7, where a tile was heated to 2300°F (1260°C) in an oven, then picked up with bare hands while the inside was still glowing white hot. Because of the very slow rate of heat flow through the tile, the outer surface had cooled in a few seconds to near room temperature while the interior was still at 2300°F (1260°C).

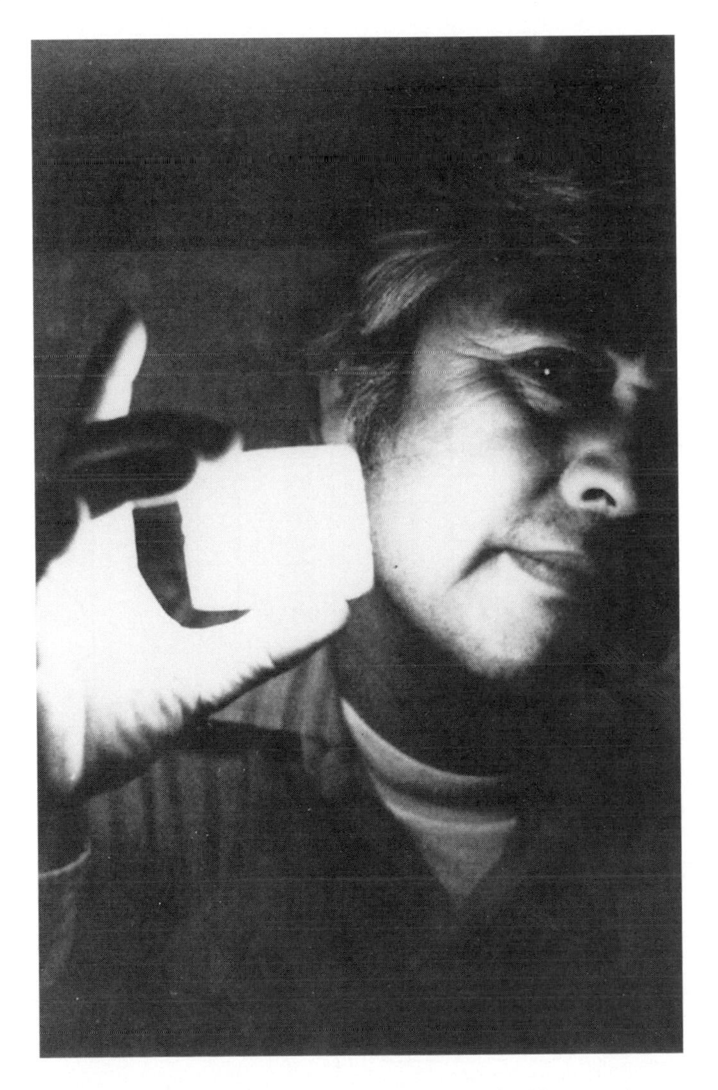

Figure 8.7 White hot tile held in bare hands. Photograph was taken by the light of the glowing tile less than 10 seconds after it was removed from the 2300°F (1260°C) oven. *Courtesy of Lockheed.*

About 70 percent of the orbiter is covered with tiles. Tiles on the bottom, front part of the fuselage, and leading edge of the tail are subject to heating in the range 1200° to 2300°F (650 to 1260°C). They are given a shiny black boron-silicate glass coating which allows 90 percent of the reentry heat to be radiated back into the atmosphere. These tiles measure about 6 inches (15 cm) square and vary in thickness from 1 to 5 inches (2.5–13 cm). On sections of the fuselage sides, tail and upper wing surfaces subject to temperatures in the range of 750° to 1200°F (370–650°C), the tiles are coated with a shiny white aluminum oxide to reflect solar radiation and help keep the spacecraft cool while in orbit. These are about 8 inches square and vary from 0.5 to 2.5 inches (0.4 to 6.3 cm) thick.

Each of the 30,000 tiles had to be cut and shaped separately to conform to the curved surfaces of the orbiter. Some of them are shown in Figure 8.8; no two are exactly alike. Groupings of about 20 tiles are set into a frame after the outer surfaces are cut and coated. Then the inner surfaces are milled to shape and bonded to a felt pad which in turn is bonded to the aluminum skin of the orbiter using a silicone resin glue. The felt pad isolates the tiles from strain due to the orbiter's vibrations as it flies into and out of orbit. Excess weight due to the absorption of rain water or condensation could be a problem, so the tiles are also given a waterproofing coat.

Each tile has a unique bar code like the ones seen on packages and cans at the grocery store. The white tiles are painted with the black lines of the bar code and the black tiles are painted with the white spaces between the bar code lines. The tiles are so important to the successful operation of the Shut-

Figure 8.8 Heat protection tiles for the orbiter. Each tile is individually milled to match the contour of a specific spot; no two are alike. *Courtesy of Lockheed.*

tle that any work done on a tile or group of tiles on an orbiter is recorded in a computer data base. The bar codes on those tiles are scanned into the computer to maintain a complete history of each one.

Although the tiles are soft and easily damaged they are also easily repaired by spraying scratches and plugging larger holes. Impregnating the tiles with ceramic shows promise as a next generation thermal protection material for new spaceflight vehicles. They are stronger and less subject to damage, yet still provide the necessary protection against the extreme heat produced during reentry.

The parts of the orbiter not subject to excessive heating are covered with Nomex felt thermal blankets, a nylon material coated with silicon, which affords sufficient protection to those areas where the temperature does not exceed 700°F (370°C). These areas include the cargo bay doors, most of the upper wing surface, lower rear fuselage sides, and pods. Some of the tile-covered areas on the orbiter do not heat up as much as was originally expected, so by replacing them with Nomex thermal blankets those areas are lighter, stronger, and less subject to damage.

Electric Power

Electricity for the Shuttle and its payload is produced by three fuel cells which use hydrogen and oxygen, combining them into water and generating electricity in the process. In a car battery, electricity is produced by the chemical reaction of lead and lead dioxide plates in a solution of sulfuric acid contained within the battery. In a fuel cell, hydrogen and oxygen are brought in from outside tanks.

Fuel cells were invented in 1959 and later used in the Apollo program. Shuttle fuel cells are about the same size as those used on Apollo missions but generate six times as much electric power. Under the floor of the cargo bay are located four liquid hydrogen tanks with a capacity of 92 pounds (42 kg) apiece for a total of 368 pounds (168 kg) and four liquid oxygen tanks holding 781 pounds (355 kg) each, totaling 3124 pounds (1420 kg). That is a sufficient supply to generate 1530 kilowatt-hours of electrical energy, generally enough for an 8-day mission. As the hydrogen and oxygen evaporate, tank pressure is built up to force the fluids out of the tanks and into the fuel cells. As the tanks are emptied, the pressure decreases so evaporation is speeded up with electric heaters immersed in the cryogenic fluids. The tanks are actually thermos bottles consisting of an inner tank and an outer shell with a vacuum between, which keeps the cold hydrogen and oxygen in a liquid state. Pressure relief valves assure that the tanks do not explode if excessive pressure builds up.

The pure water produced as a by-product in the fuel cells supplies all human needs with plenty left over. Any excess is dumped overboard into space at intervals.

A pallet with four additional oxygen tanks and four additional hydrogen tanks can be mounted in the cargo bay, essentially doubling the water and power available for missions lasting up to 16 days. The loaded pallet weighs about 7000 pounds (about 3200 kg) and is therefore carried only when a long duration flight is planned.

External Tank

The external tank (ET) serves two purposes: it carries the propellants for the orbiter's three main rocket engines and it is the support structure that connects the orbiter and SRBs together during ascent to orbit. Figures 8.9 and 8.10 show its structure; Table 8.3 tells its statistics. It is 154 feet (47 m) long and 27 feet (8.4 m) in diameter.

Because it carries both liquid hydrogen fuel and liquid oxygen oxidizer, the ET is really two inner tanks in one outer

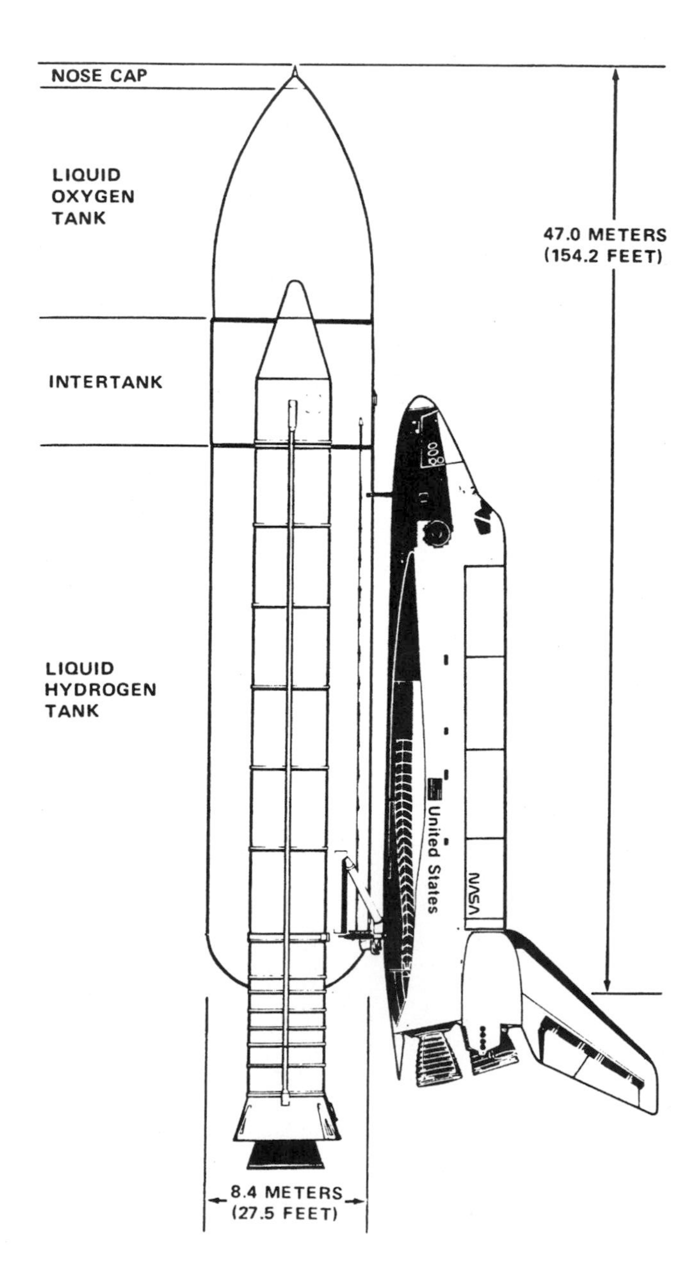

Figure 8.9 The Shuttle's external tank. *Courtesy of NASA.*

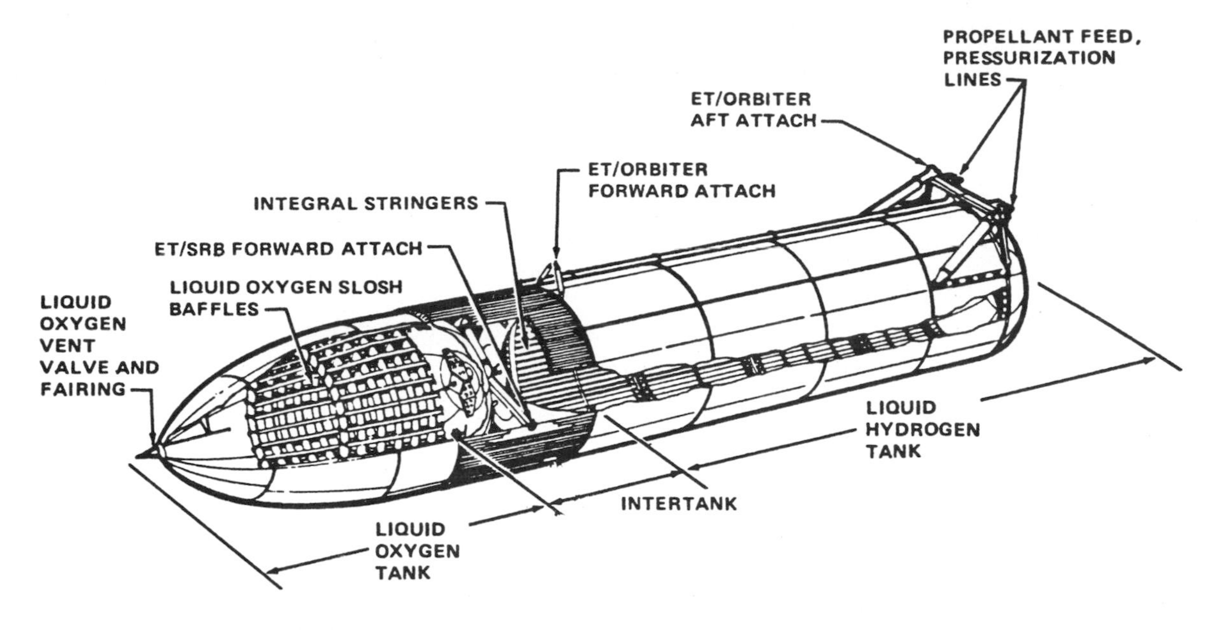

Figure 8.10 Internal structure of the external tank. *Courtesy of NASA.*

shell. The upper tank carries the liquid oxygen, 1.36 million pounds (618,000 kg) of it at −297°F (−183°C) at liftoff. Its 19,500 cubic foot (143,000 gallon) (540,000 l) volume is greater than that of a 2000 square foot house. The lower tank is about 2.5 times larger, 383,000 gallons (1,450,000 l) and carries about a quarter of a million pounds (103,000 kg) of liquid hydrogen at −423°F (−253°C). If these proportions do not seem to make sense, see MATHBOX 8.1.

Both inner tanks are constructed of aluminum and titanium alloys up to 2 inches thick. The oxygen tank contains baffles to keep the liquid oxygen from sloshing around during flight and throwing the Shuttle off course. The density of liquid hydrogen is so low that sloshing is not a problem. An intertank collar connects the two propellant tanks together and provides space for most of the electrical components. Since the first tank was built, the manufacturer has made some design changes and substituted titanium for aluminum in some components. At first the tanks were painted white. Then it was realized that several hundred pounds of weight could be saved by leaving them unpainted. As a result, the weight of the newer tanks was reduced by more than 6000 pounds (2700 kg), allowing the Shuttle to carry a heavier payload. A still lighter ET fabricated of an aluminum-lithium alloy developed later saved another 8000 pounds (3600 kg).

TABLE 8.3 External Tank Dimensions

Dimensions	
Diameter	27.5 feet (9.6 m)
Overall length	154 feet (47.0 m)
Liquid oxygen tank	53.3 feet (16.2 m)
Liquid hydrogen tank	97.0 feet (29.6 m)
Intertank	22.5 feet (6.9 m)
(Intertank overlaps propellant tanks)	
Total weight (mass)	
Empty	66,000 pounds (30,000 kg)
Loaded	1,676,000 pounds (762,000 kg)
Propellant weight (mass)	
Liquid oxygen	1,359,000 pounds (618,000 kg)
Liquid hydrogen	226,000 pounds (103,000 kg)
Total	1,585,000 pounds (720,000 kg)
Propellant volume	
Liquid oxygen tank	143,000 gallons (540,000 l)
Liquid hydrogen tank	383,000 gallons (1,450,000 l)
Total	526,000 gallons (1,990,000 l)
Propellant densities	
Liquid oxygen	71.1 pounds per cubic foot (1141 kg/m^3)
Liquid hydrogen	4.2 pounds per cubic foot (67.4 kg/m^3)

To reduce atmospheric drag, the oxygen tank curves to a point at the upper end of the structure. The entire outer surface of the external tank is insulated with a half inch thick cork/epoxy layer covered with 1 to 2 inches (2.5 to 5 cm) of spray-on foam. Insulation is necessary for two reasons. The propellants are very cold: liquid oxygen boils at −297°F (−183°C) and liquid hydrogen at −423°F (−253°C). An uninsulated tank would absorb heat from the surroundings causing uncontrolled boiling of the propellants. This poses two problems: excessive loss of hydrogen and oxygen through vent valves and buildup of excessive pressure in the tanks. Controlled boiling is necessary on the launch platform to keep the tanks pressurized for structural strength and also to assist the pumps in moving the propellants out to the engines. During flight, the tanks are pressurized by gases from the engines. In addition, because of the cold temperatures, if the tank were not insulated, water vapor in the air would readily condense as ice on the sides. At liftoff, the ice would break loose and damage the Shuttle.

Propellants flow to the engines through 17-inch (43-cm) pipes: oxygen at a rate of about 14,000 gallons (53,000 l) per

MATHBOX 8.1

Hydrogen-Oxygen Combustion

An atom of oxygen has an atomic weight of sixteen; an atom of hydrogen has an atomic weight of one. When the two react chemically (burn) they produce water. In the shorthand of chemistry, this is written as

$$2H + O \rightarrow H_2O.$$

Therefore, for the complete combustion of hydrogen it is necessary to have twice as many hydrogen atoms as oxygen atoms. This is why the hydrogen tank has more than twice the volume of the oxygen tank.

A water molecule has a molecular weight of 18 (2 for the hydrogens + 16 for the oxygen). Therefore the weight of the oxygen atoms is eight times the weight of the hydrogen atoms. If the external tank starts with 1,359,000 pounds (618,000 kg) of oxygen, then, for complete combustion it will need 1,359,000 lb / 8 = 170,000 pounds of hydrogen. (In metric units, 618,000 kg / 8 = 77,250 kg of hydrogen)

Actually, it carries more hydrogen than that, 226,000 pounds (103,000 kg) because it burns a hydrogen-rich mixture, only six parts of oxygen to one part of hydrogen, instead of the 8 to 1 ratio for perfect, complete combustion.

Thrust depends on the mass of the exhaust and its acceleration out the nozzle. So, the extra hot hydrogen in the exhaust contributes to the thrust even though it is unburned.

minute and hydrogen at nearly 50,000 gallons (190,000 l) per minute.

Assembly

At Kennedy Space Center the various parts of the Space Shuttle are brought together and assembled inside one of the largest buildings in the world, the Vehicle Assembly Building. It covers 8 acres (3.2 hectares), stands 525 feet (160 m) tall, 716 feet (218 m) long, and 518 feet (158 m) wide. The large doors operate in two sections. The upper section, 342 feet (104 m) high and 76 feet (23 m) wide, consists of seven leaves which move vertically to open. The lower section is 114 feet (35 m) high and 152 feet (46 m) wide with four leaves which move horizontally. More than 70 lifting devices are in the building, including two 250-ton cranes.

The entire Space Shuttle is put together on a mobile launch platform. First the solid rocket boosters are assembled and attached to the platform and the external tank is attached to the boosters. Horizontal payloads are mounted into the cargo bay before the orbiter is hoisted into position and connected to the external tank. Each of these components is tested individually. Then, when mated together, the entire vehicle is thoroughly checked out. A crawler-transporter (Figure 8.11) is then attached to the mobile launch platform with the Shuttle resting on top. The whole rig moves out through the doors along a 130 foot (40 m) wide roadway to one of two launch pads 3.4 or 4.2 miles (5.5 or 6.8 m) away, at a maximum speed of 1 mile per hour (1.6 km/h). See Figure 8.12 (Plate 23). The launch platform with the Shuttle is deposited on the pad (Figure 8.13) and the crawler-transporter moves back a safe distance.

Typical Mission Profile

A typical sequence of events during a Shuttle launch is tabulated in Table 8.4 and a diagram of a typical mission is shown in Figure 8.14. (Check the table and diagram as we describe the action.) By 1 hour prior to liftoff the propellant tanks are full, the crew is aboard, and computers and crew have checked out all systems to be sure they are ready. While awaiting the launch, the cold hydrogen and oxygen boil and vaporize. Excess gas must be vented, allowed to escape, so the pressure in the tanks does not build up to an unsafe level. In the final minutes of the countdown, the engines and nozzles are checked for proper operation and set into their launch position. The orbiter is switched to internal electric power supplied by the fuel cells. There are several opportunities for holding the countdown if something does not appear to function properly.

Ignition and Liftoff

The three main rocket engines in the orbiter ignite one at a time at intervals of 0.12 second, the first at T − 3.46 seconds. By time T the engines have reached 90 percent thrust. The entire vehicle lurches forward about 40 inches in the direction of the tank, straining against the eight holddown bolts

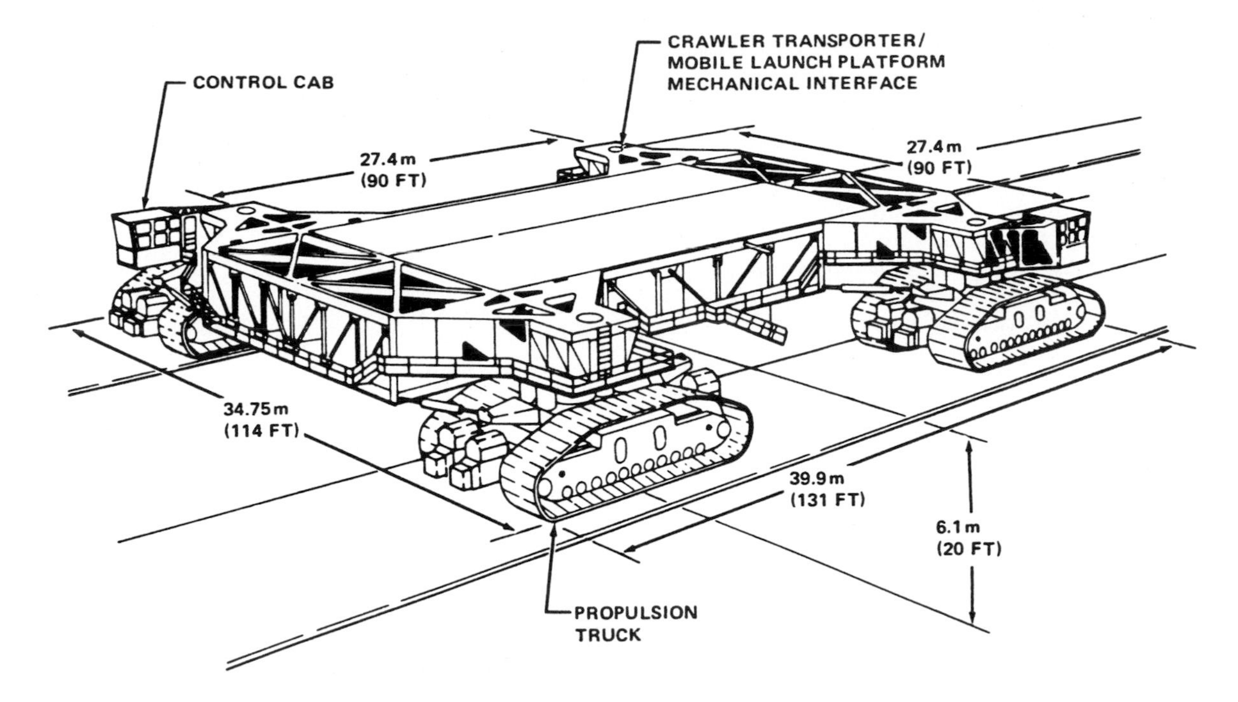

Figure 8.11 Space Shuttle crawler-transporter. Note the dimensions; the tracked drive trucks stand 10 feet (3 m) tall. The entire vehicle weighs 6 million pounds (2.7 million kg) unloaded. *Courtesy of NASA.*

which attach the solid rocket boosters and the entire vehicle to the launch platform. It takes 2.64 seconds for the vehicle to rock back to a vertical position. The booster engines then ignite and come to full power by T + 3 seconds. The hold-down bolts are severed by small explosive charges and the Shuttle lifts off. See Figure 8.15 (Plate 24).

Clouds of smoke and water vapor engulf the area. Much of it comes from water which is poured onto the launch pad area to suppress the intense sound waves produced by the rocket engines. Without sound suppression the orbiter could be damaged by the acoustic energy. The solid rocket boosters also contribute a great amount of smoke to the scene, but the three main rockets on the orbiter produce very little visible exhaust. Recall that they are burning hydrogen and oxygen, and the product of this combustion is plain water, so hot that the vapor is nearly invisible.

This sequence obviously must be computer controlled; human reaction time is not fast enough to carry it out. Sensors measure temperatures and pressures at strategic points on the main engines, continuously sending the information to the computer. As long as all readings are satisfactory, the countdown continues. If any sensor gives a reading which is out of acceptable limits, the countdown stops automatically. The liquid propellant main engines can be shut down by simply cutting off the flow of propellant, but once the solid booster engines ignite, there is no way to stop. The Shuttle must lift off the launch pad.

Up to the point when the vehicle lifts off the pad, the entire operation is under the control of the Kennedy Space Cen-

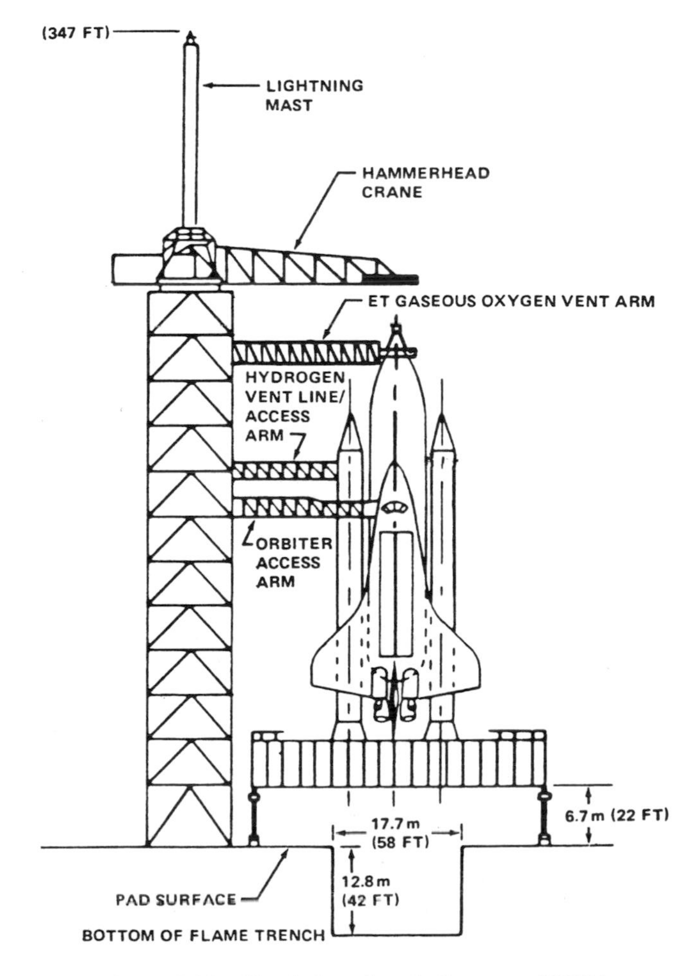

Figure 8.13 Shuttle launch pad. *Courtesy of NASA.*

TABLE 8.4 Space Shuttle Launch Events

Time	Event
T − 4 hr 30 min	Begin filling liquid oxygen tank
T − 2 hr 50 min	Begin filling liquid hydrogen tank
T − 1 hr 5 min	Crew aboard, hatch closes
T − 9 min	Automatic launch sequence starts
T − 4 min 30 sec	Orbiter on internal power
T − 3 min	Main engines move to start position
T − 2 min 55 sec	Oxygen tank at flight pressure
T − 1 min 57 sec	Hydrogen tank at flight pressure
T − 3.46 sec	First main engine starts
T − 3.34 sec	Second main engine starts
T − 3.22 sec	Third main engine starts
T − 0	Main engines reach 90% power
	Delay timer set for 2.46 sec
T + 2.46 sec	Command to start solid rocket boosters
T + 3 sec	Boosters reach 100% thrust
	LIFTOFF

Time	Event	Altitude	Speed*	Range
T + 7 sec	Begin pitchover	545 ft (166 m)	917 mph (1476 km/h)	0
T + 1 min 9 sec	Maximum dynamic pressure	8.3 mi (13.3 km)	1654 mph (2663 km/h)	4.0 mi (6.4 km)
T + 2 min 4 sec	Booster separation	29.4 mi (47.3 km)	3438 mph (5535 km/h)	23.7 mi (38.2 km)
T + 8 min 38 sec	Main engine cutoff (MECO)	73.0 mi (117.5 km)	17,500 mph (28,175 km/h)	829.0 mi (1334.6 km)
T + 8 min 50 sec	External tank separation	73.5 mi (118.3 km)	17,498 mph (28,172 km/h)	886.6 mi (1427.4 km)
T + 10 min 39 sec	OMS 1 ignition	78.3 mi (126.9 km)	17,479 mph (28,141 km/h)	1380 mi (2222 km)
T + 12 min 24 sec	OMS 1 cutoff	83.2 mi (134.0 km)	17,591 mph (28,321 km/h)	1860 mi (2995 km)
T + 43 min 58 sec	OMS 2 ignition	173.6 mi (279.5 km)	17,201 mph (27,694 km/h)	9775 mi (15,738 km)
T + 45 min 34 sec	OMS 2 cutoff	174.2 mi (280.5 km)	17,321 mph (27,887 km/h)	10,269 mi (16,533 km)

*Speed is with respect to space. Earth at Kennedy Space Center is rotating eastward at 914 miles per hour (1472 km/h).

ter. Immediately upon liftoff, control of the mission is transferred to the Johnson Space Center in Houston.

Almost immediately after liftoff the vehicle rolls over so the orbiter is beneath the external tank (Figure 8.16). At about 1 minute after launch the vehicle reaches Mach 1, the speed of sound, and experiences maximum aerodynamic pressure as it accelerates through the dense lower atmosphere. The main engines are throttled back and the regressive grain of the solid boosters decreases their thrust to keep the acceleration below the 3 g limit. MATHBOX 8.2 shows how to calculate the average acceleration during this period.

Solid Booster Rocket Operation

The boosters burn out about 2 minutes into the flight at an altitude of about 28 miles and a speed of over Mach 4. Then the boosters are jettisoned (Figure 8.17). To prevent a collision, they are moved away from the orbiter and external tank by a 1-second burn of eight small separation rocket motors, four are located in the nose frustum and four in the aft skirt of each booster. Even though the boosters have separated, they are still travelling at the same forward speed as the rest of the vehicle. But, because they are no longer under power, they follow a ballistic path arcing upward to an altitude of about 41 miles (66 km), then falling back downward into the more dense atmosphere. A barometer-altimeter switch ejects the nosecap at 15,400 feet (4690 m) altitude and the parachutes deploy, first a pilot chute which pulls away the nose cap, then a drogue chute which carries the upper conical-shaped frustum. Finally three main parachutes open to lower the booster into the ocean. The main chutes, each one 115 feet (35 m) in diameter, become fully inflated at 2200 feet (670 m) altitude. Initially they are falling at over 230 miles per hour (370 km/h), but by the time they reach the water they have slowed to 60 mph (96 km/h).

The booster rockets hanging from their parachutes strike the water bottom end first, forcing water into the empty interior and trapping air in the upper portion. Thus, they float upright in the ocean. The parachutes disconnect automatically, beacon lights are turned on, and a radio transmitter begins sending a signal to assist aircraft and ships in finding and recovering the spent motor. A ship tows the rocket to shore where it is disassembled and sent back to Utah to be refurbished and refueled for another flight.

Main Engine and External Tank Operation

Meanwhile, the orbiter and external tank (ET) continue on their way, the three main engines accelerating the vehicle upward and downrange. During the first part of the flight, the emphasis is on gaining altitude, getting through the dense part of the atmosphere. Now it is necessary to increase hori-

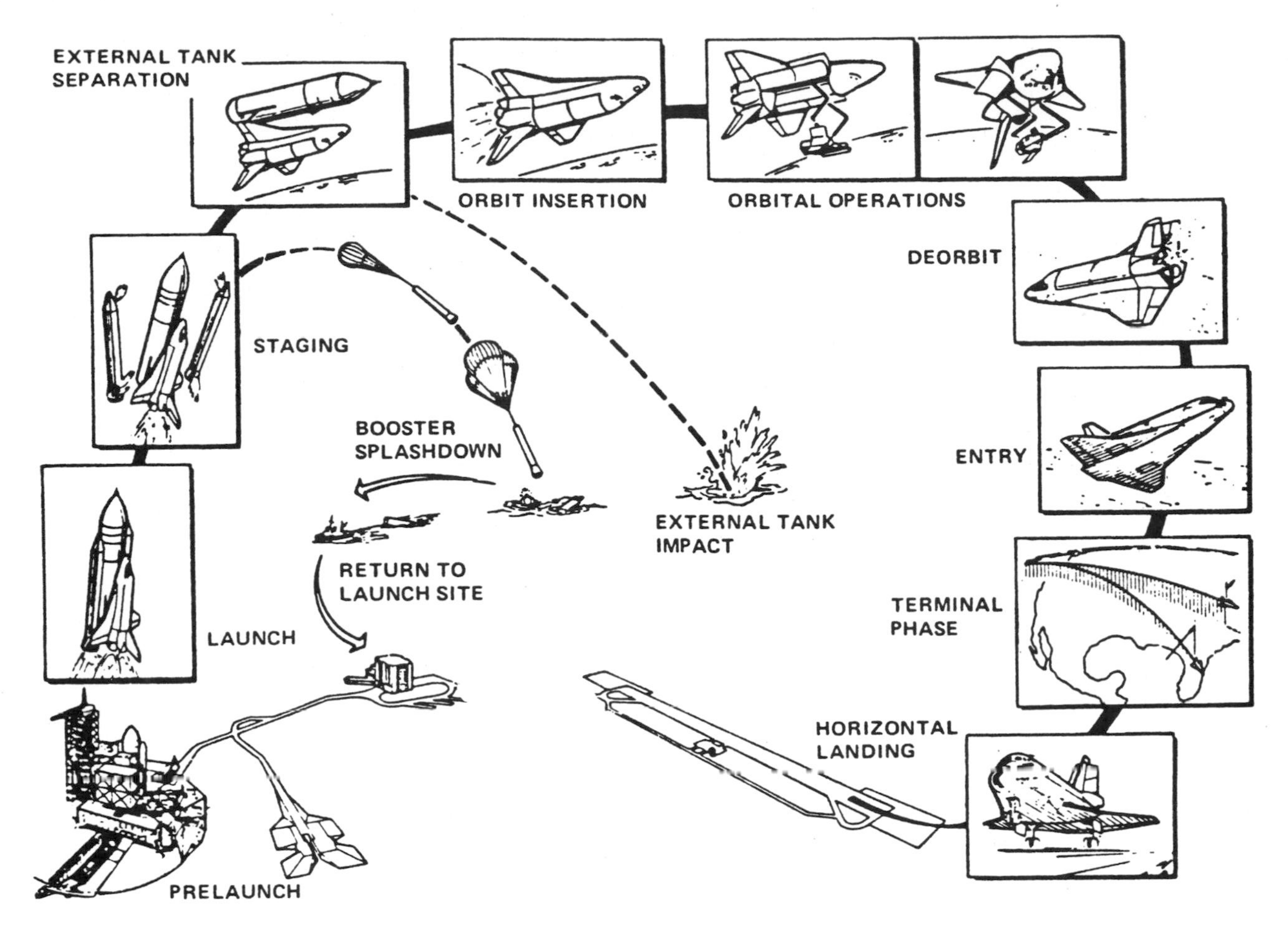

Figure 8.14 Typical launch-to-landing mission profile. *Courtesy of NASA.*

zontal velocity to reach orbital speed. When the Shuttle reaches an altitude of about 80 miles (128 km), it begins a long shallow dive to 73 miles during which its speed approaches the 17,500 mph (28,200 km/h) needed for orbit.

When the hydrogen and oxygen in the ET are nearly consumed and the vehicle is just short of orbital velocity, the orbiter's main engines cut off (MECO). A vent valve at the top of the external tank opens and oxygen escapes through the nose cap. A few seconds later the tank is disconnected (Figure 8.18), the venting oxygen causes the ET to pitch away from the orbiter and start to tumble. Tumbling assures that atmospheric drag will cause it to break up as it falls back to Earth and lands in the Indian Ocean. There is some uncertainty as to where it will land because of the way it tumbles. The designated impact area is an oval 2100 miles (3400 km) long by 62 miles (100 km) wide.

Incidentally, many people feel this is a waste of a valuable resource. External tanks are large, airtight, and structurally capable of being used as living quarters, workshops, or storage tanks. NASA has been urged to stockpile them in a parking orbit for possible future use. They already reach 99 percent of orbital velocity and only a little extra boost would inject them into orbit. There are several objections. One, the payload of the Shuttle flight would be reduced by several thousand pounds. Also, if left in very low Earth orbit, the tanks would have to be periodically boosted to higher altitudes to keep them from becoming a hazard to space traffic

Figure 8.16 During ascent the orbiter rides beneath the external tank. *Courtesy of NASA.*

MATHBOX 8.2

Space Shuttle Acceleration

Recall MATHBOX 2.2. We made a rough calculation of the acceleration of the Space Shuttle at takeoff based on Newton's second law of motion. Now let us find the actual acceleration based on the data in Table 8.2.

In the first 1 minute and 6 seconds (66 seconds) after liftoff, the Shuttle accelerates from rest on the launch pad at Kennedy Space Center, an initial speed of 914 miles per hour (1472 km/h) with respect to space, to a final speed of 1654 miles per hour (2663 km/h).

Average acceleration is the change in speed divided by the time interval required to make that change. If we use S_f to represent the final speed and S_i for the initial speed, then

$$a = \frac{S_f - S_i}{t}.$$

We want our results to be in feet per second so we multiply by 5280 feet in a mile and divide by 3600 seconds in an hour.

$$S_f = \frac{1654 \text{ mi/h} \times 5280 \text{ ft/mi}}{3600 \text{ s/h}} = 2426 \text{ ft/s}.$$

Similarly, $S_i = 1341$ ft/sec. The average acceleration of the shuttle during its first 66 seconds of flight is, then

$$a = \frac{2426 \text{ ft/s} - 1341 \text{ ft/s}}{66 \text{ s}} = 16.4 \text{ ft/s}^2$$

In metric units

$$S_f = \frac{1472 \text{ km/h} \times 1000 \text{ m/km}}{3600 \text{ s/h}} = 409 \text{ m/s}$$

and $S_i = 740$ m/s. Thus the average acceleration is

$$a = \frac{740 - 409}{66} = 5.0 \text{ m/s}^2$$

Why is this different from the acceleration we calculated in MATHBOX 2.2?

and from eventually burning in. A costly alternative is to strap on rockets and boost them to a higher stable parking orbit. Some planners envision them clustered together as a space station, fitted with rockets and launched to the Moon for a lunar colony, or refilled a little at a time and used as orbiting "gas stations" for vehicles heading to the outer reaches of the Solar System.

Orbit Insertion

A couple of minutes after the external tank separates, the orbital maneuvering system (OMS) engines are ignited to inject the vehicle into orbit. The two OMS engines, located in the pods just above the main engines on either side of the tail (Figure 8.19), produce 6000 pounds (27,000 N) of thrust

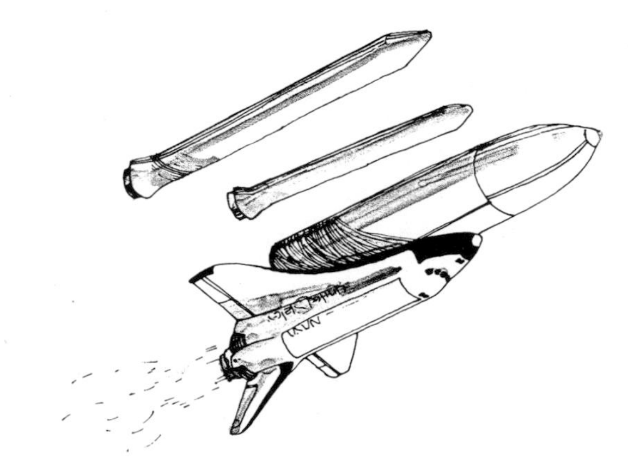

Figure 8.17 Solid rocket booster separation. *Courtesy of NASA.*

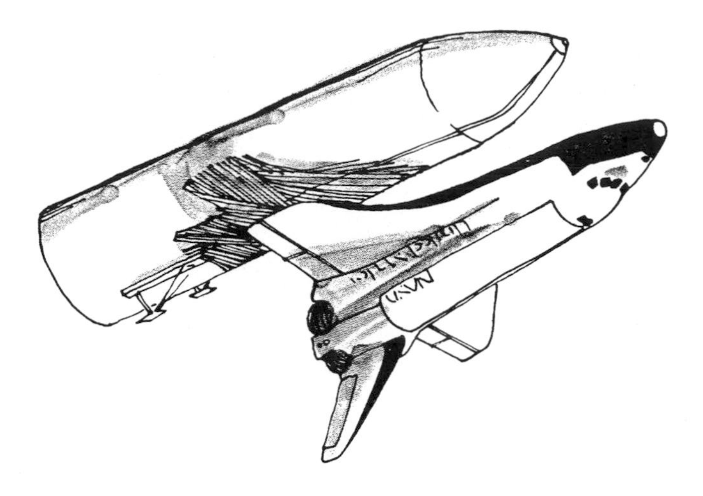

Figure 8.18 Orbiter separates from external tank. *Courtesy of NASA.*

each. They are propelled by monomethyl hydrazine fuel and nitrogen tetroxide oxidizer stored in the pods. These propellants are hypergolic; that is, they ignite and turn to hot gas on contact. Because no ignition system is required, the system is highly reliable. The propellants are forced out of their storage tanks and into the engines under pressure supplied by helium stored in liquid form in nearby containers.

The length of this first OMS burn determines the height of apogee; a longer burn means a higher apogee. Typically the engines are fired for about 2 minutes which increases the speed by 100 miles per hour (160 km/h). With engines off, the orbiter coasts to apogee, near 174 miles (280 km) in the example of Table 8.4. The orbit is now completely determined. After passing apogee the orbiter will return to the point where the OMS engines cut off, about 83 (134 km) miles. This will be perigee.

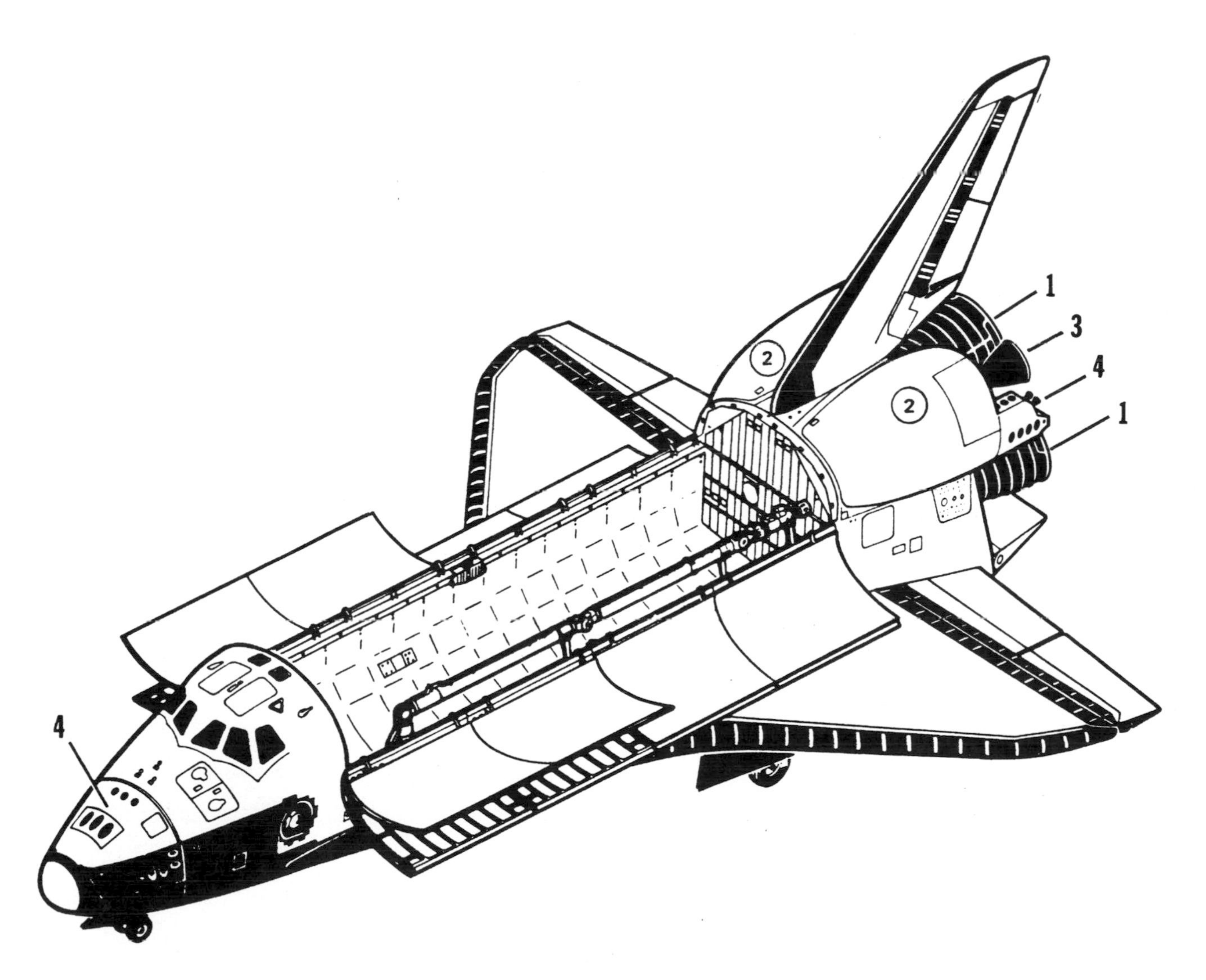

Figure 8.19 Orbiter's propulsion systems: three main engines (1), two orbital maneuvering system (OMS) engines (3), and the reaction control system (RCS) clusters of thrusters in the nose and tail sections (4). Two pods in the rear of the orbiter (2) hold the OMS engines and their fuel tanks. *Courtesy of NASA.*

Atmospheric drag on the vehicle at such a low perigee is sufficient to rapidly decay the orbit. Energy (speed) is lost to friction each time the orbiter passes perigee. With less speed at perigee, the vehicle will not be able to reach as high at apogee. Thus, apogee lowers with each orbit, increasing the drag even more, until finally the vehicle no longer has sufficient speed to remain in orbit. To prevent this from happening, the OMS engines are ignited again at apogee, a procedure called "apogee kick." This increase in energy raises the height of perigee above the altitude of excessive atmospheric drag. If the apogee kick is of just the right duration, perigee is raised to the same height as apogee and the orbit becomes a circle. Table 8.4 shows an OMS 2 burn of 96 seconds to circularize the orbit at about 174 miles (280 km).

On some flights, if the payload is less than maximum, the Shuttle main engines can provide enough energy to insert the orbiter directly to the desired apogee. Then the OMS 1 burn is unnecessary and only an OMS 2 burn at apogee is needed in order to circularize the orbit.

Attitude Control Thrusters

Thrusters are small rocket motors which change the attitude (not altitude) of the orbiter, that is, its orientation with respect to Earth. The reaction control system (RCS) consists of 44 thrusters as shown in Figure 8.19. Fourteen primary thrusters and 2 vernier thrusters are located in the nose; 24 primary and 4 vernier thrusters are located in the rear, on the back of the two pods. Primary thrusters produce 870 pounds (3900 N) of thrust; vernier thrusters, used for fine adjustment, produce just 24 pounds (107 N).

Firing the proper combination of thrusters causes the orbiter to pitch (nose up or nose down), yaw (swing nose left or right), or roll (Figure 8.20). Recall Newton's law of action and reaction. For example, firing a thruster pointed upward from the nose causes the nose to pitch down. Firing a rear end thruster toward the right causes the tail to move left and the nose to yaw to the right. The orbiter can automatically keep itself in any position, such as tail pointed straight down toward Earth and nose straight up at the sky. Different jobs require different orientations. For tasks which must be carried out without vibration, the orbiter is allowed to drift freely without firing any of the thrusters. If no particular attitude is necessary for the job at hand, the orbiter is usually oriented with the white top side toward the Sun to reduce solar heating and reduce the energy consumption of the air conditioners. The attitude which can be maintained with the least amount of fuel is the nose up, tail down attitude. Because the engines at the tail end are more massive than the cabin at the nose end, gravity tends to pull the tail end toward Earth. This is called a gravity-stabilized attitude and few or no thruster burns are needed.

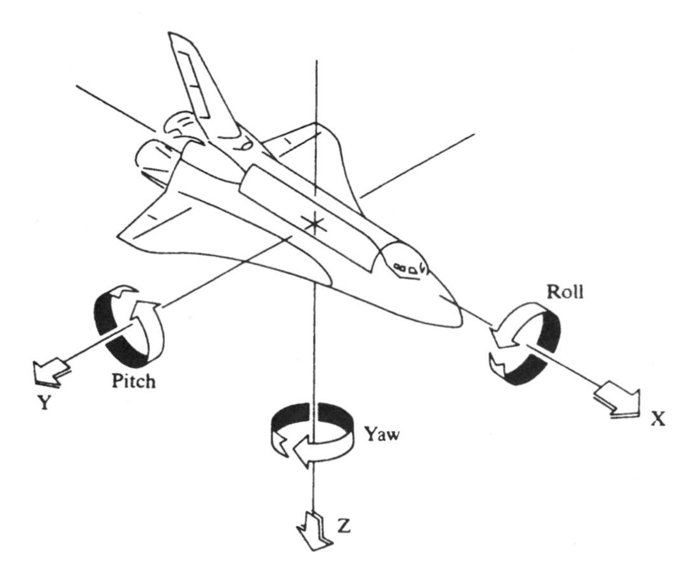

Figure 8.20 Pitch, yaw, and roll. *Courtesy of NASA.*

Propellants for these thrusters are monomethyl hydrazine and nitrogen tetroxide, the same propellants as the OMS. In fact, propellants from the OMS tanks can be fed to the reaction control system if needed. See Figure 8.21 for a diagram of one of the pods. Hydrazine is also the propellant for the auxiliary power units in the orbiter and for the boosters' hydraulic power units.

Because their thrust is so small, operating the reaction control system has little effect on the size or shape of the orbit. Small orbital adjustments in velocity are possible with the reaction control system, but if a large orbital change or rendezvous with another spacecraft is needed, the OMS engines are used. Each provides 6000 pounds thrust (27,000 N).

Reentry and Landing

To leave orbit and return to Earth, the orbiter must again fire the OMS engines. Over the Indian Ocean, about half an orbit, 45 minutes, before touchdown, the orbiter is rotated by the reaction control system so the engines are pointed in the direction of motion. Engines are fired in that position, against the motion, in order to slow the vehicle down and transfer it to a new orbit in which perigee lies inside Earth, that is, the orbit intersects the ground. As the orbiter begins to "fall" toward Earth, its speed once again increases by 200 or 300 miles per hour (300 to 500 km/h). The thrusters turn the vehicle again so it is falling underside first with a nose-high attitude.

Thirty-five minutes later, at an altitude of 50 miles (80 km), the orbiter encounters sufficient atmospheric drag to act as a brake. Maximum heating occurs at around 42 miles (67 km) while moving at 15,000 mph (24,200 km/h). During the next 20 minutes, it drops to 15 miles (24 km) altitude and slows to about 1700 mph (2700 km/h), as its kinetic energy of motion is transformed to heat energy. This would seem to be hitting the brakes quite hard, yet it is only about one-half g deceleration. See the calculations in MATHBOX 8.3.

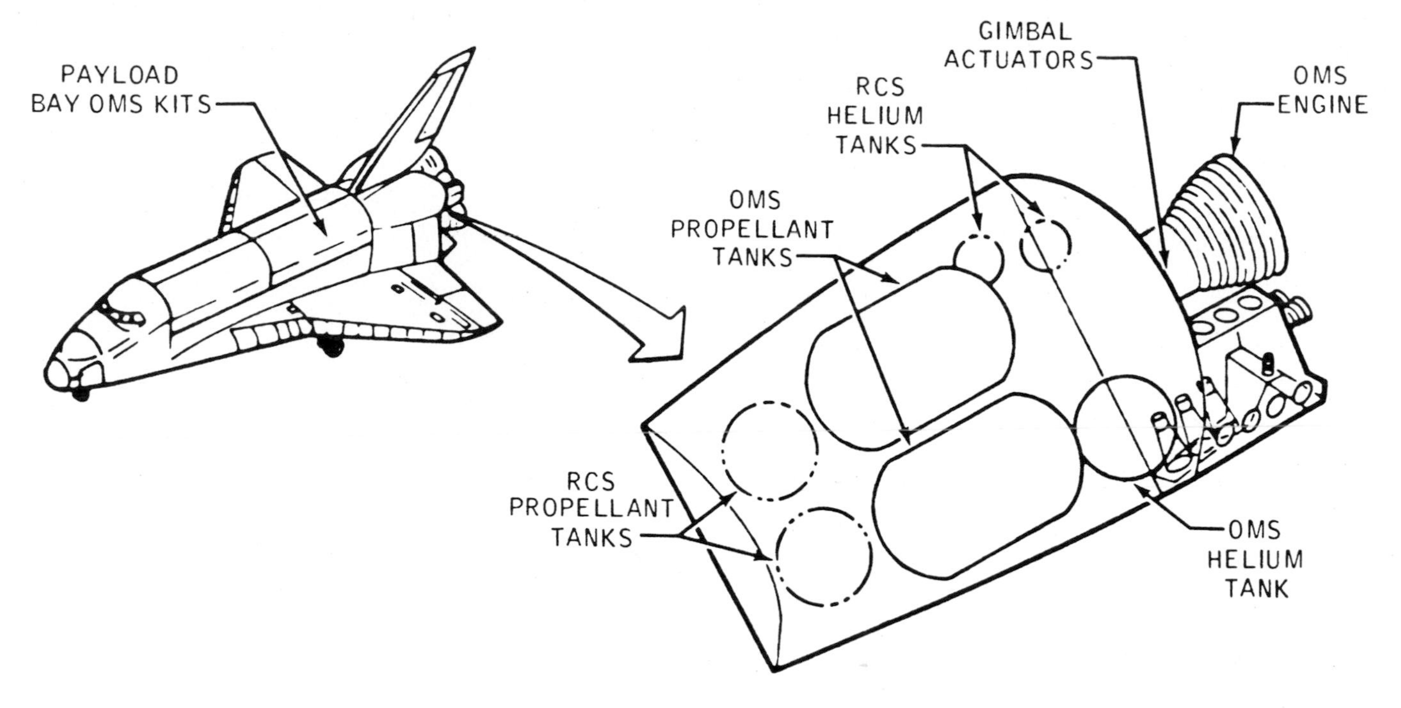

Figure 8.21 Left pod at rear of Space Shuttle. Propellants for the OMS and RCS engines are forced from their tanks under pressure supplied by the helium tanks. OMS and RCS are interconnected so the propellants from either system can feed the other. *Courtesy of NASA.*

MATHBOX 8.3

Reentry Deceleration

How large is the g force acting on the orbiter and its occupants during reentry into the atmosphere? According to the text the speed of the orbiter drops from 15,000 mph (24,000 km/h) to 1700 mph (2700 km/h) in 20 minutes. The average acceleration (deceleration in this case) is the change in speed divided by the time period. The most difficult part of this problem is handling the units. To convert mph to feet per second we multiply by 5280 feet in a mile and divide by 3600 seconds in an hour. To convert minutes to seconds we multiply by 60.

$$a = \frac{v_f - v_i}{t} = \frac{(1700 - 15000) \times 5280}{3600 \times (20 \times 60)} = -\ 16 \text{ ft/s}^2.$$

The answer comes out negative because the orbiter is slowing down. The acceleration of gravity, g, is 32 ft/s^2 so the average force on the orbiter and its occupants is a tolerable one-half g.

In metric units we convert kilometers per hour to meters per second by multiplying by 1000 meters in a kilometer and dividing by 3600 seconds in an hour. As before, we convert minutes to seconds by multiplying by 60.

$$a = \frac{(2700 - 24000) \times 1000}{3600 \times (20 \times 60)} = -\ 4.9 \text{ m/s}^2$$

In metric units, g is 9.8 m/s^2, so of course the answer is the same. The average deceleration is one-half g.

The heat ionizes the surrounding air by tearing electrons from the atoms so that a sheath of oxygen ions, nitrogen ions, and electrons, called a *plasma,* encloses the descending orbiter. It is not possible to transmit radio signals through the plasma sheath, so for 13 minutes the orbiter has no communications with the ground. It can easily be tracked by radar during that time, however, because radar pulses readily reflect off the plasma sheath.

While in the vacuum of space the wings, elevons, body flap, vertical stabilizer, and rudder (Figure 8.4) are "excess baggage"; without air they serve no useful purpose. During reentry the orbiter's altitude and speed decrease as it enters the denser atmosphere and these aerodynamic control surfaces begin to operate. Unlike the earlier manned spacecraft of the Apollo era which followed a ballistic trajectory like a cannonball, the Space Shuttle orbiter can maneuver to the left or right of its entry path by over 1200 miles (1900 km). The rudder can split in half, one side moving left and the other side moving right, to act as a speed brake. The body flap is covered with tiles and doubles as a heat shield for the engines during reentry.

Once back in the atmosphere, the orbiter flies like an airplane with one important difference, it has no engines running. At this point it is better called a glider! The approach to the runway and the touchdown must be done precisely the first time. Without power, there is no second chance. As it approaches the runway, the orbiter's glide angle is about six times steeper than a commercial jet liner. At touchdown its speed is a little more than 200 miles per hour (320 km/h).

A crew of technicians meets the orbiter on the runway and checks the exterior of the vehicle for safety before the astronauts emerge. In particular, they use instruments to "sniff" around the OMS engines and thrusters to be sure there is no leakage of toxic propellants. Also, it takes time for the orbiter to cool off so it can be touched. The temperature of the top surface of the orbiter is still about 200°F (93°C) and the hatch cover is 300–350°F (150–180°C) after landing!

The preferred landing site is at Kennedy Space Center in Florida. If bad weather prevents a Florida landing, an alternative location is at Edwards Air Force Base in the California desert. From there the orbiter is flown back to Florida, piggyback on a specially adapted Boeing 747. See Figure 8.22. At Kennedy Space Center the orbiter is refurbished, mated to a pair of reconditioned boosters and a new external tank, checked out, and made ready for another flight.

Figure 8.22 A modified Boeing 747 transports the orbiter piggyback.

Safety

If an emergency occurs while the crew is in the orbiter on the launch pad, the crew can leave the vehicle to a safe location by means of the escape system shown in Figure 8.23. However, the Presidential Commission that investigated the *Challenger* accident was concerned that there was no way for the crew to escape after liftoff. On the initial test flights, the orbiter *Columbia* was equipped with ejection seats for the two pilots. When testing was completed, they were replaced with the operational seats to save space and reduce weight. It would not have been possible to supply an entire seven person crew with individual ejection seats and exit hatches. In case of failure of one or more engines, the plan is to ride the orbiter to a landing back at the launch site, in southern Europe, in north Africa, or, as a last resort, to ditch in the ocean, depending on the speed of the vehicle at the time of the emergency. Even with an ejection system, it is doubtful that the crew could survive a catastrophic explosion. There was no warning of trouble either at mission control or in the *Challenger* until the external tank ruptured. Then it was too late.

Several experimental escape systems have been tried. In one, a small rocket is expelled from its storage canister by gas pressure. When it reaches the end of a 10-foot (3-m) lanyard, the 2000 pound thrust (8900 N) rocket motor fires for a tenth of a second to pull the crew member, parachute, and survival pack free of the orbiter, clearing the wingtip by 30 feet (9 m). Another type of escape system makes use of a 21-foot (6.4-m) telescoping rod extending from the side of the hatch opening. The crew member clips a ring onto the rod and slides down and out away from the vehicle before opening the parachute. In case of a horizontal emergency landing, the top windows of the orbiter can be removed and the crew can escape using a cable over the top and side. A 34-inch (86-centimeter) inflatable sphere is also available for a crew member to crawl into with a portable oxygen system for life support if a spacecraft-to-spacecraft rescue mission is feasible. None of these has been used, of course, but under certain specific conditions they may save an astronaut's life.

Other Winged Vehicles

The single most important problem in opening up the space frontier is the cost of getting off Earth to low orbit. Without

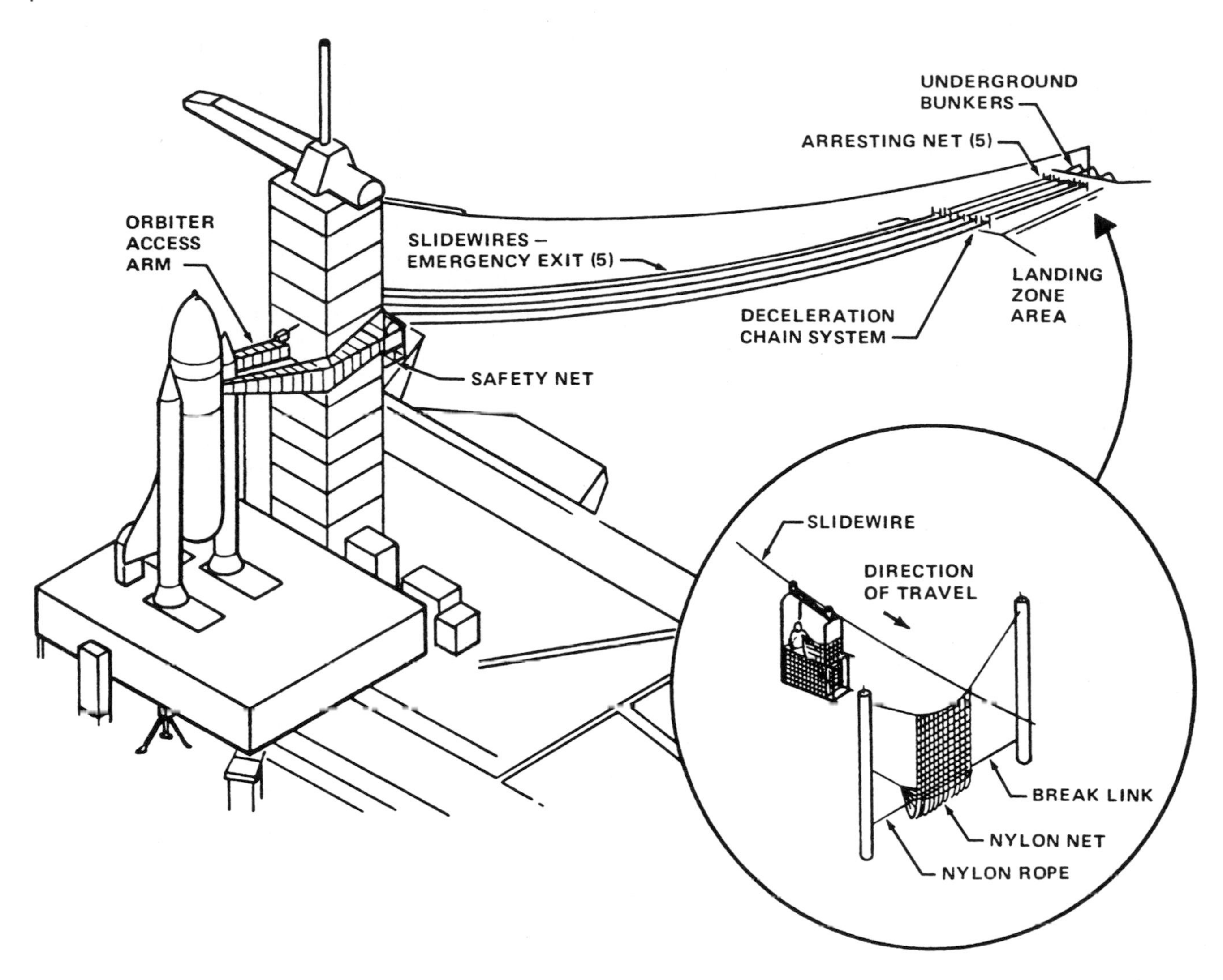

Figure 8.23 Emergency escape while on the launch pad. Crew leaves orbiter, enters baskets on slide wires, two people per basket. Near the bottom, the baskets catch and drag chains to decrease their speed before they reach the nets which stop them completely. Crew then takes cover in bunkers. *Courtesy of NASA.*

cheap access to low Earth orbit, we cannot afford to carry the large masses needed for manned flights to Mars, a colony on the Moon, or any other giant step in space. Developing a reusable launch vehicle that significantly reduces the cost of access to space is, at the moment, NASA's number one priority.

All space launches are now done in the Space Shuttle or on high-powered disposable rockets. Research efforts have been aimed toward making these more efficient, reusable, and cheaper. Taking a payload to orbit costs about $10,000 a pound. The largest part of this cost lies in the fact that most of the launch vehicle is thrown away during the launch.

Only the Space Shuttle is even partly reusable. Major parts are dropped into the ocean and the cost and time involved in turning it around for another flight are excessive. The Space Shuttle is basically 1970s technology and has proved to be very expensive to operate. Research is underway for the next generation of manned spaceships.

The goal of this research is a tenfold reduction in the cost of putting payloads into orbit, from $10,000 to $1,000 per pound ($22,000 to $2200 per kg). Currently, the best hope for reaching this goal is by using composite materials instead of metal to reduce the weight dramatically and improved rocket engines with higher thrust at lower weight. By taking advantage of wings while in the atmosphere, the fight against gravity is not as severe as it is in a vertical takeoff.

New experimental craft called the X-vehicles hopefully will be reusable, simpler to build, and cheaper to operate, and will be turned around more quickly by a small ground crew the way airlines operate. The X-vehicles are a continuation of the research done by the X-15 rocket planes that flew in the 1950s. Engineers are studying and testing new technologies that must be perfected in order to build a true spaceship. Significant advances must be made in more efficient propulsion systems, durable composite materials, lightweight thermal protection, cryogenic fuel storage, and electronics. The only alternative is to continue flying the Space Shuttle beyond 2012, the currently proposed phase-out date.

DC-X

The first in this new series of X-vehicles was the DC-X shown in Figure 8.24. It was a vertical takeoff, vertical landing vehicle using rocket engines to lift off the ground, maneuver, and slowly set back down upright in its launch position. The U.S. Air Force had been experimenting with a similar vehicle. A composite epoxy liquid hydrogen tank reduced the weight by 1200 pounds (550 kg). Four successful launches showed feasibility of the concept. In the last flight one of the landing gears failed to open and lock into place and the unmanned vehicle fell over on its side while landing. Much important information for follow-on planes came from these flights: how to build a vehicle with composite materials instead of metal, how to handle liquid oxygen and liquid hydrogen in such a vehicle, and how quickly a small team could prepare it for another takeoff.

Figure 8.24 On its maiden flight, the DC-X, Delta Clipper, lands vertically on four legs. *Courtesy of McDonnell Douglas Aerospace (now a division of Boeing).*

X-34

The X-34 is a suborbital, unmanned hypersonic rocket plane capable of flying to 250,000 feet (76 km) at speeds up to Mach 8, eight times the speed of sound. Its purpose is to test the new technologies being developed for hypersonic flight. It is shown in Figure 8.25 (Plate 25) and its statistics are given in Table 8.5. The X-34 is carried under a L-1011 jet airplane to high 40,000 feet altitude (12 km) where it is dropped in free fall for 6 seconds and the rocket engines are ignited.

The 60,000 pound thrust (270,000 N) Fastrac engine is an aerospike LOX-kerosene rocket. A test of its thrust chamber is shown in Figure 8.26. The X-34 is the first space plane to try out such an engine.

The wings are made of aluminum framework with a graphite/epoxy composite covering which maintains its strength under high temperatures. The nose and leading edge of the wings are covered with 2-inch (5-cm) thick graphite/epoxy tiles which keeps their temperature below 1845°F (1010°C). Thermal blankets protect the outer skin from the heat of high-speed flight in the atmosphere, keeping the temperature of the structure below 350°F (177°C). To prevent the LOX from boiling and building up excessive pressure, the tank is insulated with sprayed-on polyurethane foam.

The X-34 is designed for a 2-week turnaround time with a small maintenance team for up to 25 flights per year. It should also be capable of two flights in a 24-hour period. Cost goal is $500,000 per flight. Three airframes will be

TABLE 8.5 X-34 Statistics

Structure	Aluminum and composite
Gross vehicle weight	45,000 pounds (mass 20,500 kg)
Empty weight	15,000 pounds (mass 6,800 kg)
Propellant weight	30,000 pounds (mass 14,000 kg)
Payload	400 pounds (180 kg)
Length	58 feet (18 m)
Wingspan	27 feet (8 m)
Fastrac engine:	
Propellants	RP 1 (refined kerosene) and LOX
Thrust	60,000 pounds (267,000 N)
Burn time	up to 150 seconds
Propellant tanks:	
LOX	Aluminum, Diameter: 54 inches (1.4 m) Length: 13 feet (4 m)
RP-4	Composite, Diameter: 54 inches (1.4 m) Length: 9 feet (2.7 m)
Altitude	At least 250,000 feet (76 km)
Maximum speed	Mach 8
Touchdown speed	250 miles per hour (400 km per hour)
Braking parachute diameter	24 feet (7.3 m)

Figure 8.26 A test of the Fastrac aerospike engine's thrust chamber. The exhaust is a ring of fire around a spike or plug in the nozzle. It will be used on the X-34. *Courtesy of NASA / Marshall Space Flight Center*

built—one for tests on the ground and while attached to an L1011. Two of the airframes will first be tested in unpowered drop tests. Then the engines will be installed and powered flights will commence.

Some possible future research and development projects for the X-34 include the following. The Russian NK-39 engine has higher thrust, can be throttled, is fully reusable, and can be adapted to the X-34. Atmospheric oxygen could be used for atmospheric flight and stored LOX in space, reducing takeoff weight considerably. An experimental pulse thrust engine fills its cylinders with propellants, then ignites it to give a pulse of thrust, repeated as many as 60 times per second, to produce greater efficiency.

X-33 and Venture Star

The X-33 will be a one-half scale unmanned craft capable of reaching altitudes of 60 miles (100 km) at speeds of Mach 15. Its shape is that of a lifting body, like a wing or parafoil. It will take off vertically and land horizontally using technology developed and proven in the X-34 test flights. It is shown in Figure 8.27 (Plate 25) and its statistics are given in Table 8.6.

In building a single stage rocket there are two prime considerations. The engines must have maximum efficiency and the vehicle must have minimum weight. Two linear aerospike engines will power the X-33. Their efficiency was discussed in Chapter 2. See Figures 2.9 and 2.10. Their ramps can be seen in the rear of the vehicle in Figure 8.27 (Plate 25). The vehicle can pitch, yaw, and roll by throttling the combustion chambers differently on the two sides of the engines. For example, greater exhaust flow on the right side will cause the vehicle to turn toward the left, greater exhaust flow on the underside will cause it to pitch up, and so on. It is not necessary to gimbal these engines. Two body flaps are also used as control surfaces at hypersonic speeds.

Much of the structure of the X-33 is lightweight carbon/carbon and graphite epoxy composite materials which are not as heavy as metals. Using composites for building cryogenic fuel tanks is experimental but is essential to reducing the weight of the vehicle. About 8 pounds of propellant is required to lift 1 pound of payload to orbit (8 kg to lift 1 kg to orbit). See MATHBOX 8.4. For each pound of weight taken off the vehicle, about 8 pounds (3.6 kg) of propellant is saved. If less propellant is needed, the tanks can be made smaller which further reduces the weight. That further reduces the fuel requirement, reducing the weight even more. Engineers calculate that reducing the weight of a component part by 1 pound saves an overall 40 pounds of weight (mass of 18 kg) in fuel, tanks, and structure. The X-33 program suffered a setback when the outer skin of a liquid hydrogen tank fractured during a pressurization test in November 1999. Such setbacks are certain to occur in any research program that tries to push the limits of technology.

If the X-33 is successful, it will show the way to build a full-scale piloted vehicle, the Venture Star, which is designed to fly into orbit with a 45,000 pound (20,000 kg) payload. It will look very much like the X-33, but twice as big and 10 times the weight. Its statistics are given in Table 8.7.

Venture Star will be powered by seven linear aerospike engines. A single stage vehicle, it will have all engines run-

TABLE 8.6 X-33 Statistics

Structure	Lifting body; metal and composite materials
Weight	
Gross vehicle weight	285,000 pounds (mass 129,000 kg)
Empty weight	75,000 pounds (mass 34,000 kg)
Propellant weight	210,000 pounds (mass 95,000 kg)
Payload	None. A technology demonstrator
Dimensions	
Length	63 feet (19 m)
Wingspan	68 feet (21 m)
Engines	Two linear aerospike
Propellants	LH and LOX
Thrust, total at liftoff	410,000 pounds (267,000 N)
Propellant tanks	
LOX	Aluminum-lithium, Diameter: 54 inches (1.4 m) Length: 13 feet (4 m)
LH	Graphite-epoxy, Diameter: 54 inches (1.4 m) Length: 9 feet (2.7 m)
Reaction control system	
Thrusters	Eight at 500 pounds each (2200 N)
Propellants	Gaseous oxygen: 360 pounds (mass 164 kg) Gaseous methane: 340 pounds (mass 155 kg)
Altitude	At least 50 miles (76 km)
Maximum speed	Mach 13–15

MATHBOX 8.4

Propellant Weight vs. Dry Weight

How many pounds (kilograms) of fuel is required to lift a pound (kg) of payload to orbit by the Space Shuttle? Refer to the text and Tables 8.1 and 8.3. Add together the dry weights of the orbiter, payload, external tank, and solid rocket boosters. Then add together the weight of all propellants. (We will do it in pounds. Try it in kilograms yourself. The result will be the same.)

Component	Dry Weight (pounds)	Propellant Weight (pounds)
Solid Rocket Boosters	188,400	1,112,000
	188,400	1,112,000
External Tank	66,000	LOX 1,359,000
		LH 226,000
Orbiter	170,000	
Payload	65,000	
Totals	677,800	3,809,000

Now divide the propellant weight by the dry weight: $3{,}809{,}000 \div 677{,}800 = 5.6$. On this basis it takes 5.6 pounds of fuel to lift 1 pound to orbit.

This estimate is too small, however, because the booster rockets drop off along the way and are not carried all the way to orbit. If we make another estimate by ignoring the dry weight of the two boosters, 376,800 pounds, we get 12.7. This estimate is too high because the booster rockets are carried part of the way. The actual value is between these two estimates, about 8 pounds of propellant to lift 1 pound of payload. The external tank is carried all the way to burnout before it drops off and falls back to Earth.

ning and checked out before liftoff which increases the safety and reliability of the vehicle. A hatch in the top gives access to the cargo bay.

The heat of reentry will not be as extreme on Venture Star as it is on the Space Shuttle. Because Venture Star will be a relatively lightweight lifting body, its reentry trajectory can be much less steep so it can decelerate at a higher altitude where the friction of the thin air does not produce as much heat. It will take about 35 minutes to reach the ground from orbit compared to the Space Shuttle's 20 minutes. Because it experiences less heating, the thermal protection panels are metallic, mounted over an insulating material that prevents the heat from reaching the interior of the vehicle. This construction saves additional weight and gives a sturdier longer-lasting surface than the Shuttle's ceramic tiles.

If all goes well, Venture Star will be a commercially produced reusable vehicle flying routine flights into space.

TABLE 8.7 Venture Star Statistics

Structure	Lifting body, metals and composites
Weight	
Gross vehicle weight	2,186,000 pounds (mass 994,000 kg)
Empty weight	257,000 pounds (mass 117,000 kg)
Propellant weight	1,929,000 pounds (mass 877,000 kg)
Payload	
To LEO	45,000 pounds (20,000 kg)
To Space Station	25,000 pounds (11,000 kg)
To GEO transfer orbit	18,000 pounds (8,200 kg)
Dimensions	
Length	127 feet (38.7 m)
Wingspan	128 feet (39.0 m)
Cargo Bay	45 by 15 feet (14 by 4.6 m)
Propulsion	7 linear aerospike engines
Propellants	LH and LOX
Thrust, total at liftoff	3,010,000 pounds (13,390,000 N)
Propellant tanks	Graphite epoxy
Altitude	To orbit
Maximum speed	Orbital

X-37

The X-37 is a small 4 foot by 7 foot (1.2 by 2.1 m) unmanned spacecraft weighing 500 pounds (225 kg) to be carried to orbit in the cargo bay of the Space Shuttle orbiter. Deployed from the cargo bay, it will operate autonomously, conducting experiments, rendezvousing with satellites for close operations or recovery, then returning to Earth. Eventually it could be launched on a booster rocket. A rocket engine using a kerosene type of fuel with hydrogen peroxide for an oxidizer will provide the power to change orbits and maneuver and a second stage could carry it to geosynchronous orbit. Electric power is provided by solar panels which fold back inside before deorbiting. Like the

Shuttle orbiter it is covered with ceramic tiles, thermal blankets, and graphite carbon composites for protection during reentry. This little research craft will first fly in 2003. A full mission will last 21 days.

X-38, Crew Return Vehicle

The X-38 is a prototype crew return vehicle to be attached to the International Space Station for emergency evacuation or medical emergency, or if the Space Shuttle fleet is grounded. It is a lifting body design similar in appearance to a small X-33 with a parafoil for steering to a landing site. As a single stage vehicle, it is intended for a one-way trip back to Earth, but a booster stage could be attached to launch it back to the Station. The European Space Agency is particularly interested in fitting it to a French Ariane booster rocket and using it for two-way crew transportation.

X-43, Hyper-X

The X-43 is a small 12-foot long unmanned vehicle designed around a supersonic ram jet (scramjet) engine. It will be carried to high altitude by a B-52 and dropped off to fire its engine and proceed through several tests and maneuvers before diving into the Pacific Ocean. A scramjet has the potential for hypersonic speeds beyond Mach 5. Instead of carrying an oxidizer like a rocket, it uses atmospheric oxygen for combustion. A previous program had attempted to develop a scramjet-powered vehicle, the X-30 National Aerospace Plane, that would reach orbital speed. The X-43 is continuing that research.

Many other concepts are under study or development by private companies or under government sponsorship, too numerous to describe here.

DISCUSSION QUESTIONS

1. Why are the Space Shuttle main engines started first? Why not fire the solid rocket boosters first?
2. How is the orbiter steered when it is on its way to orbit? when it is in orbit? when it returns to Earth?
3. Why do you suppose the orbiter flies beneath the external tank and solid rocket boosters on its way to orbit? (See Figure 8.16.)
4. Why are these such complex joints between segments of the solid rocket boosters, shown in Figure 2.17? Why not just bolt the segments together?
5. At what points in a Shuttle flight would it be feasible to bail out of a disabled orbiter? When would it not be feasible?
6. What limits the altitude the orbiter can reach?
7. At liftoff, what percentage of the total weight of the Space Shuttle is propellant weight?

ADDITIONAL READING

Allaway, Howard. *The Space Shuttle At Work, NASA SP-432,* Government Printing Office, 1979. Easy reading, written before first flight.

Beardsley, Tim. "The Way to Go in Space." *Scientific American,* February 1999. Discusses a dozen far-out concepts for spaceships.

DiChristina, Mariette. "Highway Through Space." *Popular Science,* November 1999. Simplified article about several concepts for plasma propulsion.

Joels, Kerry M., and Gregory P. Kennedy. *The Space Shuttle Operator's Manual,* Ballantine Books, 1982. Very detailed and complete, popular reading level.

Kaplan, Marshall H. "Are RLVs Economically Viable?" *Launchspace,* October/November 1999. A discussion of the pros and cons of reusable launch vehicles from an economic point of view.

Kerrod, Robin. *The Illustrated History of Man in Space.* Mallard Press, 1989. Profusely illustrated, easy reading.

Kincaid, Jeffrey S. "Inside the Linear Aerospike." *Launchspace,* July/August/September 1999. Fundamentals of aerospike engines.

Mattingly, T. K. "A Simpler Ride into Space." *Scientific American,* October 1997. Excellent understandable discussion of the X-33 and Venture Star.

NASA. *Space Shuttle News Reference,* Government Printing Office. Everything you always wanted to know about the Shuttle.

NOTES

Chapter 9

Living in Space

For millions of years the human body has adapted to living on Earth with the normal force of gravity, where up and down are clearly defined, in an atmosphere of 21 percent oxygen and 78 percent nitrogen at 14.7 pounds per square inch (101 kPa) pressure, with food and water in abundance. All these things are absent in space. If people are to survive in space, they must take a suitable environment with them, particularly air at a proper temperature and pressure. If they are to remain for an extended period of time they must have food and water. In weightlessness the body begins to deteriorate in a number of ways and some arrangement must be made to keep physically fit by exercising or perhaps by creating artificial gravity. Protection from cosmic and solar radiation is necessary. In addition, being cooped up in a small space with a few other people for a long time can lead to sociological and psychological difficulties. All these things must be considered in designing spacecraft and in planning missions.

Cabin Atmosphere

The Space Shuttle orbiter cabin is shown in Figure 9.1. Living space is in the middeck and life support equipment is mostly in the lower deck, accessible through hatches in the floor of the middeck.

The orbiter has a "shirt-sleeve environment," that is, the atmosphere is a duplicate of the atmosphere on Earth at sea level: 21 percent oxygen, 78 percent nitrogen, 14.7 pounds per square inch (psi) (101 kPa) pressure. An automatic system maintains the proper oxygen-nitrogen ratio. It senses the amount of oxygen in the air and adds more oxygen, more nitrogen, or both as necessary, from storage tanks.

By contrast, the Apollo spacecraft used a pure oxygen atmosphere at a pressure of 5 psi (34 kPa). Pure oxygen is very reactive and the launch pad fire, discussed in Chapter 1, showed the danger of using it unless the spacecraft is made absolutely fireproof and the pressure is kept lower than normal atmospheric pressure. Diluting the oxygen with an inert gas makes it less hazardous. The Skylab space station had an atmosphere of 75 percent nitrogen and 25 percent oxygen at a pressure of 5 psi (34 kPa). Other inert gases such as helium have been considered as a substitute for nitrogen in large space vehicles, primarily because helium is a very light gas and less mass would have to be transported to orbit. For example, if the volume of a space station is 16,000 cubic feet (453 m^3) and the pressure is 14.7 psi (101 kPa), 787 pounds (357 kg) could be saved by using helium. See MATHBOX 9.1. However, helium gives the voice a Donald Duck-like quality that could be very disturbing after a time.

Besides providing a proper mixture of oxygen and nitrogen, the life support system must remove exhaled carbon dioxide and other gases, and maintain a comfortable temperature. Poisons can accumulate in the spacecraft atmosphere. Excess methane and ammonia gas from human and food wastes could be a problem if the air purification system failed. More serious would be an increased level of carbon dioxide. At first this would cause a headache and the astronaut's breathing rate would increase noticeably. Then hearing ability would be reduced, followed by dizziness, nausea, mental confusion, convulsions, and unconsciousness.

On Earth, air circulates because warm air is less dense than cold air. Gravity causes the cooler, denser air to sink while the warm air rises. In orbit, when weightless, air will not circulate of its own accord. In the Space Shuttle orbiter, blowers are necessary to keep the air from stagnating. First the blowers force the air through filters that catch debris such as bits of paper, hair, and crumbs. Then about 10 percent of the air is diverted through purifying canisters containing lithium hydroxide which absorbs and removes carbon dioxide, and activated charcoal which removes other noxious gases and odors. The canisters handle about 320 cubic feet (9 m^3) of cabin air per minute and must be changed at regular intervals as they become saturated with the gases.

Maintaining a comfortable temperature, in the range of 65° to 80° F (18°C to 26°C) requires cooling, not heating. Body heat, solar heat, and heat generated by the equipment are removed by a refrigerator air conditioner. Excess heat is dumped into space via radiators located on the inside of the

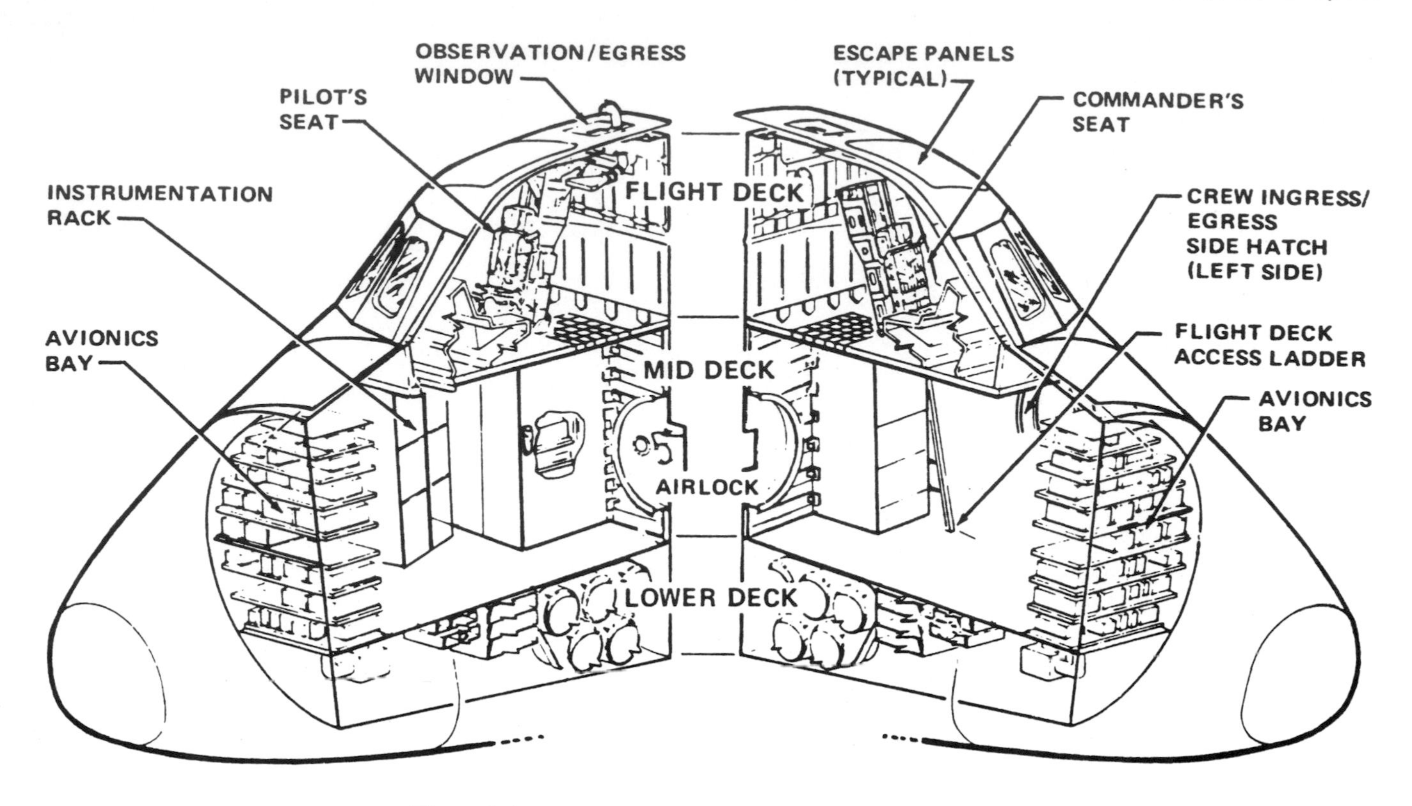

Figure 9.1 Space Shuttle orbiter cabin. *Courtesy of NASA.*

orbiter's cargo bay doors. The doors must be opened and the radiators exposed to space for the system to work. Your refrigerator does the same job, removing heat from the box and dumping it into the kitchen via a radiator on the back or bottom. Notice how hot the radiator gets when the refrigerator is running.

High humidity is not necessarily a hazard, but can be very uncomfortable. Excess moisture is removed by the air conditioning system. As the air passes over cold metal plates the water vapor condenses out, much as water droplets collect on your bathroom mirror when you take a shower. The droplets are then blown off into a waste storage tank. Cosmonauts use the condensed water, but American astronauts found that it had a peculiar taste, probably acquired from dust and other impurities on the metal plates.

After all this recycling and processing are finished, the cabin atmosphere is cleaner and more comfortable than the air in your home. Because machinery sometimes fails, systems are designed with redundancy. Life-critical equipment generally has two backups in case the primary one goes out.

Loss of Atmospheric Pressure

Early in the space age there was concern that a meteoroid could puncture the spacecraft and let the air out. Complete loss of the atmosphere from the cabin would, of course, kill everyone. Although the concern over meteoroid impacts proved to be exaggerated, an underlying and realistic concern is that mechanical failure of some component of the spacecraft, space suit, or helmet could be disastrous. Three Russian cosmonauts died in 1971 when a valve in their Soyuz II failed.

Death comes quickly and quietly when the atmospheric pressure is suddenly reduced to almost zero as it is in the near vacuum of space. First, air in the lungs would be expelled in one quick exhalation. As blood flowed through the lungs, the oxygen in the blood would also be released to space. Symptoms of hypoxia (lack of oxygen)—dizziness and blurred vision—would be evident. Without a continuing fresh supply of oxygen, the brain would die in a few minutes.

Early in the space age it was thought that the blood would boil and a body would literally explode when exposed to near zero atmospheric pressure. Water boils at 212°F (100°C) at sea level, but at an altitude of 6000 feet above sea level (1800 m), where atmospheric pressure is only about 80 percent of sea level pressure, water boils at about 190°F (92°C). Ascending to higher altitudes, one would eventually reach the point where the pressure is so low that water would boil at 98.6°F (37°C), body temperature. This fact led to the false conclusion that, in the near vacuum of space, blood would boil and a body would explode. Skin is a strong membrane, however, and it is more likely that the body fluids would evaporate slowly rather than explosively. The skin would blister and some blood vessels would rupture producing bruises. But the person would have been brain-dead long before.

MATHBOX 9.1

Space Station Atmospheres

Air weighs a lot more than you may think. A 2000 square foot (186 m^2) house with ceilings eight feet high contains 16,000 cubic feet (453 m^3). Air has a density of 0.0755 pound per cubic foot (1.21 kg/m^3) at sea level pressure of about 15 pounds per square inch (101 kPa). The total weight of the air in the house would be $16{,}000 \times 0.0755 = 1208$ pounds. Consider a space station the size of a house with a normal sea level atmosphere, and suppose we substituted helium for nitrogen.

First calculate in common units. The nitrogen, which constitutes 76 percent of the air, would weigh $1208 \times 0.76 = 918$ pounds. Helium has a molecular mass of about 4 while nitrogen has a molecular mass of about 28. Therefore, the weight of helium would be only one-seventh the weight of nitrogen, or 131 pounds. This provides a saving of $918 - 131 = 787$ pounds which can be carried as extra payload.

If the atmospheric pressure in a space station is only 5 pounds per square inch, one-third normal sea level pressure as in Skylab and Apollo, then the total weight of the atmosphere in the house-sized space station would be 403 pounds, the nitrogen would weigh 306 pounds, the helium 44 pounds. Therefore, substituting helium for nitrogen would save 262 pounds.

We can do these same calculations in metric units with the following results. The mass of the air in the house is $453\ m^3 \times 1.21\ kg/m^3 = 548$ kg. This contains $548 \times 0.76 = 417$ kg of nitrogen. An equivalent volume of helium would have a mass of one-seventh of 417 kg, or 60 kg. By substituting helium for nitrogen, we save $417 - 60 = 357$ kg of payload mass. If the pressure is one-third of normal sea level pressure, 34 kPa, then the nitrogen would have a mass of 139 kg, and substituting helium would save 119 kg of mass.

Decompression Sickness—The Bends

In extravehicular activity (EVA), decompression sickness could be a problem. Decompression sickness, technically called dysbarism and commonly called the bends, is well known to deep sea divers. When descending in a diving suit, the pressure in the suit must be increased to keep it from collapsing under the great pressure of the surrounding water. After working under the higher air pressure for a time, the body absorbs an excessive amount of nitrogen. Oxygen is used up in metabolic processes, but nitrogen in the air is simply absorbed by blood and body tissue. On rising to the surface the pressure reduces to normal atmospheric pressure and the nitrogen forms expanding bubbles that move along with the blood in the veins and arteries until they cannot go any farther because of their size. There the bubbles form a blockage, preventing the flow of blood to the surrounding tissue. This condition is not only extremely painful, but can cause death.

Divers cope with this problem by ascending very slowly, reducing the pressure slowly. In contrast, astronauts are subject to the bends when they *leave* the orbiter in space. In the cabin the pressure is 14.7 psi (101 kPa), but pressure in the space suit is only 4.3 psi (30 kPa). Thus the nitrogen in the astronaut's body will expand as it does in a rapidly ascending diver's body. To prevent this from happening, in anticipation of an EVA the pressure in the cabin is reduced to 10.2 psi (70 kPa) for 24 hours prior to the planned EVA so astronauts become acclimated to a lower pressure over a period of time. Also, they breathe pure oxygen for a time before leaving the cabin to allow the body to eliminate most of the nitrogen. NASA has been developing a space suit that will hold an atmosphere at 8.2 psi (56 kPa). If one can be successfully built, bends will be less of a problem. It has been noted that women are at a greater risk than men.

Food

The first American to eat in space was John Glenn in 1962, squeezing applesauce through a tube into his mouth. Before the flight, some experts were concerned that, in weightlessness, food would be difficult to swallow and would collect in the throat. The concerns were unfounded, however, and Glenn had no problem except that the food was not very appetizing. When another astronaut heard about the unappetizing meal, he smuggled a ham sandwich aboard his flight.

For present day spaceflights, all food must be taken along. No running down to the corner convenience store for something you forgot! In the future, on large space stations and interplanetary spacecraft, food will undoubtedly be grown in greenhouses aboard the craft. It would be difficult to carry all provisions necessary for very long trips without resupply from Earth. The weight at liftoff would be enormous. Cosmonauts have already attempted to grow food on their space stations with mixed results. Concepts for doing this will be discussed further in Chapter 11.

Five problems and considerations arise in preparing food for Shuttle flights.

1. Achieving proper nutrition. Astronauts often experience a loss of appetite in space.
2. Preserving food from spoilage. Food must have a shelf life of 6 months at 100°F (38°C). Five basic methods are described below.
3. Preparing and eating in zero g. Bite-sized cubes carried aboard the Mercury flights crumbled and crumbs floated about the cabin. The cubes were later coated with an edible gelatin to control the crumbling. Freeze-dried foods were difficult to rehydrate. Improved plastic packs were developed with a means for inserting the nozzle of a water gun into the package.
4. Making the food and its packaging lightweight and compact. At first, some foods were pureed and placed in aluminum tubes to be squeezed out like toothpaste. However, the tubes weighed more than the food and that concept was abandoned.
5. Developing meals that are appetizing and pleasing to eat. There were early complaints that the food tasted bland.

Water supply aboard the Space Shuttle is not a problem. It is produced by the fuel cells which generate electricity from hydrogen and oxygen. Water is a by-product, made at a rate of nearly 2 gallons (8 liters) per hour. This is four times the amount needed for a crew of seven, so much of it is dumped overboard. With an abundance of water, dehydrated foods are a natural choice. The food is rehydrated as needed.

Food Preparation

Normally, no refrigerator is carried in the orbiter because of the weight, although occasionally one is brought aboard for life science experiments. Therefore, food must be preserved in other ways. Five basic methods are used:

1. Rehydratables: More than 100 foods—including cereals, vegetables, beef patties, soups, spaghetti, and some fruits—can be dehydrated by freeze drying, air drying, spray drying, or other methods. This is the most commonly used method and is likely to remain so. The astronauts rehydrate some of them by simply putting them in their mouths, such as bananas and strawberries. Others, such as shrimp cocktail, scrambled eggs, and beverages are rehydrated with water. Many instant drinks are dehydrated, like tea, coffee, and chocolate. Orange juice, grapefruit juice, and milk do not rehydrate well. They get lumpy and do not mix well with the water.
2. Intermediate moisture: Some foods are partially dehydrated, such as dried apricots and dried peaches. Most of the water is removed, 70 to 85 percent of it, and the remaining water is in chemical combination with sugar or salt so bacteria cannot grow in it.
3. Thermostabilized: Foods are sealed in cans or aluminum-laminated pouches and cooked at temperatures that destroy harmful bacteria and enzymes. Foods processed this way can be stored while retaining their original moisture. These foods include canned fruit in heavy syrup, ground beef with relish and mushrooms, grilled chicken, ham, and stewed tomatoes.
4. Irradiated: As with thermostabilized food, bacteria are destroyed to allow food to be kept at room temperature with its natural moisture. But in this case, the bacteria are killed by exposure to ionizing radiation. Beefsteak and smoked turkey are the only foods on the Shuttle that are treated this way.
5. Natural form: Foods that are naturally low in moisture, such as nuts, hard candy, gum, peanut butter, cookies, and candy bars, can be taken on the spaceship in the same form that they are purchased at the supermarket.

Menus

The menus aboard the Space Shuttle are strictly regulated to give 3000 calories per day, consisting of 16 to 17 percent protein, 30 to 32 percent fat, and 50 to 54 percent carbohydrate, and include the Recommended Dietary Allowances (RDA) of vitamins and minerals. While weightless, the body tends to lose essential vitamins and minerals, so adequate supplies of calcium, nitrogen, potassium, and other minerals are included in the diet. A typical day's menu may include:

Breakfast: Dried pears, beef patty, scrambled eggs, vanilla instant breakfast, and orange juice.

Lunch: Peanut butter, apple or grape jelly, tortilla, fruit cocktail, and peach-apricot drink.

Dinner: Hot dogs, macaroni and cheese, green beans with mushrooms, peach ambrosia, and tropical punch.

Astronauts can make up their own menu which is then checked by a dietitian for proper nutrition. Each package of specially requested food is color coded for that person. There are nearly 100 foods and beverages to choose from. Seven different menus are prepared for 7 days. If the flight is longer than 7 days the menus are repeated. Total weight is limited to 3.8 pounds (1.7 kg) per person per day which includes a pound (0.45 kg) of packaging.

Tortillas are a favorite and they avoid the bread crumb problem, but they must be specially prepared and packaged in an oxygen-free atmosphere to avoid mold. The pantry carries an extra 2-day supply of food in case the mission is extended. Additional snacks and beverages are also stored in the pantry, a supply of favorites requested by the astronauts. Tobacco and liquor are not allowed.

Food tends to taste bland in space, perhaps because of the nasal congestion often experienced by the astronauts or perhaps because the air circulation is low in microgravity. Therefore, a number of condiments are provided for seasoning such as ketchup, mustard, liquid salt and pepper, and hot sauce.

The galley aboard the orbiter is shown in Figure 9.2. Each package is labeled with a flight day and meal number and contains a meal for each crew member. Crew members take turns preparing the meals. The first step is to add water to the dehydrated food by sticking a hollow needle through a small opening in the pouch and kneading it thoroughly. If it is to be heated, it is placed in the 160–170°F (71–77°C) oven. A fan circulates the hot air in the oven so the food heats evenly. Trays are stuck to the wall with clamps or magnets. The food packages are set into the tray with tape or magnets and cut open. Scissors are an essential part of tableware. Containers are cut open and the food is eaten directly from them. The Shuttle food tray is shown in Figure 9.3 and a meal in weightlessness aboard the orbiter *Columbia* is shown in Figure 9.4.

Sticky foods are easier to handle. Foods in gravy, heavy syrup, or sauce stick together and to the container; the forces

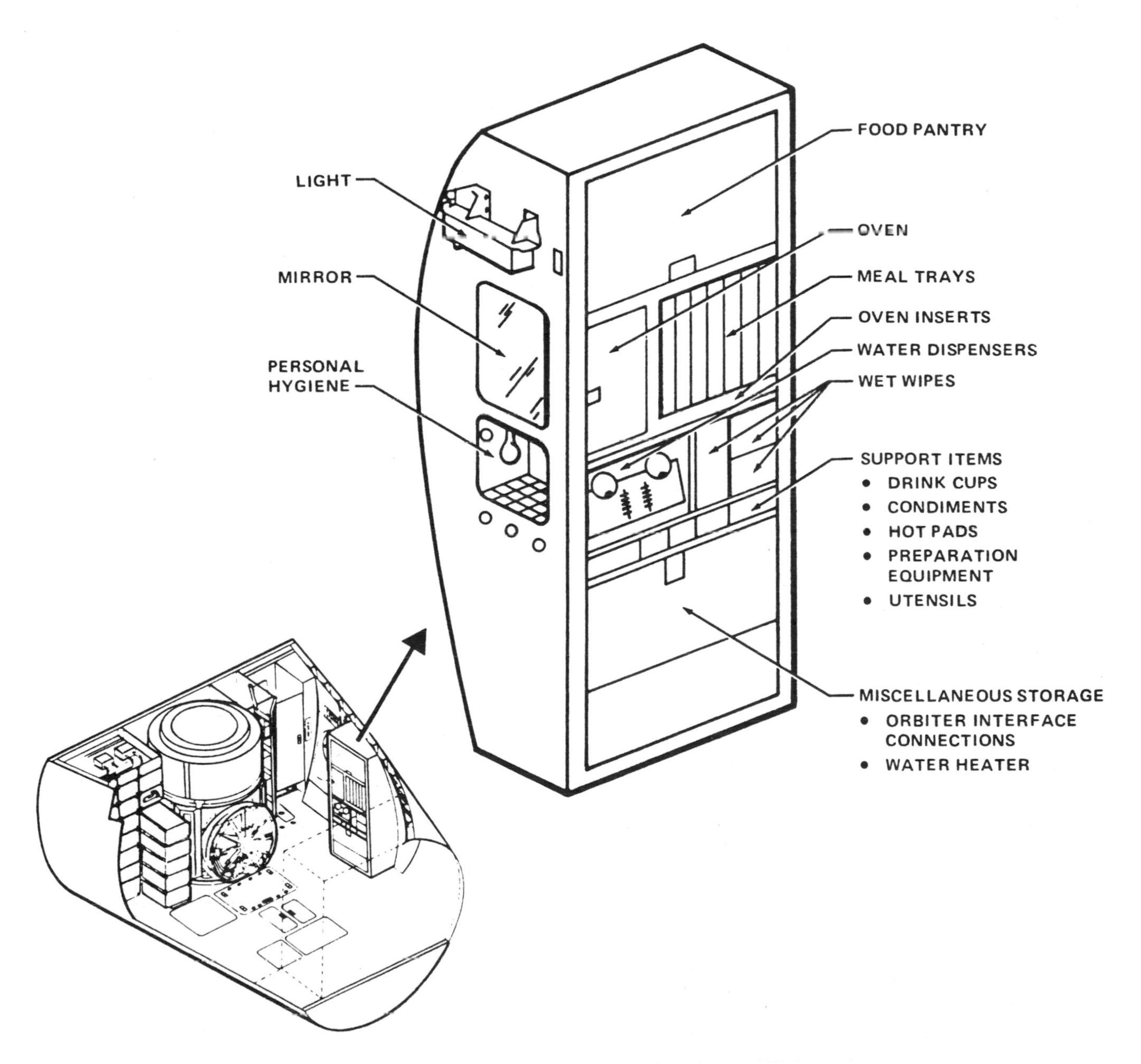

Figure 9.2 Space Shuttle pantry and personal hygiene station. *Courtesy of NASA.*

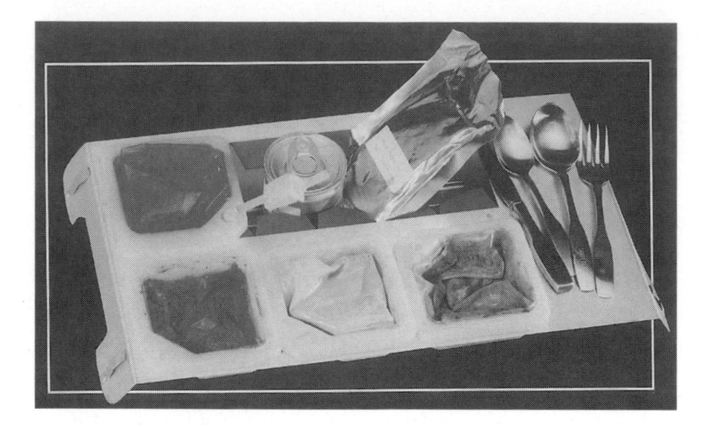

Figure 9.3 Shuttle food tray. This meal consists of (left to right, top row) fruit punch, butterscotch pudding in the can, smoked turkey in the foil bag, (bottom row) strawberries, mushroom soup, and mixed vegetables. *Courtesy of NASA.*

Figure 9.4 Mealtime aboard the orbiter *Columbia*. A full meal tray floats at the right while an empty pudding can floats at the center. *Courtesy of NASA.*

of adhesion and cohesion keep them from floating away. They also stick to forks and spoons so eating is almost normal. There are some cautions, however. Food sticks to the side and bottom of the utensils as well as the top, so a forkful is twice as big. Sudden moves will send the food floating off into space. A straw inserted into a covered container must be used for beverages. Otherwise, when you bring an open container to your lips and stop, the liquid keeps going. In fact, a favorite sport in the orbiter is playing with one's food. Spilled water does not fall down, but gathers into floating globules. Salt sprinkled on the food may scatter into the air. Bread crumbs scatter everywhere, so crumbly food is avoided. A trash compactor takes care of the waste.

Sleep

When you are weightless you do not have to lie down to sleep. Some astronauts have slept while simply floating in the cabin. The problem is, of course, that they drift around and get in the way of the rest of the crew.

There was concern in the early days that a person would not be able to sleep while weightless, and some astronauts have had difficulty. After sleeping for a short time, some have tried to roll over and woke themselves up flailing their arms and legs. Waking with a dizzy feeling resulted from the weightless head bobbing around. Others could not sleep well without the firm feeling of a mattress under them.

Figure 9.5 shows the arrangement of padded bunks in the orbiter. One person sleeps on the top bunk, a second on the lower bunk, a third on the bottom side of the lower bunk facing down, and the fourth vertically at the end of the other bunks. Side panels close for privacy. Fireproof sleeping bags are attached to the bunks to keep them from floating away. Additional sleeping bags can be attached to the locker face either horizontally or vertically for additional crew members. Of course, the words "top," "down," "horizontal," and "vertical" are only in reference to the orbiter's top and bottom. They are meaningless when you are weightless. A waist strap presses the body against the bunk to get the sensation of lying on a mattress. Slipping the hands under the waist strap keeps them from floating in space.

In a 200-mile (320 km) orbit, the Sun rises and sets every hour and a half, so there is no long dark night for sleeping. Eye shades and earplugs reduce disturbing light and noise for those who want to use them. If the entire crew sleeps at the same time, two must wear communications headphones in case an emergency warning signals or the ground controllers call.

Hygiene

The personal hygiene station is part of the orbiter's galley, shown in Figure 9.2. A globular-shaped chamber with two openings is used for washing hands without getting water all over the place. The Skylab space station was equipped with a

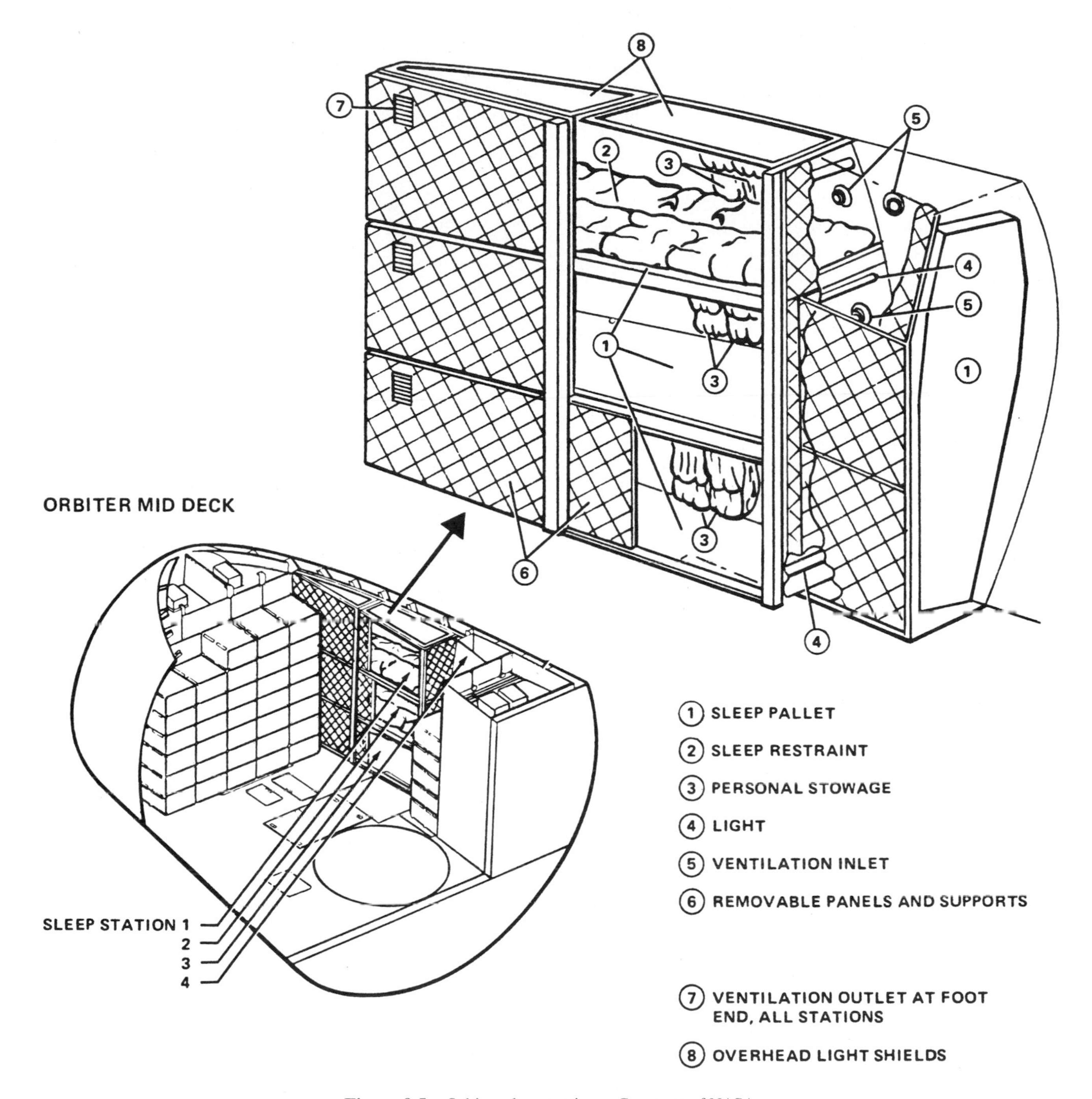

Figure 9.5 Orbiter sleep stations. *Courtesy of NASA.*

shower. A curtain enclosed the astronaut. He used a spray to squirt water on his body and lather up, then used a vacuum cleaner device to remove it from his body, the curtain, and walls. It did not work very well. Some water escaped from the shower stall and had to be chased around and vacuumed up.

There is no shower on the Space Shuttle, so astronauts must wash by hand. At the wash station, a handgun supplies water at a comfortable temperature between 65°F and 96°F (18°C and 36°C). Water clings to the body in weightlessness and makes the job easier than you may expect. Each member of the crew gets two washcloths and two towels per day. One washcloth is used to soap up and a second to rinse off. Waste water goes into a storage tank beneath the floor. For privacy, a curtain can be pulled across the front of the personal hygiene station. Towels, washcloths, and other personal hygiene items are equipped with Velcro so they can be stuck to the wall without floating away.

Dry shaving with an electric shaver in weightlessness caused problems because the whiskers floated off around the cabin causing eye and lung irritation and getting into equipment. A wind-up electric shaver with a vacuum cleaner attached to suck up the whiskers helped, but didn't give a comfortable shave according to at least one astronaut. A depilatory cream or gel can be used to melt off the whiskers. Shaving cream and safety razors seem to work best. Some astronauts feel it is simply less trouble to grow a beard.

Cleaning up immediately after a meal is more important on the Shuttle than at home. Odors and bacteria from food wastes are unwelcome in the small enclosed space of the orbiter. Garbage and trash are sealed in plastic bags in the trash compactor. Wet wipes containing a strong disinfectant are used on utensils and trays as well as around the eating area. Used clothing is also sealed in airtight plastic bags.

A question frequently asked astronauts on their public appearances is, "How do you go to the bathroom in space?" Figure 9.6 is a diagram of the Space Shuttle's commode for collecting human waste. It had to be carefully designed for use in weightless conditions by both male and female astronauts. Toeholds, handholds, and a waist restraint, like an automobile seat belt, hold the occupant firmly in place to assure a good seal between the user and the seat. A reading light, a nearby window, and privacy curtains are part of the commode area. Female astronauts have continually had trouble with the toilet. While it easily collects urine from a man using a tube, women have a more difficult time. It has been a constant annoyance and better designs have been sought.

Then comes the problem of what to do with the waste. The commode has separate receptacles for feces and urine. In the absence of gravity, high speed air streams carry the solid and liquid waste into their respective receptacles. Solid waste is vacuum dried, chemically treated with germicides to prevent odor and bacteria growth, and stored until return to Earth. Liquid waste joins other waste water in the storage tank beneath the floor. When the tank fills up, it is dumped overboard.

Waste is not recycled during Space Shuttle flights. Everything is saved in the orbiter and returned to Earth. On longer trips to other planets and beyond, it will be advantageous to recover whatever can be reused in order to reduce the weight of the vehicle on the launch pad. Whether or not there is an advantage to recycling depends on the weight of the recycling equipment plus the initial supplies compared to the weight of use-once-and-throw-away materials. Present recycling technology gives no weight advantage for flights less than 2 years in duration. This will be discussed in more detail in Chapter 11.

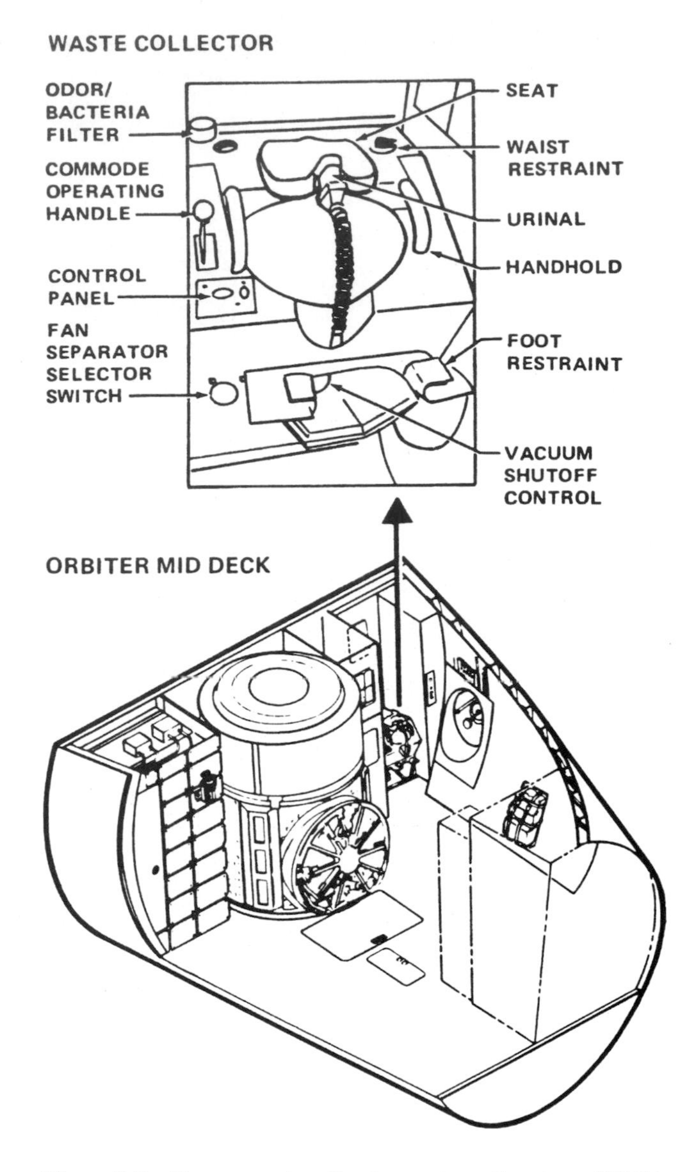

Figure 9.6 Human waste collection system. *Courtesy of NASA.*

Radiation

Radiation in space consists of electromagnetic waves from the Sun, high speed protons and electrons in the solar wind, cosmic rays from the Sun and beyond the Solar System, and trapped particles in the Van Allen belts. (Recall Chapter 4.) Some of these pose a serious threat to human life. On the ground, we are protected from this radiation by Earth's atmosphere and magnetic field. The cabin of a spacecraft provides protection from electromagnetic waves and some protection from the particles. However, when a primary cosmic ray strikes the vehicle walls it produces secondary radiation by ionizing the materials of which the walls are made. This secondary radiation may penetrate into the cabin. (Secondary radiation is produced in the upper atmosphere of Earth also, when cosmic rays strike the molecules of atmospheric gases. Few primary cosmic rays reach the ground, but the secondaries do.)

When high energy protons, cosmic rays, x-rays, or gamma rays enter a living cell they ionize the atoms and split apart the molecules of the cell, disrupting its operation and perhaps killing the cell. Recall that cosmic rays are the nuclei of atoms that have lost all their electrons. They are mostly hydrogen and helium nuclei (protons and alpha particles) with a few nuclei of heavy atoms such as carbon, oxygen, and iron. One of these heavy cosmic ray particles does more damage than a proton because it is more massive, carries more positive electrical charge, and can cause more ionization in the cell. A single heavy cosmic ray particle will destroy a cell. In total, though, protons do more damage simply because there are so many more of them.

The human body is continuously manufacturing new cells and the loss of a few is no big thing, but exposure to high dosages of radiation results in the destruction of many cells in a short time. Then the body has a problem, particularly if the cells are in a vital organ.

A particular problem of a woman exposed to high radiation dosages is that one or more of her latent egg cells may be damaged. A woman is born carrying all the latent egg cells she will ever have. They develop at the rate of about one per month during her reproductive years. If one is damaged and later is impregnated, the fetus may be malformed. Male sperm cells are produced frequently and are not subject to the same lifelong concern as a woman's egg cells. However, a man can become sterile if he receives a large dose of radiation in a short time, causing the sperm-creating organs to stop functioning.

Radiation dosage is measured in rads (radiation absorbed dose), a quantity of energy absorbed per unit of body tissue. For humans the lethal dose is about 600 rads in a day. Exposure to 100 to 200 rads per day may cause nausea, vomiting, and fatigue. Biological damage from radiation is difficult to predict and the effects vary from person to person, depending on body weight and general health. To further complicate the situation, different particles have different abilities to damage biological tissue. Alpha particles do more damage than x-rays. A "quality factor" is applied to each type of particle to specify how much damage that particle can do. The quality factor is 1 for x-rays and electrons, 2 to 25 for neutrons and protons, and more than 15 for alpha particles and heavy nuclei. When the dose in rads is multiplied by the quality factor, the result is called the dose equivalent in rems, which is a better indicator of the damage done.

However, except in a research laboratory, there is no easy way to tell just what kind of particle is being intercepted. Under normal conditions, the radiation in a spacecraft cabin is lower than first anticipated. During the Skylab's longest mission of 84 days, total radiation received was less than 8 rads. NASA sets a limit of 75 rads per year whole body dosage and a maximum career limit of 400 rads. Astronauts carry dosimeters to monitor their exposure. Instruments in the spacecraft cabin also monitor the radiation level.

Solar Proton Storms

Large solar flares may produce cosmic ray proton storms. The radiation intensity from these events occasionally reaches levels lethal to humans outside the protection of the Earth's atmosphere. It would be particularly hazardous in a polar orbit inside the magnetosphere because the Earth's magnetic field deflects the particles toward the north and south pole. Consequently, the Space Shuttle has never been launched into a polar orbit. During a proton event an astronaut could be exposed to more radiation in half an hour than NASA standards allow for a year. For the most energetic of these events, even the cabin walls of the orbiter would not provide sufficient shielding.

A solar proton storm in October 1989 and one which lasted longer than a week in August 1972 were intense enough to have killed an astronaut on an EVA outside the spacecraft in polar orbit. On August 4, 1972, the proton flux as measured by unmanned satellites translated to over 100 rads per hour for about 8 hours. Exposed to such a high dose, the astronaut would have lived 4 to 6 days in severe pain with diarrhea and constant vomiting. Exposure to the smaller dosages inside the cabin might not have killed the astronauts, but they might just wish they were dead! Vomiting and diarrhea, leading to dehydration and emaciation, and the destruction of blood cells would have led to death for some of them within a month.

The Space Shuttle was not flying during those two storms, but an event did occur in August 1989 while a Shuttle was in orbit. NASA considered bringing the orbiter home early, but it was at the end of the mission anyway, and increased particle radiation was observed only on the last two orbits.

Solar activity was discussed in Chapter 4. It is highly variable and not completely understood. Scientists have only a rudimentary knowledge of the causes of solar proton events, and forecasting their occurrence has only limited accuracy. However, given that a flare has occurred, the forecaster can state with some reliability whether or not high energy protons have been blown into space and spacecraft at the L1 point can observe them leaving the solar corona. The forecaster can then issue a warning to the astronauts who would have from 20 minutes to perhaps 3 hours to take cover in the best shielded part of the spacecraft or to return home. Since reentry and landing at a particular point on Earth are possible only from a certain starting point in orbit, this may not be a feasible emergency procedure.

The inhabitants of a space station would have to ride out the storm. The only practicable protection from radiation is shielding by a thick mass. Some materials are more effective shields than others, but any kind of mass will afford some protection. Lead is perhaps the best commonly available material, but is not practical on a spacecraft because of its weight. Any habitat designed for a long duration flight should include shielding against normal day-to-day radiation and a heavily shielded compartment, a "storm cellar," for protection during a large solar storm.

Low Level Radiation

We have a pretty good understanding of the effects of high radiation dosage on people based on studies of the atomic bomb explosions in Japan during World War II. However, the effects on the body of continuous exposure to low radiation levels such as galactic cosmic rays are not well understood. We on Earth are continually exposed to low level radiation from cosmic rays entering the atmosphere, radioactive rocks and gases, and dental and medical x-rays. We often hear statistics such as "exposure to such radiation will lead to two deaths from cancer per 100,000 people in 50 years." The question is, who are those two and would they have died from something else in those 50 years? Certainly driving an automobile, smoking, and a dozen other human activities are more hazardous than low level radiation. But here the discussion gets emotional rather than analytical. People can limit many of their hazardous activities if they want to. Radiation is frightening to many people because it cannot be seen and it is difficult for the average person to limit radiation exposure.

The normal level of intensity of galactic cosmic rays is rather constant, but researchers differ widely in their estimates of the hazard. One source estimates that continuous exposure to galactic cosmic rays for a year would exceed the safe dosage by some 60 times. Another says there is little danger, that the total annual dose is from 6 to 20 rads. The main reason for this disparity is that researchers disagree as to the biological effect of protons. Estimates of their damage range from one to ten times the damage done by a similar dose of x-rays. While it is possible to accurately estimate the number and intensity of proton encounters, there is wide disagreement as to how to convert that to rads. We do know that for short trips the dosage is not excessive. The highest total dose received by an Apollo crew on a trip to the Moon was a little over 1 rad. The greatest hazard on a long journey, such as a 2-year trip to Mars, comes from the proton storms following large solar flares. Some sort of protection other than the spaceship itself will have to be provided.

The trapped particles are also a serious hazard to people in space, but the Van Allen belts are well mapped and generally avoidable.

One interesting effect of cosmic rays is the occurrence of light flashes in the eyes, first reported by Buzz Aldrin on the Apollo 11 flight to the Moon. Other astronauts have also reported seeing flashes of light with their eyes closed. The most obvious explanation of this phenomenon is that cosmic rays penetrate the eye and deposit in the retina, sending a signal down the optic nerve which the brain interprets as a flash of light. However, a variety of shapes were observed such as streaks, commas, and diffuse clouds. These are difficult to explain.

In summary, people in space are exposed to higher levels of radiation than people on Earth even if the Sun is quiet. The short duration of Space Shuttle flights keeps the exposure to tolerable levels. The long-term effects on any particular individual cannot be predicted. This question will become more demanding of an answer as people live for longer periods in the space station, on the Moon without an atmosphere, and on Mars with only a thin atmosphere.

Gravity and Weightlessness

A person on Earth, along with everything else, experiences a constant acceleration due to gravity of 32 feet per second per second, called 1 g. Just as a car increases speed when the driver presses down on the accelerator, any unsupported object would fall toward the center of the Earth with an ever increasing speed. An object accelerating at 1 g would be falling at a speed of 32 feet per second after 1 second, 64 feet per second after 2 seconds, 96 feet per second after 3 seconds, and so on. In metric units, the acceleration of gravity, 1 g, is 9.8 m/s^2. An unsupported object in free fall would be falling at a speed of 9.8 m/s after 1 second, 19.6 m/s after 2 seconds, 29.4 m/s after 3 seconds, and so on.

The 1 g acceleration of gravity is always with us on Earth. There is no way to "turn off" gravity. We are born with it and we live and die with it. The human body is designed and conditioned to operate in a 1-g environment.

The term "g force" or "a force of 3 g" is somewhat misleading because g is not a force. Rather it is the acceleration due to the force of gravity. However, in common usage, "a force of 3 g" means any acceleration that gives a sensation of weighing three times more than normal. It makes no difference whether the acceleration is due to gravity or due to a rocket engine. The sensation is the same and the effect on the body is the same. We will use the term "g force" in that context.

When going into space, astronauts experience dramatic changes in g forces. First, while the engines are running and accelerating the spacecraft, they experience more than the normal 1 g of gravity. Then when the engines shut down, the spacecraft is in freefall and they are instantly weightless. This does not happen gradually. The thrust suddenly stops and they immediately float! When the rocket motors are retrofired to return to Earth, they experience an acceleration again and its accompanying feeling of weight. When the thrusters or OMS engines are fired in orbit, the astronauts experience an acceleration and if they are strapped in, would feel as if weight had returned momentarily.

High g

Let us first examine the effects of a greater than normal g force on the body. Experience in high speed jet aircraft and experiments in a centrifuge and on a rocket-powered sled in the 1950s provided a good idea of the tolerance of the human body to higher than normal g forces. The record was set by Col. John Stapp when he decelerated at 35 g on a rocket sled at White Sands, New Mexico, in 1954.

While the general outer body structure is relatively rigid, the fluids within are free to move. Internal organs in the chest and abdomen are not rigidly fixed in place and can also move. Human bodies are well adapted to function in 1 g. Flexing muscles in the legs forces blood upward in the veins against normal gravity, helping the heart to keep the blood circulating. One-way valves in the veins also help prevent the blood from pooling in the lower body. At normal 1 g the internal organs hold their proper positions.

Problems occur if a person is sitting or standing erect and an acceleration greater than normal 1 g acts lengthwise, head to feet. Leg muscles cannot supply enough force to prevent blood from leaving the head and pooling in the lower part of the body. Lack of blood in the brain causes dizziness and possible blackout. You feel this sensation briefly when an elevator starts suddenly upward or when a roller coaster reaches the bottom of a curve. Col. Stapp in his rocket sled showed that very high g forces can be tolerated if the acceleration is directed front to back, perpendicular to the chest. The key factor in how much a body can withstand is the direction of the acceleration.

In the early manned flights into space, each astronaut had a contour seat specially fitted to his form. During launch into orbit he was lying back in the seat facing forward. During reentry he rode backward, again with the acceleration directed back to front, perpendicular to his chest. This allowed him to withstand more than 8 g without blacking out.

The Space Shuttle is designed to operate at lower accelerations during launch. The engines are throttled so that the acceleration reaches a maximum of only 3 g twice during the rise to orbit, once just before the solid rockets burn out and separate, and a second time for about 5 minutes just before the external tank empties and drops away. This is for the safety of the spacecraft as well as the comfort of the crew. For comparison, 3 g is less than may be experienced on some carnival rides. The crew sits up in seats like riding in an airliner and the Shuttle accelerates nose first so the g forces are perpendicular to the chest. Just after liftoff, the Shuttle has greater acceleration than the Saturn boosters that carried men to the Moon. However, the Saturn attained much higher acceleration just before the first stage rocket burned out.

During reentry Shuttle crews encounter 1.5 g. Because the orbiter enters the atmosphere bottom side first, the force on the crew acts lengthwise as they sit upright in their seats. After several days or more in weightlessness a person could be especially susceptible to pooling of blood and blackout. Therefore, crew members wear antigravity suits during reentry. An antigravity suit looks like a pair of pants with inflatable bladders. As the g force increases, the bladders are inflated and apply pressure to the lower part of the body, forcing blood toward the upper body and brain.

Weightlessness

Many of the problems a human body confronts in space are related to what is commonly called weightlessness or microgravity or the absence of gravity or zero g (Figure 9.7). None of these terms is precisely correct. The force of gravity does decrease with distance from Earth. However, at orbital altitude an orbiter is not beyond Earth's gravity. (Recall MATHBOX 2.1.) At 300 miles (180 km), barely off the ground, g is only 10 percent less than it is on the ground. A spacecraft could not remain in orbit if gravity were zero. (Recall Newton's first law of motion: an object moves in a curved path only if some force deviates it from a straight line path. In the case of an orbiting spacecraft, gravity is the force.) If the astronauts were not in free fall, their weight would be nearly the same as it is on Earth. So the terms *zero g* and *microgravity* are somewhat misleading, although they do communicate the intended meaning.

As discussed in Chapter 3, an object in orbit is in free fall, falling toward Earth, but with sufficient horizontal velocity that it never strikes Earth and therefore remains in orbit. On Earth, the supporting upward force from the chair we sit on or the floor we stand on prevents us from falling downward. It is this force that gives us the sensation we call "weight." But in orbit, the floor is in freefall, too. It does not provide any support because the entire spacecraft is falling the same as the crew is. Since there is no upward supporting force as there was on Earth, the astronauts feel weightless.

People on Earth feel the same sensations and experiences as astronauts, but only for a fraction of a second at a time while in freefall in elevators, off diving boards, and on trampolines. Astronauts, without support from beneath, are in freefall for days. Imagine that you jump off a diving board with an apple. You let go of the apple and it falls alongside you. To you it appears that the apple is floating in space.

As described earlier, the problems of high g forces on the human body are well understood and taken into account in the design of the spacecraft and in its mission profile, In contrast, problems with weightlessness are not yet solved.

Under weightlessness these are some of the effects on a human body:

a. Fluids shift from the lower part of the body to the upper. Lower body muscle structure is such that, under normal gravity, blood is forced upward to keep it from pooling in legs. Even in weightlessness these muscles continue to force blood and other body fluids upward.

b. Dehydration, loss of body fluids, and loss of lymphocytes. When the fluid shifts upward, the body attempts to eliminate some of it.

c. Cardiovascular deconditioning and red blood cell loss. The heart does not have to work so hard to move weightless blood through the arteries and veins. Total quantity of body fluids is reduced, including blood.

d. Muscular deconditioning. Muscles do not have to work at normal levels so they begin to atrophy. It takes days or weeks to recover, depending on the length of time the per-

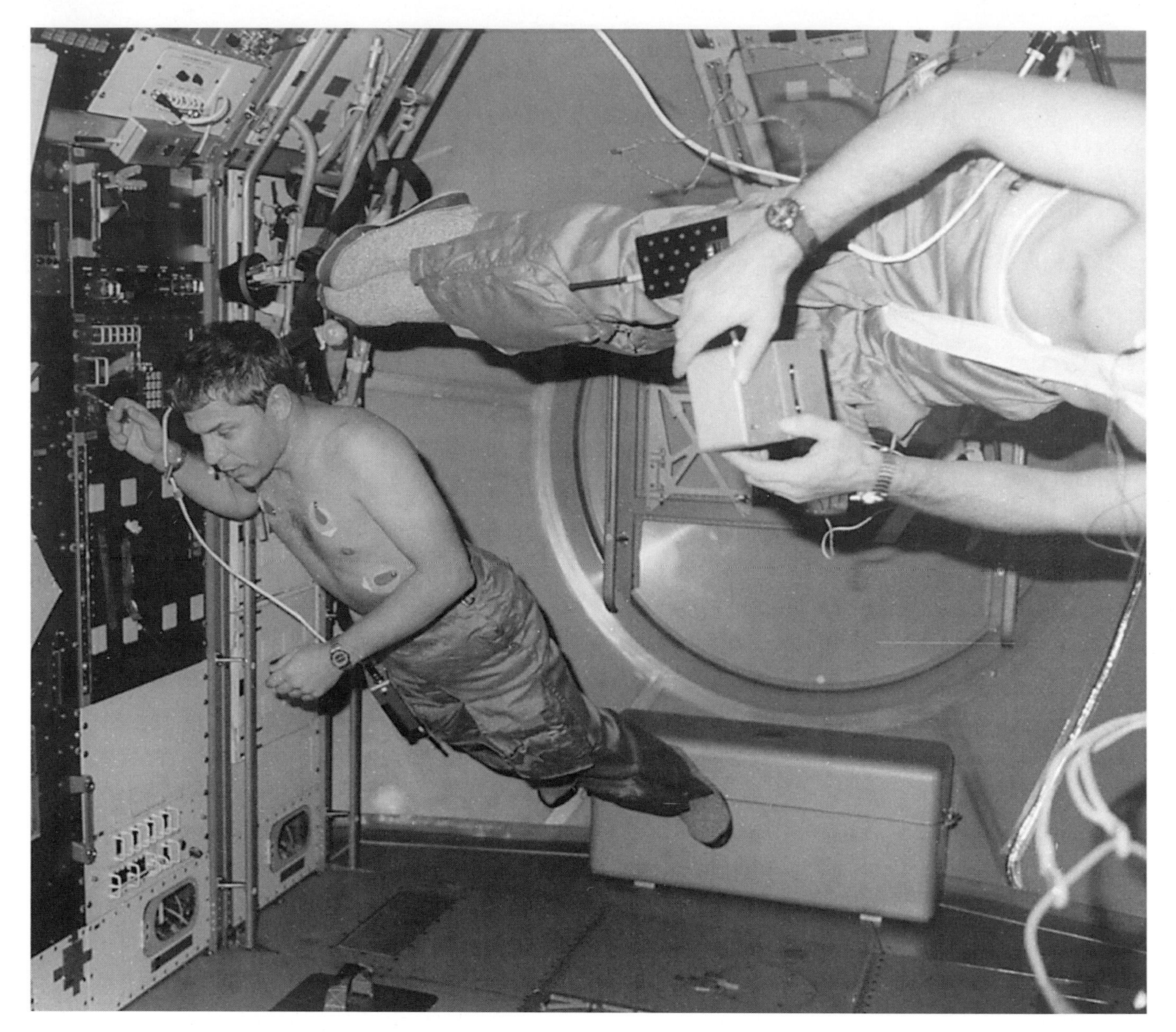

Figure 9.7 Weightless astronauts working in Spacelab aboard orbiter *Columbia*. In a normal relaxed position, knees bend slightly, head raises or tilts back, shoulders hunch up, and arms float upward. It requires some effort to bend at the waist or to hold arms down to waist level. *Courtesy of NASA.*

son was weightless and how much exercising was done during the flight.

e. Calcium loss from bones. The loss continues throughout the entire flight. The longer the flight, the more calcium is lost. Bones do not repair themselves as they do with normal gravity.

f. Space adaptation syndrome. Space sickness lasts a day or two, affecting about half the people who go into space.

g. Changes in the number of neural connections. This increases in weightlessness and decreases in high g.

When the spacecraft reaches orbit and the engines stop, the vehicle and its contents are weightless, in freefall. At first there is a great feeling of euphoria, an absence of force on the body, and the experience of freely floating about in the cabin. Without the downward force of gravity, the space between vertebrae in the backbone stretches and the average astronaut gains an inch or two in height. As a result, many astronauts have complained of back pains. The natural posture of a weightless astronaut has knees and hips bent like a crouch, with arms floating out in front. Straightening out takes considerable effort. Of course, while in orbit there is no reason to straighten out.

After a short time, body fluids begin to shift. Muscles in the lower limbs that work against gravity on Earth, pushing blood upward in the body, continue to work. Blood vessels in the legs are soon emptied because of the "absence of gravity." Other fluids from the surrounding tissue flow into them. The upper body becomes somewhat swollen as the blood and other fluids move in that direction. The face becomes swollen in appearance as the normally loose flesh fills with fluid and "floats" on the cheekbones. Wrinkles and double chins disappear and eyes appear smaller. At the same time, the lower body, having lost the fluid, becomes thinner. Waist measurements decrease by an inch or two. Feet and legs become noticeably smaller; boots and pants may not fit as well. Astronauts called this "bird legs" and "wasp waist."

All this is accompanied by head congestion, a full stuffy head, throbbing in the neck, and perhaps a severe headache. The sensation is similar to that of hanging upside down on a gym bar.

Meanwhile, the body misinterprets what is happening. Sensors in the upper body indicate excess fluid and the kidneys begin action to eliminate it in the form of urine. One or two quarts of fluid and the dissolved salt are released this way in the first few days. The individual may not feel thirsty and does not replace the lost fluid by drinking water. Dehydration could be a problem.

Most of the symptoms described above disappear in about the first 3 days in orbit, either by the body self-adjusting its functions or by the mental attitude of the individual simply "getting used to it." The cardiovascular circulatory system settles down to a more normal operation but with a smaller volume of blood. Loss of calcium in the bones is an exception. It continues throughout the flight. A recent study suggests that this may be due to a reduced ability of the pituitary gland to secrete certain hormones which control growth. Over a period of several years, researchers sent both rat pituitary cells and live rats on flights of the Space Shuttles and the Russian biosatellites. Pituitary cells produced only half as much active hormone as similar cells on Earth. This is an interesting study, even though no connection between the hormone loss and calcium loss has been proven.

Prior to reentry on the last day of the flight, astronauts take salt tablets and drink about a quart of water to begin the process of renewing the blood volume to its normal Earth level. Recently a researcher concocted a drink she called "Astroade" that requires much less volume for rehydration. In one test by an astronaut on a Shuttle mission, the drink appeared to help the dehydration problem.

On returning to Earth after a flight, about one in ten astronauts is unable to stand upright and walk off the orbiter without help. Because of the low blood volume, the symptoms are similar to giving several pints at the blood bank. It may take several days for the cardiovascular system to readjust to normal gravity.

Space Adaptation Syndrome

About half the people who have ventured into space have reported becoming sick on the first or second day of the flight. Earliest onset was during the first hour. The symptoms are similar to motion sickness, being seasick or airsick. From one-fourth to one-half of the astronauts have reported vomiting, headache, malaise, fatigue, or loss of appetite, or a combination of those symptoms. The symptoms subside suddenly after 2 or 3 days and do not recur on that flight. Virtually all astronauts recover. Apparently the body adapts to whatever causes it.

This illness is not the same as motion sickness. All the early astronauts had been test pilots and were not susceptible to airsickness. Yet many of them suffered from space sickness. Tests on Earth do not identify those who may be susceptible. It becomes apparent only when they reach orbit. Those individuals who get sick on one flight will likely get sick if they make another flight. Conversely, those who do not get sick on one flight probably will not on another. The common motion sickness drugs such as dramamine do not work against space sickness. Another drug has been developed to counter space sickness, but many astronauts prefer not to take it.

Space sickness has been the subject of considerable research effort. Because the symptoms are similar to motion sickness, it is thought that the cause lies in the vestibular apparatus, the semicircular canals in the inner ear that give us a sense of balance and provide information to the brain on accelerations and rotations. The relationship between vomiting and this apparatus has been shown by experiment. Monkeys and people who have defective semicircular canals or have had them removed cannot be made to vomit when subjected to unusual acceleration or rotation.

A person on Earth can tell up from down and estimate direction and speed of motion from visual observation and from feeling the force of gravity. When weightless, however, information sent from the eyes and from the vestibular apparatus may conflict. One may indicate you are upside down while the other says you are tumbling. After a day or two in orbit, a person adapts to the confusing signals and can function normally (almost). When they return to Earth, crew members have the same problem, causing difficulty with balance. It may take a few days to readapt to the gravity environment.

One NASA researcher, Dr. Patricia Cowings, equipped a laboratory with motion equipment designed to make people sick. One of the symptoms described to her by astronauts is that they have the feeling they are standing upside down. This seems related to the shift of fluids to the upper part of the body. Dr. Cowings suggested that biofeedback might help solve the space sickness problems. The subjects in her experiments learn to control their own heart rate, blood flow to their hands, facial pallor, and perspiration.

A simulator has been tested at Johnson Space Center that gives different visual cues than the vestibular system would expect. This causes confusing relationships to be fed to the brain, similar to those experienced in weightlessness. The purpose of the simulator is to teach the individual to rely only on what is seen, not on what the vestibular system is feeding to the brain. This type of training should help inexperienced astronauts to adapt more quickly to weightlessness on their first flight and may enable them to control their space sickness symptoms as well.

Space Adaptation Syndrome did not show up in Mercury and Gemini astronauts, perhaps because their spacecraft were so small that their visual perception did not conflict with their vestibular system. As spacecraft became larger, it was easier for astronauts to see how they were moving and oriented in the space. Many of the Skylab astronauts suffered space sickness in their large space station. In Skylab, equipment, switches, and labels were mounted without respect to up and down. One piece of equipment was sideways to others. Even the commode was attached to a wall where one would expect a window to be. This disregard for the normal human expectation to have a floor, ceiling, and walls in a room may play a part in the onset of space sickness. Designs for the new space station do make the living quarters and laboratories more earthlike with a familiar up, down, and sideways orientation.

Exercise

A regular regimen of exercise as part of the astronaut's daily routine is essential to keep the bones, heart, and muscles in condition. A stationary bicycle was used on the Skylab space station. Treadmills, stationary bicycles and rowing machines are excellent means of exercising in weightlessness. Spring-type exercise devices also work in weightlessness, but weight lifting is useless! The treadmill shown in Figures 9.8 and 9.9 uses shoulder and waist straps to keep the astronaut in contact with the machine. Originally, the equipment was attached directly to the floor or bulkhead of the orbiter. Experience revealed that an astronaut peddling a stationary bicycle can produce as much as 100 pounds (445 N) of force which can cause severe vibrations of the entire orbiter. A platform for the exercise equipment was then designed with special stabilizers to absorb vibrations, reducing the force to less than 1 pound (4.45 N), so astronauts do not disturb sensitive microgravity experiments when they work out.

Sweat does not run off in weightlessness. It clings to the body and builds up to a thick layer. If it breaks loose, the sweat forms spherical globules floating around the cabin. If one strikes a solid surface, it spreads out like pancake batter and sticks to the surface. A strong air current from a nearby duct helps evaporate the sweat so it does not become a problem.

Figure 9.8 Exercise treadmill.

For Shuttle astronauts, at least 15 minutes of vigorous exercise per day is recommended for flights up to 2 weeks and 30 minutes a day for missions up to 30 days. The Russians have far more experience in spaceflight than the United States. While they are working, cosmonauts wear an exercise suit, nicknamed the penguin suit, laced up with elastic cords that create a force against the body and require exertion to do nearly any task. They eat a special diet to combat physical deterioration and keep a more rigorous exercise schedule including daily 2-mile runs on a treadmill. Even with a vigorous exercise program, calcium loss continues and muscles do not remain as strong as they would on Earth. The Russians report that they have reduced calcium loss to 5 percent at most, compared to as much as 30 percent in early flights. They also report a wide variety of reactions to weightlessness among their cosmonauts. Following a flight of several months by two cosmonauts, one showed a 10 percent de-

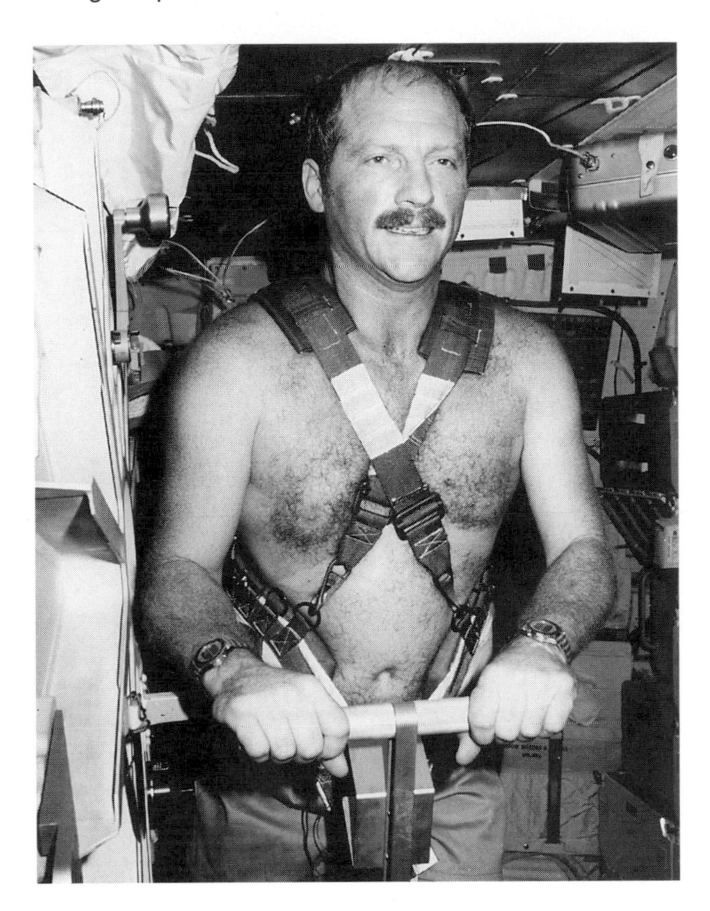

Figure 9.9 Astronaut Fred Hauck exercises on the treadmill. Straps are needed to keep the weightless astronaut in contact with the machine. *Courtesy of NASA.*

crease in bone density in his vertebrae while his companion showed no loss at all.

Still, depending on the length of the flight, it may take days or even weeks to recover after returning to Earth. After his record 366 days in space ending in December 1988, cosmonaut Vladmir Titov reported that he was able to operate the controls of the spacecraft during reentry in a semiprone position. Back on Earth, however, he had difficulty sitting up for very long. But a few months later, Titov and Musa Manarov, the other cosmonaut on the yearlong flight, were traveling the world reporting on their experiences and appeared none the worse from their extended flight.

The human body may be the limiting factor in further exploration of space, particularly on such long missions as a 2-year flight to Mars. It may be necessary to produce an "artificial gravity" by rotating the space ship. The Russians believe that on long duration flights, psychological factors may become more significant than physical factors—more about these subjects in Chapter 11.

DISCUSSION QUESTIONS

1. A deep sea diver who experiences the bends on return to the surface is put into a pressure chamber. Why?
2. What are your favorite foods? How would you package them for eating in weightlessness?
3. Think of your daily habits of personal hygiene and cleanliness, such as combing hair, clipping fingernails, and dressing. What problems would have to be overcome while trying to do these things in weightlessness?
4. Newton's first law says that an object does not change its motion unless a force is applied. Why, then, does liquid float out of a cup and food float off a plate in an orbiting spacecraft?
5. Describe exercises that could be done in weightlessness.
6. Describe what it might be like to play basketball or handball in weightlessness.

ADDITIONAL READING

Compton, W. David, and Charles D. Benson. *Living and Working in Space, A History of Skylab.* NASA SP-4208, Government Printing Office, 1983. Very comprehensive hardback.

Kerrod, Robin. *The Illustrated History of Man in Space*. Mallard Press, 1989. Profusely illustrated, easy reading.

NASA. *Food for Space Flight.* NASA Facts, NF-133/6–82, NASA, 1982. Easy reading pamphlet.

Oberg, James and Alcestis. *Pioneering Space: Living on the Next Frontier*. McGraw-Hill, 1986. Popular book for the interested layperson.

Page, Lou and Thornton. *Apollo-Soyuz Pamphlet No. 6, Cosmic Ray Dosage*. NASA, 1977. Technical description of the experiments.

Pogue, William. *How Do You Go to the Bathroom in Space?* Doherty Associates, 1985. Delightful, fun to read; question and answer format.

Possehl, Suzanne Rene. "Keeping Life Sciences Alive in Russia." *Aerospace America,* March 1997. The state of research in Russia.

Rambaut, Paul C. *Space Medicine*. Carolina Biological Supply Co., 1985. Excellent, readable pamphlet.

Rogers, Terence A. "Physiological Effects of Acceleration." *Scientific American,* February 1962. Early high-g experiments; out-of-date conclusions on weightlessness.

Steinberg, Florence S. *Aboard the Space Shuttle*. NASA EP-169, Government Printing Office, 1980. Easy reading pamphlet.

White, Ronald J. "Weightlessness and the Human Body." *Scientific American,* September 1998.

NOTES

Chapter 10

Working in Space

The Space Shuttle has made it rather easy for people to get into orbit and to do various jobs there. A list of the many impressive things done by Shuttle astronauts includes:

1. Drop off satellites in orbit.
2. Recover, service, and repair satellites in low Earth orbit, such as the Hubble Space Telescope and Solar Max.
3. Build large structures outside the orbiter.
4. Measure the plasma environment around the orbiter.
5. Manufacture perfect spheres and crystals.
6. Prepare pharmaceutical products by separating biological cells from one another.
7. Study effects of weightlessness on people, plants, and animals.
8. Refuel a satellite.
9. Use photography, radar, and infrared to make high resolution images of the ground.
10. Measure solar and cosmic radiation.
11. Ferry crews, supplies, equipment, and waste between Earth and space stations.

Some of the work has direct commercial application, some is for developing new techniques for working, manufacturing, and processing materials in space, and some is basic research for a better understanding of the world around us.

Tools of the Trade

Good tools make any job easier. When working in weightless conditions, jobs become easier in some ways and more difficult in others. Just because an object is weightless and floating does not mean that it requires no work to move it. It takes some force to get it started and an equal but opposite force to stop it. A force must also be applied to change its direction of motion (Newton's laws again).

Drilling a hole or tightening a screw is no easy task. When a weightless astronaut inserts a screwdriver into the head of the screw and applies a twisting force, the screw may remain firmly in place while the applied force causes the astronaut to rotate instead. Remember Newton's third law. This does not happen on Earth because the person with the screwdriver is firmly fixed in place by gravity and friction with the floor. Electric drills, screwdrivers, and other tools were developed to prevent this from happening. When the tool is turned on, a disk inside the tool is geared to rotate in a direction opposite to the motor's rotation. Thus the two rotations cancel each other out.

Foot restraints are located at strategic spots in the cabin and in the cargo bay, and they can be easily moved to other locations if needed. Working with hand tools is much easier when the feet stay in one place.

The orbiter is equipped with foot restraints and Velcro in strategic locations where people and things can be attached to keep them from drifting off. An astronaut is firmly planted in one spot by slipping both boots into a foot restraint. Small objects can be stuck to the Velcro or tethered to straps on the wrist so they cannot float away.

Remote Manipulator Arm

Much of the work outside the Shuttle is done from inside the orbiter cabin using the *remote manipulator arm.* It can recover satellites from orbit and stow them in the cargo bay; lift scientific equipment from the cargo bay, drop it overboard and later grasp it and place it back into the cargo bay; hold scientific experiments out away from the orbiter to study the space surrounding the vehicle; hold astronauts in space suits and move them from place to place while working in the cargo bay; assemble together pieces of a large structure; and make video images of inaccessible places in and around the cargo bay.

This remarkable robot device, designed and built in Canada for the orbiter, is powered by six electric motors. In many ways it resembles a human arm with a shoulder, elbow, wrist, and hand. In robotics terms, the hand is called an *end effector*. See Figures 10.1 to 10.3. When not in use, the arm is stowed against the side of the cargo bay. Heaters keep the joints warm so they do not freeze up. From a control panel on the upper deck of the cabin an astronaut lifts the arm from its cradle and moves it to any desired position and location. Movement is controlled by a joystick, similar to those used for video games. A computer reads the joystick motion and translates it to motion in the arm. The operator can see what is happening through windows into the cargo bay and in the ceiling of the orbiter, and by television monitors next to the control panel. TV cameras are mounted at the wrist and the elbow of the manipulator arm and can be focused and zoomed by the operator to give the best view.

The end effector does not have fingers, but uses an ingenious rotating wire arrangement to capture and hold a payload. See Figure 10.4. A mechanical finger is a complicated device with many joints; the wire device is simple by comparison and less likely to fail. Any payload to be manipulated by the arm must be equipped with a *grapple,* a rod-like protrusion. The end effector is slipped over the grapple and the snare wires tighten around it to make a firm connection. A platform with foot restraints can be attached to the end effector so an astronaut can ride the arm (Figure 10.5).

The robot arm works only in orbit. In weightlessness it can manipulate any payload the orbiter can carry, but it cannot support its own weight on the ground. In this respect, it is something like a fish out of water. A fish is nearly weightless in water and needs only small flips of its tail and fins to move and maneuver, but on dry land it is helpless.

Figure 10.2 Remote manipulator arm as seen from the flight deck of the orbiter. Notice the TV camera at the elbow. *Courtesy of NASA.*

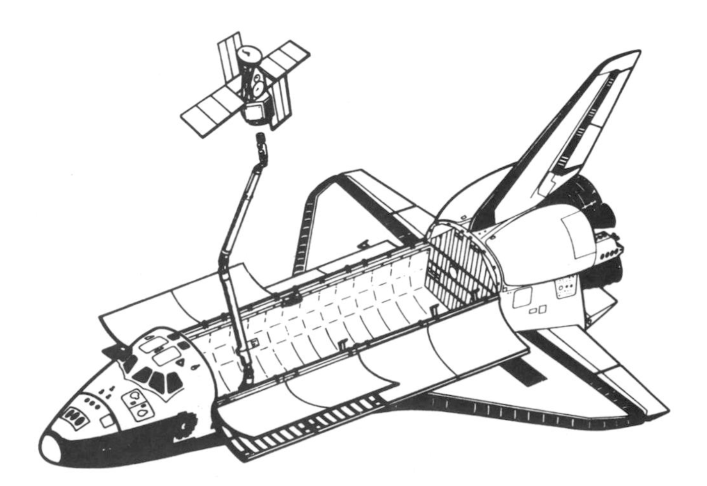

Figure 10.1 Remote manipulator arm. Fashioned after a human arm, it has a shoulder attached to the cargo bay, an elbow, a wrist, and a hand called the end effector. In this diagram it has just dropped off a satellite. *Courtesy of NASA.*

Figure 10.3 The remote manipulator arm moves in over the tail of the orbiter to grasp a small payload. The end effector is clearly seen. *Courtesy of NASA.*

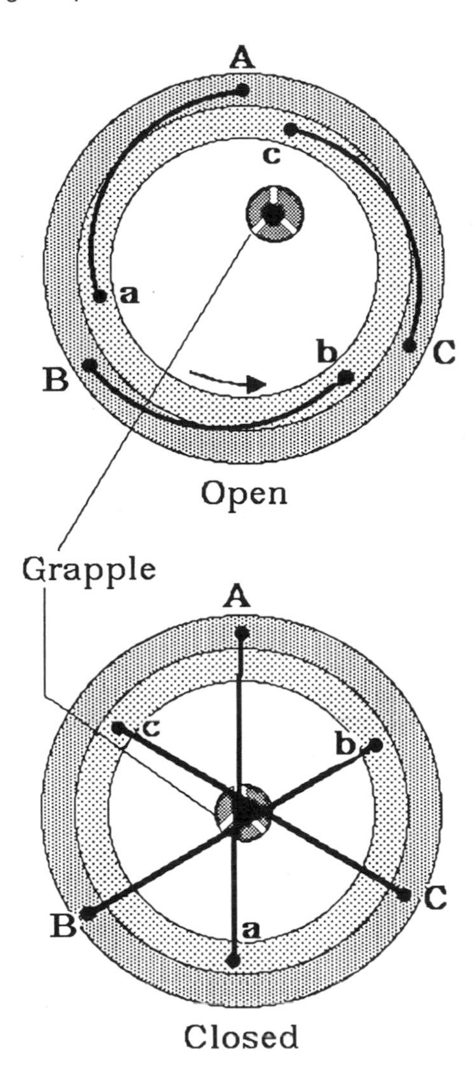

Figure 10.4 Manipulator arm's end effector. One end of each wire (A, B, and C) is fixed to the outer ring forming a triangle. The other ends (a, b, and c) are attached to a rotating inner ring; as it rotates the wires are drawn together, shrinking the triangle, to take hold of the grapple attached to the payload.

Figure 10.5 Astronaut Jerry Ross is firmly attached to the manipulator arm which moves him toward a tower he and Woody Spring have just constructed in the cargo bay of the orbiter *Atlantis* to test methods of constructing large objects in weightlessness. Earth is in the background. *Courtesy of NASA.*

Space Suits

When work is to be done outside the cabin, in the cargo bay, or away from the orbiter, the astronaut wears a space suit equipped with a self-contained life support system. It also provides some protection from radiation, both particles and electromagnetic radiation. The structure of the Shuttle space suit is shown in Figure 10.6.

The Shuttle space suit is the outgrowth of many years of experience in constructing suits for high altitude aircraft and spacecraft. As early as the 1930s, aviation pioneer Wiley Post had a suit molded in the shape of a man in the sitting position with a helmet something like a diver's helmet, and pressurized by the engine supercharger. In it, Post set cross-country speed and altitude records with his plywood airplane.

Mercury astronauts wore space suits as backup in case the cabin lost pressure. The suits were not pressurized except in an emergency, which never happened. Gemini space suits were designed specifically for working in empty space outside the cabin, called *extravehicular activity* (EVA). They were made in several layers to give greater flexibility than the Mercury suits. An inner bladder was made of nylon coated with neoprene rubber. Over that was a net woven of nylon and Teflon strands which acted as a casing to keep the bladder from ballooning out, just as a bicycle tire acts as a casing over an inner tube. Flexibility was improved by making the outer casing slightly smaller than the inner bladder. In 1965, Ed White was the first American to leave his vehicle while in orbit, connected to the spacecraft's life support system by a 25-foot (7.6-m) long umbilical cord. Cosmonaut Alexei Leonov had been the first human to "walk in space" just 3 months before.

Apollo space suits presented other design problems because they had to work on the surface of the Moon as well as in space. The inner structure was similar to the Gemini suit, but the outer layers had to stand up to jagged rocks, the heat of the lunar day, micrometeoroids falling onto the Moon, and rough

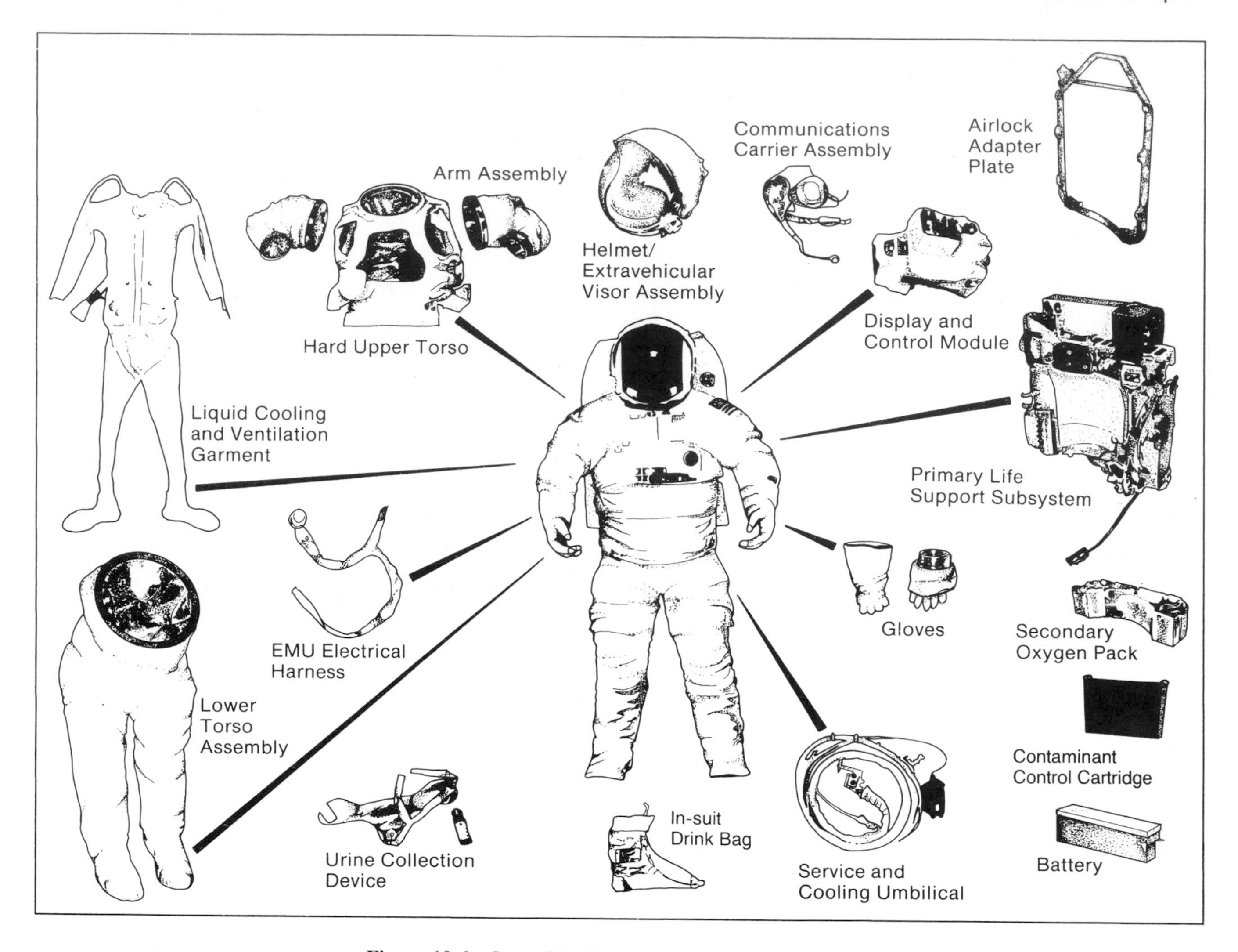

Figure 10.6 Space Shuttle space suit. *Courtesy of NASA.*

handling while climbing in and out of the lunar lander. The suit also had to be flexible enough to allow stooping, bending, adjusting instruments, and sitting on the lunar rover "dune buggy." Bellows-like joints at the knees, elbows, and waist gave greater freedom of movement. In earlier suits, when an astronaut wanted to bend a joint, an elbow, knee or shoulder, that entire section of the suit had to compress, reducing the volume of the enclosure and increasing the air pressure inside. Fighting that increased pressure quickly tired the astronaut. Bellows joints do not increase the pressure. When they bend, the outer side of the bend expands while the inner side compresses and the volume remains constant.

Like the Gemini suit, the Apollo suit was made with an inner bladder of neoprene rubber enclosed in a casing to keep it from ballooning out. Multiple layers of thin plastic alternating with fiberglass, followed by several layers of Mylar and Teflon cloth provided the needed insulation and outer protection. Lunar boots fit over the space suit feet like overshoes. Backpacks carried communications and life support equipment. Each suit was custom made to fit its occupant at a cost of $400,000 each.

The Space Shuttle space suit is a direct descendent of the Apollo suit. At the shoulders, elbows, wrists, and waist are ball bearing joints which rotate easily with little force, increasing the flexibility and reducing the fatigue of a working astronaut. It is easier to get into than previous space suits. In preparing for an EVA from the orbiter, the astronaut first dons a urine-collection device designed for either men or women with a bag attached at the groin to hold up to a quart of urine. Next comes a suit of longjohns made of spandex mesh. Air can move freely around and through this undergarment. A sealed-up suit holds in the body heat which must be removed, so plastic tubes running through the undergarment carries cooling water to keep the suit and its occupant at a comfortable temperature.

Twenty-one ounces of drinking water is stored in a bag at the torso. A microphone and headphones are built into a tight-fitting cap, the Snoopy Cap, and are connected through the electrical harness to a radio transmitter-receiver for communicating with the cabin and with other astronauts. Instruments for monitoring heart rate and respiration are built into the suit and connected to the radio.

The space suit itself consists of a hard upper torso with an aluminum shell, a lower torso, gloves, helmet, and visor. All components are multilayered like the Apollo suit. In the orbiter, the upper torso part of the space suit is mounted on a wall so an astronaut can slip easily up into it. The astronaut puts on the lower half of the suit, then slips up into the upper torso and aligns the many connections between the two parts. The two halves lock together to make an airtight seal. Helmet and gloves are then sealed in place, also with metal rings. A visor snaps over the helmet to provide protection from solar ultraviolet and from micrometeoroids. Work lights and a TV camera can be mounted on the helmet.

The suits are not custom made for each astronaut as in previous manned space programs. Each component comes in several sizes and all are interchangeable so any astronaut, male or female, can be properly fitted. The only things that are custom made are the gloves. They pose a special engineering problem, because fingers will tire quickly if the astronaut must work against the space suit's higher pressure in ill-fitting gloves. Casts are made of the astronaut's hands in several positions, and patterns for the gloves are made from the castings.

A built-in backpack contains two oxygen tanks; a fan to keep the air circulating; a contaminant control cartridge which contains filters, lithium hydroxide, and charcoal to remove carbon dioxide and other gases; cooling water and a pump to circulate it through the suit; a unit to cool the water and remove water vapor; and batteries for power. Sufficient oxygen is carried in the main tank for 7 hours; a 30-minute reserve tank provides for an emergency. The batteries power the suit for 7 hours also. Completely equipped, the suit weighs nearly 260 pounds (118 kg) on Earth.

A control panel is attached to the hard chest part of the suit. Seventeen sensors monitor the levels of carbon dioxide, water vapor, and other contaminants, as well as the temperature, the pressure, and the supplies of oxygen, water, and power. A caution and warning system in the suit notifies the occupant of a malfunction by sounding a tone in the headphones and flashing a light on the control panel. Because a higher pressure would make the space suit less flexible and difficult to work in, pressure is kept at 4.2 psi (29 kPa), compared to 14.7 psi (101 kPa) in the orbiter cabin. This gives rise to the potential problem of decompression sickness, the bends, which was discussed in Chapter 9.

Through the Airlock

Of course, astronauts cannot just open the cabin door and step out; their friends inside would not appreciate them letting the air out! Leaving the cabin is done through an airlock, a drum-shaped compartment 63 inches (1.6 m) in diameter and 7 feet (2 m) tall. See Figure 10.7. It may be mounted in the orbiter in one of several places shown in Figure 10.8, depending on what is being carried in the cargo bay. If a laboratory module is carried in the cargo bay, then a tunnel connects it to the middeck and if the mission calls for EVA, the airlock would be attached to the tunnel. If the cargo bay does not hold a laboratory, the airlock may be connected to the hatch either inside or outside the cabin.

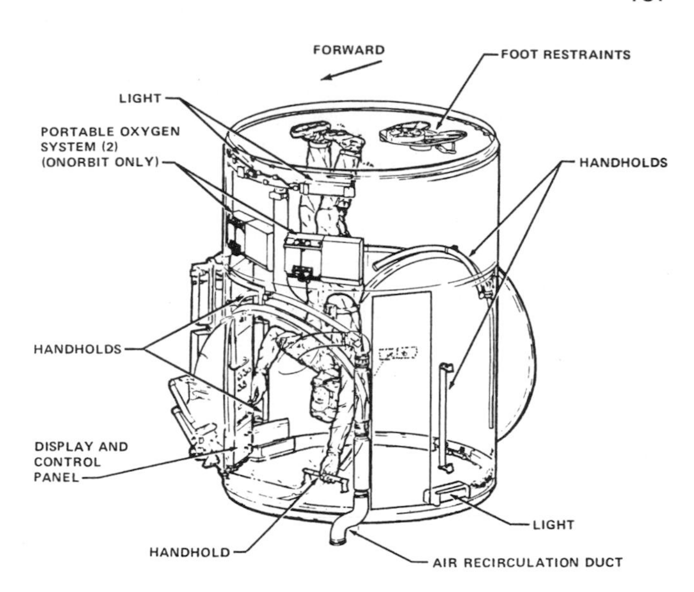

Figure 10.7 Space Shuttle orbiter airlock. *Courtesy of NASA.*

The day before an EVA, the pressure inside the orbiter is allowed to drop from 14.7 psi to 12.5 psi through normal breathing by the crew. An hour before the EVA begins, the cabin pressure is further reduced to 10.2 psi where it stays until the EVA is complete. The astronauts who are going outside then breathe pure oxygen for 45 minutes to purge nitrogen from their blood stream. The astronauts enter the air lock, don their space suits, and attach a tool kit and flashlight. Then the air is allowed to escape slowly into space and the outer hatch is opened for exit into the payload bay.

Handrails are conveniently located along the full length of the cargo bay near the door hinges and in several other places. See Figure 10.9. Long wires run the full length of the cargo bay just above the handrails on each side. A tether attached to the space suit and clipped to these wires keeps the astronaut from floating away, yet allows freedom of movement throughout the area. Perhaps you have tethered your dog that way, using a long wire in the back yard with the leash looped around it so the dog can run over a wide area but cannot run away.

Manned Maneuvering Unit

When there is work to be done beyond the reach of a person on the manipulator arm, the manned maneuvering unit

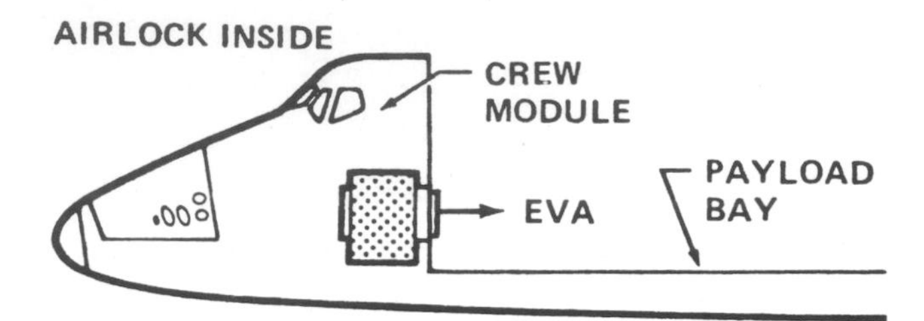

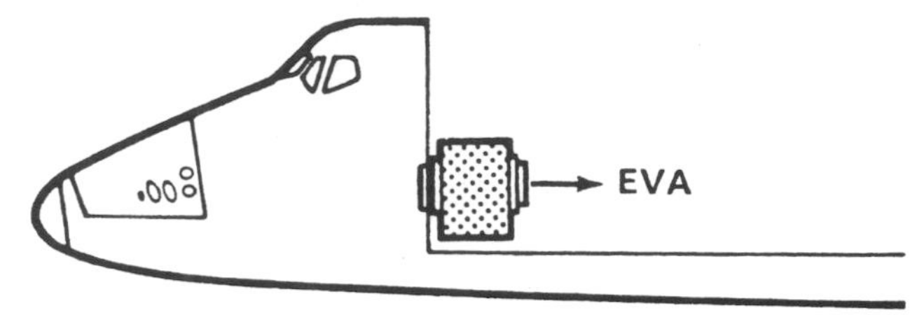

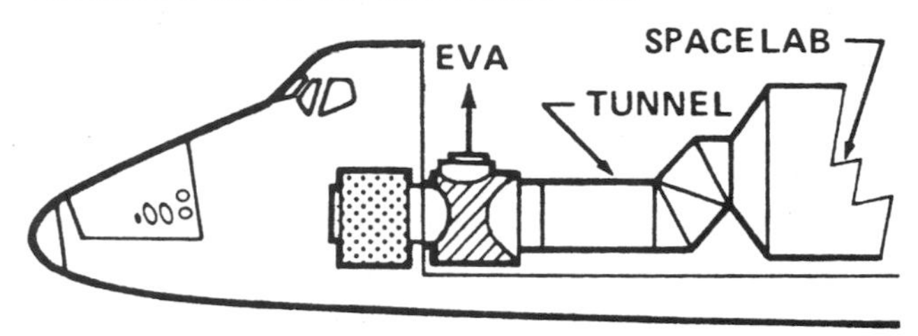

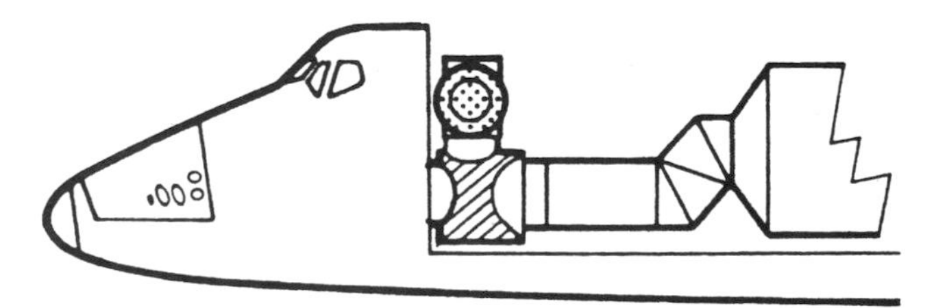

Figure 10.8 Mounting positions for the orbiter's airlock. *Courtesy of NASA.*

Figure 10.9 Astronaut moves along the side of the payload bay using handholds and a wire tether. Earth is in the background. *Courtesy of NASA.*

(MMU) converts the astronaut into the world's smallest completely self-contained spacecraft. Figure 10.10 shows Bruce McCandless on the first untethered spacewalk. He and his partner Bob Stewart flew more than 300 feet from the orbiter and returned several times. On one return McCandless flew around to the nose of the orbiter, looked in and asked if they wanted the windows washed.

The MMU is stowed in the payload bay against the front wall. An astronaut in a space suit simply backs up into it, closes two latches, and flies off.

Mounted on the MMU are 24 thrusters which produce 1.6 pounds (7.1 N) of thrust each. Compressed nitrogen gas is used for propellant. Supply tanks mounted on the back carry enough propellant for about 6 hours and can be refilled from the orbiter's nitrogen supply. Flight controls are located on the ends of the armrests. At the left hand is the "throttle" and "brake" which control straight line motion: left or right, forward or backward, up or down. At the right hand is a joystick to control attitude: up and down for pitch, left and right for yaw, and rotation for roll. If the astronaut wants to stay in a

Figure 10.10 Bruce McCandless in the manned maneuvering unit, the first man to fly untethered in space. *Courtesy of NASA.*

particular attitude, the MMU can be put on "autopilot," freeing both hands for other tasks.

There are actually two redundant sets of controls and propulsion systems operating at all times. The thrusters work as two sets of twelve, operating independently, yet together. If thrusters or controls on one set should fail, that set can be shut down and the other can bring the astronaut back home to the orbiter.

Normal flight speed is from 0.3 to 1.0 mile per hour (0.5 to 1.6 km/h) with respect to the Shuttle, in addition to orbital speed of about 17,500 miles per hour (about 28,000 km/h) with respect to Earth, of course.

To help the astronaut monitor what is happening, a tone is heard in the headphones anytime a thruster fires. Also, a fiber-optic cable connects the MMU to the space suit control panel so the nitrogen supply and battery condition can be monitored. MMU silver-zinc batteries are interchangeable with those that power the space suit and can be recharged in the orbiter.

MMUs were used to capture and repair several satellites on three missions in 1984 before they were retired. Recently a new, more compact maneuvering unit called the Simplified Aid For EVA Rescue (SAFER) has been tested in the Space Shuttle and Mir. SAFER is designed primarily as a self-contained "life vest" for emergency use by astronauts who might accidentally float off while working outside. It fits over the life support system on the back of the space suit. SAFER is shown in Figure 10.11. Compare it with the MMU in Figure 10.10. It is a miniature version of the MMU with 24 thrusters, each producing 0.8 pounds (3.5 N) of thrust. The attitude control system operates essentially the same as the MMU. The control panel and joystick are pulled out of the backpack and brought around to the front for operation. Three pounds (1.4 kg) of high pressure nitrogen gas for propellant are stored in a tank in the backpack. The tanks can be recharged and the battery can be replaced in the orbiter.

Figure 10.11 Astronaut Mark Lee test-flies the SAFER backpack mobility unit. *Courtesy of NASA.*

Care and Feeding of Satellites

One of the original jobs of the Space Shuttle was to act as a first stage booster taking large satellites to low Earth orbit. Some surveillance satellites remain in low orbit to do their jobs; others, such as communications satellites, must be fired to higher orbits by an attached rocket engine. Certain satellites may be recovered and either brought back to Earth for maintenance or repaired by astronauts in orbit. The Hubble Space Telescope described in Chapter 7 is a prime example of the Shuttle's ability to launch, repair, and maintain satellites while in orbit. The HST was designed with that in mind.

Launching Satellites

Figure 10.12 shows a satellite just leaving its protective cradle in the orbiter's cargo bay. The clamshell doors are kept closed to shield it from the Sun until launch time. The spacecraft rests on a turntable, and just prior to launch, a motor starts the turntable and satellite rotating. Spinning the satellite serves two purposes. First, heat from the Sun is more evenly distributed over the entire surface. Intense solar heating on one side can damage the satellite. Second, a spinning object stays pointed in the same direction. Consider how a rotating top or gyroscope remains upright while spinning and topples over as the rotation slows. Similarly, a spinning satellite remains oriented in the same direction in space.

When the satellite is rotating at about 50 revolutions per minute, springs eject it out of the cradle into space. Their slight difference in speed causes the orbiter and the satellite to slowly drift apart. In 45 minutes they will be separated by 20 to 30 miles (32 to 48 km). The Shuttle's maximum altitude is about 600 miles (960 km) so if a satellite is to be sent to a higher orbit, an attached solid booster rocket is fired by a command from the ground at the proper time. This transfers the satellite from the Shuttle's orbit to an elliptical orbit with apogee at the desired altitude. When it reaches the desired altitude, an apogee motor in the satellite fires to circularize the orbit.

Figure 10.12 Communications satellite being deployed from the orbiter's payload bay. White insulated clamshell doors cover the satellite to protect it from overheating while in the Sun or getting too cold while in the dark. The egg-shaped rocket attached to the satellite will boost it to geosynchronous orbit. *Courtesy of NASA.*

Retrieving Satellites

Communications satellites are not designed to be repaired in space because the Space Shuttle cannot reach that altitude. However, in February 1984, two communications satellites, Westar VI and Palapa B-2, had been dropped off by the Shuttle, but their booster engines failed to carry them to geosynchronous altitude and they remained in low orbits, useless for communications satellites. In November 1984, two more communications satellites were deployed from the Shuttle and then Westar VI and Palapa B-2 were recovered and returned to Earth for repair.

A number of new untested procedures had to be devised. Astronaut Joe Allen flew the MMU to the spinning Palapa satellite and, using the MMU thrusters, adjusted his rotation so it matched that of Palapa. He then inserted a device called a "stinger" into the nozzle of the apogee kick motor. Attached to the satellite by the stinger, Allen fired his MMU thrusters to stop their combined rotation. He then backed away so the remote manipulator arm could take hold of the stinger and move it into the cargo bay. The next step would have been to attach a bracket to the top of the satellite for the manipulator arm to grab onto and set it into a cradle for the trip back to Earth. However, the bracket would not fit, so Allen held the satellite for a complete trip around the Earth while another astronaut, Dale Gardner, prepared it for berthing. The two together then pulled the satellite into its cradle. The satellites weighed over a thousand pounds on Earth. Although they are weightless in orbit, some force must be exerted to start them moving into position and then to stop them.

Discovery caught up with Westar the following day. Gardner and Allen traded jobs and followed the same procedures to capture it and berth it in the cargo bay. See Figure 10.13. Insurance companies had paid out about $180 million for the lost satellites. By paying NASA about $10 million to bring them back to Earth, they hoped to refurbish and relaunch them, recouping up to $40 million of their settlement.

Repairing Satellites

Syncom IV-3 is another satellite that failed to reach geosynchronous orbit when the booster rocket engines would not fire on command from the ground. It too was stuck in a low orbit. In August 1985, Shuttle *Discovery* rendezvoused with the disabled Syncom. Astronaut James van Hoften rode the manipulator arm to a position where he could attach a handle to the satellite and pull it down to astronaut Bill Fisher. See Figure 10.14. In van Hoften's words, "This is me, . . . wondering what the hell I was doing out there. . . . The shuttle was very active, with 44 reaction control jets shooting out fireballs every which way, and I'm standing on the end of a 50-foot diving board, getting moved around, while ap-

Figure 10.13 Astronauts Gardner and Allen display a "For Sale" sign for the two disabled satellites they recovered from orbit. The two satellites were brought back to Earth for repair and re-launch. *Courtesy of NASA.*

Figure 10.14 Van Hoften manually started Syncom rotating at about three revolutions per minute, then shoved it off away from the orbiter. The Moon is seen below the satellite's antenna. *Courtesy of NASA.*

Figure 10.15 Astronaut William Fisher peers into the nozzle of the perigee-kick motor of Syncom IV-3. The remote manipulator arm (visible at the top of the picture) holds the satellite in place. The grapple fixture had been attached to the satellite by Astronaut van Hoften. *Courtesy of NASA.*

proaching a 15,000 pound satellite, then trying to get hold of it with my hands. It was *very* colorful."

While Fisher held it steady, van Hoften attached a grapple to the satellite. The manipulator arm then took over the chore of holding it in place while the two astronauts repaired the trouble. See Figure 10.15. First they shorted across some of the wiring and attached a new electronics box to bypass the defective part. Next they connected a battery power supply and sent a command to unfold the radio antenna. The satellite had been floating in space for 4½ months. Therefore, they placed a cover and temperature sensor over the nozzle of the perigee-kick motor to help warm up the solid rocket propellant. Finally, van Hoften stood on the manipulator arm and, using a handle he had attached, started the satellite rotating at three revolutions per minute and shoved it off to drift slowly away from the orbiter. A 2-month warmup period was planned before it would be safe to ignite the solid propellant and head for geosynchronous orbit.

Repairing the Hubble Space Telescope

The Space Shuttle makes periodic visits to the Hubble Space Telescope for repair and maintenance. On-orbit maintenance was included in the original concept for the telescope and it was designed to be worked on by astronauts in space suits. Original plans also included bringing the instrument back to Earth for a complete overhaul every few years. That idea was pretty well abandoned because of possible damage and contamination on a return to Earth. Maintenance and repair is done in orbit, and only when needed.

A Shuttle orbiter can rendezvous with the telescope, pull it into the cargo bay with its manipulator arm, and dock it to a fixture in the cargo bay as shown in Figure 10.16 (Plate 26). Astronauts open the large doors on the side of the telescope for easy access to the interior and upgrade the sensors and computers, repair worn or damaged equipment, and install new, modified, improved, and updated replacement parts. See Figure 10.17. When they are finished, the telescope is gently undocked and released over the side of the cargo bay by the manipulator arm. The orbiter slowly drifts away to a safe distance before firing engines or thrusters to avoid any possibility of a collision and to prevent exhaust gases from accumulating on the telescope.

The Hubble Space Telescope was placed in orbit by the shuttle *Endeavour* in April 1990. Subsequently three visits have been made to the telescope. On the first visit in December 1993 astronauts inserted corrective optics in the light path to compensate for an error made in grinding the mirror (See Chapter 7). They also replaced shaky solar panels, installed a second-generation wide field planetary camera, replaced two magnetometers and two gyros, added a 386 computer processor, and made other repairs.

On the second mission to Hubble in February 1997, astronauts updated a camera and a spectrometer, and installed a 12 gigabit digital data recorder to replace the old tape recorder which could hold only 1.2 gigabits and was subject to mechanical failure. In December 1999, new gyros for stabilizing the telescope, a new computer to replace the 1970s model, a backup transmitter and another 12 gigabit solid-state data recorder were installed.

Figure 10.17 One astronaut sticks his head inside the Hubble Space Telescope while another looks on. Both are firmly attached to the remote manipulator arm. Notice the rack attached to the manipulator arm in front of the astronaut which holds tools and parts so they do not drift away. *Courtesy of NASA.*

Figure 10.18 Ross floats freely while constructing the Ease tetrahedron. *Courtesy of NASA.*

Construction of Large Structures

Large structures, such as solar power satellites and space stations, cannot be boosted into orbit completely assembled. They must be built in space.

Developing elements of a structure which fit together easily, Tinkertoy fashion, while weightless, required several years of engineering and experimentation. Astronauts wearing space suits in a tank of water to simulate weightlessness tried working with various struts, tubes, beams, and joints until they had two designs that seemed to work well. Pieces locked together using only one tool, the astronaut's hands.

The designs were then tested in space during a flight of the Shuttle *Atlantis* in November and December 1985. Two structures were erected, a tower-like truss called Access shown in Figure 10.5 and a tetrahedron called Ease shown in Figure 10.18.

With their boots clamped into foot restraints, astronauts Woody Spring and Jerry Ross assembled the three-sided truss into 4.5-foot (1.4-m) sections using a rotating jig on the floor of the cargo bay. One side of a section was assembled, then the jig was rotated and another side built, then rotated again and the third side finished. As each section was completed, it was pushed up the jig and another section assembled beneath it. Using 93 struts and 33 joints, the astronauts built ten sections, a total of 45 feet (14 m), the height of a four-story building, and tore it down in an hour and 25 minutes.

Astronaut Spring, riding the manipulator arm, picked up the tower by one end and moved it in several directions to see how easily it could be maneuvered and repositioned as might be done when building a space station. He found it easy to point the structure at some point on Earth and keep it in that position. He also found it easy to hold it at its center and spin it without undue stress on the manipulator arm. Later he removed and replaced some struts in the truss to simulate repairing a damaged part. That too was not a difficult task. Finally, he ran a cable along the inside of the tower and then removed it to see how difficult a job it is in weightlessness.

Looking something like a pyramid, a tetrahedron has four triangular sides. (A pyramid has four triangular sides and a square base.) The astronauts built the Ease tetrahedron while floating freely without foot restraints, connected to the orbiter only by a tether. The entire structure consisted of six beams, each 12 feet long, connected together with only four joints.

Trying to maintain a fixed position in space while rotating the 12-foot (3.6-m) sections into place and snapping the corners together proved to be more difficult than building the Access tower. On Earth the beams weighed 60 pounds (27 kg) and even though weightless in orbit, they resisted the force applied to move them (Newton, again). It was particularly tiring on the hands and wrists. The time needed to built the tetrahedron decreased with practice, of course. The first time took Spring 12 minutes; the last time required only 9 minutes. At one point he disconnected the structure from the orbiter to see how well it could be moved into various positions and found it to be more difficult to maneuver than the tower.

Microgravity Research

Research in physics, life sciences, and materials sciences, the testing of prototypes of equipment for spacecraft and the space station, and the development of commercially useful processes are all done in the microgravity environment of the

Space Shuttle orbiter. This kind of work does not usually receive the media attention that dramatic spacewalks do because basic research is not as exciting to watch and much of the work is difficult to appreciate without some understanding of the sciences.

The experiments depend on weightlessness and require a very smooth ride without bumps or jolts. One method of avoiding disturbing motions is to let the orbiter drift free in orbit with the tail pointed down toward Earth. This attitude is called gravity gradient stabilization. Gravity acting on the engines, the most massive parts of the orbiter, keeps it oriented in that position without firing the thrusters. See Figure 10.19.

Spacelab

Major scientific research was done in the Spacelab. The European Space Agency developed the laboratory module which fit into the Shuttle cargo bay and connected to the cabin through the airlock hatch. This configuration can be seen in Figure 10.8. Astronauts had to float through the tunnel to enter Spacelab from the orbiter middeck. Experiments could be placed either in the Spacelab or outside on pallets toward the rear of the cargo bay and operated from inside as seen in Figure 10.20. Figure 10.21 (Plate 27) is a view inside Spacelab during a Shuttle flight.

Figure 10.19 Shuttle in gravity gradient attitude. *Courtesy of NASA.*

Hundreds of experiments were carried on Spacelab flights in biological sciences, materials processing, Earth sensing, astronomy, solar physics, and a variety of other areas. The astronauts had voice, data, and video communications with ex-

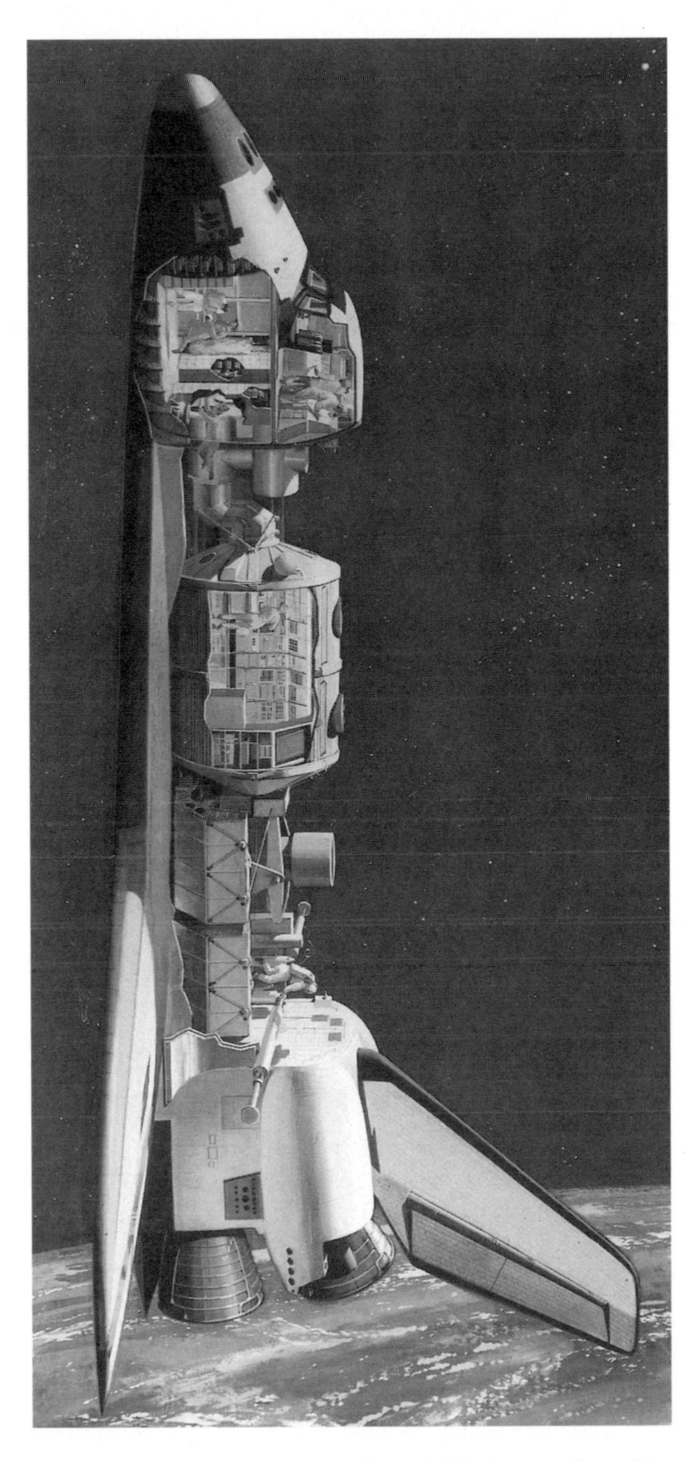

Figure 10.20 Artist's concept of Spacelab in the cargo bay of the orbiter. Scientists work inside the pressurized laboratory in a shirt-sleeve environment; some of the instruments, located just to the rear, are tended by a space suited astronaut. A tunnel connects Spacelab with the orbiter cabin and the airlock for exit to space is seen attached to the top of the tunnel. (Refer to Figure 10.8.) The orbiter is in the gravity gradient attitude. *Courtesy of NASA.*

perts and advisors on the ground so their work could be discussed as the experiments progressed. If a problem came up or if a new idea arose from the results, immediate modifications to the experiment could sometimes be made.

Spacelab flew 10 times before it was retired in April 1998. It had hosted scientist-astronauts from many countries working together in a "shirtsleeve" environment. The International Space Station, which will be discussed in Chapter 11, will accommodate the Spacelab experiments and a multitude of others as well.

Life Sciences

The effects of weightlessness on the human body were described in Chapter 9. Researchers are trying to find the root causes of these effects, and numerous experiments have been done on Space Shuttle missions to that end. Some measurements that are simple to do on Earth are impossible in weightlessness unless special equipment is constructed. For example, how do you weigh a weightless astronaut? A seat attached to springs is the answer. If an object is connected to a spring and the spring is stretched out, it will snap back and oscillate a number of times like a pendulum before coming to a rest. The period of the oscillations depends on the mass of the object. The device in the orbiter has a seat which is attached to a frame by means of springs at both the front and back. An astronaut sits in the seat and begins the oscillations. A timer records how long it takes to make one oscillation; the mass of the astronaut is calculated from the period of oscillation. A sample calculation is shown in MATHBOX 10.1.

Astronauts use a bicycle ergometer for aerobic exercise. The exercising person is wired to sensors that measure physical stress, workload, and heart rate. A rotating chair is used for vestibular experiments. The subject is held tightly in place with straps and electrodes. Cameras and accelerometers record the motions while the chair is rotated in three dimensions. Also, the subject may try to keep focused on a rotating disk mounted in front to test susceptibility to dizziness, nausea, or other symptoms of space adaptation syndrome.

Internal body temperature can be monitored by a miniaturized thermometer-transmitter, about the size of a vitamin pill, small enough to be swallowed by the astronaut. An antenna and receiver are strapped on the subject and internal body temperature is recorded for 1 to 3 days until the transmitter passes out of the body. This recently developed device offers great promise on Earth for monitoring patients in intensive care, in pregnancy and labor, and during surgery. Measurements from these and other devices are recorded for later study by physiologists who are trying to understand the response of the body to life in space.

Protein Crystal Growth

Proteins are complex compounds contained in all living cells. They provide structural support for cell walls, transport materials from place to place, promote chemical reactions, contract muscles, regulate metabolism, and do a host of other functions. When they float, immersed in body fluids, protein cells growing inside a body are nearly weightless. Similar cells growing outside the body in a laboratory are usually deformed because gravity pulls them against the bottom of the container. In the weightlessness of an orbiter or a space station, however, the cells can grow perfectly in three dimensions as they would inside a body. See Figure 10.22.

A perfect crystal shows the molecular structure of the protein, how the atoms are arranged in the molecule. This is important in medical research because the molecular structure determines how the molecule behaves in a living organism. With that information from, say, a disease causing molecule, researchers can design a drug that will counteract the action of the disease molecule.

Trays for growing protein crystals aboard the orbiter have 20 experiment chambers, each of which is equipped with a syringe with two barrels; one contains the protein solution and the other contains a solution which encourages crystallization. Several hours after the Shuttle reaches orbit, an astronaut starts the process running by turning a wheel which mixes the two solutions and forces out a drop which will hang on the tip of the syringe. At that point, the hanging drop is surrounded by a reservoir material saturated with a liquid. Water evaporates from the drop on the tip of the syringe and collects in the reservoir liquid which causes the protein solution to become more concentrated and the crystal to grow. This process is called vapor diffusion. Some proteins grow to large sizes more easily if they are first seeded with a small protein crystal brought from Earth. Before landing, the chambers are photographed and the drops retracted back into the syringes. Because they are perishable, the experiments must be removed from the orbiter as soon as possible after returning to Earth.

Protein crystals that have been grown in space include an antibiotic for treating infections; an antibody and a drug related to AIDS; a protein used to treat diabetes; an iron-containing enzyme associated with liver functions; and dozens of others. Cultures of cancerous tumors, such as breast and ovarian tumors, grown in space can be used to learn how they form and grow, perhaps leading to methods for treating them growing inside a body.

Biological Science Experiments

Numerous small scientific experiments have been carried into space on Shuttle missions. Only a few will be mentioned here. A colony of 3300 honeybees in a 9-inch cube was carried on a 6-day mission to see if they would go about their usual bee-business in weightlessness. They did. They built a normal honeycomb of hexagonal cells and the queen laid 35 eggs. That experiment had been suggested by a Tennessee high school student.

In another insect experiment, flies, moths, and bees were observed while weightless. Moths flew around in the cham-

MATHBOX 10.1

Weighing a Weightless Astronaut

The period of oscillation of a mass hanging on a spring is given by the equation

$$T = 2\pi \sqrt{\frac{m}{k}}$$

where T is the period in seconds, m is the mass of the object, and k is the spring constant. Notice that the equation does not include the acceleration of gravity, g. If we want to find the mass of an oscillating astronaut, we solve this equation for m:

$$m = \frac{kT^2}{4\pi^2}.$$

The spring constant is a measure of the stiffness of the spring, as indicated by the force that must be applied to stretch it a certain length. It can be found experimentally on Earth by hanging the spring from a hook, hanging a mass from it, and measuring how far the spring stretches. Say the spring stretches 8 centimeters (0.08 meter or 0.26 foot) when a 10 kilogram mass (22 pounds weight force) is hung on it. On Earth, Newton's second law tells us that the 10 kilogram mass exerts a downward weight force of $F = W = mg = 10 \times 9.8 = 98$ newtons. Then the spring constant is

$$k = \frac{F}{d} = \frac{98 \text{ N}}{0.08 \text{ m}} = 1225 \text{ newtons per meter.}$$

The spring constant is just that, a constant for any particular spring. It is the same in orbit as it is on the ground.

An astronaut sits on the seat and oscillates with a period of 1.5 seconds. Because he is weightless, it makes no difference if he oscillates up and down or back and forth. His mass is, therefore,

$$m = \frac{kT^2}{4\pi^2} = \frac{1225 \times (1.5)^2}{4\pi^2} = 69.8 \text{ kg}$$

In common units, a 22-pound force stretches the spring 0.26 foot, so the spring constant is

$$k = \frac{F}{d} = \frac{22}{0.26} = 84 \text{ pounds per foot.}$$

The mass of the astronaut is

$$m = \frac{kT^2}{4\pi^2} = \frac{84 \times (1.5)^2}{4\pi^2} = 4.8 \text{ mass units}$$

Mass units are converted to Earth-weight by multiplying by the acceleration of gravity, g. In common units, $W = mg = 4.8 \times 32.2 = 155$ pounds.

ber rather well, bees let themselves tumble without flapping their wings, and flies spent most of the time walking on the walls of the chamber.

The death of 16 chicken embryos after a 5-day ride in the Shuttle raised questions as to whether higher life forms could reproduce in weightlessness. Sixteen eggs were fertilized 9 days before the flight, and another 16 were fertilized 2 days before liftoff. The first 16 embryos survived, but of the second group 8 were dead when the eggs were opened on arrival back on Earth. The other 8 died soon after.

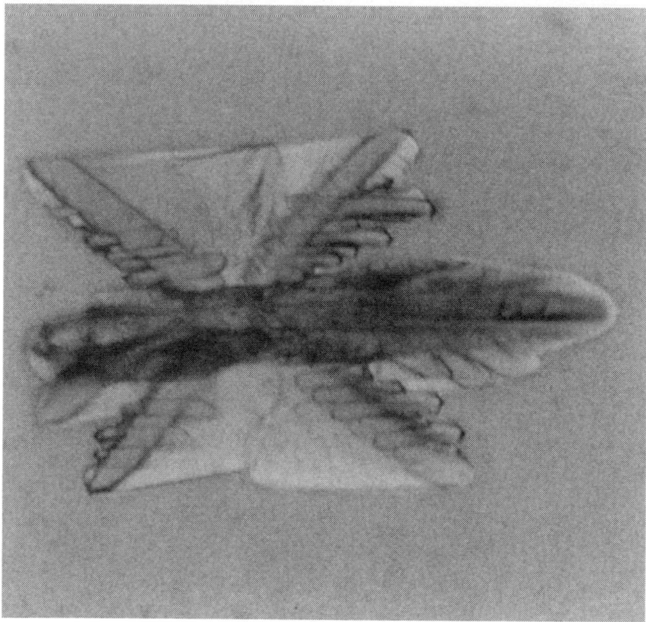

Figure 10.22 Isocitrate lyase crystal grown in the Spacelab (top) compared with one grown on Earth (bottom). *Courtesy of NASA.*

Sunflower seeds were carried in sealed plant containers with measured amounts of soil and water. On Earth, gravity pulls water down into the soil. In weightlessness the water collects around the seed and drowns it. Previous experiments have failed because of this problem. The quantity of water in the soil often determines whether a seed will germinate or not.

Plants produce a substance called *lignin* which gives them structural strength to stand up against gravity. Although lignin is vital to the plants, commercially it is a relatively useless material which interferes with paper manufacturing. Seedlings have been carried on Shuttle missions to see how they grow while weightless, particularly how much lignin they produce. Better understanding of lignin production may help in more efficient utilization of wood products.

Floating algae, *phytoplankton,* in the oceans are a basic food for fish. Find the phytoplankton and you find the fish. In a Shuttle experiment, light reflected from the eastern sides of the Atlantic and Pacific Oceans was directed through a telescope to a diffraction grating which broke the light into its spectrum of colors (wavelengths). Chlorophyll in phytoplankton absorbs certain blue wavelengths and reflects certain green wavelengths. If the bright green wavelength was present while blue was absent, a concentration of phytoplankton and, therefore, fish were probably at that location. Routine methods for leading commercial fishermen to good fishing spots could come from these experiments.

Meteorology

At any given time on Earth, about 2000 thunderstorms are occurring, most of them in the tropics. On two flights, nighttime lightning strokes were photographed through a diffraction grating which spread the light into its spectrum of component wavelengths. From these spectrograms the temperature, pressure, types of molecules, electron density, and amount of ionization in the lightning stroke can be calculated. Daytime lightning was detected by a device which created an audible pulse and recorded it on a cassette recorder while a 16-millimeter camera simultaneously photographed the clouds. This information may lead to a better understanding of the development of severe storms.

Industrial Research

The low gravity environment of an orbiting vehicle offers possibilities for conducting manufacturing processes in space that are more difficult and less efficient to do on Earth. For example, very large pure crystals have been grown on Skylab and on Shuttle flights which would have been impossible to grow in normal Earth gravity. Gravity causes the crystal to sag and become malformed. In addition, impurities from the container can become part of the crystal's structure. A large percentage of the silicon crystals grown for computer chips must be thrown out because of imperfections. Gallium arsenide chips are faster and more powerful than silicon chips, but are even more difficult to manufacture. In weightlessness, crystals can be processed without a container, eliminating the impurity problem, and can be grown much larger with a near perfect molecular structure as we saw in Figure 10.22. In spite of the high cost of transportation to and from space, it may become profitable in the future to manufacture such crystals in a continuously occupied space station.

Manufacturing high quality alloys is another possible commercial venture in space. An alloy is a mixture of two or more different metals to produce a material with different characteristics than any of its components. For example, titanium steel is a material of exceptional strength even at high temperatures. Some aluminum alloys are as strong as steel but weigh much less. When mixing an alloy in Earth gravity, the more dense metal tends to sink to the bottom before the mix cools and hardens, producing a nonuniform material. That separation would not happen in the microgravity of an orbiting spacecraft. Materials of unprecedented purity and uniformity could be prepared by allowing the mix to float in a vacuum without a container until it hardens.

Electrophoresis

One of the manufacturing processes that was tested for commercial use in space is electrophoresis. Electrophoresis is a method of separating materials from one another while they are immersed in a liquid or gel. Every substance has an inherent electrical charge because its molecules are made up of atoms which contain negatively charged electrons and positively charged protons. If the molecule has equal numbers of protons and electrons, then it is electrically neutral. But if one or more electrons are missing, as is often the case, the molecule bears a net positive charge, and it can be made to move through the fluid by applying an electric field. Positively charged particles migrate toward the negative side of the electric field. The speed at which the particles move in the fluid depends on their mass and net charge, and on the strength of the electric field.

In one type of electrophoresis machine, substances to be separated are injected into a fluid flowing continuously downward through the chamber. See Figure 10.23. An electric field is applied perpendicular to the flow. As the various substances move downward with the flow, they split into several particle streams, depending on their charge and mass, and migrate across the flow at various rates in the electric field. The substances are collected at outlets at the bottom of the chamber. Shawn Carlson, in a *Scientific American* article, (December 1998) describes how to set up a home demonstration of electrophoresis.

A most important application of electrophoresis is in the separation of medically useful biological substances, enzymes and hormones. For example, insulin is produced by beta cells in the pancreas. Beta cells can be a source of natural insulin if they can be separated from the pancreatic cells. Pure natural insulin is used for research and treatment of diabetes. Similarly, kidney cells produce an enzyme, urokinase, which can help in treatment of blood clots, phlebitis, and strokes. Pituitary cells produce growth hormone. The problem comes in separating the pure hormones, enzymes, cells, and other proteins from one another.

Electrophoresis works on Earth, but two problems occur, both due to gravity. First, if the biological materials are much more dense than the moving fluid, the injected stream collapses under gravity around the inlet and the purity of the separation is reduced. If they are diluted to a lower density,

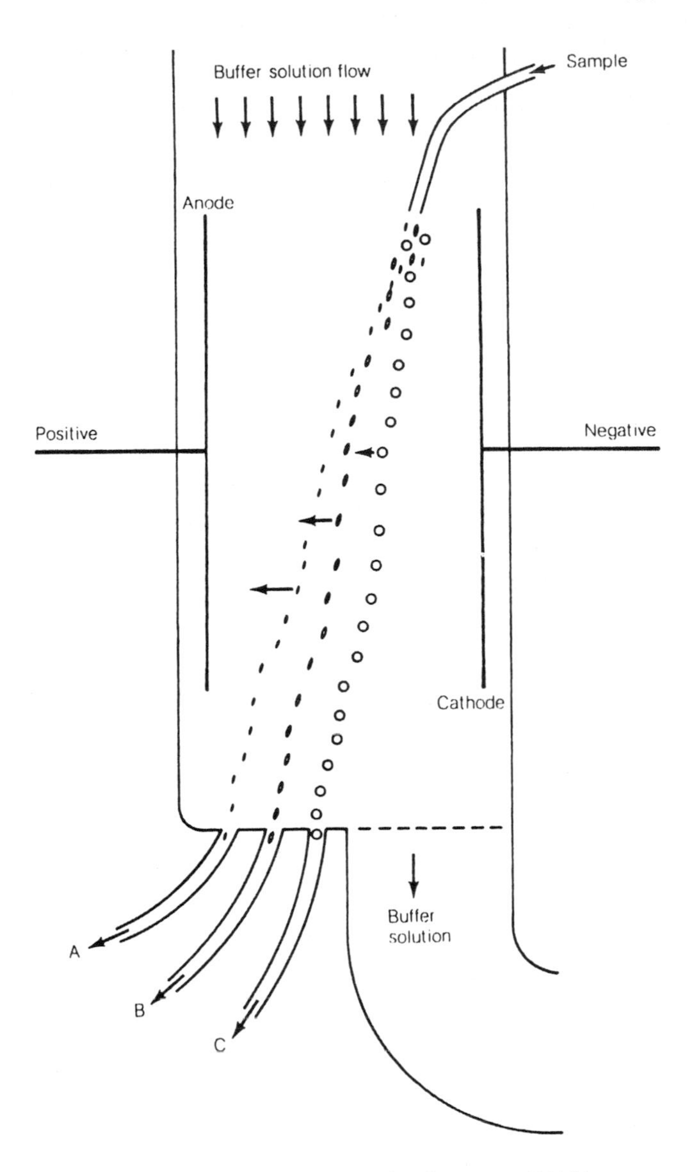

Figure 10.23 Electrophoresis. *Courtesy of NASA.*

the purity is improved, but the quantity that can be separated in a given time is reduced.

There is a second gravity problem. If the temperature of the moving fluid is not uniform throughout, the cooler parts, being more dense, will sink while the warmer parts rise, producing convection currents. This phenomenon can be observed when looking into a pot of water on a stove. The water at the bottom of the pot is hotter because it is in contact with the burner while the top surface of the water is cooler. Convection currents can be easily observed. In an electrophoresis machine, convection cells disrupt the moving streams, reducing their purity. Using a smaller chamber slows the convection currents, but also reduces the output of the separated substances.

In a weightless environment these problems disappear. High volume, high density samples can be injected into large volume chambers and large quantities of pure substances can be produced. Two experiments were flown in the Apollo-Soyuz project in 1975, one a continuous flow system from Germany and the other a static flow system by NASA. A mixture of horse, rabbit, and human blood cells were successfully separated into the three constituents. McDonnell Douglas Corporation built electrophoresis machines which were operated by astronauts on five Shuttle missions from 1982 to 1985. The results were remarkably successful. On one of the flights the unit separated more than 700 times the material with four times the purity than was achieved in a similar machine operating on Earth.

Plans have been developed for an electrophoresis factory in space, either manned or unmanned. An unmanned factory would be visited several times a year by astronauts to bring up fresh materials for processing, return the separated materials back to Earth, and perform any necessary repairs and maintenance on the equipment. A manned version could be operated aboard the International Space Station. Cost effectiveness would be greatly improved because the machine would have to be launched only once and would operate continuously while in orbit.

SPACEHAB

SPACEHAB, Inc., is a privately financed company that built laboratory modules for commercial research which can fly in the Space Shuttle cargo bay. The module is 10 feet (3 m) long so it occupies only part of the cargo bay, leaving space for satellites or other payloads. University and commercial researchers may lease space in the laboratory in the form of lockers or equipment racks. All the arrangements for electrical connections, data handling, temperature control, and crew-tending the experiments are provided so the researcher needs only to be concerned with the experiment itself.

The first flight of SPACEHAB was in June 1993. It carried 13 experiments, specifically designed to further the commercial development of space, including such things as high temperature melting of metals, cell splitting, and soldering in weightlessness. By the end of 1999, six other Shuttle flights had carried SPACEHAB.

DISCUSSION QUESTIONS

1. If objects are weightless in the payload bay, what is the advantage of using the manipulator arm to move them around?
2. What kinds of jobs can robots do better than people? What jobs can people do better? What limits the usefulness of robots in space?
3. What limits the length of time an astronaut can work outside the cabin on an EVA?
4. Why does the pilot warn the crew when the thrusters are about to be fired to change the attitude of the orbiter in space?
5. Does it make any difference which way the orbiter is oriented when a satellite is released into orbit?

ADDITIONAL READING

Carlson, Shawn. "The Amateur Scientist, Sorting Molecules with Electricity." *Scientific American,* December 1998. How to make an electrophoresis apparatus.

Compton, W. David, and Charles D. Benson. *Living and Working in Space, A History of Skylab.* NASA SP-4280, Government Printing Office, 1983. Hardback, complete, readable.

Joels, Kerry M., and Gregory P. Kennedy. *The Space Shuttle Operator's Manual.* Ballentine Books, 1982. Easy reading, many diagrams.

Lebedev, Valentin. *Diary of a Cosmonaut: 211 Days in Space.* Bantam Books, 1990. Experiences of two Russian Cosmonauts during their record-breaking flight on Mir.

NASA. *Microgravity . . . A New Tool for Basic and Applied Research in Space.* NASA EP-212, Government Printing Office, 1985. Semitechnical booklet.

Pogue, William R. *How Do You Go to the Bathroom in Space?* Tom Doherty Associates, 1985. Question and answer format, by a Skylab astronaut.

Roland, Alex, "The Shuttle, Triumph or Turkey," and Overbye, Dennis, "Success Amid the Snafus." *Discover,* November 1985. The other side of the coin.

Science, July 13, 1984. Special issue on Spacelab results.

NOTES

Chapter 11

Space Stations

When Europeans set out westward across the Atlantic, national governments provided the necessary support It is said that Spain's Queen Isabella even sold her jewels to finance Columbus's expeditions to the "New World." When Americans moved westward in the nineteenth century, the government provided land to settlers, incentives for building railroads, and military protection. In modern terms, the government provided or encouraged the development of an *infrastructure* to allow the westward expansion of the country. Development of air transportation also required a government-sponsored infrastructure of airports, flight control centers, radar sites, weather stations, and communications systems. Now the time has come for developing an infrastructure for expansion into the new frontier of space: a transportation system and a permanently manned space station are the first steps.

The United States had a space station in the 1970s called Skylab. See Chapter 1 and Figure 11.1. Skylab was constructed from the large fuel tank of a Saturn rocket, the type that had been used to carry men to the Moon. It weighed 100 tons, the equivalent of seven cross-country buses, the heaviest object put into orbit by the United States until the Space Shuttle. Skylab was occupied by three crews of three men, the first for a month, the second for 2 months, and the third for 3 months.

Russia has far greater experience with space stations than the United States. The seven Salyut space stations were a continuing presence in orbit from 1971 to 1986. They were mentioned in Chapter 1. The *Mir* space station was launched in 1986 and was permanently occupied until mid-1999. In 1995 cosmonaut Valeri Polyakov set a record for continuous time in space, 14 months onboard *Mir*. American astronauts lived and worked on *Mir* from 1995 to 1998.

Why a Space Station?

The International Space Station, under construction in orbit, is said to be "the next logical step." It is described below. First, let us consider why.

The International Space Station will be a center of scientific research in the unique weightless environment in such areas as biomedical science, combustion physics, materials processing, astronomy, Earth resources, atmospheric science, and solar and space physics. Many of the tasks now being done by satellites and by astronauts during Space Shuttle missions can be done better aboard a permanently manned space station. Shuttle flights last 17 days at the most. On a space station, experiments can be continued as long as desired and the equipment does not have to be hauled to and

Figure 11.1 Skylab space station, photographed by the last crew to work there as they departed for Earth. The shadow of the Apollo command module can be seen on the solar array. *Courtesy of NASA.*

from Earth each time it is needed. For example, some viruses and proteins take longer than 17 days to crystallize.

Unmanned satellites must be proven 100 percent perfect before launch. If something goes wrong in orbit, unless the Space Shuttle can reach and repair it, that is the end of it. Ground testing to certify the equipment ready for space is expensive and time consuming. Such perfection is not so necessary if the equipment is aboard a space station with someone tending it. When an experiment goes wrong on the Shuttle, it has often been repaired on the spot by astronauts.

Also, if the results of a satellite experiment or a measurement suggest a new or modified approach to the experiment, a new satellite has to be designed, tested, and launched. On the other hand, people tending experiments in or near a space station can modify the experiment and test, repair, or replace the equipment. Spacelab, carried in the orbiter's cargo bay, gave a hint of this. Radio, television, and data link interactions between the on-board astronaut and the scientists on Earth proved fruitful in many ways.

SPACEHAB, carried in the orbiter's cargo bay, was a facility for doing biological and medical experiments in weightlessness while being monitored from the ground. An understanding of the long-term effects of weightlessness on people is needed before prolonged flights can be safely carried out. The 17-day maximum duration of Shuttle flights is not long enough to compile the necessary data. Continuous research can be done for extended time periods aboard a space station.

Manufacturing in space has been done experimentally, electrophoresis for example (see Chapter 10). To be commercially profitable, however, manufacturing must be a continuous operation. Idle machinery makes no profit. Experiments and manufacturing aboard a space station could continue indefinitely.

Mir–Shuttle Program

To gain experience in long-term space station operations, the United States sent astronauts and experiments to and from *Mir* nine times from March 1995 to May 1998. The Shuttle flights are listed in the chronology in the Appendix. The first docking by *Atlantis* is shown in Figure 11.2. Astronauts stayed from 3 to 6 months. The interior of *Mir's* living quarters is shown in Figure 11.3 (Plate 27).

Experiments similar to those in Spacelab were done in combustion physics, fluid physics, materials science, protein and virus crystal growth, and tissue culture. However, they could be carried out for months on *Mir* instead of days on a Shuttle orbiter. Crew members also learned about how to keep fit during extended weightlessness, how to make best use of available space for living and working, and how to interact with people from different cultures and with different languages. Astronaut Shannon Lucid set a record for both

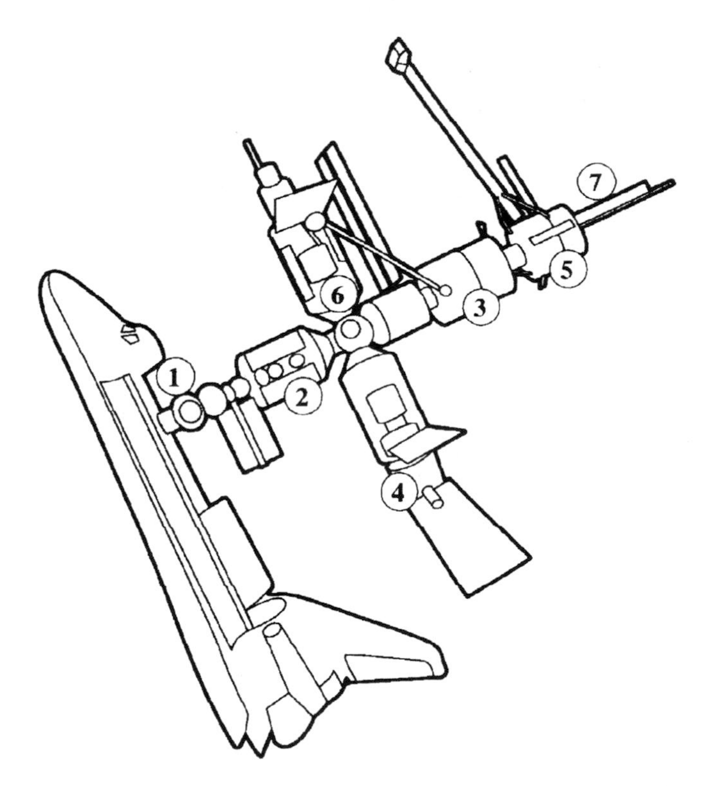

Figure 11.2 Space Shuttle *Atlantis* docked to the Russian space station *Mir*. The component parts of *Mir* are identified in the diagram. *Courtesy of NASA.*
1. Docking system
2. Krystall, material processing research module
3. Core module, living quarters with life support and power systems.
4. Spektr, Earth monitoring module
5. Kvant 1, astrophysics module
6. Kvant 2, biological research and Earth monitoring with airlock
7. Docking port for Soyuz vehicle

Americans and women with her 188-day tour of duty with two cosmonauts on *Mir*. She had previously flown on four Shuttle missions.

Two serious accidents occurred during the program. When more than three people were onboard, additional oxygen was generated by burning lithium perchlorate "candles." On February 23, 1997, the burning went out of control and until it burned itself out 14 minutes later, it filled the space station with smoke, and set off fire alarms. Some of the *Mir* hardware was damaged but no one was injured.

Also in 1997, a collision between *Mir* and an approaching unmanned Progress supply spacecraft damaged the solar panels and punctured the hull of the Spektr module. The small puncture was sealed before too much air was lost. Lessons have been learned from both of these serious incidents for safe operations on the International Space Station.

At the time of the astronauts' visits, *Mir* had been in operation for 10 years; it was designed for a 5-year lifetime. Cosmonauts were spending most of their time just maintaining the station, repairing or replacing broken parts. Upkeep is expensive. Left unattended, its orbit will decay and it will fall back to Earth. It would be better to deorbit *Mir* deliberately to be certain it lands in an ocean. A Holland-based company, MirCorp, has leased *Mir* from Russia and proposes to raise the money to boost *Mir* to a higher orbit, resupply it, and renovate it as a laboratory and tourist attraction. If MirCorp cannot raise the money, the Russian government will probably deorbit it.

The International Space Station (ISS)

In his State of the Union Message of January 5, 1984, President Reagan directed NASA to develop a permanently manned space station and to do it in within a decade. It didn't happen in that time frame. International partners—Canada, the European Space Agency, and Japan—joined in the project, but the U.S. share of the projected cost of designing, constructing, and launching the components into orbit was far beyond what the U.S. Congress was willing to authorize. With the end of the Cold War and the disintegration of the Soviet Union, the entire picture changed. Russia, whose space resources exceeded many other countries, especially in heavy lift launch vehicles and space station experience, could be included. With the participation of Russia, the International Space Station is now under construction.

The International Space Station is truly an international endeavor. The United States, Canada, Russia, the nine member countries of the European Space Agency, Japan, and others are working together toward a common goal in its construction and operation. Hopefully it will prove that diverse nations can work together successfully on a major scientific and technical undertaking, and that peaceful cooperation can replace hostile competition. Economic and scientific gain and major leaps in research and technology are anticipated. It can be a model for other large-scale space ventures such as a colony on the Moon and a manned expedition to Mars. If this project succeeds, it certainly bodes well for the future of international affairs.

Design

The fully operational configuration of the ISS is illustrated in Figure 11.4. It consists of one long keel truss with attached modules. The main truss is 365 feet (111 m) long, the length of a football field. Connected to it are six laboratories, two habitat modules, and two logistics modules, totaling more than 42,000 cubic feet (1200 m^3) pressurized space, nearly equal to the passenger compartments of two 747 airliners. An average of 110 kilowatts of electric power is provided by 24 solar arrays: 16 main panels and 8 panels on the Russian modules.

The ISS is designed to accommodate an international crew of six engineers, scientists, and technicians. An artist's conception of the interiors of a laboratory module and a habitat module is shown in Figures 11.5 and 11.6. Figure 11.7 is a cutaway drawing of the ESA laboratory, Columbus, showing its attachment to the ISS forward *node*. A node is a module to which other modules are attached.

The central truss and its attached modules maintain a constant orientation with respect to Earth. The solar arrays, however, must be continuously aimed toward the Sun. This is achieved by means of a rotating joint at the point of connection to the central truss. While ISS is in Earth's shadow, electricity is supplied by nickel-hydrogen batteries which are recharged when the station returns to the sunlit side of Earth. This charge-discharge cycle repeats every orbit.

The laboratory modules are located near the center of the complex where microgravity is at a minimum. Any activity that may perturb the microgravity experiments, such as docking or undocking the resupply and passenger vehicles, will be scheduled so that quiet microgravity conditions are maintained for 30 to 60 days at a stretch. Astronauts will spend about 200 hours per year of extravehicular activities (EVA) outside the station for maintenance. Such external activities will also be restricted during the microgravity quiet periods to avoid disturbing the experiments.

Construction

Much of the ISS consists of prefabricated modules designed to be hauled to orbit in the cargo bay of the Space Shuttle or aboard unmanned Russian Proton and Soyuz launch vehicles. In November 1998 a Proton rocket launched the Russian Zarya module into orbit. In December, Space Shuttle *Endeavour* carried a connecting node to rendezvous with the Zarya module. The assembly sequence involved using the orbiter's robotic arm first to mount the node on the orbiter's docking hatch, then to grapple the Zarya and attach it to the node, then to attach and connect power and data cables between the two modules during an EVA, and finally to release it into space. The result of this first step in the construction of the ISS is shown in Figure 11.8 (Plate 28).

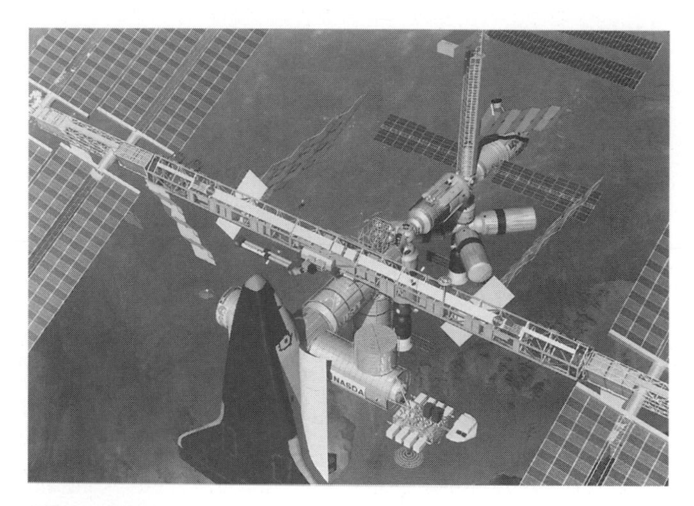

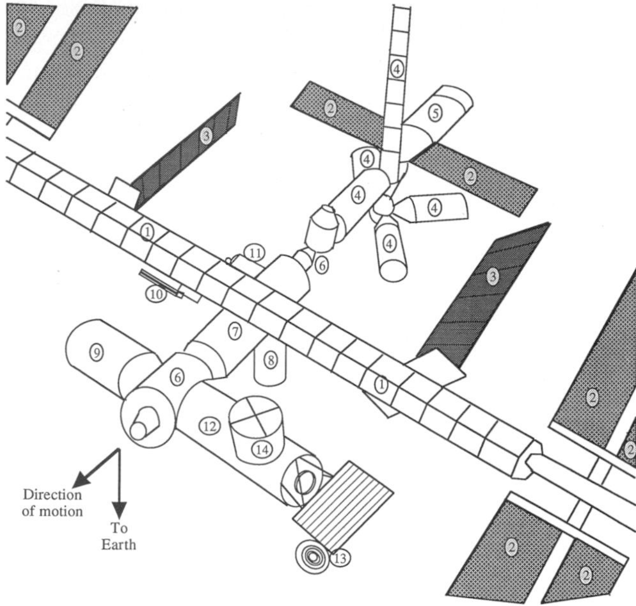

Figure 11.4 Overhead view of center section of completed International Space Station complex. The segments are identified by number in sketch. *Courtesy of NASA.*

1. Main truss structure
2. Solar arrays
3. Radiators
4. Russian components
5. Russian *Mir*-derived service module
6. Node for attaching modules and Soyuz return vehicles
7. U.S. laboratory module
8. Habitation module
9. European Space Agency laboratory module
10. Canadian-made robotic manipulator arm
11. Italian-made logistics module
12. Japanese laboratory module
13. Japanese exposed experiment
14. Japanese logistics module

A Shuttle flight in May and June 1999 carried supplies and equipment to the fledgling ISS. Figure 11.9 (Plate 28) shows an astronaut with a crane which was attached to the outside.

The subsequent sequence of assembly includes:

1. Attaching the Russian service module to the aft end of Zarya. The service module contains life support and living quarters, propulsion for boosting the station to a higher orbit, and solar panels for additional power.
2. Carrying supplies and equipment to Zarya to prepare for occupancy.
3. Ferrying the first three-person crew, two Russians and one American, to the station on a Soyuz rocket for a 5 to 6 month stay. The Soyuz remains attached for a return vehicle.
4. Attaching the first segment of the trusses.
5. Attaching additional solar panels to the new truss segment to supply power for the U.S. laboratory module.
6. Attaching the U.S. laboratory and two loads of supplies and equipment.
7. Ferrying the second three-person crew to the station.

These nine flights have to be carried out in this order. Any slippage in one flight will delay all the rest in line. At this point the station will look like Figure 11.10. More than 30 additional Space Shuttle, Proton, and Ariane flights along with about 160 spacewalks by space-suited astronauts over another 4 years will be needed to complete the ISS assembly.

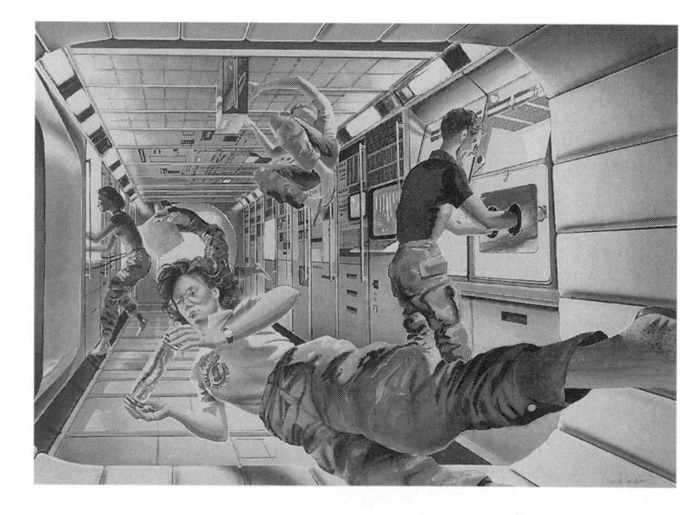

Figure 11.5 Artist's concept of the interior of the U.S. laboratory module. The astronaut in the foreground is inspecting the results of a fluids experiment. Behind her another astronaut manipulates a life science experiment. Near the ceiling an astronaut pulls out a utility box. The two in the rear are working at a computer and removing a rack for servicing. *Courtesy of NASA; Harold Smelcer, artist.*

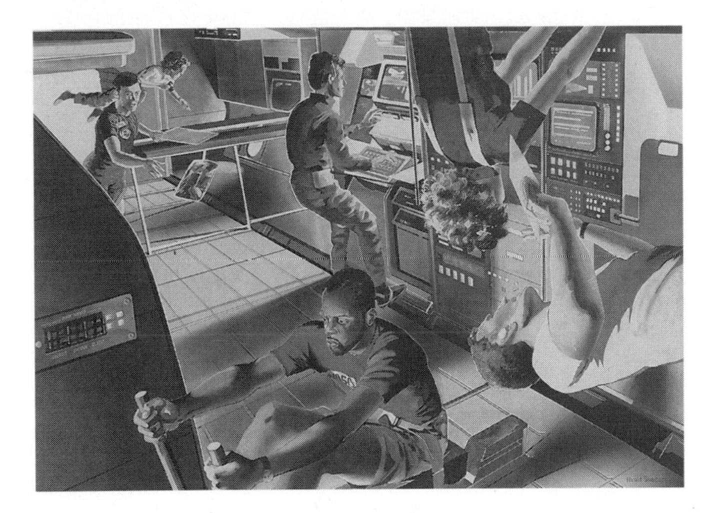

Figure 11.6 Artist's concept of the living quarters. The astronaut in the foreground is working up a sweat on a rowing machine. Another appears to hang from the ceiling as she is strapped to a treadmill for a jog. In the center, an astronaut checks a computer monitor. In the rear, one prepares to heat his dinner in a microwave oven while another looks out a window at the Earth, camera in hand. *Courtesy of NASA; Harold Smelcer, artist.*

Robotic Arms

Canada's two advanced remote manipulator arms are the primary machines used in the assembly of the ISS. Each is 56 feet (17 m) long. Using new advanced robotics technologies, the arm is longer and smarter than its predecessor on the Shuttle. An enhanced video system sends three-dimensional imagery to the operator. Thus, the operator can steer the arm without a direct view of the end. A small two-armed robot attached to the end effector will handle delicate tasks currently done by astronauts during spacewalks, such as tightening and loosening bolts. In advanced robotics, any movement of the operator's hands and arms is reproduced exactly by the remote machine.

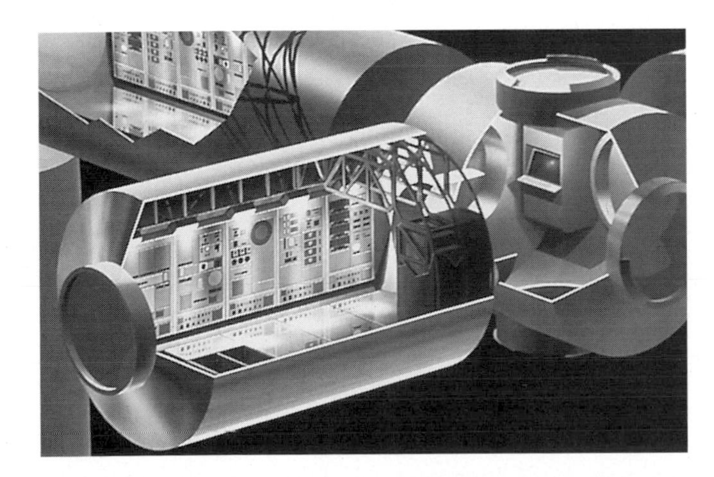

Figure 11.7 Cutaway view of Columbus, the European Space Agency's laboratory module showing its attachment to the node at the front of the space station. *PHOTO ESA.*

Figure 11.10 The International Space Station by 2001. Russian and American laboratories are on opposite ends, each with their own three-person crew. *Courtesy of NASA.*

The arm will also feed back to the operator the feel of the force and torque being applied. Newly developed proximity sensors attached to the robotic arm will be able to detect a nearby object and stop the arm or move away to avoid a collision.

When construction is complete, the arms will be used in many station activities such as assisting with docking the Space Shuttle orbiter, loading and unloading materials to and from the cargo bay, supporting space-walking astronauts, servicing outside payloads, and moving equipment and supplies around the exterior of the station. One of the arms will be attached to a platform which can move along a rail attached to the central keel truss as shown in Figure 11.11. The other, equipped with two end effectors, will move end over end around the exterior. With one end effector connected to a point on the station, the arm will stretch out and connect the other end effector to another point. The first end effector will then release and connect to a third point. Thus, moving like an inchworm, the arm will be able to reach nearly any place on the station.

EVA

Spacewalks will be a frequent occurrence in the operation of the ISS. Equipment on the outside will need tending and perhaps repairs will be called for. Many of these tasks will be done by the robotic arms, but some will necessarily be done by astronauts in space suits. Unless a suit is developed with internal pressure greater than the Shuttle suit, astronauts will have to condition themselves for the lower pressure to avoid

Figure 11.11 A Canadian remote manipulator arm similar to that on the Space Shuttle, but redesigned for the ISS. Its base slides along the black rails attached to the truss, visible at the bottom of the picture. It can reach any part of the completed station. *Courtesy of NASA.*

the bends as discussed in Chapter 9. The SAFER backpack maneuvering system discussed in Chapter 9 will be standard equipment for EVA from the ISS. At the present time, Shuttle space suits are refitted and reconditioned after each mission. Space suits will have to be maintained aboard ISS so they do not have to be sent to Earth for reconditioning. The SAFER units will have to be maintained for immediate use, also. However, pressurized nitrogen is not available aboard the ISS as it is on the Space Shuttle unless special equipment is provided.

Orbit

The ISS orbit is circular at about 275 miles (445 km). The area of the completed station is about equal to the area of two football fields. Because of this large size and the low altitude, atmospheric drag will cause the orbit to lower steadily. A boost back up to 275 miles will be needed every 3 months or so. The aft end Russian module provides propulsion, guidance, and attitude control.

The inclination of the orbit is 51.6 degrees for two important reasons. First, one of the major tasks aboard the ISS is Earth observation. At that inclination it will overfly 85 percent of the Earth's surface and 95 percent of Earth's population. This is easily seen in Figure 11.12, the simulated ground track for 26 hours. Second and equally important, all international participants will be able to launch directly into an orbit at that inclination. Originally, the United States intended to use an orbit at 28.5 degree inclination, the latitude of the Kennedy Space Center launch site. However, the Russian launch complexes are located at higher latitudes from which

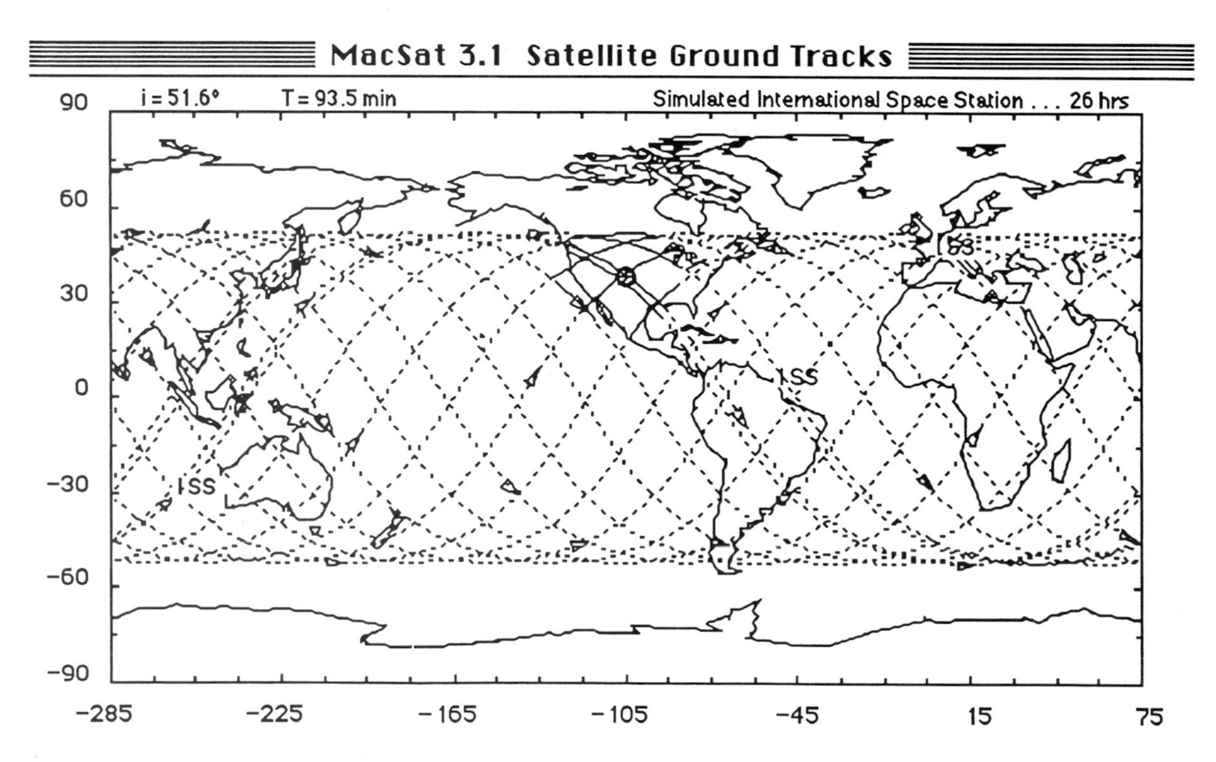

Figure 11.12 Ground track of the International Space Station for 26 hours at an altitude of 275 miles. At this altitude, the entire pattern shifts a little more than 3° eastward each day. The first 2 hours of the second day are seen over the central United States, South America, and in the far western Pacific near the Philippines. Thus, the station passes over different areas of the Earth each day and covers the entire area between 51.6° S to 51.6° N in about a week.

it is not possible to launch directly into a 28.5 degree orbit. Vehicles launched from the Russian sites would need additional fuel to change inclination after launch which would reduce the payloads they could carry to orbit.

Life Support

At first, food water, clothing, and other expendable supplies will be hauled up from Earth and trash will be returned to Earth. Later it may prove more efficient to recycle at least some of the expendables. An economical means of recycling air and water in the ISS is the first possibility. The Space Shuttle now carries all necessities for its relatively short flights. In the Shuttle, water is manufactured by the fuel cells; hydrogen and oxygen supplies are brought along to power them. Since electricity for the ISS will come from solar panels, not fuel cells, that source of fresh water will not be available.

In the Shuttle, excess humidity, carbon dioxide, and other noxious gases are removed from the air and oxygen and nitrogen are added as needed from tanks carried aboard. As described above, the Russians burn lithium perchlorate to produce fresh oxygen in *Mir*. This process has been made safer since the fire in 1997 and will be used in the Russian ISS modules. For a continuously operating space station it may be less expensive in the long run to install machinery to recycle air and water rather than replenish the supplies every few weeks. The scientific principles involved are well known, but the technology to do it reliably is not perfected.

What happens to a closed and sealed atmosphere after a few years? For periods of weeks or months, we have the experience of nuclear submarines to draw on. But for longer times, research is just getting underway. Toxic chemicals tend to build up in a closed environment and must be carefully monitored. For example, nickel has been detected in fuel cell water. Excessive intake could be a problem for astronauts.

Beyond recycling air and water, the science and technology of recycling solid waste are even less well understood. It will probably not be feasible on the ISS, but will be a necessity for very long term ventures such as large colonies in space or on other celestial bodies. We will look at this subject again later in this chapter.

Bus Services to Orbit

Once the station is completed and occupied, U.S. Shuttle flights and the Russian Soyuz and Progress vehicles will ferry fresh supplies and crews to the station and return crews and trash to Earth about every 3 months. Crews will rotate every 6 months. The Italian Space Agency has designed and developed two pressurized logistics modules for transporting equipment and supplies to the ISS and returning necessary items. One will always be docked to the station while the other is on the ground being reloaded. The Shuttle carries them back and forth.

At first the Russian Soyuz capsule will be used as a “lifeboat” to provide a fast safe way to return to Earth in an emergency. Because a Soyuz carries only three passengers, two vehicles would have to be continuously docked on the International Space Station ready for instantaneous use. The United States is developing the X-38 which will serve the same function when it becomes operational. It was described in Chapter 8. The European Space Agency is considering using the X-38 as a round trip vehicle, not just a lifeboat, to carry its astronauts both to and from the station.

Future Innovations

Once ISS is up and running, modifications will be relatively easy. Because of its modular design, new equipment, even entire new laboratories and living facilities, can be added. Later the station can be expanded by attaching additional trusses and payloads. More power will then be required. The drag from additional solar panels could cause the station’s orbit to decay quickly, so concentrating solar collectors, which produce more electricity with a smaller area, might be used to reduce the drag.

Each time a Shuttle or Soyuz docks, the clunk could shake the whole station. One solution involves free flying platforms, nearby but not attached, to carry machinery and experiments that must be free of vibrations, protected from contamination, or require precise pointing. Nearby free flying platforms could be teleoperated, that is a video camera would send a view to the person in the station who operates the machine.

Using orbital maneuvering vehicles (OMV), satellites and perhaps people will be carried to and from higher orbit, particularly geosynchronous orbit. The OMV will be able to return satellites to the ISS when necessary for repair and maintenance. It, too, would be remotely controlled from the station.

An example is the space telescope, described in Chapter 7. An OMV could boost the telescope to higher orbits as needed to decrease drag. If it transported the telescope to an orbit higher than can be reached by the Shuttle, boosts would not be required as often. An OMV could also bring the telescope back to the Station for routine maintenance and overhaul; doing these jobs in a “drydock” at the Station would save the expense and possible damage to the instrument inherent in capturing it for maintenance in the Shuttle orbiter. Because the OMV will use hydrogen-oxygen fuel, the contamination problem will be alleviated.

An advanced space station could also become a way station for construction and fueling of spacecraft headed for interplanetary space.

City-Sized Space Colonies

Perhaps the best known concept for a large space station is the “big wheel” that appeared in the movie *2001: A Space Odyssey* based on a story by Arthur C. Clarke. Such a structure was first presented to the public by Wernher von Braun in an article in *Colliers* magazine, April 30, 1954. See Figure 11.13.

The big wheel habitat of Wernher von Braun and Arthur C. Clarke may become a reality in the far future. NASA study groups in 1975 and 1977 did serious research into large space colonies and much of the material in this section comes from their work. Although NASA's greatest effort in the following decade was the Space Shuttle and the International Space Station, some research on space colonies has continued and many questions have been answered. At least as many other questions have been raised.

Imagine 10,000 people riding in the rim of a rotating wheel a mile in diameter (Figures 11.13 and 11.14). The rim is a tube, like the tire on an automobile wheel, with the occupants on the inside. We live on the surface of Earth; living in the wheel would be as if Earth were turned inside out and the atmosphere and people were put on the inside. Figures 11.15 and 11.16 are artists' conceptions of what the interior might look like.

To get the funding necessary to build and support a large colony in space, there has to be some economic payoff. Certainly it would be part of the infrastructure necessary for further exploration of the Solar System. But colonists could not expect a continuous flow of supplies to come from Earth unless they offered something in return. What is in space to entice governments to build a colony and people to live there? The International Space Station has many useful and profitable jobs to do; that is part of the answer. But that won't support 10,000 people.

What else is there? One major industry of a space colony could be exploitation of the Moon's natural resources to manufacture very large satellites, such as solar power satel-

Figure 11.13 Wernher von Braun and his conceptual space station. *Courtesy of David Christensen.*

Figure 11.14 One-mile diameter big wheel space colony. Ten thousand people live inside the rim of the wheel. The round object above the wheel is a mirror which reflects sunlight into the colony. The long "rod" at the lower left extends to a manufacturing facility separated from the main colony for safety. *Courtesy of NASA.*

lites. In Chapter 5 we discussed satellites in geosynchronous orbit which could harvest sunlight and beam it to an energy hungry Earth in the form of microwaves. The enormous cost of transporting the materials and construction crews from Earth for such a huge structure dampened the initial enthusiasm for energy satellites. To ship cargo to geosynchronous orbit requires an acceleration, called *delta-V* (Δv) to nearly 7 miles per second (11 km/s) from the surface of the Earth. Now suppose ore from the Moon is sent to a space colony where it is processed and fabricated into satellites. These large structures could be transferred to geosynchronous orbit, completely built, by a Δv of only 2.9 miles (4.7 km) per second. As you will remember from Chapter 3, the amount of fuel consumed depends on how much you must accelerate the vehicle. The saving in fuel costs occurs because lunar gravity is only one-sixth Earth gravity and the Moon has no atmosphere.

Using lunar materials, the colony could be the site for construction of large space vehicles, "star ships," for long-duration flights to the outer Solar System and beyond. The

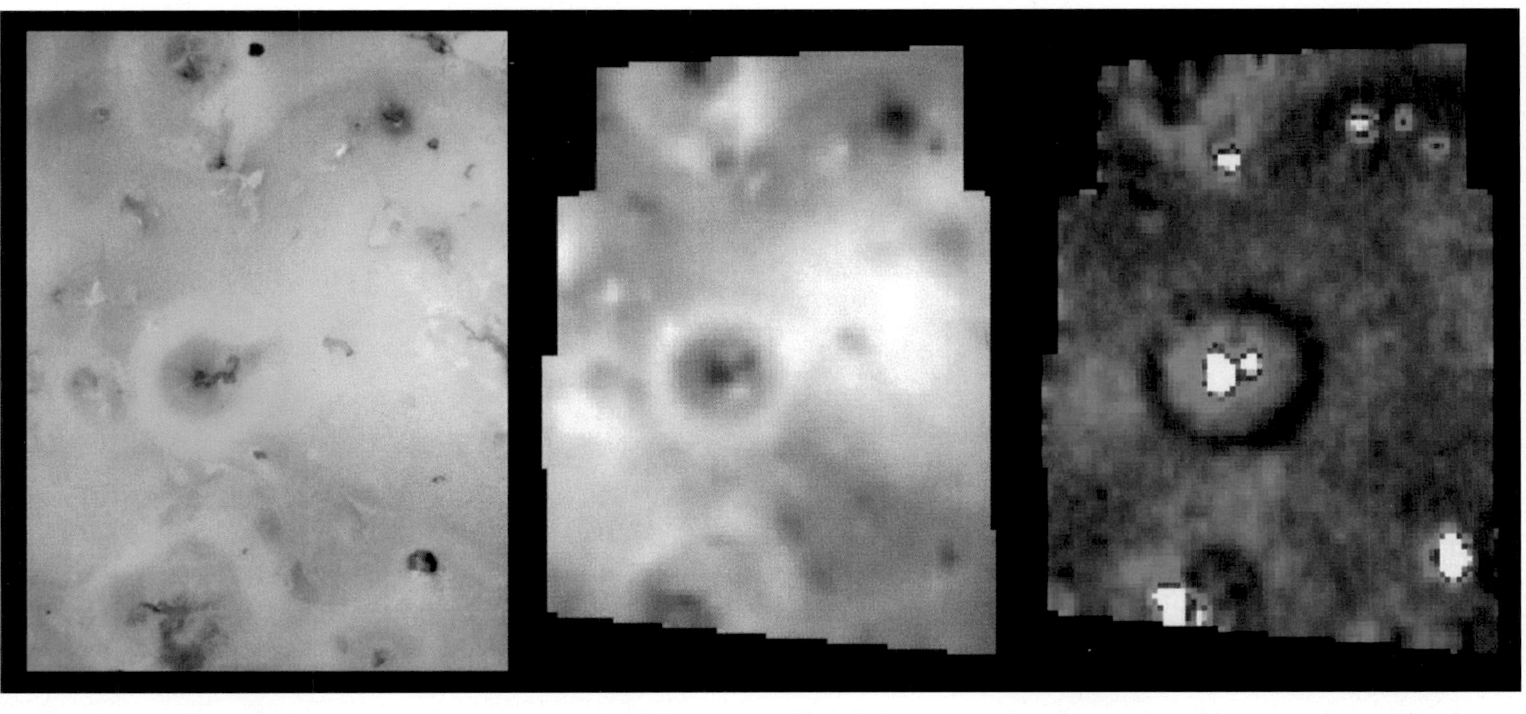

Figure 7.29 A volcano on Io, which is called the pizza planet for obvious reasons. The left picture shows the volcano as it would appear to the eye. The other two show the same area with infrared instruments. The center image at 1.3 micrometer wavelength shows variation in composition of the surface material. The right image at 4.2 micrometer wavelength records heat, producing a thermal map. The white spots in the right image are hot lava which appears dark in the visible wavelengths seen in the left picture. *Courtesy of NASA.*

Figure 7.32 Cassini ejects the European Space Agency's Huygens probe to investigate the atmosphere of Saturn's moon Titan. *Courtesy of NASA.*

Figure 7.33 Artist's rendering of the Huygens probe arriving at the surface of Titan after being released from the Cassini spacecraft. *Courtesy of NASA.*

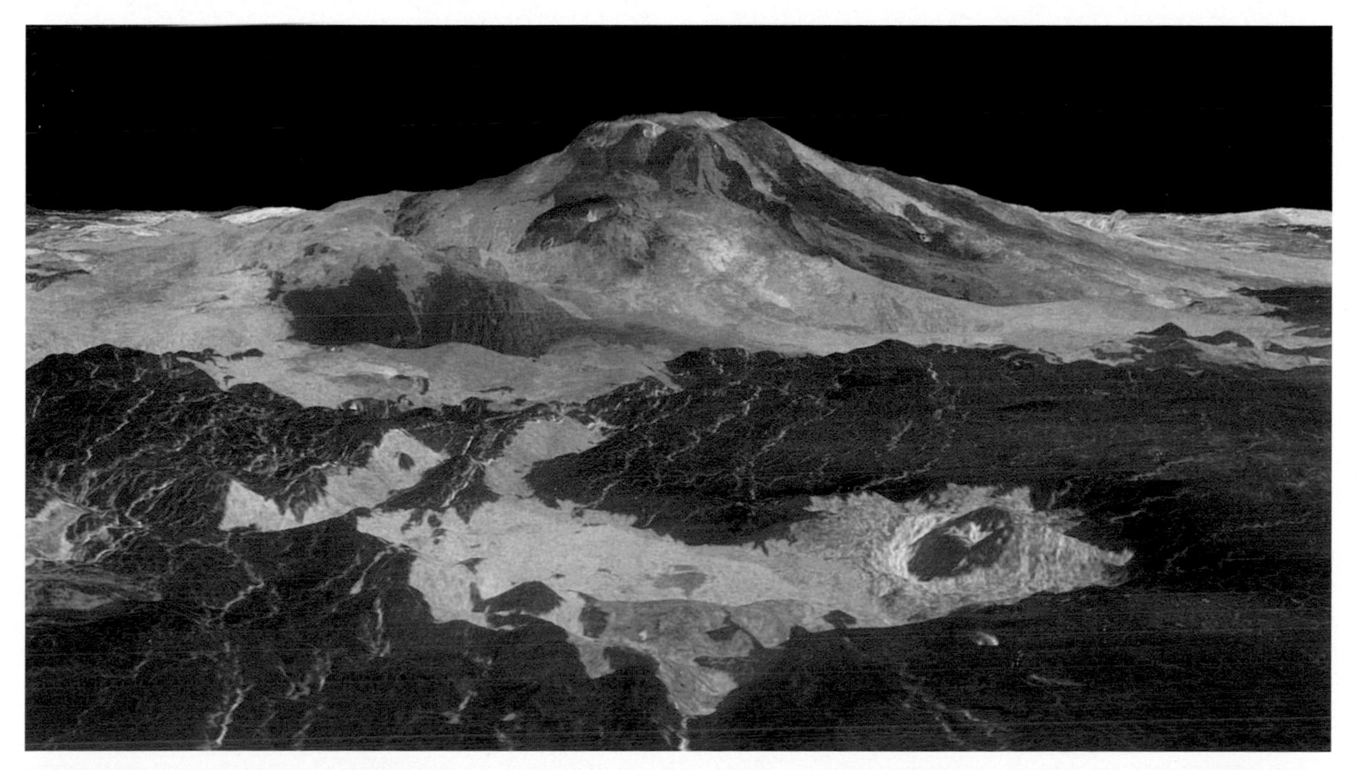

Figure 7.36 Maat Mons, Venus's largest shield volcano built up of overlapping lava flows. This three-dimensional view was produced by a computer using a combination of radar and altimeter data. Vertical height is exaggerated about 10 times. The colors are approximately true, based on the color pictures returned from the Russian Venera spacecraft that landed on Venus in the 1980s. *Courtesy of NASA / JPL.*

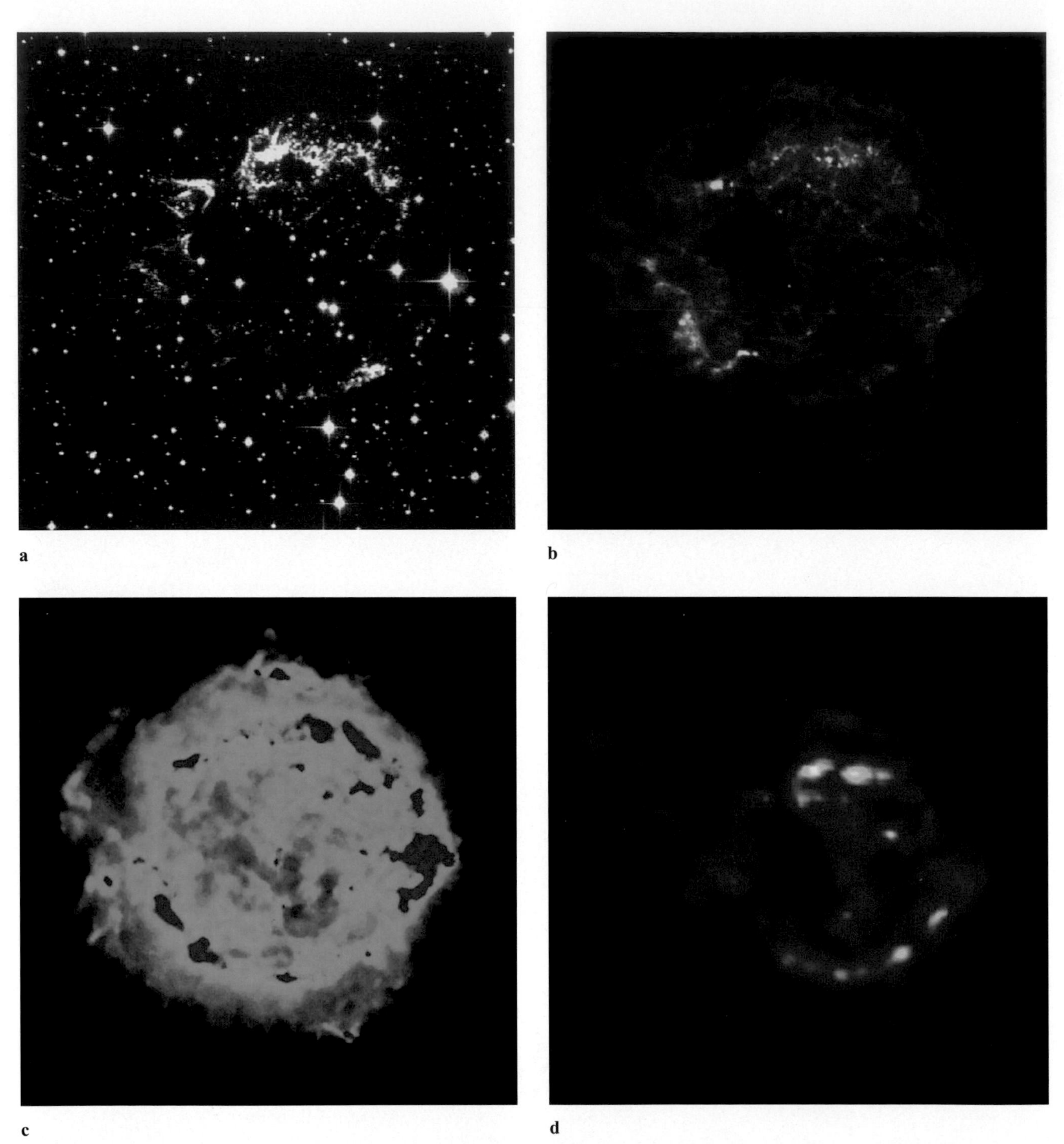

Figure 7.39 Four images of CasA, a bubble of hot gas in the constellation Cassiopeia created by a supernova explosion about 300 years ago, 9700 light-years away. (Actually, the star exploded 10,000 years ago, but it took 9700 years for the light to reach Earth.) The bubble is about 10 light-years in diameter and exploding at a rate of 10 million miles (16 million km) per hour. Image **a** was taken with visible light as the eye would see it. The x-rays emitted by the hot 50 million degree Fahrenheit (28 million degree C) gas are seen in image **b**. The greatest intensity x-rays come from the bright spots. Radio emissions are displayed in image **c**. The red areas are the most intense radio sources. Image **d** shows the infrared emissions which represent the high temperature spots. Notice that the locations of intense emissions at one wavelength do not necessarily correspond to intense emissions at another wavelength. For example, the lonely little red spot near the center of the x-ray image may be a neutron star or a black hole, the remains of the original star that went supernova. It does not show up in the other wavelengths. For this reason, astronomers are intent on examining interesting objects in the sky at different wavelengths. Each wavelength represents some particular physical phenomenon occurring in the object. *Courtesy of NASA / MSFC / CXC / SAO.*

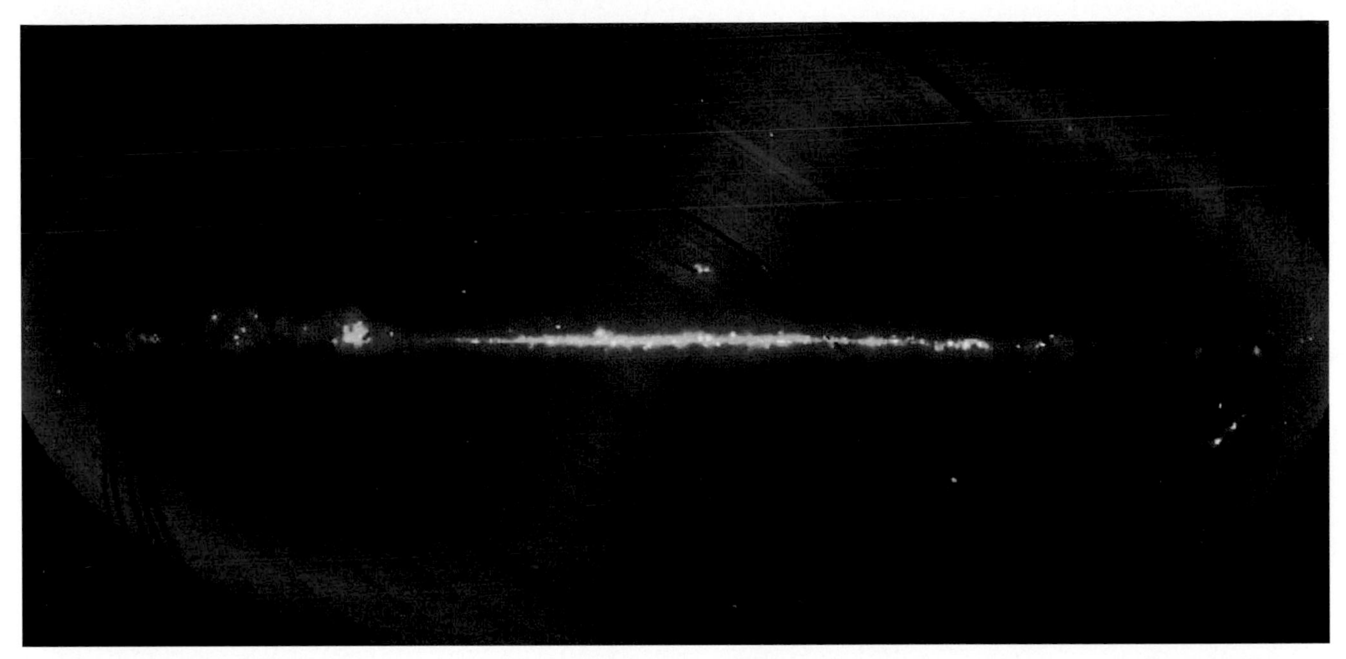

Figure 7.42 Infrared image of entire sky made from IRAS data. The bright band is our view of the Milky Way Galaxy from our position on the inside. The blue and white areas are hot 12-micrometer wavelength radiation and the red areas are cooler 100-micrometer radiation. The black streaks are areas of no data from the satellite. Directly above the center of the plane of the Milky Way is the constellation Ophiuchus and just below the plane on the far right is Orion; new stars are forming from hot gas in both of these regions. The Large Magellanic Cloud, a small galaxy outside the Milky Way, is below the plane to the right of center. *Courtesy of NASA / JPL.*

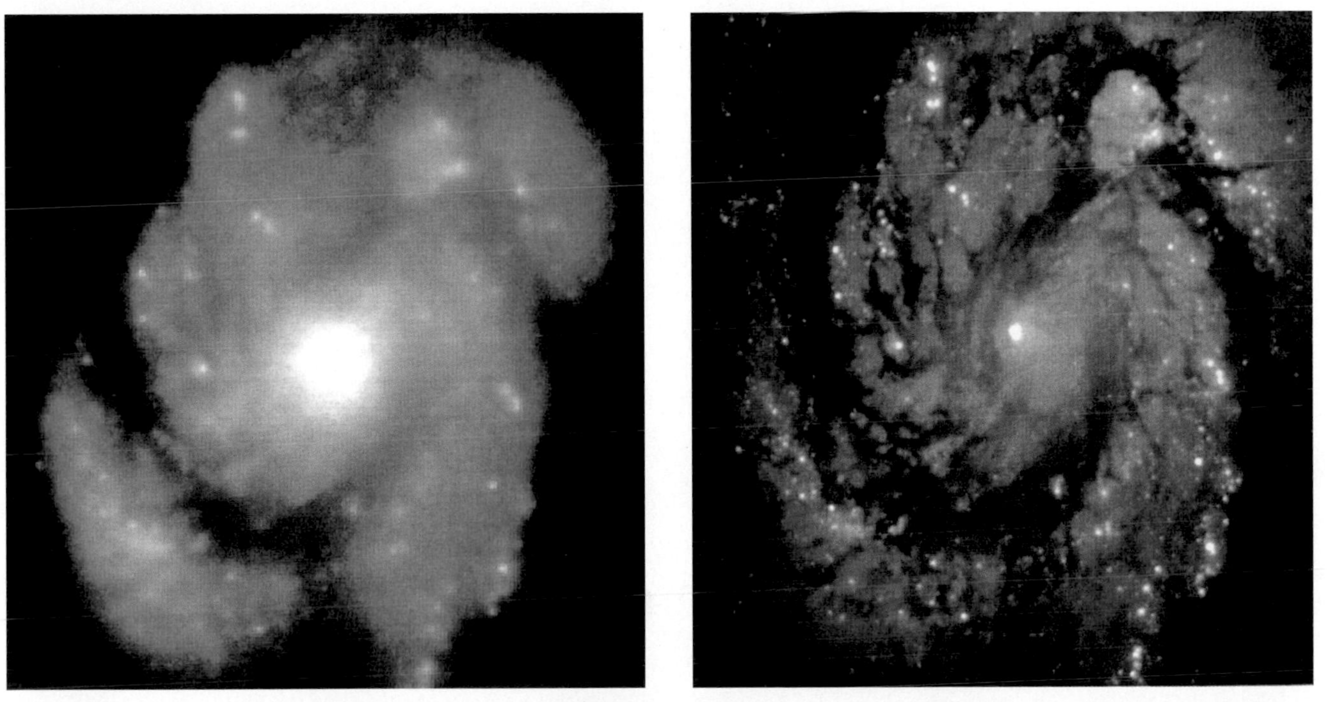

Figure 7.45 Two images of the same galaxy made by the Hubble Space Telescope. The one on the left was made with the mirror in its original flawed condition. The right image was made after astronauts inserted corrective optics into the telescope. *Courtesy of NASA.*

Figure 7.46 Core region of the Orion Nebula. The brightest star is θ^1C-Orionis and the next two brightest in the upper right corner are stars in the Trapezium Cluster. About a dozen low mass stars and their protoplanetary disks surround θ^1C-Orionis. Intense radiation pressure from θ^1C-Orionis sweeps the protoplanetary disks out into teardrop shapes pointing away from the star. The vertical streak is caused by overexposure of the star in the CCD. *Courtesy of Robert O'Dell and NASA.*

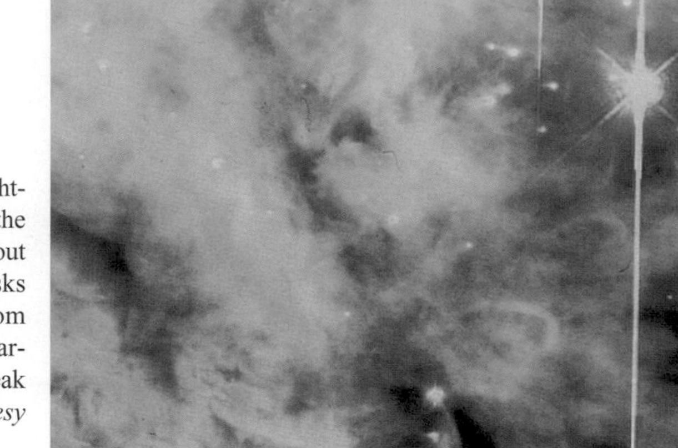

Figure 7.47 The Eagle Nebula photographed by Jeff Hester and Paul Scowen of Arizona State University using the Hubble Space Telescope. The nebula consists of clouds of hydrogen gas and dust particles. Brilliant hot stars out of the picture at the top illuminate and heat the cloud. New stars are likely to form within the cloud. *Courtesy of NASA.*

Figure 7.48 The Hickson Compact Group 87, a cluster of four galaxies. Two of the galaxies are spirals and one is elliptical. The fourth one is a disk-shaped galaxy seen on edge with dark dust lanes stretched through its center. *Courtesy of the Hubble Heritage Team (STScl / AURA / NASA).*

Figure 8.12 A Space Shuttle on the mobile launch platform atop the crawler transporter speeds at 1 mile per hour (1.6 km/h) along the road to the launch pad. For size comparison, note the van along the side of the road. *Courtesy of NASA.*

Figure 8.15 Space Shuttle *Atlantis* lifts off from the mobile launch platform. *Courtesy of NASA.*

Figure 8.25 The X-34 technology test-bed vehicle. It will not fly into orbit; its purpose is to test new concepts for building vehicles that will. *Courtesy of NASA.*

Figure 8.27 The X-33 advanced technology demonstration vehicle. Note the lifting body and the two linear aerospike engines in the rear. *Courtesy of NASA.*

Figure 10.16 With the Hubble Space Telescope firmly docked in the cargo bay, one astronaut checks the tools and replacement parts while another rides the remote manipulator arm toward the access doors. *Courtesy of NASA.*

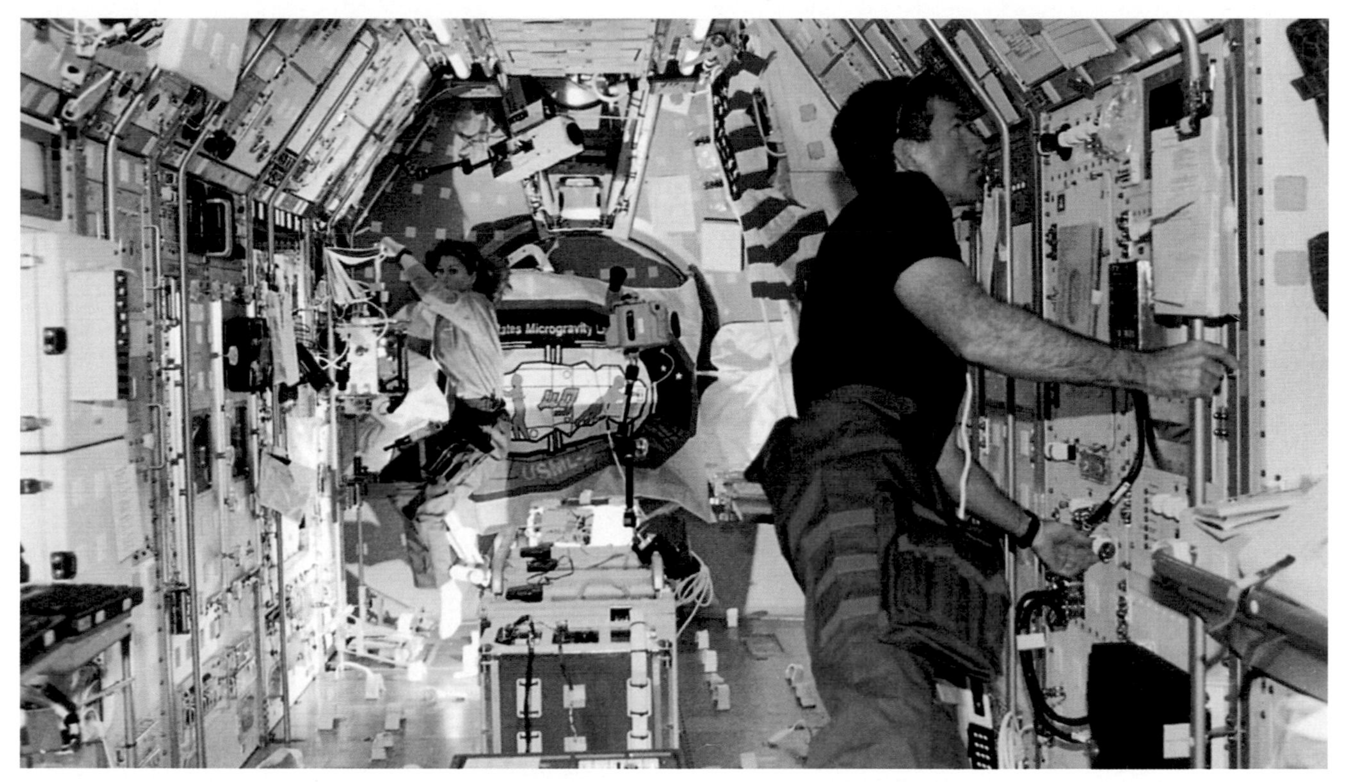

Figure 10.21 Astronauts Catherine Coleman and Fred Leslie conducting microgravity experiments in Spacelab in November 1995. Because there is no up or down in weightlessness, experiments are mounted on floor and ceiling as well as walls. Notice the foot restraints on the floor to keep a person in place while working. *Courtesy of NASA.*

Figure 11.3 Three astronauts and a cosmonaut in the *Mir* living quarters in the Base Block, number 3 in Figure 11.2. Notice the maze of wires, ducts, and equipment that accumulated during 13 years in orbit. *Courtesy of NASA.*

Figure 11.8 The International Space Station as construction began in 1998. At the top is a U.S.-built node for attaching other modules and as a temporary docking port for the Space Shuttle. At the bottom with solar panels deployed is the Russian Zarya module, number 5 in Figure 11.4. *Courtesy of NASA.*

Figure 11.9 Flying over Greece and Turkey on May 30, 1999, astronaut Tamara Jernigan holds a crane to be attached to the International Space Station at the top of the picture. She is attached to the manipulator arm of Shuttle *Discovery*. *Courtesy of NASA.*

Figure 12.2 Artist's concept of a lunar mining base. Lunar soil is scooped up (red endloaders) and ground up (red structure to left); oxygen is extracted from the soil and liquefied (yellow buildings), then piped to the yellow storage tanks. Waste tailings are carried away by a conveyer system. A rocket takes off in the background carrying several tanks of oxygen to a rendezvous point in lunar orbit. The habitat for a dozen technicians is just right of the communications antennas, covered by a mound of soil. Power lines come from a nuclear generator beyond the horizon. *Courtesy of Eagle Engineering for NASA; artist Pat Rawlings.*

Figure 12.8 Later stage in the construction of a lunar base using inflatable structures. The low flat building with the flag and NASA logo is the first in place. The dome is set over a crater. An inflatable hallway connects the two. Bags of lunar regolith are stacked around and over the inflatables for shielding against the solar wind. Solar panels for power are next to the low building, at the upper right, and in the upper left. The road leads to the landing pad. *Courtesy of NASA.*

Figure 12.10 The Lunar Games. New out-of-this-world records are made here. Antennas at the right broadcast the games to Earth in the background sky. With the low gravity, people fly like birds in the dome at the top right. At the top center of the dome is a hologram projection of a basketball game next door. *Courtesy of NASA; artist Pat Rawlings.*

Figure 12.11 Mars mosaic made from Viking 1 orbiter pictures. Major features are identified in the drawing. Valles Marineris is over 3000 miles (4800 km) long, up to 150 miles (240 km) wide, and 5 miles (8 km) deep. Three of Mars's largest volcanoes are on the left. *Courtesy of NASA.*

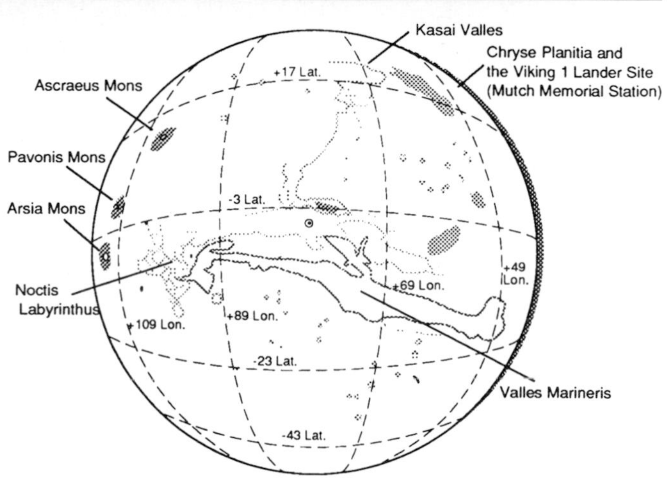

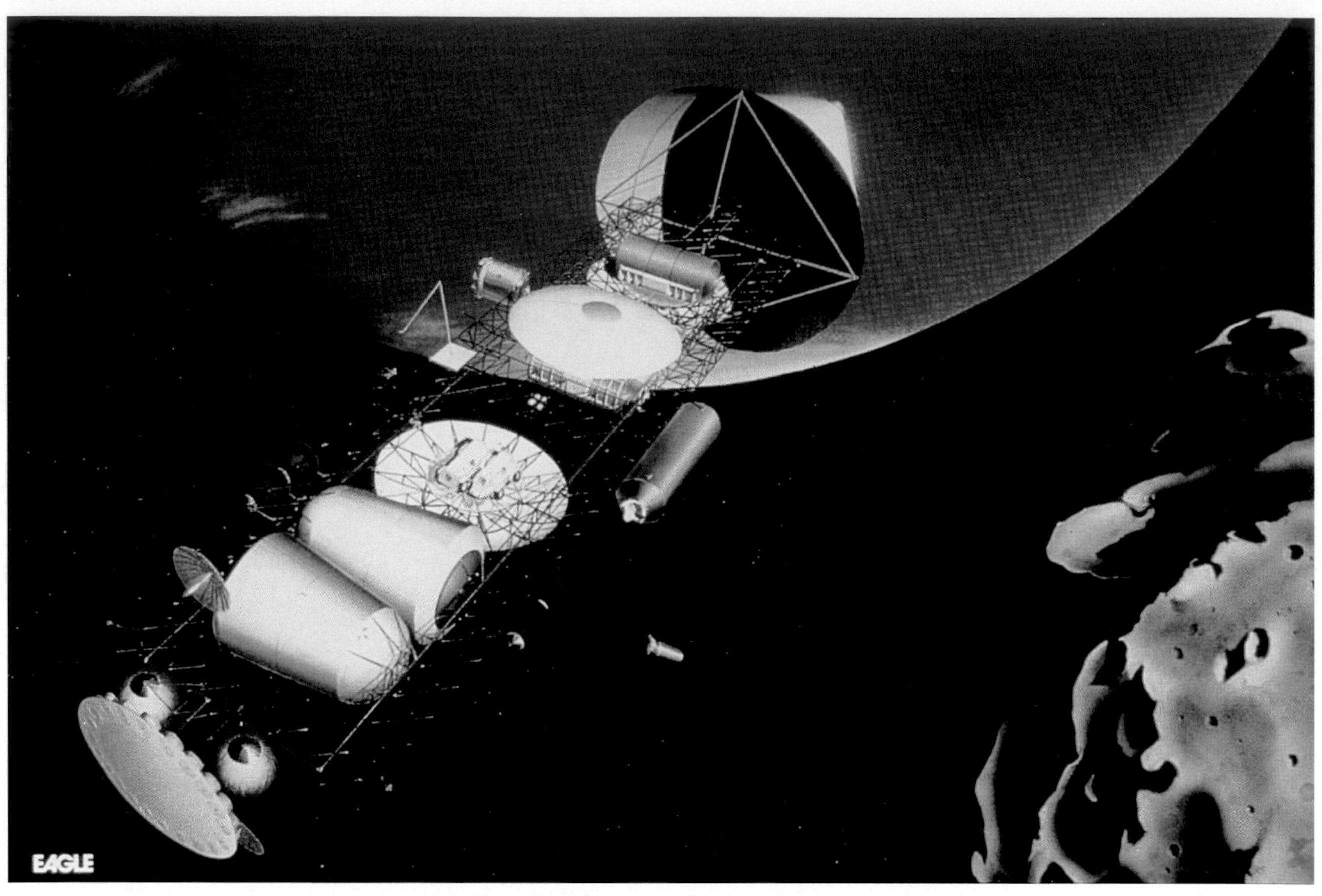

Figure 12.19 A space barge hauls a massive load of cargo from low Earth orbit to low Mars orbit. The Martian moon Phobos is visible at the lower right. *Courtesy of Eagle Engineering for NASA; artist Mark Dowman.*

Figure 13.6 A piece of Mars, meteorite ALH84001.0. *Courtesy of NASA.*

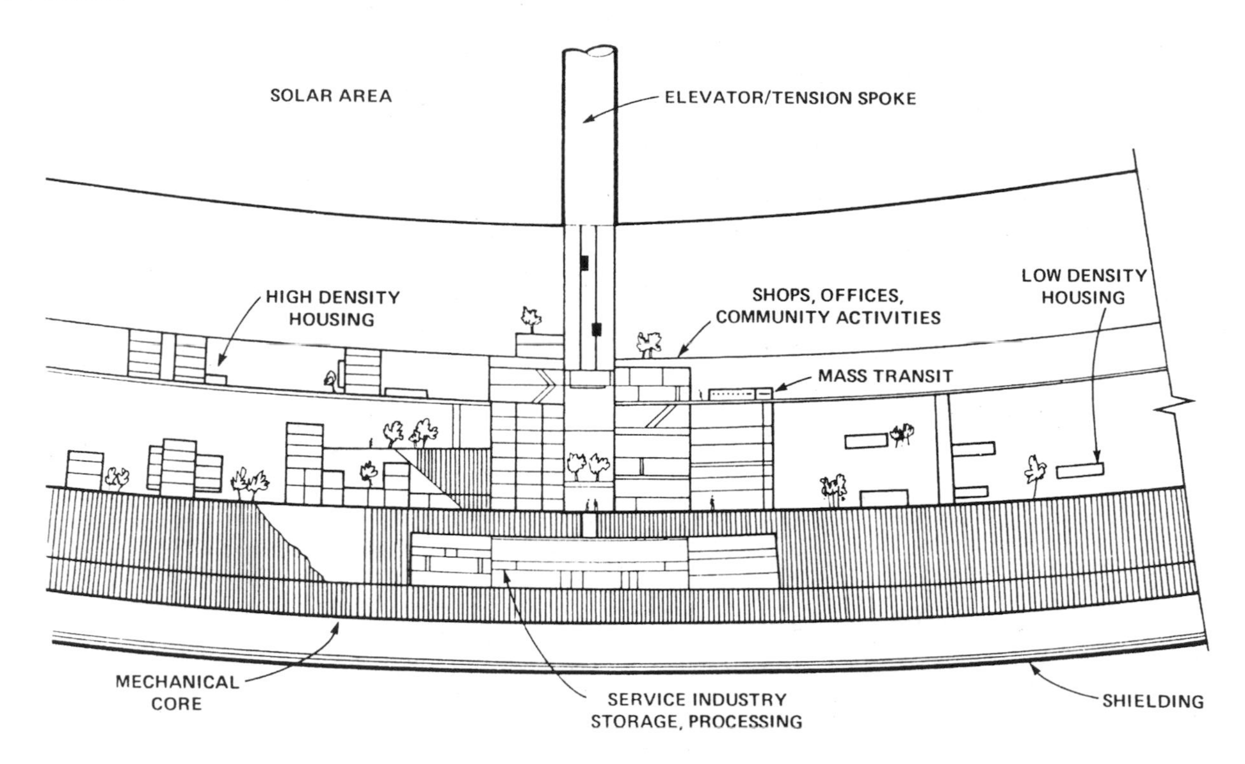

Figure 11.15 Partial cross section of the interior of the big wheel space colony. *Courtesy of NASA.*

cost of an interplanetary vehicle to carry people on an expedition to Mars, for example, is greatly reduced if it originates from a space colony. The spaceship is built and stocked with supplies at the colony; crews come aboard, check it out, and blast off. Escape velocity from Earth is about 7.5 miles (12 km) per second; from the Moon it is only 1.5 miles (2.4 km) per second, and from an orbiting colony it would be even less. Both mining the Moon and human exploration of Mars are discussed in Chapter 12.

The initial processing plant, solar power equipment, and construction crew would have to come from Earth, but the habitat itself would be built with a supply of lunar ore. When the habitat and manufacturing facility are completed, other projects, such as the two described above, could be undertaken. Building additional new colonies is another logical step. Of course, the initial investment is high. An industrial plant in space and a mining camp on the Moon would not be cheap. However, once in place, they could become a profit-making venture.

Figure 11.16 Artist's concept of the interior of the big wheel space colony. *Courtesy of NASA.*

Location

Where would you locate such a city in space? Low Earth orbit is not the best choice because some of each orbit is spent in darkness, and the colony will depend on a continuous supply of solar energy. The Van Allen belts must be avoided. But if the colony's industry is to operate in conjunction with a lunar mining colony, then a location near the Moon may be advantageous. L5 is often suggested as the best location. The L-points were discussed in Chapter 3.

Energy

Because the space colony is in direct sunlight, solar energy would be continuously available for smelters, mills, and fabrication factories as well as for light and agriculture inside the colony. Solar energy falling on Earth averages about 25 watts per square foot (270 watts/m^2); the maximum reaching a square foot with the Sun directly overhead is about 70 watts (750 watts/m^2). In space, without an intervening atmosphere, a continuous 130 watts falls on a square foot surface (1400 W/m^2). The only limit to the amount of energy available to the colony is the size of the array of solar cells or concentrating mirrors.

Materials

To build the L5 habitat, excluding a massive protective shield, will require an estimated 200,000 tons of materials (Earth weight). The proportions of the primary elements are shown in the center pie diagram of Figure 11.17. The main structure would be made chiefly of aluminum. That plus atmospheric oxygen and soil constitutes most of the mass of the habitat. Analysis of lunar soil brought back by Apollo 16 shows it to be composed mostly of oxygen, silicon, and aluminum. See the right pie diagram in Figure 11.17. This is not too surprising; those three are the most common elements in Earth's crust, too, although the Moon has a higher percentage of aluminum, at least at the Apollo 16 site. These materials, then, would come from the Moon. Because of the Moon's lower gravity, it is much easier to lift mass to L5 from the Moon than it is from Earth. Only hydrogen to make water; nitrogen to dilute the atmospheric oxygen; carbon in the plants, animals, and humans; initial structures; and machinery would have to be brought up from Earth. As seen in the left pie diagrams of Figure 11.17, these amount to a small percentage of the total mass of the habitat. As discussed in Chapter 7, there is evidence of water ice in craters at the Moon's poles. If water is present in large quantity, hydrogen from Earth would not be necessary.

Artificial Gravity

As pointed out in Chapter 9, weightlessness even for a few days has adverse effects on the human body. Permanent weightlessness may destroy a body, so a large colony in space must have some way of producing an artificial gravity. Rotating a big wheel space colony does just that. When you go around a corner in your car you can feel a force pushing you toward the outside of the turn. On a revolving carnival ride you get the same sensation; you need to hold on to keep from being thrown off. Objects in the tube of the rotating big wheel would feel a similar force, but would interpret it as "gravity."

What is really happening follows Newton's first law of motion. Remember Chapter 2: objects in motion will continue to move in a straight line unless acted on by an unbalanced force. Imagine you are in the tube of the big wheel, as in Figure 11.18. Because the wheel is rotating, carrying you with it, you are in motion in the direction of A. According to

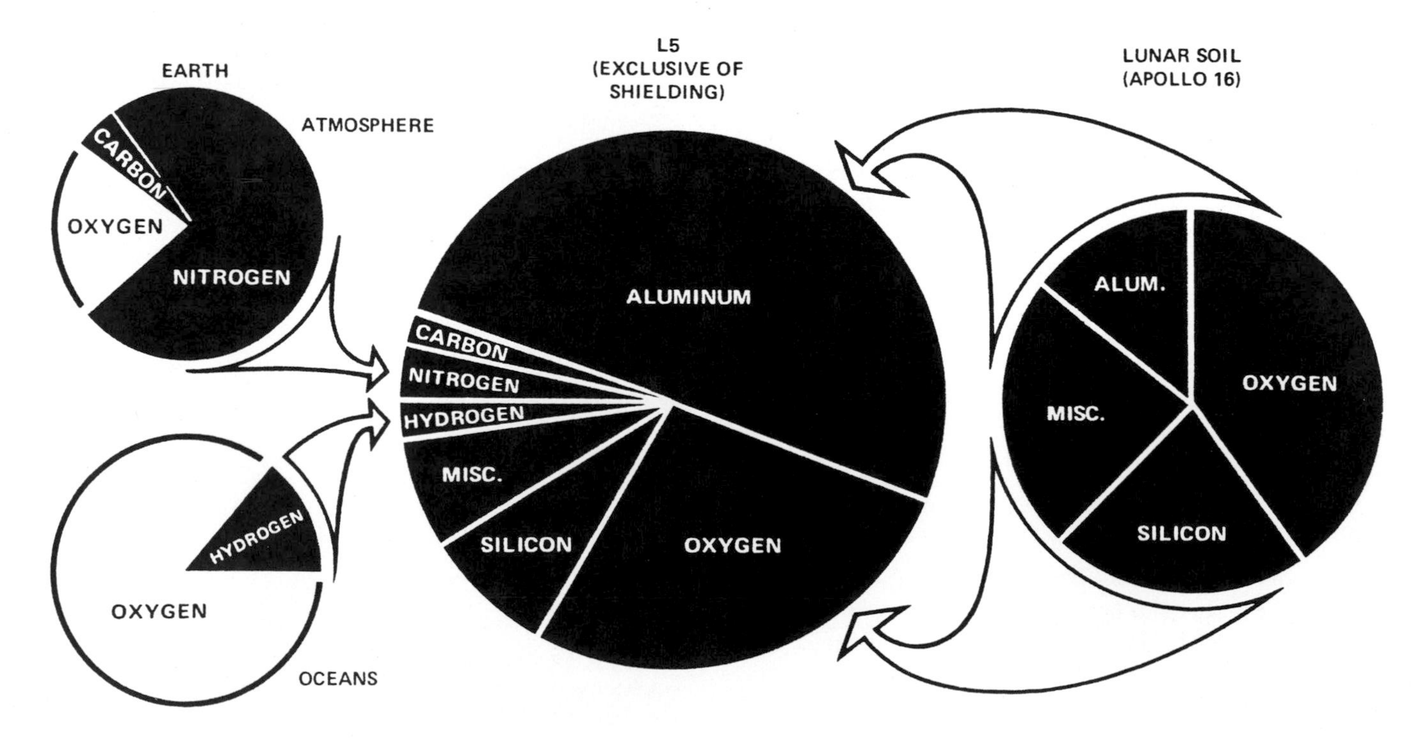

Figure 11.17 Sources of materials required for construction of the L5 habitat. Center diagram shows proportion of various elements needed; left and right diagrams show what could be supplied from the Moon and from Earth. Shielding can be the slag and waste materials from processing lunar ore. *Courtesy of NASA.*

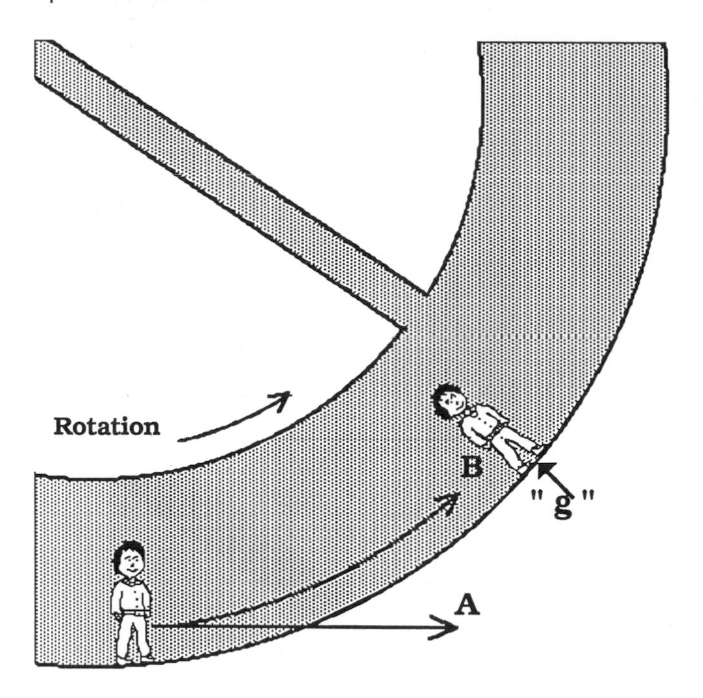

Figure 11.18 Rotation of wheel to produce artificial gravity.

Newton, you would continue to move in direction A except that the rim of the wheel curves upward, forcing you to follow the curved path B. In doing so, the rim pushes up on the bottoms of your feet, toward the center of the wheel. This force toward the center is called, appropriately, *centripetal force*. You get the same sensation that you feel on Earth when gravity pulls you down but the floor pushes up on the bottoms of your feet keeping you from falling through. Centripetal force becomes an artificial gravity. In the rim of the rotating wheel, heads point in toward the central hub while feet point out. "Down" is away from the hub.

The amount of artificial gravity, that is, centripetal force, depends on how large the wheel is and how fast it is turning. A 1-mile (1.6-km) diameter wheel rotating at one revolution per minute produces an artificial gravity at the rim approximately equal to Earth gravity. See MATHBOX 11.1 for the mathematics involved. It is possible to obtain any *g* force desired by changing the radius or the speed of rotation.

Living in a rotating vehicle can cause motion sickness. The problem arises when a person's head moves out of the plane of rotation of the vehicle. Physiologists have found that most people can adapt to rotation rates up to four revolutions per minute. To be conservative, most space colony designs limit the rotation rate to one revolution per minute.

The open question is: how much gravity, or rather, how little gravity can the human body tolerate permanently? Although Russian cosmonauts have spent as much as a year at a time in weightlessness, no one as yet has a conclusive answer to that question. The answer is obviously crucial to the design of the L5 habitat. For economy and ease of construction, it would be desirable to keep gravity as low as tolerable. One g in the living area means that the buildings would weigh the same as they do on Earth and would have to be built to same strength as if they were on the surface of Earth. The habitat shell, in turn, would have to be built to support that weight. If humans could live at one-half g then everything would weigh only half its Earth weight, allowing the shell and buildings to be built with less structural strength, thus greatly reducing the mass of material that must be shipped out to build the colony. One of the experiments to be carried out on the ISS will be an artificial gravity lab which can be rotated at various speeds to find out how much gravity humans require to stay in good physical condition.

Visitors to the colony would dock their spacecraft at a port on the hub shown in Figure 11.19, enter through an airlock and take an elevator in one of the spokes out to the rim of the wheel. The hub is an interesting place because there the g force is zero, that is, weightlessness. Try to imagine that ride to the rim. You start out weightless, but as the elevator carries you "down" you feel your weight returning until, at the end of the ride, your weight is normal. The hub would make an exciting recreation center. Imagine a game of basketball in weightlessness!

Meteoroid and Radiation Protection

Possible collisions with meteoroids must be considered. A big wheel colony could lose 60 percent of its air in 1 day through a hole 1 yard (1 m) in diameter. A hole that size would be discovered and patched quickly, however. Based on the density and size distribution of the meteoroid population in the Earth-Moon environment, catastrophic loss of atmosphere in a collision appears to be only a small hazard. The probability of a collision creating a 1-yard (1-m) hole has been calculated at one in 10 million years. Meteoroids do occur in clusters and multiple collisions in a short time are possible. Certainly, collisions with micrometeoroids would occur often and a routine program to repair small leaks would be necessary.

A more serious aspect of the meteoroid problem is the explosion overpressure caused by an impact. Even a small (1 gram) meteoroid would create a pressure wave of killing force for a distance of about 7 feet (2 m) from the point of impact. A meteoroid shield or bumper separated from the main shell of the space colony would provide protection from such pressure waves.

Protection from cosmic rays and solar protons would also be needed. Solar flares of sufficient magnitude to cause a serious radiation sickness problem occur during the high part of every 11-year solar cycle. We on Earth are protected from particle radiation by Earth's magnetic field. Similarly, a magnetic field around the space station would deflect the high energy particles. However, building a machine to produce a magnetic field of sufficient strength to protect a big wheel space colony is beyond present technology.

MATHBOX 11.1

Artificial Gravity

Centripetal force is the force directed toward the center that makes an object move in a circular path. If you stand inside the rim of a rotating big wheel space station with your head toward the center, you feel this force on the bottom of your feet pushing you into a circular path. The sensation is the same as the force of gravity produces on your feet when standing on Earth, therefore it is called artificial gravity.

The formula for centripetal force is

$$F_c = \frac{mv^2}{r}$$

where m is the mass of the object, v is its velocity and r is the radius of the circle. Remember Newton's second law

$$F = ma$$

which relates the acceleration of an object to the applied force. Compare the two equations above and note that the acceleration due to centripetal force must be

$$a_c = \frac{v^2}{r}.$$

Solving for v we get

$$v = \sqrt{a_c r}.$$

Our rotating big wheel space colony is 1 mile (5280 feet or 1609 m) in diameter; the radius, r, is 2640 feet (805 m). If we desire 1 g artificial gravity, a_c must be 32 feet per second per second, the acceleration of gravity on Earth. Then the velocity of an object at the rim must be

$$v = \sqrt{32 \times 2640} = 290 \text{ ft/sec} = 198 \text{ mi/h}.$$

That is less than half the speed of a jet airliner.

Next, let us calculate the period of rotation. First find the circumference of the wheel, that is, the distance travelled in one rotation.

$$5280 \text{ feet} \times \pi = 16{,}590 \text{ feet}.$$

Next, divide that distance by the speed to get the period of rotation.

$$\frac{16{,}590}{290 \text{ ft/sec}} = 57 \text{ seconds}.$$

So a big wheel space colony 1 mile (1.609 km) in diameter rotating approximately once per minute will have 1g artificial gravity at its rim.

In metric units, the acceleration of gravity at the surface of the Earth is (9.8 m/s^2). Therefore, the velocity at the rim must be

$$v = \sqrt{9.8 \times 805} = 89 \text{ m/s} = 320 \text{ km/h}.$$

The circumference of the wheel is 1609 m $\times$ π = 5055 m, so the time it takes for one rotation at that velocity is

$$\frac{5055 \text{ m}}{89 \text{ m/s}} = 57 \text{ s}$$

Of course, we get the same result no matter what system of units we use.

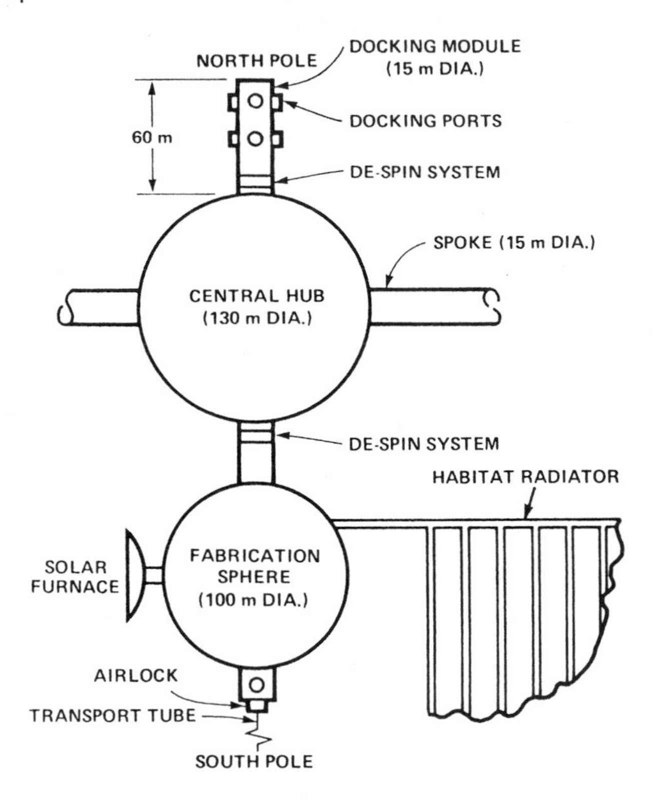

Figure 11.19 Hub configuration. The de-spin system keeps the docking ports, fabrication sphere (manufacturing plant), and radiator from rotating with the main part of the habitat. Elevators in the spokes carry passengers and materials to the living areas in the rim. *Courtesy of NASA.*

A massive shield could provide protection from both high energy particles and meteoroids, but would not hold together if rotated at one revolution per minute. It would have to be nonrotating, and therefore not attached to the rest of the space station. However, a shield placed between the Sun and the habitat to block solar particles would also block the needed sunlight. One proposed design calls for a mirror on the opposite side of the big wheel, not connected to it, reflecting sunlight into the colony. The mirror can be seen in Figure 11.14. This design would allow a massive shield to be located on the sunward side. Ten million tons of mass would be required. The slag from the industrial processing of the lunar soil would provide part, but not all of it.

Life Support

The life support question rises again as it does for any inhabited spacecraft. The design of the habitat is greatly dependent on solutions to the problems of life support: air, food, water, and protection from radiation. In Chapter 9 we discussed creating a comfortable and healthy environment in the Space Shuttle. All the supplies needed for the relatively short mission times are carried along, oxygen and nitrogen are added to the atmosphere as needed, carbon dioxide is removed by canisters of lithium hydroxide, excess water is dumped overboard, and the wastes are returned to Earth. Now, however, we must design a city which will be completely self-sufficient, independent of Earth. Continually shipping supplies to L5 from Earth is out of the question. It would be much too expensive. The life support system must be a closed loop, everything recycled.

Atmosphere

Can a person live in an atmosphere different from normal Earth atmosphere for long time periods? Space Shuttle carries a normal Earth atmosphere, 21 percent oxygen and 79 percent nitrogen. Most previous manned spacecraft had pure, low pressure oxygen atmospheres. The long-term effect of a pure oxygen atmosphere on people, plants and animals is unknown. Pure oxygen also increases the fire hazard. Until more definitive research is done, the oxygen should be diluted with some inert gas such as nitrogen. Oxygen would come from lunar ore, but the initial supply of nitrogen would have to come from Earth. Helium has been suggested as an inert dilutent instead of nitrogen because it is not as dense. See MATHBOX 9.1. There would be less mass to bring from Earth, and the rotating habitat would have less weight to support. However, helium gives a strange Donald Duck-like quality to voices.

Sea level pressure is 14.7 psi (101 kPa), but some people, plants, and animals live their entire lives at altitudes where atmospheric pressure is only two-thirds that at sea level. Pressure in Skylab was only 5 psi (34 kPa), one-third normal. It would be advantageous to equip a space colony with an atmosphere at lower pressure to reduce the strain on the shell of the habitat, thus requiring less structural strength. One difficulty is that sound does not travel as far in a gas at low pressure. Skylab astronauts reported that a loud speaking voice carried only about 15 feet (about 5 meters).

For transportation, no cars would be allowed, only an electric railway or moving sidewalks, roller skates, and bicycles. The distance around the colony is only a little more than 3 miles (5 km), and the elevator shafts through the central hub are shortcuts to the other side. Thus, the air would only have to be cleansed of dust, excess water vapor, carbon dioxide, and other gases. Plant photosynthesis is a primary factor in closing the atmosphere loop, but will not do the complete job. Cold plates and coils can condense the water vapor into droplets to remove it from the air, making it available for reuse.

Food and Water

Although much is understood about human nutritional needs, there is no certainty that all the necessary nutrients have been identified. An important question is whether loss of bone calcium, loss of fluids, muscle deterioration, and other effects of living and working in space can be compensated for by diet adjustments. Proper diet is not only necessary for good health, it is also a factor in psychological well-being.

Initial food supplies will be brought from Earth, but farming and raising livestock will begin as soon as they can be established. Plant agriculture is practical and beneficial for two reasons: food and oxygen. While animals inhale oxygen and exhale carbon dioxide, plants intake carbon dioxide and release oxygen. That is a great oversimplification, but is basically true. Photosynthesis is the chemical and metabolic process by which plants use water from the ground and carbon dioxide from the air with sunlight as an energy source to manufacture carbohydrates. Oxygen is released as a by-product. In fact, the original atmosphere of Earth probably contained carbon dioxide and water vapor, but no oxygen. According to the geological fossil record, plants inhabited Earth before animals did. It wasn't until plants came along that oxygen began to build up in the atmosphere, which made higher forms of animal life possible.

The large variety of foods available on Earth may be impractical to produce in the colony simply for lack of space. Meats are a good example. What kind of livestock should be raised? Cattle for beef? Hogs for pork? People have different preferences. Hogs, chickens, and turkeys have been suggested because they can consume table scraps and noncellulose plant wastes. Turkeys are particularly efficient in protein production. However, these choices may not be the best because their main diet, corn, could also be eaten by humans. Energy is lost when the animal converts the food to meat and it may be more energy efficient for humans to eat the food directly.

Ruminant animals would be an excellent choice in a closed environment because they can digest cellulose plant wastes inedible for humans. Cellulose parts of plants include the structural parts, roots, husks, and hulls. Cows are ruminants and yield both dairy products and meat. A beef steer grows to about 900 pounds in16 months. Dairy products are a particularly important source of nutrition because of the high calcium content, and dairy cows convert food to milk with greater efficiency than they convert food to meat.

Small animals such as goats, rabbits, ducks, and geese may be good choices. They reproduce frequently, mature rapidly, and would only have to be fed during their short rapid growth period, then slaughtered for their meat. Rabbits reach 7.5 pounds (3.4 kg) in about 4 months. Table scraps and by-products from vegetable production would provide the bulk of their diet. Goat milk and duck eggs would add variety to the human diet, also.

Aquatic food production is another likely possibility and has been carried on in some parts of Earth for many centuries. Animal and human solid waste is effective fertilizer for aquatic production. In some countries, human waste is often dumped directly into lakes or ponds for cultivating water plants and fish. Fish can be harvested in a year weighing in at about 4.5 pounds (2.0 kg).

Space Agriculture

"Weather" in a space habitat would be under complete control. In the agricultural areas, light, temperature, humidity, and carbon dioxide content would be kept at proper levels to promote most rapid growth and maturation of crops. Agricultural and animal production can be highly efficient in a small area providing diseases are eliminated or at least kept under careful control.

In one series of experiments simulating the controlled environment of a space habitat, Bruce Bugbee from Utah State University grew a common breed of wheat in hydroponic tubs with a yield five times greater than the highest productivity ever achieved in a wheat field. The most important factor is providing enough light. Plants need a light intensity 100 times greater than people need for work light. Bugbee used sodium vapor lamps and left them on 24 hours a day. Wheat does not need a rest period. The range of useful wavelengths of light is 400 to 700 nanometers. (A nanometer (nm) is a billionth of a meter.) There is a sharp cutoff for photosynthesis: 695 nm works, but 710 nm does not.

Given enough light the next most important thing is an adequate supply of water and nutrients. Seeds were planted in a 3/4 inch (2 cm) thick layer of sterile, inert rock wool, similar to house insulation, to give support to the plant. The rock wool was fastened over the tops of 16-inch by 20-inch tubs, 4 inches deep, (40 cm by 50 cm by 10 cm) filled with about 5 gallons (20 l) of nutrient solution. Plants in each tub would take up and transpire more than 50 gallons (200 l) during the wheat's 80-day life cycle. The solution contained carefully measured amounts of the 13 essential elements for plant growth. In that time the tub full of wheat would absorb about 2½ ounces (70 grams) of the mineral. To provide oxygen to the roots, the solution flowed through the tubs continuously.

Wheat is most productive at 63°F (17°C); at 73°F (23°C) yield is cut in half. Relative humidity of 80 percent is optimum. A higher percentage of carbon dioxide is desirable to a point, but wheat is more sensitive than people are to excess carbon dioxide. At more than 0.2 percent, it becomes toxic to wheat. A slight breeze is needed to waft carbon dioxide to the leaves and oxygen away.

The wheat used in this experiment was not a special breed. A breed could be genetically engineered to be better adapted to a closed environment. For example, a shorter plant would save space and have less waste stalks, stems, and leaves. While these experimental results apply to wheat, other plants would have similar needs and limitations.

A long-running attempt to grow dwarf wheat in the Kristall module on *Mir* had mixed results. Shannon Lucid, during her 6-month stay, planted seeds in a bed of Zeolite. A computer program controlled the light and moisture. The wheat stalks were monitored and photographed daily. A few plants were harvested at intervals and preserved for later examination. After about 40 days the plants began to produce seed heads. Several months later Lucid's replacement, John Blaha, harvested the wheat and brought more than 300 seed heads back to Earth only to find that they were empty. Apparently a low level of ethylene in *Mir* interfered with polli-

Figure 11.20 Artist's view of agricultural area in the L5 space colony. Compare with Figure 11.21, a diagram of the area. A residential section is in the background. *Courtesy of NASA.*

nation. Later, another astronaut on Mir, Michael Foale, planted rapeseed that successfully pollinated.

About 150 acres of lunar soil would be enough to produce sufficient food for 10,000 colonists under such controlled conditions. Grown hydroponically, only half that space may be needed. As to which medium to use, the tradeoff is the high cost of delivering rock wool or other base material from the Earth versus the available space. Transporting lunar soil to L5 would certainly be cheaper but it would have greater mass and occupy more space. For a colony on the Moon, lunar soil would be the obvious medium of choice, but the cost of constructing a large lunar greenhouse must also be considered.

See Figure 11.20 for an artist's conception of an agricultural area on an L5 space colony. A cross section is shown in Figure 11.21. Using a tiered arrangement yields more usable growing area in a smaller space. Water cascades down from the aquaculture ponds on the top tier to irrigate the lower tiers, then is pumped back up to the top to start over again. Plants must also be protected from harmful radiation. They can be as sensitive to this as people are.

The colony is divided into six sections, three for living and three for agriculture, alternating around the circle. See Figure 11.22. Each section can be isolated from the others. By

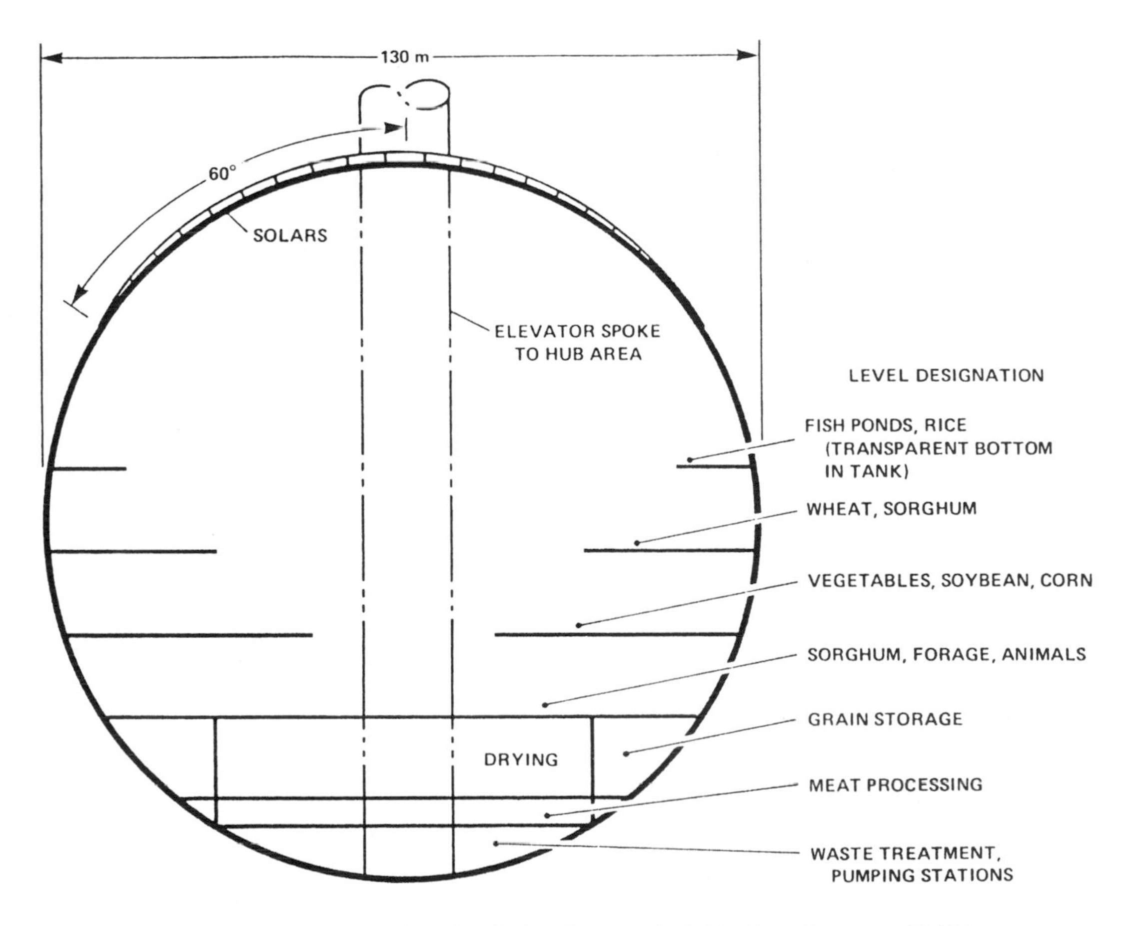

Figure 11.21 Cross section of agricultural area in the L5 habitat. *Courtesy of NASA.*

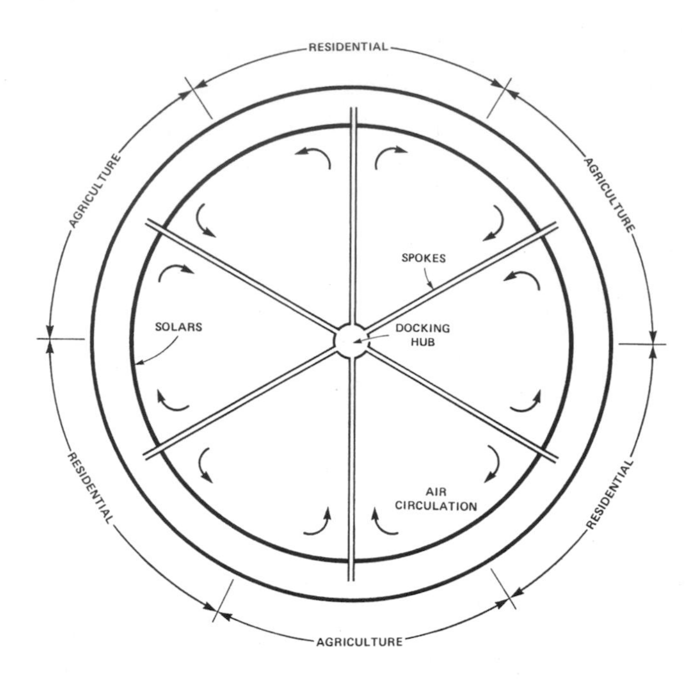

Figure 11.22 Distribution of residential and agricultural areas in the L5 habitat. *Courtesy of NASA.*

placing the crops and livestock into three separate areas, crops can be started at various times to assure a continuous food supply. Alternating residential and agricultural areas would also create a more pleasant living environment for the inhabitants. Fruit would grow in the residential areas for their beauty as well as their fruit.

Careful screening of everything and everybody brought to the colony would be essential to keep out infectious diseases and parasites. Infrared sensing technology from remote sensing satellites could be used in the agricultural areas to monitor the health of crops. If disease should occur in one area, it could be isolated and hopefully the other two would not be affected.

Common human ailments may be impossible to keep out. Immunization and quarantine are the usual tactics against the spread of infection. Using chlorine or other disinfectants to control microbes in a closed environment creates another problem, the accumulation of toxic substances. Hospitals have similar problems, but can bring in fresh air from outside to remove toxic wastes.

Closing the Loop

The upper part of Figure 11.23 shows on the left side what humans, plants, and animals need to live, and on the right side what they put back into the environment as a result of living. The object of a closed loop life support system is to assure that all the outputs on the right side of the diagram are converted to usable inputs. If everything is not recycled, there will be a gradual buildup of waste. Mechanical, chemical, and biological methods to convert the waste back the other way must be perfected. The lower part of Figure 11.23 shows the complexity of a closed loop life support system. Following the arrows from place to place will give an idea of what is involved. Circles represent chemical and mechanical devices which recycle the waste into usable substances.

Chemical and mechanical means for converting outputs to inputs have been devised for some but not all of the waste products. Animals and humans output water in respiration and perspiration which are recoverable mechanically from the air. Purifying urine into drinking water is easy, although there is a psychological barrier to its acceptance by many people. Wash water is also easy to reclaim and purify using mechanical and chemical processes.

Notice that biological activity partially closes the system. For example, the respiratory output of animals and humans is partially carbon dioxide, which must be removed from the air. That is the job of plants. The atmospheric input to plants is carbon dioxide and the respiratory output of plants includes oxygen which is the atmospheric input to humans and animals. Plants use the carbon from carbon dioxide in building their leaves, stems, roots, and fruits. These become food for animals and humans. Animals are also food for humans. Human and animal feces makes fertilizer for plant growth.

The great difficulty in completely closing the loop is that *everything* must be recycled. If it is not, there will be a gradual buildup of unusable waste and a decrease of usable materials in the system. Just imagine the accumulation of unrecycled fingernail or hair clippings of 10,000 people! Some of the waste may be toxic, causing additional problems. If the toxic substance is not removed by the recycling process, then it will accumulate, its concentration increasing with each pass through the cycle.

In the final analysis, for any particular space endeavor there is a break-even point in the means of supplying food. Whether to recycle in a closed loop system or to bring a bag lunch depends primarily on the weight at liftoff. If the recycling equipment and initial food supply weigh more than the bag lunches, then the bag lunch is the way to go. With present technology, a 2 to 3 year mission is the break-even point.

Medical and Psychological Well-Being

Until the Space Shuttle era, astronauts were a homogeneous bunch of healthy, tough test pilots, who could rough it for a week or two cramped together in a small space. They were well screened in advance. Now there is a changing crew composition, less homogeneous, less screened. In the future, space travel will be for everyone. Missions are more routine. As a consequence there is a growing interest in the medical and psychological aspects of spaceflight. *Ergonomics* is the study of human factors.

Astronauts receive emergency medical training, including a limited knowledge of preventative and diagnostic procedures. In case of an emergency they could act to stabilize a

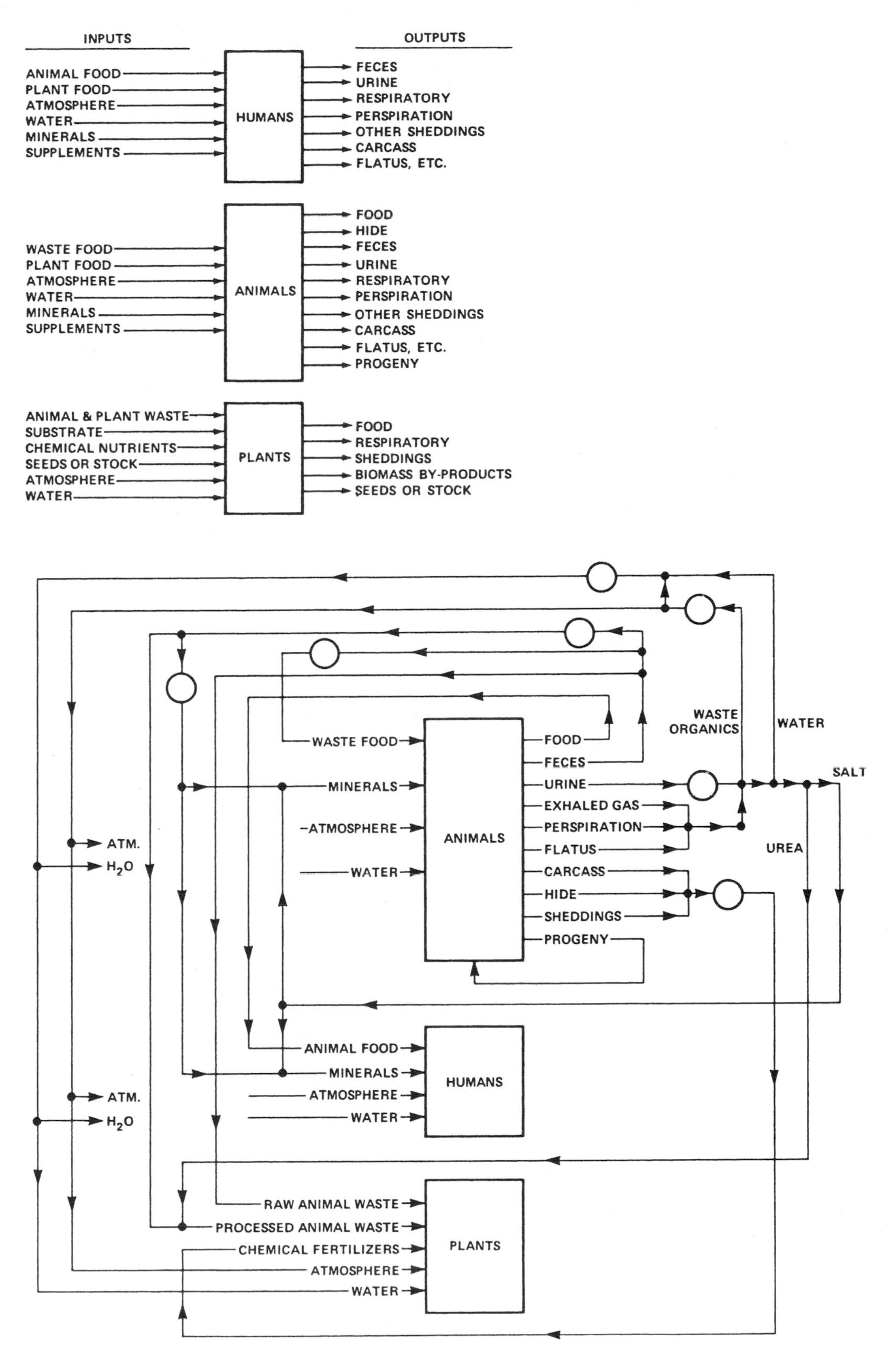

Figure 11.23 Closed loop life support system. The problem is as complex as the diagram appears. *Courtesy of NASA.*

situation until the person could be returned to Earth, although returning from weightless to normal gravity might cause more problems for a patient. It would be advisable to have a doctor, perhaps a surgeon, on long missions and on space colonies. At this time, no one knows whether it is possible to perform a surgical operation in weightlessness.

Data on human behavior in space are being better analyzed and understood. During long-duration flights in confined spaces, motivation is high at first, but irritation sets in and small annoyances become big ones. Personal hygiene, use of the toilet, and lack of privacy become irritants. Astronauts also become less efficient in their work after a few months and cannot concentrate as well. Problems arise with attention span in performing cognitive tasks. On Earth during long winters in the far northern countries, it has been noted that irritability increases and a higher suicide rate is seen.

Submarine studies also bear out this reaction to long confinement. There is a decrease in morale and mood, accompanied by higher stress, after the first 10 days. Things get better about a week before the final day of the mission when the end is in sight. People forget their petty grievances in the excitement over ending the mission. Group cohesion is important. Years ago, the Russians put their cosmonaut crews in a small car and sent them on a 30-day trip around the country to see how well they got along! It is said that cosmonauts do not play chess aboard their space station because it brings out aggressiveness, especially among the Russians who take the game quite seriously!

Underground missile crews are given training as to what to look out for in their coworkers, things which indicate they may "crack." Certain changes in mood or changes in activity levels are some of the indicators. Psychological tests are available to determine profiles of candidates to predict if they might not be able to take it. Schizophrenia, delusional disorders, and hysteria are relatively easy to detect. Anxiety or emotional disorders are difficult to diagnose. Buildup of excess carbon dioxide, toxic substances, noxious gases, or bacteria in a closed environment can also lead to mental or emotional disorders or confusion which may be difficult to recognize.

Simple things become significant to humans during a long stay in a relatively small volume. Even the colors of the walls and equipment have a bearing on the feelings and moods of the occupants. Some Skylab astronauts complained about the lack of color and texture in their habitat. At that time the only colors on board were the color bars used to check the cameras. Everything else was dull earth tones. Work areas need to be well illuminated so lighter colors are logical: greys and whites with touches of blue. Also, lighter paints are more reflective and therefore reduce the amount of electric lighting needed. This reduces both power requirements and heat production, important considerations in space vehicles. Warmer, homey colors are best in living areas and darker shades in sleeping areas. Glossy paint is easy to keep clean, but may cause unacceptable glare. Semigloss and flat paints are preferable. Japanese studies have shown that volume in a habitat is more important than surface area for psychological well-being.

During Skylab activities the astronauts sometimes complained of having too much to do and not enough time to relax and just look out the window at the Earth. On one occasion the crew refused to do any work for a day. They just relaxed and did what they pleased. In longer flights the Russians have noted that comrades in a space station tend to stick together, but get irritated with ground personnel in the command center. Interesting, varied tasks help relieve stress and boredom. Free time is important. Relaxation training may help relieve stress. From the Soviet experience, important factors are communication with families, tapes of music of preference for each crew member, and just being alone once in a while.

Astronaut Shannon Lucid says she learned some important lessons from her 6-month experience on *Mir* that should be applied to the International Space Station. First, the crew must be selected for compatability. She and her two cosmonaut crewmates got along very well. They worked together, did not criticize one another, and most important, she says, they laughed a lot together. She said that if the crew members do not enjoy working together, the flight will be a miserable experience. She also said that the work schedule should not be so precisely and totally rigid as it is for the 2-week Shuttle flights. For long-term flights the astronaut-scientists should have some flexibility to set their own schedules for operating the experiments. Therefore, their training should be skill based rather than step-by-step procedure based so they thoroughly understand the science involved. She added, "An intellectually engaged crew member is a happy crew member."

One possible problem is the disruption of diurnal circadian rhythms. A spacecraft in low Earth orbit has 16 sunrises and sunsets a day. Fortunately humans are very adaptable. Only one-tenth of the variability in most psychological factors tested are accounted for by genes. Studies of adaptation show no significant differences between men and women. Edward Teller says women are smaller and smarter than men, so we should send more women.

The bottom line is that spaceflight is expensive and performance must be maximized. Getting the job done depends on people. Psychological factors become more important as the duration of the flight increases.

Biosphere II

The most complete closed environment experiment yet devised was the privately funded Biosphere II. (Biosphere I is Earth itself.) A crew of eight people, four men and four women ranging in age from 30 to 70, lived in a completely closed ecological system for 2 years in 1991–1993. Constructed of a framework with 8,000 panes of glass, the building covers 3 acres (1.2 hectare) and has a volume of 7 million cubic feet (200,000 m^3). It was divided into seven biomes: desert, savannah, tropical rainforest, marsh, ocean, and human habitat. Over 3,800 species of plants and animals were collected worldwide to occupy Biosphere II.

Two thousand sensors connected to more than 150 computers and an array of video cameras and monitors were scattered throughout the building. Careful monitoring from both inside and outside enabled the research scientists to determine how the environment was acting and reacting at any time.

Much of the crew time was spent in agriculture, growing their own food, and caring for the animals. Because of an unusually cloudy 2 years in Arizona, they were able to produce only about 80 percent of their needs. The home-grown food was supplemented by a supply previously grown in the building. Although adequate, it was not as much or as varied as some of the experimenters would have liked. Most of the crew lost weight. Near the end they said they would give almost anything for a pizza and a beer.

Several things went wrong. In the 2-year closure of the building, the supply of oxygen ran unexpectedly low and the carbon dioxide level was unexpectedly high. Fresh oxygen had to be added from time to time. Microbes in the excessively rich soil were responsible. Some species of plants grew rampant threatening to crowd out other desirable plants unless the crew intervened to keep them under control. Some insect populations also grew more rapidly than anticipated, notably ants and cockroaches.

There were some personality conflicts and quarreling among the crew members. This increased as time went by, but decreased as the end of the 2 years approached. The project was ultimately discontinued but a great deal has been learned about living in a small, closed environment. The building has been converted to an environmental education and research center by Columbia University.

This chapter has covered only one possible configuration and location for an orbiting space colony. Many other configurations have been considered. You can perhaps think of alternative ways to live, work, and play in space. New ideas come from the imaginations of creative people who understand enough of the basic science to know what is possible.

DISCUSSION QUESTIONS

1. Do you think the International Space Station will be worth the expense to the world? Why?
2. Discuss the advantages and disadvantages of manned vs. unmanned space missions. What jobs are best done by each?
3. Describe some recreational activities which can be conducted in a nonrotating space station.
4. Describe the artificial gravity and living conditions in a spherical or a cylindrical shaped rotating space colony.
5. Would it be possible and desirable to completely sterilize all bacteria and viruses in a space colony and its population as a means of eliminating disease?
6. Compare Biosphere II to an orbiting space station. In what ways are they alike? How are they different?
7. What personal adjustments would an astronaut have to make before working with astronauts of another country?
8. What technical modifications would have to be made before docking with a spacecraft of another country?

ADDITIONAL READING

Allen, John, and Mark Nelson. *Space Biospheres.* Orbit Book Co., 1986.

Beardslley, Tim. "Science in the Sky." *Scientific American,* June 1996. About the International Space Station.

Billingham, John, et al. *Space Resources and Space Settlements,* NASA SP-428, 1979. Report of NASA summer workshop, technical but readable.

Brown, David. "Manning the Lifeboats on the International Space Station." *Launchspace,* April 1998. About the X-38.

Brown, David A., and others. "International Space Station." *Launchspace,* June/July 1997. Several articles about the ISS.

Bugbee, Bruce. "Hydroponics on the Moon," *Fine Gardening Magazine,* November/December 1989. Popular article.

Foust, Jeff. "Is a Space Station the Ultimate Greenhouse?" *Final Frontier,* July/August 1997. Changes in chemical processes by plants while weightless.

Grey, Jerry. *Beachheads in Space, A Blueprint for the Future.* Macmillan Publishing Co., 1983. Popular, nontechnical summary of where we are and where we are going.

Johnson, Richard D., and Charles Holbrow, eds. *Space Settlements, A Design Study.* NASA SP-413, 1977. Report of NASA summer workshop, technical but readable.

Lucid, Shannon W. "Six Months on Mir." *Scientific American,* May 1998. A fascinating personal account by the record-breaking astronaut.

National Commission on Space. *Pioneering the Space Frontier.* Bantam Books, 1986. Recommendations to the president on future space programs.

O'Leary, Brian. *Project Space Station.* Stackpole Books, 1983. Easy reading, nontechnical.

Ride, Sally K., et al. *Leadership and America's Future in Space.* NASA, 1987. Recommendations to NASA administrator on future space programs.

Van Allen, James A. "Space Science, Space Technology and the Space Station." *Scientific American,* January 1986. Arguments against the space station.

Woodcock, Gordon R. *Space Stations and Platforms*. Orbit Book Co., 1986. Technical aspects of engineering space stations; comprehensive.

PERIODICALS

Final Frontier, The Magazine of Space Exploration. Final Frontier Publishing Co., Minneapolis, MN 55408. Bimonthly magazine, easy reading.

Launchspace, The Magazine of the Space Industry. P.O. Box 97238, Washington D.C. 2007-7328. Free to qualified space professionals.

SSI Update, The High Frontier Newsletter. Space Studies Institute, P.O. Box 82, Princeton, NJ 08542. A bimonthly newsletter for members of the Space Studies Institute describing research sponsored by the Institute founded by Gerard O'Neill.

NOTES

Chapter 12

Colonies on Other Worlds

Why go to the Moon or to Mars? Several often heard phrases may be appropriate. "Because it's there" or "To explore is human." But there is probably a more pragmatic reason, perhaps "The grass is greener on the other side of the fence." Humans are always looking for greener pastures. Other species do it, too. Consider the African bees migrating northward into Texas or the Mediterranean fruit flies spreading into California. Each form of life on planet Earth finds an appropriate environmental niche into which it settles, perhaps moving with the seasons, perhaps suspending animation during harsh times. When times get tough, space becomes overcrowded, and food gets short, then some of the individuals leave for a better place to procreate and populate. Humans move for these social, biological, and economic reasons, too. When one leader or group makes life unbearable for other individuals or groups, the oppressed may pick up and leave.

Expanding human presence from Earth into space is sometimes compared to the great age of exploration from the 1500s to the 1800s when Europeans expanded into the Western Hemisphere. There are indeed similarities—Columbus was sponsored by the government, Queen Isabella of Spain,—and there are some important differences. Columbus knew that he would find food, water, and air at the end of his journeys though there were shortages during the long voyages at sea. Unlike astronauts and cosmonauts, Columbus did not have continuous contact with home port.

One of the signs of human intelligence is that, rather than search out a hospitable niche, we have learned to survive in any environment by building habitats to maintain a comfortable temperature and bringing in the necessities of life: fresh air, food, and water. All that is needed is raw materials, energy, and intelligence.

Earth has many places where people cannot survive without a habitat to protect them from an inhospitable environment. Antarctica, perpetually covered with ice and snow, provides plenty of water and air, but temperatures are above freezing only during the short summer and little or no plant or animal life is available for sustenance. The side of a mountain may be hospitable and inviting in summer, with abundant streams of clean water, fresh air, warm sunshine, plants and animals for nourishment—all the necessities of life. But it may be impossible to survive in that same spot in winter; temperatures drop below freezing even during the day, plants wither and die, and streams freeze and stop flowing.

Just like people living in Antarctica or on mountains, the first colonists on other bodies of the Solar System will need to build habitats to create environments in which they can survive comfortably. The most likely candidates for imminent human habitation are the Moon and Mars. Their situation is something like that of Antarctica. Prehistoric people migrated and settled on all other continents, but Antarctica wasn't colonized until this century when the technology became available to make it livable and governments were willing to provide support for the colonists. We are just reaching that point in space exploration. Mars may be our next Antarctica.

Permanent Colonies on the Moon

Why live on the Moon? It is difficult to think of a more hostile environment. Only 12 humans have ever set foot on the Moon, all of them American astronauts. In Figure 12.1 one of them takes a sample of Moon rock to bring back to Earth. Since the moon has no atmosphere, the most obvious drawback is the lack of air to breathe. Colonists would have to live in sealed and pressurized astrodomes or semiburied habitats and would have to wear space suits while outside. Second, without an atmosphere there is no protection from ultraviolet, x-rays, gamma rays, cosmic rays, solar particles, and meteoroids. Third, you cannot run down to the corner convenience store when you run out of milk. All food and other supplies would have to be brought from Earth.

This description sounds like space itself; the Moon appears to have little to offer. In a space station, on the Moon, or on Mars the needs are the same: raw materials and energy. Living conditions on the Moon would be similar to living on a space station with two important differences. First, the Moon has natural gravity, about one-sixth the force of Earth gravity, and second, there are resources to be tapped. There are two pragmatic reasons to colonize the Moon: mining and science, particularly astronomy.

Moon's Natural Resources

If it were cheap and easy to go to the Moon, colonists would have done so long ago. The lunar *regolith,* the surface material, is an enormous source of valuable raw material. Chemical analysis of lunar rocks and soil returned to Earth by Apollo astronauts shows them to be rich in helium, aluminum, titanium, iron, calcium, and silicon. Pure iron is found everywhere as a fine powder from the bombardment of the Moon by meteorites during the past eons; it can be easily separated by a magnet as soil passes by on a conveyor belt. Aluminum and titanium are found in ores not commonly refined on Earth, so new methods of extraction would have to be developed for use on the Moon. Silicon is just what is needed for photovoltaic solar cells.

Lunar Water

Water is one of the essentials for living. Many comets contain a high percentage of water and comets undoubtedly crashed all over the Moon's surface in the past. However, the rocks and soil brought back by the Apollo astronauts showed no sign of water, not even in chemical combination. Those Moon rocks were collected in the equatorial regions which have been desiccated by the intense solar radiation in the absence of a protective atmosphere. Water remains in the polar regions where the sun never shines in the deep fissures and craters. The Moon's axis is not tipped as Earth's axis is, so some deep polar craters and crevices never see sunlight and temperatures may never rise higher than 40°F (22°C) above absolute zero. Not only water, but carbon dioxide, methane, and other gases are frozen there.

The polar regions are not visible from Earth and only recently have spacecraft looked into them. Clementine (1994) detected signs of water near the south pole, and Lunar Prospector (1999) detected hydrogen ions at both poles. (See Chapter 7.) Estimates range from 11 million to 300 million tons of water are spread over the north polar region and about half that amount near the south pole.

At the end of its mission in August 1999, Lunar Prospector was programmed to crash into a crater near the south pole that may contain water, with the hope that "spray" from the impact could be observed from Earth and studied spectroscopically for signs of water. None was seen in this experiment.

Extracting water from the soil is not difficult. Just heat the soil and the water evaporates. Then condense the vapor by running it through pipes in the shadow of a crater where it is always cold and collect the liquid.

Lunar Oxygen and Hydrogen

Oxygen and hydrogen are particularly important for a colony. Oxygen is, of course, essential for a colony's atmosphere. Oxygen and hydrogen are the best rocket propellants we have and they can be combined in fuel cells to produce electricity and a supply of pure water for the colony. Over 40 percent of the surface soil mass is oxygen compounded with other elements. Oxygen would be easy to extract from surface soil simply by heating it. Hydrogen is relatively scarce on the Moon, however. Its original supply escaped the low lunar gravity long ago, and solar wind protons are the source of the hydrogen now found in the soil (recall that protons are nuclei of hydrogen atoms). At least half of the top 20 to 30 feet of lunar surface materials consists of fine dust. Solar protons and other particles striking the Moon become attached to the dust. Meteorite impacts stir them into the soil to a depth of at least several feet. They can be removed by heating the regolith to 1200°F (650°C). It has been estimated that there is enough hydrogen stored in the lunar regolith to manufacture a ton of water per square kilometer (0.4 square mile).

If a large supply of water can be extracted from the lunar regolith it could be electrolyzed into hydrogen and oxygen for rocket fuel. The *electrolysis* of water can be demonstrated by dropping a 9-volt battery into a glass of water with a pinch of salt dissolved in it and watching the bubbles. The water, H_2O, is broken down into oxygen and hydrogen gases by the electric current. Hydrogen bubbles up at the negative terminal, the cathode, and oxygen appears at the positive terminal, the anode. (If you try this, don't leave the battery in the water for more than a few seconds or it may be destroyed by the immersion.)

Mining the Moon's Helium

Harrison Schmitt, the astronaut shown in Figure 12.1, recently pointed to the helium found on the Moon as a potential fuel for nuclear fusion, an energy source for both Moon and Earth. Nuclear fusion in the Sun was described in Chapter 4 and MATHBOX 4.1. On Earth, nuclear fusion has been produced on a

Figure 12.1 Apollo 17 astronaut-geologist Harrison Schmitt studies the rocks of the Taurus-Littrow Valley of the Moon. *Courtesy of NASA.*

very small scale in laboratories and on a very large scale in thermonuclear bombs. Scientists have been trying for several decades to find a practical way to control the energy so it is released slowly over a period of time instead of instantaneously as in a bomb. Controlled fusion for fractions of a second has been experimentally successful in laboratories in both the United States and the former Soviet Union. It isn't easy to build a container that will hold something at 30 million degrees Fahrenheit (17 million degrees Celsius)! The trick is to use a magnetic field to keep the hot particles suspended in a vacuum away from the container walls. High temperatures can be achieved by initiating the reaction with high-powered lasers focused on small pellets of fuel.

Several different fusion reactions are possible involving hydrogen (H) and its isotopes, deuterium (^2H) and tritium (^3H). One reaction involves tritium with deuterium. An isotope of helium, ^3He has two protons and one neutron in its nucleus. The most common helium is ^4He which has two protons and two neutrons. The ^3He-^2H fusion yields a ^4He and a proton. Theoretically it produces less waste heat and radioactive by-products, two major concerns of environmentalist opponents of nuclear energy. Deuterium is found in ocean water, but ^3He is not found in any quantity on Earth, so this reaction was not considered to have any practical value. However, ^3He, a component of the solar wind, is found on the Moon in large enough quantities to have real value as a fuel.

As described in the paragraphs on *Moon's Natural Resources* above, hydrogen and other particles from the solar wind, including ^3He, strike the Moon and become attached to the dust. The concentration of ^3He is low in the lunar regolith, less than 30 parts per billion in the top 10 feet (3 m), but it can be removed easily by heating. When controlled fusion becomes practical it would be shipped to an energy hungry Earth for generating electricity in fusion power plants. The hydrogen and oxygen separated in this process could be used to produce electricity in fuel cells with water as a byproduct.

It has been estimated that the Moon has at least a million tons of ^3He. Theoretically, a ton of ^3He can produce 10,000 megawatts of electricity for a year, enough for a city with a population of 10 million; 25 tons would produce a year's supply of electricity for the entire United States. The United States now spends about $75 billion a year to fuel its electric generators. At that rate, ^3He is worth about $3 billion per ton! A lunar mining colony could easily become self-sufficient and profitable within half a century selling ^3He to Earth and using the by-products for its own use.

Of course, no fusion-powered electric generators have been built to use ^3He. Up to now, controlled fusion in the laboratory has not turned a profit. Far more energy must be put into a laboratory machine than is produced by the machine although researchers are approaching the break-even point. Some researchers expect eventual breakthroughs and predict that nuclear fusion will be the major energy source of the new millennium. Mining the Moon's ^3He could make it so.

Mining the Moon's Metals

Consider the Moon's environment: rocks and soils, plentiful solar energy for 2 out of 4 weeks, and a near perfect vacuum. That has some advantages and some disadvantages. Many processes used for extracting metals from ores on Earth require chemical agents which are not likely to be found on the Moon. For example, gold is extracted from ore by dissolving it in a sodium cyanide solution. Electrochemistry is the most likely approach for adaptation to the lunar environment.

The simplest example of an electrochemical process is the electrolysis of water described above. Similarly, if you run an electric current into a molten mineral oxide, a compound of metals and oxygen, the metals will collect at the cathode and oxygen will be given off at the anode. For example, anorthite, found in abundance in the lunar highlands, is a compound of calcium, aluminum, silicon, and oxygen. Its chemical formula is $CaAl_2Si_2O_8$. By weight, it is 14 percent calcium, 19 percent aluminum, 20 percent silicon, and 46 percent oxygen. (See MATHBOX 12.1.) This means that if you completely process 1000 pounds of pure anorthite, you will get 190 pounds of aluminum. Aluminum is a most useful metal. It can be used in the construction of buildings, satellites, mirrors, vehicles, and electrical wiring. It is easily rolled into sheets, drawn into wire, welded together, and with a low melting point, about 1220°F (660°C), it can be melted and cast into almost any shape.

The difficulty is that all three metals in the anorthite—the silicon, aluminum and calcium—would be collected at the cathode. Preprocessing the mineral by heating it in the vacuum of the Moon would evaporate the silicon oxide first. Electrolysis would then produce calcium and aluminum at the cathode to be later separated.

Anorthite is only one of a number of minerals which could be mined profitably. Ilmenite is a mineral compound of iron (Fe), titanium (Ti), and oxygen with a chemical formula $FeTiO_3$. It is relatively abundant on the Moon. To process the ore, it is first heated with hydrogen to 1560°F (900°C). The oxygen combines with the hydrogen to form water and leave a solid mixture of iron and titanium oxide. The mixture can then be ground into a fine powder and a magnet used to separate the iron from the titanium, both valuable construction materials.

Mining the Moon's Oxygen

The electrolysis of 1000 pounds (455 kg) of anorthite will also produce, as a by-product, 460 pounds (210 kg) of oxygen, approximately the amount of oxygen in a 1400 square foot (130 m^2) house. There may be an easier way to extract oxygen from other minerals on the Moon that need only to be heated and oxygen is released. For example, some oxygen would be released while processing the regolith for ^3He as described previously. If lunar mining and processing could supply the oxygen for rocket engines plying *cislunar space* (Earth-Moon space), much less mass would have to be lifted

MATHBOX 12.1

Molecular Weights

The chemical formula for anorthite is $CaAl_2Si_2O_8$ which says a molecule consists of one atom of calcium, two atoms of aluminum, two atoms of silicon, and eight atoms of oxygen. Let us find the percentage by weight of these four elements. First find the atomic weights of each in a table of the elements:

calcium (Ca)	40.080	aluminum (Al)	26.981
silicon (Si)	28.086	oxygen (O)	15.999

Next find the total atomic weight contribution of each to the molecule of anorthite:

calcium	$1 \times 40.080 = 40.080$
aluminum	$2 \times 26.981 = 53.962$
silicon	$2 \times 28.086 = 56.172$
oxygen	$8 \times 15.999 = 127.992$

The total molecular weight of anorthite is the sum of these four: 278.206.

Now we can find the percentage weight of each of the elements as follows:

$$\text{calcium: } \frac{40.080}{278.206} \times 100 = 14.4\%$$

$$\text{aluminum: } \frac{53.962}{278.206} \times 100 = 19.4\%$$

$$\text{silicon: } \frac{56.172}{278.206} \times 100 = 20.2\%$$

$$\text{oxygen: } \frac{127.992}{278.206} \times 100 = 46.0\%$$

The sum of these must be, of course, 100%.

Notice that nearly half the weight is oxygen, which is one of the most needed elements for a colony on the Moon.

---

from Earth. At least three-fourths of the takeoff weight of most launch vehicles is propellant.

Figure 12.2 (Plate 29) is an artist's concept of a lunar mining colony and Figure 12.3 shows a cargo vessel hauling liquid oxygen from the Moon to a "filling station" in low Earth orbit. The oxygen tanker is equipped with an inflatable heat shield for aerobraking in Earth's upper atmosphere. Therefore it does not have to carry as much fuel as would be needed if the orbital change were done by rockets alone.

Automatic robotic mining machinery could be operated from inside the habitat so people would not be exposed to the harsh environment outside. In fact, it has been suggested that a completely automated mining operation could be controlled remotely from Earth and only a small maintenance crew would have to be sustained on the Moon.

Plenty of solar energy is available during the solar day for processing ores to extract the useful materials such as aluminum, silicon, and oxygen. Just spread out an array of photovoltaic cells and tap in. Since the Moon rotates on its axis once a month resulting in a 2-week day and a 2-week night, another source of energy would be needed during the long lunar night. There are several possibilities. First, use orbiting mirrors to concentrate and reflect sunlight onto the collectors like miniature suns. With four of them equally spaced in lunar orbit, one would always be in the right position throughout the 2-week night. Second, store some of the electricity produced during the day by spinning up massive flywheels to high speed using electric motors. After dark, connect the flywheel to a generator to generate electricity. In fact, the motor could be designed to act as the generator. Third, the brute force method, bring in a nuclear power generator from Earth.

Figure 12.3 Ferry vehicle transports liquid oxygen manufactured on the Moon to low Earth orbit. Spherical tanks are assembled at the facility in the lower right; the rocket engines and heat shield are attached. The ferry heads for Earth (upper left), the heat shield inflates to several times its illustrated size, and the vehicle aerobrakes in the atmosphere to reduce the fuel required for orbital transfer. On the return trip the ferry brings a supply of liquid hydrogen from Earth. In the upper right is a lunar space station; a lunar lander has just arrived with more oxygen from the surface. *Courtesy of Eagle Engineering; artist Pat Rawlings.*

And finally, mine the ore on the Moon, and ship it for processing to a space colony which is in continuous sunlight at L5, as we discussed in Chapter 11.

The Mass Driver

Material can be lifted from the Moon to space with only 5 percent of the work needed to lift material from Earth to space. Instead of using chemical rockets to haul the ore from the Moon to the L5 space station, an electrically powered launcher on the surface could accelerate the ore to escape velocity. Such a launcher, called a *mass driver,* has been designed and tested on Earth.

The principle of a mass driver is simple. Put the ore into a bucket, then accelerate the bucket and its cargo to escape velocity, stop the bucket, and let the ore go flying off into space. One approach to accelerating the ore uses the attraction and repulsion of magnets. When an electric current flows through a coil, it becomes an electromagnet with a north pole and a south pole. If another magnet is brought near, with its north pole pointing toward the north pole of the coil or its south pole pointing toward the south pole of the coil, the two will repel one another. However, if a north pole and a south pole are facing one another, they will attract each other.

We could not use such a device to launch objects off Earth directly into orbit. Escape velocity from the Earth's surface is about 7 miles per second (11 km/s) and, at that speed, atmospheric friction would quickly slow the object, converting its kinetic energy of motion to heat, and destroy it. However, one could be used as a first stage booster. Referred to as *magnetic levitation* tracks, they are being tested as a means of giving a spacecraft an initial boost into orbit from the surface of Earth. A magnetic field levitates the spacecraft a short distance above the track to nearly eliminate friction and accelerates the vehicle rapidly to high speed. One 50-foot (15-m) test track accelerates a scale model craft from zero to 60 miles per hour (96 km/h) in less than a half-second. See Figure 12.4. A full scale 1.5-mile (2.4-km) track could accelerate a 25,000-pound (54,000-kg) vehicle to about 400 miles per hour (640 km/h) in 9 seconds. Then the rocket engines would ignite to carry the craft into orbit. This initial acceleration on the track takes the place of a first stage

Figure 12.4 A magnetic levitation track with a model space plane resting on a sled which hovers a few inches (cm) off the track and accelerates electromagnetically. *Courtesy of NASA.*

booster rocket, dramatically reducing the starting mass of the vehicle.

A mass driver is entirely practical on the Moon because lunar escape velocity is only about 1.5 miles per second (2.4 km/s) and there is no atmosphere. It could launch to escape speed directly from the surface. A container fitted with superconducting magnets is inserted into a tube and a coil is energized just ahead of the bucket to attract it and accelerate it forward. Just as the container reaches it, the coil is turned off and the next coil a little farther down the track is turned on to accelerate it further. Passing through coil after coil the loaded container is accelerated to escape velocity. Figure 12.5 is an artist's concept of a mass driver in operation on the Moon.

In one design about 25 pounds Earth weight (11 kg mass) of lunar soil is *sintered* (heated and compressed) to form a cylindrical slug about 18 inches (46 cm) long and 4.5 inches (11 cm) in diameter. The slug, firmly attached in its container, enters the mass driver where it accelerates at 1,000 g to escape velocity, 1.5 miles per second (2.4 km/s), in the first thousand feet (300 m). It then moves through guide rails for a distance of 6500 feet (2000 m) to damp out any sideways motion. At that point the slug is released and the container is suddenly decelerated by two magnetic loops and deflected sideways away from the ore. The slug of ore, now separated from the bucket, continues on its way into orbit, while the container is directed onto a return track back to the loading point for another run.

Figure 12.5 Advanced lunar base equipped with a mass driver which accelerates ore to escape velocity and sends it to a collection site at L2. Ore buckets enter the driver at the near end and, after releasing their load at the far end, return for another load on the track behind the mass driver. In the right foreground an astronaut sits with his son at the future site of the Apollo Museum. © *Lunar and Planetary Institute; artist Pat Rawlings, Eagle engineering.*

The slug continues on its escape trajectory toward a catching device located at L2. (Refer back to Figure 3.29.) Although the processing and manufacturing plant may be located at L5, it does not seem advisable to launch the ore directly toward the colony. The best location for the mass driver would be on the front side of the Moon, near the equator, so the slug would follow a half-orbit to L2, similar to a Hohmann transfer orbit. The actual trajectory is more complicated than a simple ellipse because of the balance between Earth gravity and Moon gravity in that part of space. Recall, an elliptical orbit has only one central body, located at one focus, which supplies the gravitational force for the orbit. The velocity of the ore slugs must be very exact as they leave the mass driver; otherwise they would scatter over a wide region of space near L2. Two additional "trim stations" are located downrange from the launcher to precisely adjust the final velocity.

At L2 a device is positioned to catch the slugs as they arrive. Several designs have been suggested, from a metal mesh net to a 10-mile (16-km) wide piece of styrofoam. There are two main problems to overcome. First, there is sure to be some scatter in the arriving pieces of ore, and second, the catcher would be accelerated backward from the impacts of the ore slugs. Solutions have been suggested, but both of these problems are difficult to deal with.

A space tug would haul an accumulated stock of the slugs to the L5 colony for processing. While the L5 habitat is under construction, raw ore would be sent. The waste slag from ore processing would be used for the L5 shield. Later, preliminary separation could be done on the Moon so only the useful minerals such as aluminum and titanium oxides are shipped out. Operating at full capacity, this design of a mass driver would launch 10 ore slugs per second, a total delivery of 600,000 tons a year to L2.

Once construction of the L5 space colony is completed, its citizens can go into business manufacturing and selling solar power satellites to Earth and oxygen to passing spaceships.

Astronomy on the Moon

Astronomy from a lunar base has tremendous advantages. Optical telescopes on the Moon would be unhampered by atmospheric refraction and absorption, pollution, and city lights. Low gravity and the slow rotation of the Moon would allow construction of large telescopes and simplify the construction of telescope mounts. Radio telescopes located on the far side of the Moon, pointing away from Earth, would not suffer interference from man-made radio noise. A lunar crater would be a perfect natural bowl-shaped depression in which to build a large dish antenna like the one at Arecibo, Puerto Rico (Figure 12.6). Telescopes designed for observing other parts of the electromagnetic spectrum such as infrared, ultraviolet, or x-ray, would be unimpeded by atmospheric absorption. Over a billion dollars was spent on the Hubble Space Telescope, an indication of the value astronomers place on an observing site in space.

Luna City

Besides offering the possibility of becoming a profitable venture, the Moon is an ideal place to learn to live and work on another celestial body. It is close to Earth, within a few days flight. Not much is saved in launch costs, however, because the greatest expense in launching a space vehicle is in the massive booster rockets needed to reach low Earth orbit.

Figure 12.7 is an artist's concept of what a lunar colony may look like while in the first stages of construction. The prefabricated buildings, similar in design to space station modules, are set into trenches and several feet of lunar soil heaped over the top for shielding from solar and cosmic radiation and from meteorites. The soil also insulates the structures from extreme temperature variations during the 2-week lunar day and 2-week lunar night. A deeper "storm cellar" module is buried beneath the main habitats to provide protection during solar proton storms. Automating most of the machinery for the mining camp and operating it remotely from Earth would require perhaps as few as 10 maintenance people at the site. With additional processing and manufacturing equipment, lunar materials could be used to build and expand a permanent city. Construction would be simpler than on Earth because of the low gravity.

A later stage of development is shown in Figure 12.8 (Plate 29). Buildings are inflatable to save space and weight in shipping. They are interconnected by hallway-like tunnels; all doors are airtight so if one section loses pressure it can be sealed off until the leak is repaired.

Luna City might even become self-sustaining. Once established, it could engage in interplanetary commerce, selling its products to space stations and serving as a way station for spacecraft leaving the vicinity of the Earth. As was pointed out before, it is much less costly in energy to launch supplies from the Moon with its one-sixth g than it is to launch from Earth. Only small rocket engines and a short fuel burn were needed to launch Apollo's LM from the Moon. Only small shuttle craft would be needed to deliver fresh supplies and oxygen propellant to customers parked in orbit around the Moon before they proceed to the farther reaches of the Solar System. A possible cislunar infrastructure is shown in Figure 12.9. Each part has been discussed separately in this and the previous chapter; this diagram brings it all together.

Although the inhabitants could not go outside without spacesuits, they could lead enjoyable lives with a beautiful view of Earth, fixed in the sky but spinning once every 24 hours. Earth's news broadcasts and entertainment programs can be received directly. Athletic games would be quite different in only one-sixth of Earth's gravity. Figure 12.10 (Plate 30) shows some of the possibilities, played in a large transparent dome. Imagine pole vaulting over 130 feet (40 m) and weightlifting over 2500 pounds (1100 kg). Gymnasts may stay in the air several times longer than on Earth.

Figure 12.6 The world's largest radio-radar telescope antenna hangs in a natural bowl in the mountains of Puerto Rico. The 1000-foot (300 m) dish, made of nearly forty thousand 3 by 6 foot perforated aluminum panels, is precisely spherical in shape to an accuracy of less than an eighth of an inch over the 20-acre surface. The triangular-shaped 600-ton platform, suspended 435 feet above the big reflector, supports the antenna feed horns and lines. The telescope either receives natural signals from space or connects to a transmitter to act as a radar. The antenna feeds can be moved to study various parts of the sky in a wide range of wavelengths. *Courtesy of Arecibo Observatory.*

As good stewards of our environment, we should be careful not to devastate the Moon with our activities. Although some people might consider a large mining scar on the Moon, visible from Earth, to be a sign of the wondrous achievements of the human species, many would consider it a desecration. By careful concern for the Moon's environment at the very beginnings of our colonization, we can avoid costly repairs later. Just as we now insist that we do all we can to return Earth to a more natural state, future lunar citizens may want their home planet to remain as natural as possible.

Site Selection

Of the 15 million square miles of lunar surface, how do you choose the best spot to build a colony? The decision depends primarily on the colony's function. If exploration is the primary purpose, than a scientifically interesting site would be the first choice, one with a variety of terrain, surface composition, and geological features. If mining is the chief objective, then a location near profitable mineral deposits is essential. If the colony is to be an astronomical research station, then a location near the equator gives a view of the entire sky

Figure 12.7 Unloading a habitat module just delivered to the lunar base. The flimsy looking crane can do the job easily because the module weighs only one-sixth its Earth weight. Other modules, delivered by other landers scattered around the scene, are already in place in trenches at the base in the background. The assembled habitat is then heaped over with lunar soil for protection and insulation. Earth hangs low in the sky. *Courtesy of Eagle Engineering; artist Pat Rawlings.*

in a month's time, although Earth would always be in the scene at a site on the "front" side of the Moon. Radio astronomers would select a site on the back side of the Moon, free from Earthly interference.

Life support considerations are essential. The site should have loose soil and regolith to cover the habitat for protection from the normal continuous radiation as well as severe solar proton storms and small meteoroids. A nearby supply of oxygen-containing minerals would assure adequate oxygen for habitat atmospheres. Locating at a water supply would be immensely valuable. It may be that an icy comet collided with the Moon and is buried somewhere, although water is probably found only in the polar regions. Locating on high terrain on the top of a crater rim at either the north or south pole also has the major advantage of having the Sun continuously on the horizon. There solar cells would produce a continuous supply of electricity.

Terrain features are important in choosing a site. Because heavy soil-moving equipment may not be immediately available, relative flat terrain around the site would be safest, although it may not be interesting to miners and scientists. The bottom of a small crater nearby would make a useful launch site. Its walls would keep debris kicked up by rocket engines from spreading very far.

Access to orbit for both coming and going is an important consideration. An equatorial orbit passes over a site on the equator every time around. Similarly, a spacecraft in a 90 degree orbit passes over a polar site on every orbit. Sites at other latitudes are not accessible as frequently from an inclined orbit. Because Earth rotates on its axis once every 24 hours, a spacecraft in an inclined orbit overflies locations on Earth at latitudes less than the orbital inclination twice every 24 hours, once headed northward and once headed southward. But the Moon rotates on its axis once in about 29 days, so access to inclined orbits is not as frequent.

Not everyone agrees that the Moon will become a permanent home to a large group of colonists. Other ideas are discussed in the additional readings listed at the end of this chapter. A lunar outpost will undoubtedly be established in the new century, but how far it will expand is the question.

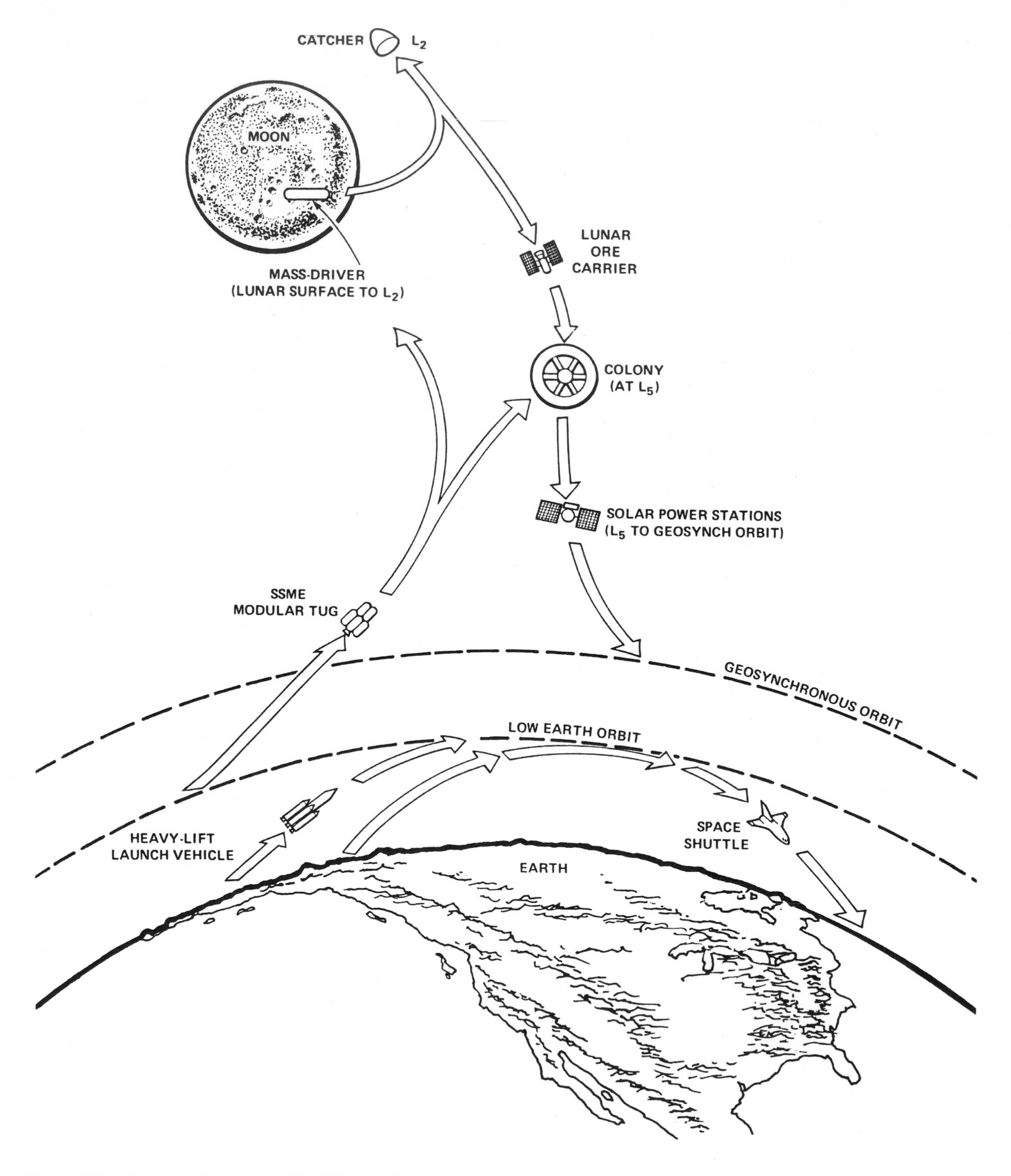

Figure 12.9 Cislunar infrastructure. (An SSME modular tug is a cargo vessel assembled from Space Shuttle main engines.) *Courtesy of NASA.*

Mars, the Red Planet

Mars as a place to live has been an appealing notion for centuries. Science fiction writers from Edgar Rice Burroughs to Ray Bradbury and many others have described the native Martians or human settlers who became "Martians" when they colonized the red planet.

As early as 1659, markings were seen on the surface of Mars and observers could estimate the period of rotation of the planet by watching the markings. In 1877 Schiaparelli reported seeing linear markings which he said were like the fine threads of a spider web. He called them *canali,* an Italian word for lines or grooves.

In the 1890s, astronomer Percival Lowell built an observatory in Flagstaff, Arizona, where he devoted much time and effort to observing Mars. His observing site was better than most and he was in possession of a fine instrument. Much of the time the image of the planet was blurry and shimmering, even when looking through the clear desert sky. Sometimes the image would clear and become quite sharp and detailed.

Lowell, too, saw the markings which had been observed by Schiaparelli, but he saw much more. He saw seasonal changes in the markings and in the polar caps. The markings seemed to widen and become darker in the spring and summer. At the same time the polar caps retreated in the summer and advanced in the winter. Lowell mistakenly interpreted these seasonal variations as being due to vegetation along canals which he thought had been constructed by Martians to carry water from the polar caps to their agricultural land in the warmer equatorial regions. He was convinced that intelligent beings inhabited the red planet, struggling to keep their civilization going as their planet was cooling and drying up.

Later Earth-based observations using more advanced telescopes equipped with spectrographs and other equipment told a different story. Then Mariner spacecraft were launched from Earth for close inspection of Mars. Mariners 4, 6, and 7 flew by for closeup looks during 1964 and 1969. Mariner 9 was injected into orbit around Mars in 1971. Then, in 1975, two spacecraft called Viking were sent. The Viking orbiters took photographs of almost the entire planet; see Figure 12.11. (Plate 31) The Viking landers settled softly onto the surface of Mars, sniffed the air, tested the soil, searched for signs of life, photographed the surroundings, and sent the findings back to Earth. They found no canals, no vegetation, no Martians. Mars is almost as inhospitable as the Moon. Almost. Several differences make it a more appealing place, a potential site for human communities. The most important is an atmosphere.

Because it is 1.5 times farther from the Sun, its year is 687 Earth-days long, 1.88 Earth-years. Recall Kepler's third law? A day on Mars is 24.6 hours long, just 37 minutes longer than Earth's day; so the Martian year is 669 Mars-days long. The tilt of its axis is 25 degrees, compared to Earth's 23.5 degrees. So Mars has seasons, but because of the long year, each season is about twice as long as on Earth; see Table 12.1. By comparison, Earth seasons last about 90 days. Also, the eccentricity of its orbit is 0.093, greater than Earth's 0.017. At perihelion it is 26 million miles (42 million km) closer to the Sun than it is at aphelion. Mars is at perihelion during the southern hemisphere summer so that season is shorter and warmer than northern hemisphere summer. Recall Kepler's second law?

TABLE 12.1 Length of Martian Seasons

Season		Length	
Northern Hemisphere	Southern Hemisphere	Earth Days	Mars Days
spring	autumn	199	194
summer	winter	183	178
autumn	spring	147	143
winter	summer	158	154
Mars year		687	669

The diameter of Mars is about 4000 miles (6400 km), half that of Earth, and its mass is only one-tenth that of Earth. Therefore, its gravity is much lower, about 0.4 g on the surface. Escape velocity from the surface of Mars is 3 miles per second (4.8 km/s) compared to nearly 7 miles per second (11 km/s) from Earth.

Mars has two tiny moons. Phobos is potato shaped, about 6 by 8 miles (10 by 13 km) in size, and therefore has very low surface gravity, only 0.001 g. It orbits 5600 miles from the planet in about 7.6 hours. Deimos is even smaller, about 3.5 by 4.5 miles (5.6 by 7.2 km), and its surface gravity is 0.0005 g. Escape speed is about 22 miles per hour (about 14 km per hour); a good pitcher could throw a baseball into orbit. It is about 14,000 miles (22,500 km) from Mars and completes an orbit in 30.3 hours. To an observer on Mars it would look like a bright star moving rapidly across the sky.

In July 1988, the former Soviet Union launched two probes to Mars and its moons, both named Phobos. Because of a computer software error, Phobos 1 lost lock on the Sun. With the solar panels pointed in the wrong direction, its batteries became depleted and contact with the spacecraft was lost. Phobos 2 reached Mars in January 1989 and was injected into an orbit matching that of the moon Phobos. It took 37 visual images of Phobos and thousands of infrared data samples of the little moon and of Mars itself. Phobos appears to be covered with a dark dust; it reflects only 7 percent of the light falling on it. A number of grooves, craters, and some rock outcrops appear in the pictures. Phobos's density is about twice that of water; by comparison, typical Earth rock is five to six times as dense as water. This would indicate that Phobos either has a porous interior or is mostly ice. The spacecraft also carried two landers which were equipped with stereo cameras and other sensors, and had rotating rods which would make them hop over the ground in 50 to 60 foot

(15 to 18 m) leaps to observe different spots on the Phobos's surface. Unfortunately, the spacecraft never got a chance to fire the probes; contact was lost after nearly 2 months of operation.

It has been suggested that these two moons could be reconstructed into way stations for spaceships traveling between Earth and Mars. Rendezvous, docking, and then escaping from such a small moon would require little energy.

Martian Landscapes

Except for the lack of vegetation, the Martian landscape is strangely Earth-like. Viking and Pathfinder pictures of the rocks, soil, and sand dunes have a familiar look about them, giving the impression that they were taken in a barren desert on Earth. Figure 12.12 shows the scenery surrounding Viking; you would almost expect to see a camel walk by. All of the rocks examined were found to be igneous, that is the basic rock of which planets are made. Igneous rocks originate from molten material when planets first form and in subsequent volcanic activity. Some few rocks appear to contain layers like those seen in *sedimentary rocks* on Earth. Sedimentary rocks, as the name suggests, are formed from sediments deposited on ocean bottoms, flood plains, and other places where water stagnates and allows the suspended material to drop out. As the thickness of the sediments increases, the weight of overlying sediments causes the layers below to *lithify*, that is, solidify into rock. Sedimentary rocks on Mars would be substantial evidence of running water at some time in the past. However, the presence of layers in a rock does not necessarily mean it was formed in this manner, so more study is needed before any conclusion can be drawn.

Soils originate from the erosion of the rocks. Soil analysis at the landers' sites detected the presence of many of the same minerals as found on Earth. Many of the iron-bearing rocks are oxidized, "rusted" to a red color like iron-bearing rocks on Earth. Over time, erosion has cut away sand and dust-sized grains, and wind has distributed them over the planet. For centuries Mars has been called the red planet; it really is red.

The southern hemisphere is covered by old, eroded impact craters much like the Moon, but the northern hemisphere shows evidence of geological activity including volcanoes, lava flows, canyons, and what appear to be dry river valleys and glacier deposits. This geological activity along with wind erosion has obliterated most of the craters. The largest of the dormant, perhaps extinct, volcanoes is Olympus Mons, measuring 15 miles (24 km) high, 325 miles (523 km) across at its base, and sporting a 40-mile (64-km) wide caldera crater at the top. It is the largest volcano so far discovered in the Solar System.

A huge canyon, Valles Marineris, stretches along the Martian equator nearly 3000 miles (4800 km), about the distance across the United States from New York to California. Parts of it have walls which appear to have been deeply eroded by running water. It is up to 150 miles wide and 4 miles deep (240 km wide and 6 km deep). The Grand Canyon of Arizona would fit into one of the side tributaries.

It is natural to compare the features seen on Mars with familiar features on Earth. Rock formations can take on many interesting shapes and many of them are given fanciful names such as the rock called Yogi in Figure 7.9 (Plate 11).

Figure 12.12 The scene at the Viking 1 lander site looks strangely Earth-like. The structure at the right is part of the spacecraft. *Courtesy of NASA.*

Figure 12.13 A Martian canyonland in the southern Elysium Planitia. The picture is about 1.9 miles (3 km) wide. The entire area was once level and flat like the upper right. Erosion created mesas and buttes like those in Utah and Arizona. Perhaps one day this too will be a Martian National Park! *Courtesy of NASA/JPL/MSSS.*

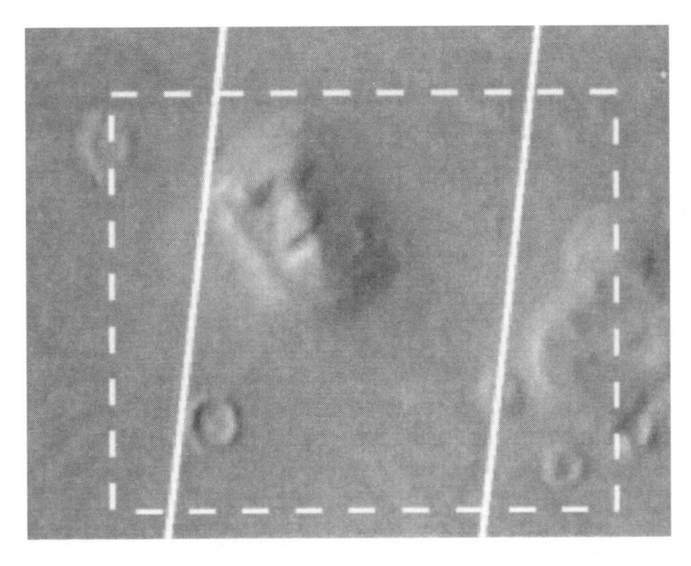

Figure 12.14 The famous face on Mars was photographed by the Mars Global Surveyor. It has a resolution of 14.1 feet (4.3 m), ten times better than the Viking picture. Another pile of rocks to the right looks like a snowman. *Courtesy of NASA/JPL/MSSS.*

Figure 12.13 shows a smaller canyonland on the other side of Mars, similar in appearance to Monument Valley in northeast Arizona. A field of sand dunes is shown in Figure 7.11 similar to those seen in Chile and Peru . Figure 12.14 shows the Mars Global Surveyor photograph of the famous face on Mars first seen in a Viking orbiter image. The Global Surveyor camera found another face, Figure 12.15.

Martian Atmosphere

Mars has an atmosphere, very thin by Earth standards, equal in density to Earth's atmosphere at about 20 miles (32 km) altitude. Pressure is less than 10 millibars, 0.15 pound per square inch (1 kPa), one-hundredth what it is at sea level on Earth. The Martian atmosphere is 95 percent carbon dioxide. The other 5 percent consists of nitrogen and argon, with traces of oxygen, water vapor, and other gases. Notice that the constituents are the same as Earth's atmosphere, just in vastly different proportions. Even so, cloud and dust patterns appear quite similar to Earthly cloud patterns. For example,

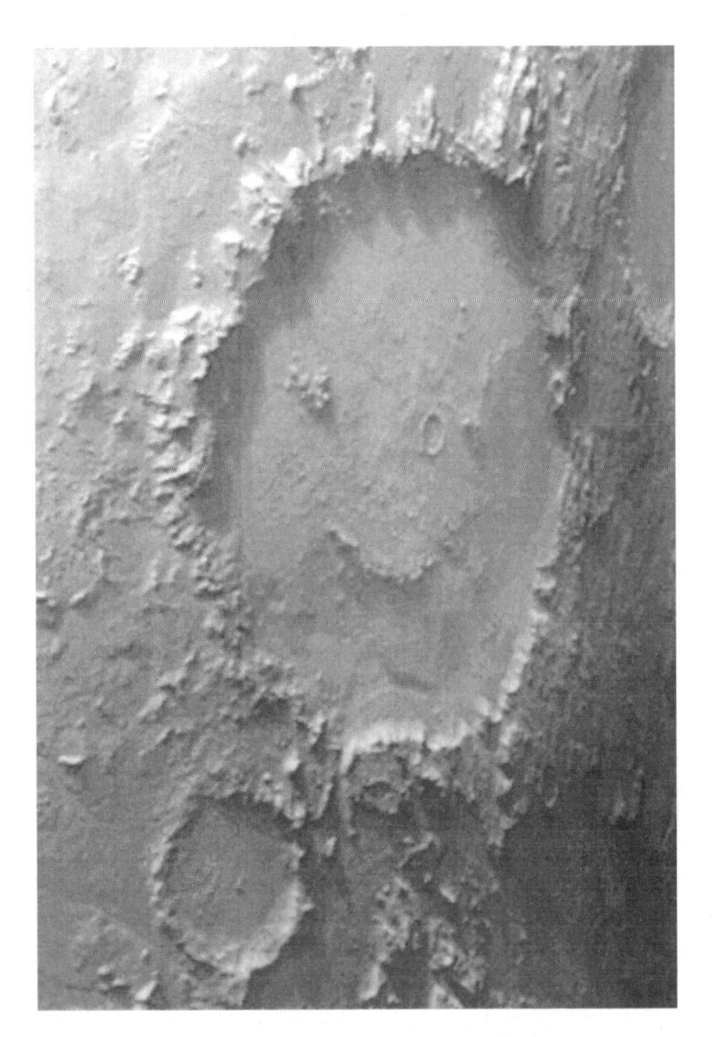

Figure 12.15 Another face on Mars, a happy face crater winking at you. It is 134 miles (215 km) across. *Courtesy of NASA/JPL/MSSS.*

look back at Figure 7.10. At the Pathfinder site, pressure dropped to a low of 6.7 millibars during the southern hemisphere winter when the south polar ice cap was at its greatest extent. At that time, 20 to 30 percent of the atmosphere had frozen onto the ice cap. Humans could not survive without space suits and pressurized habitats.

The atmosphere contains so little ozone that ultraviolet radiation reaches the surface of Mars with full intensity. Because of the distance from the Sun, the ultraviolet is much less intense than it is at Earth's distance. Nonetheless, it is sufficient to be hazardous to unprotected human beings.

Thin as it is, the atmosphere provides some protection from meteoroids. Although there are many large craters, Viking lander pictures show no small craters. Smaller meteoroids burn up as they fall through the atmosphere.

At the Viking landing sites, light southwest winds blew in the early morning and light easterly winds blew in the afternoon. Maximum recorded wind speed was 15 miles per hour (about 24 km per hour). Afternoon temperatures were typically −25°F (−32°C); at night temperatures regularly dropped to near −120°F (−84°C). At the Pathfinder site the temperature reached a maximum of 14°F (−10°C) in the afternoon and dropped to −105°F (−76°C) just before sunrise. Wind was less than 22 miles per hour (36 km/h). Because of its greater distance from the Sun, Mars receives only 43 percent as much solar energy as Earth, and, because of its thin atmosphere, the little solar heat absorbed during the day is lost to space at night. In the polar regions, summer daytime temperatures reach only −95°F (−71°C). In equatorial regions, the highest temperature recorded was a comfortable 62°F (17°C).

Dust devils passed over both Viking and Pathfinder sites, recognized by brief drops in atmospheric pressure. Mars Global Observer photographed numerous dust devils (Figure 12.16), confirming that they may be the most important meteorological feature in raising dust into the atmosphere. Each one lasts for only a few hours during the hottest part of the day. Dust storms develop as more and more red dust is picked up from the surface and suspended in the air. Major storms may encompass the entire planet for several weeks. Dust storms occur mostly during the southern hemisphere summer when Mars is at perihelion, closest to the Sun, and receiving maximum solar radiation. The energy contained in the wind is not great, however, because of the low density of the air. Very fine dust remains suspended for long periods of time causing the sky to appear pink rather than the blue we are accustomed to on Earth. It is possible that some of the features observed by Lowell were due to the red dust settling out onto the dark bedrock at some time and then being blown off the rocks during a storm.

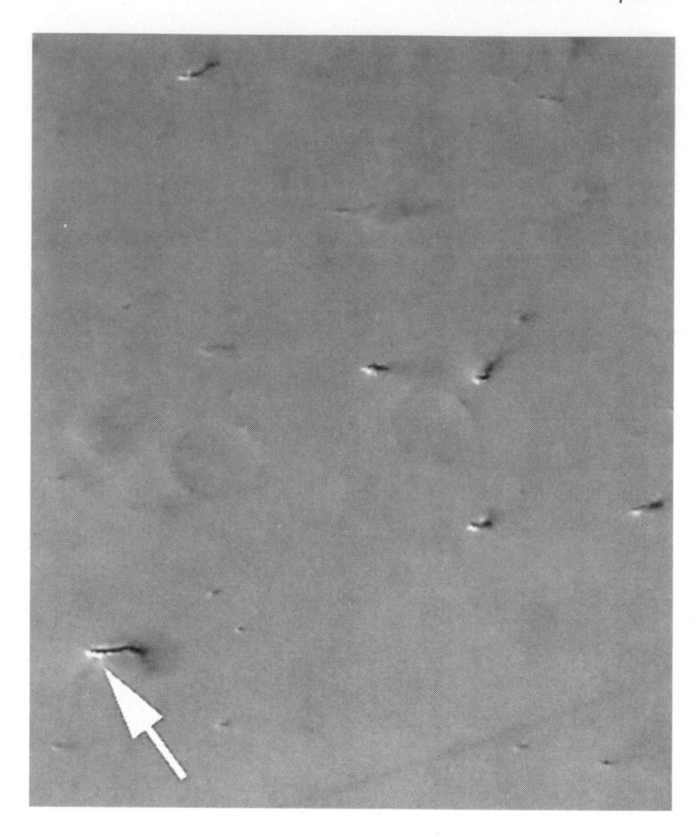

Figure 12.16 Six large Martian dust devils are seen in this image. The arrow points to the largest one, 5 miles (8 km) tall. The Sun is shining from the lower left so their shadows are pointing to the upper right. *Courtesy of NASA/JPL/MSSS.*

Water on Mars

Water must have been present on Mars in substantial quantities some time in the past. The Mariner, Viking, and Global Surveyor images show geological features that look like riverbeds on Earth and appear to have been made by running water. The Pathfinder lander pictures show rounded cobblestones and pebbles like those found in riverbeds on Earth. In some areas craters appear in the river channels and flood planes indicating that meteorite bombardment occurred after the running water disappeared. In other areas the river channels cut across craters suggesting that the craters formed first. See Figure 12.17. From this, geologists suggest that water ran on Mars somewhere between 1.8 billion and 3.5 billion years ago.

A laser altimeter on the Global Surveyor mapped the *topography* of Mars and found that the lowest elevations are in the northern hemisphere. If water flowed on Mars it would have drained in that direction. The terrain in the north is also very flat, suggesting that the area may once have been an ocean floor. However, the volume of both the polar ice caps taken together is only about 40 percent of that needed to fill an ocean that size. Where did all the liquid water go?

There is no running water on Mars now. Temperatures are too cold and the pressure is too low for water to remain in the liquid state very long. It can exist only as vapor or ice. In the Martian environment liquid water is unstable and either freezes or evaporates.

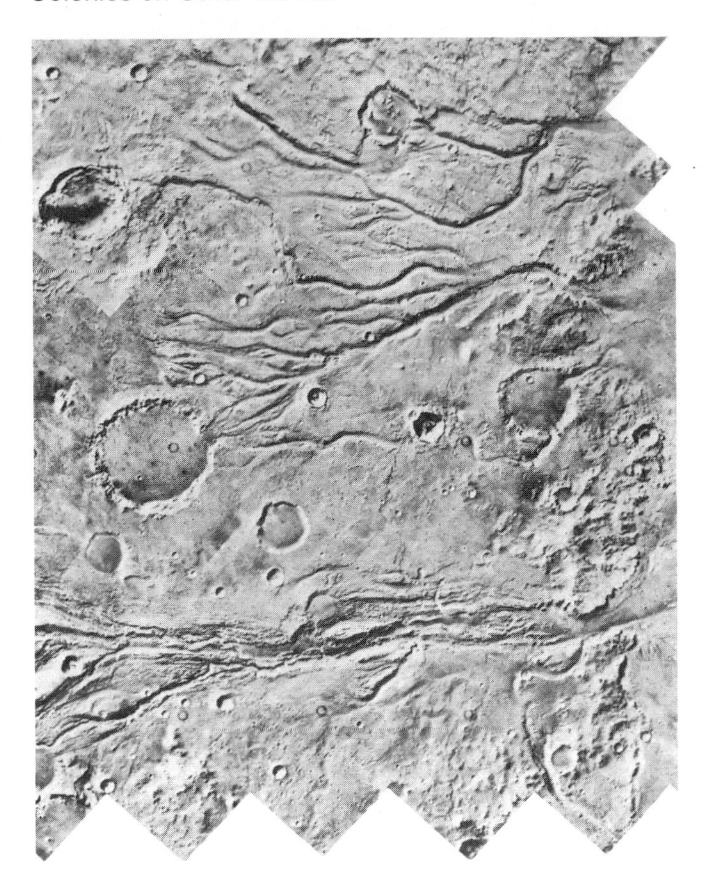

Figure 12.17 Maze of dried-up river channels. The terrain slopes down about 2 miles (3 km) from left to right. Rivers and flood waters once poured across the region in that direction. *Courtesy of NASA.*

Figure 12.18 Frost on the ground and rocks at the Viking 2 lander site. *Courtesy of NASA.*

A small amount is in the atmosphere. Clouds and fog appear on many of the Viking and Global Surveyor images, especially in craters, on the slopes of Martian volcanoes, and around the polar ice caps. Frost, either dry ice (frozen carbon dioxide) or water ice, appears on the rocks in some of the Viking lander pictures (Figure 12.18). It has been estimated that if all the water vapor in the Martian atmosphere could be condensed to liquid it would amount to about 0.2 cubic mile (0.8 cubic km) or 220 billion gallons (830 billion liters). (For comparison, Lake Mead behind Hoover Dam in Nevada holds about 10 billion gallons (38 billion liters).)

Some water is frozen in the polar ice caps. The seasonal changes observed in the ice caps are the result of the deposition of dry ice from the atmosphere during the severely cold polar winter season and the sublimation (evaporation) of the dry ice during the following spring. The ice caps do not disappear completely during the summer season, indicating that the residual is probably water ice mixed with Martian soil.

The water vapor content of the atmosphere varies widely from season to season, indicating that water is also stored in the ground in the form of permafrost, permanently frozen soil and ground water. Permafrost is found on Earth in the northern parts of Alaska, Greenland, Canada, Scandinavia, and Siberia. In the equatorial regions of Mars the top 3 feet (1 m) or so of soil has been completely dehydrated. If permafrost water is present there, it is deep in the ground. Because the interior of the planet is warmer than the surface, liquid water may be found at greater depths, perhaps a mile down. If so, it could be tapped the way groundwater is tapped on Earth, by drilling wells.

Unmanned Exploration

Although Viking, Pathfinder, and Mars Global Surveyor have given us a wealth of information, there are still a great many unanswered questions. Mars is a big place. The surface area of Mars equals the land area of Earth. Brief human visits would be an inadequate exploratory program to lay plans for colonization. EVAs covering short distances for a few hours will not do it.

We know almost nothing about the subsurface Martian rock and soil. The same is true of Martian water. We can be confident that it is there somewhere, but except for the ice caps, we do not know exactly where the water is located or in what form. No plans for human exploration and colonization of Mars can be made until further unmanned exploration reveals its composition.

In the last several years more Mars missions have failed than succeeded. The United States has been sending spacecraft to Mars every 2 years, each time the planets are properly aligned for rendezvous. In August 1993 controllers lost contact with Mars Observer 3 days before it was to reach the red planet. In 1999 the Climate Observer failed because of error in metric vs. English units and the Polar Lander was headed for a landing when it and its two penetrator probes disappeared. The only successful missions since Viking in the 1970s are Pathfinder and Global Surveyor which reached Mars in 1997.

Exploration of the Moon prior to the Apollo landings had numerous losses, too. Six of Russia's first eight Luna flights and six of nine U.S. Ranger spacecraft failed. When doing things that have never been done before, losses must be expected and accepted. As the old saying goes, "If at first you don't succeed, try, try again."

Surveyor 2001, Surveyor 2003, and Surveyor 2005 are the next in line. Each carries an orbiter and a lander with a rover. The 2001 goals include searching for water and evidence of life, gathering meteorological and climate data, and manufacturing oxygen from the Martian atmosphere. The 2005 mission will carry a return vehicle to bring back samples of the Martian regolith. A communications relay satellite is scheduled to orbit around Mars. It will speed the data flow from the landers and orbiters back to Earth and will help keep track of new arrivals on their final trajectories.

Russia had plans for several Mars probes in the 1990s but none of them happened due to lack of funding. Japan's Planet B is on its way to Mars via a complex trajectory, but it has had communication problems.

Humans on Mars

In spite of its harsh environment, Mars has the basic ingredients necessary to support a self-sufficient colony of humans. It is the most Earth-like of all the planets. Mars can be a prototype human extraterrestrial society. If we can do it on Mars, then human potential for expanding into the universe will have been proven.

Should we go? Going to Mars can be done with our present technology and there are many who would like to see it happen soon. Indeed, the research and development concomitant with a commitment to a Mars mission would undoubtedly make it happen. However, the commitment is not there. NASA has no plan for human spaceflight except the Space Shuttle and the International Space Station. In constructing the ISS we are learning how to build and operate in low Earth orbit. That is a valuable learning experience for human expansion beyond Earth.

Interest and enthusiasm for a human expedition to Mars have been steadily increasing. A particularly active group of students and teachers, dubbed the Mars Underground, held a Case for Mars conference at the University of Colorado at Boulder in 1981. Since then the group has joined with the Planetary Society in forming a Mars Institute to promote studies and research for Mars exploration. Three more Case for Mars conferences were held in Boulder, attracting more and more researchers and engineers. Studies related to the transportation systems, construction in orbit and on Mars, use of Mars resources, and human factors have been done and are continuing. The Mars Society formed in 1998 is an outgrowth of these grassroots activities. It has an increasing international membership.

With a firm commitment, a human expedition to Mars could become a reality within 5 years. We need to set a deadline and stick to it. The Apollo Moon landing took 8 years; that was all the time it was given. If 15 years had been allocated, the program would have used every minute of it. It seems to be human nature to accomplish a task in the time available to do it. Another factor is funding. It is unlikely that a Mars venture could be done with private money. Governments would have to participate, and that is where politics enters the scene.

Interplanetary Transportation

The major problem is getting off the Earth. The first hundred miles is the most difficult and the most expensive. Several ideas for getting to Mars have been explored. Once at Mars, the crew would have to wait for the planets to line up again for the return trip. An expedition would require at least 2 years. Therefore a 2-year supply of materials, food, oxygen, and fuel for the return trip would have to be available. NASA's 1980s plan called for shipping supplies and materials in small batches first to a space station in low Earth orbit. There the pieces would be assembled into a larger unmanned cargo vessel and sent ahead to Mars via a Hohmann transfer, the least-energy transfer, requiring nearly a year transit time. Figure 12.19 (Plate 32) shows what such a vehicle might look like. This has been called the "Battlestar Gallactica" approach. At Mars it would be injected into a low orbit to wait the arrival of the smaller passenger craft coming by a faster, high-energy route, taking about 5 months. An important advantage to separating the cargo from the crew is that the cargo would be in place in orbit around Mars before the crew leaves Earth, giving greater assurance of a successful mission.

A Hohmann transfer requires a minimum of fuel but takes a longer time making it a good way to deliver cargo. The slow route is not good for people for two reasons: (1) the extra weight of a year's supply of food, air, and water that must be lifted from Earth and (2) the physiological and psychological effects of a yearlong trip on the human body. On the other hand, a fast trip requires more fuel at both ends, to ac-

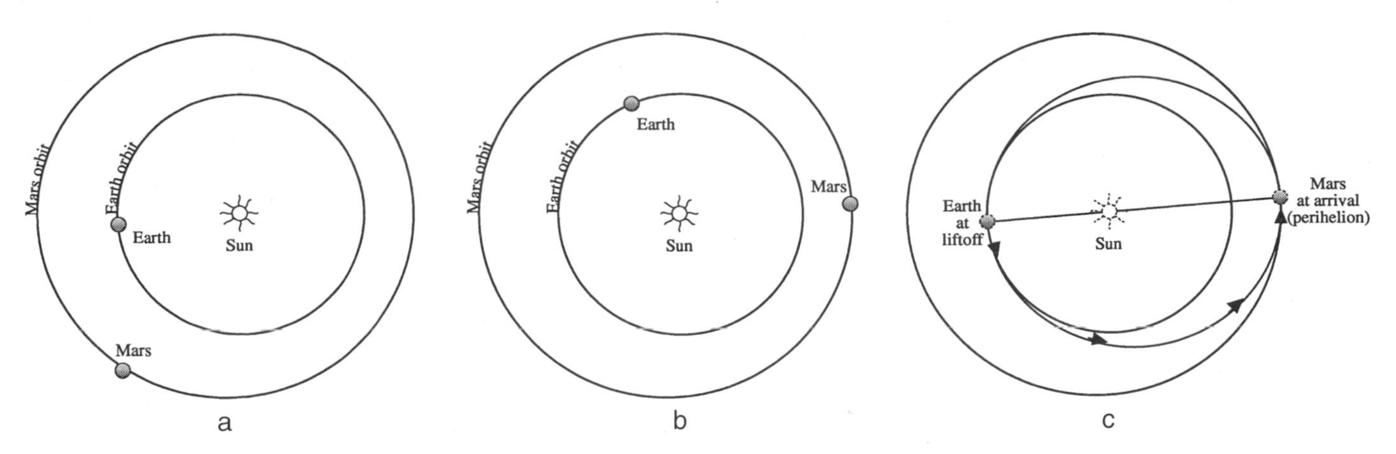

Figure 12.20 A Hohmann transfer from Earth to Mars. The position of the two planets at departure from Earth is shown in (a). Their position at arrival at Mars is shown in (b). This is a conjunction class mission as shown in (c).

celerate the craft to higher speed at departure and to slow it on arrival. Although the mass of life support supplies is less for a quick trip, mass of fuel is greater. An optimum must be determined.

The interplanetary passenger ship would rendezvous with the cargo ship and materials would be taken to the surface in small batches. Aerobraking in the Martian atmosphere reduces the fuel requirements for retrockets to inject the spacecraft into orbit around Mars.

The effects of long-term weightlessness and exposure to cosmic radiation are still undetermined. The crew must be able to function in Martian gravity immediately upon arrival, so a rotating vehicle to produce artificial gravity may be essential to keep the crew in good physical shape.

Mars Direct

Robert Zubrin, an aerospace engineer, author, and president of the Mars Society, proposed a simpler and cheaper way to go: send robotic equipment ahead to land on Mars and manufacture methane gas and water for energy and the return trip. Then it would not be necessary to haul tons of return trip fuel all the way from Earth. The equipment, including a 100-kilowatt nuclear power generator and 5.8 tons of hydrogen, sets itself up, and starts running automatically. It takes in carbon dioxide from the Martian atmosphere and combines it with the hydrogen to produce methane and water. The chemical reaction is $4H_2 + CO_2 \rightarrow 2H_2O + CH_4$. The water is electrolyzed into hydrogen and oxygen. The oxygen and methane are stored and the hydrogen is used to make more methane. The machinery runs for 2 years and, if all is working well and a good supply of methane and oxygen are in storage, the crews are sent. Methane is then the primary energy source for Mars surface activities and for launching back into low Mars orbit for return trips to Earth.

Zubrin and his colleagues have successfully built a prototype unit which produces about 1.5 pounds (0.7 kg) of methane per day, is more than 94 percent efficient, yet weighs only about 45 pounds (20 kg). The engineers think they can cut the weight of the unit in half. To prove the concept, they suggest that their machine be sent to Mars on a future sample return mission, perhaps Surveyor 2005. In 2 years of operation on Mars it would produce enough fuel to bring back 9 pounds (4 kg) of Martian rock and soil. Just as important, it would demonstrate the feasibility of scaling the machinery up to a size capable of bringing a crew back from the red planet.

Trajectories

Every 2 years Earth and Mars are properly aligned in their orbits for a voyage. Because Earth is traveling faster in its smaller orbit closer to the Sun, it overtakes and passes Mars every 780 days. Astronomers call this an opposition because, viewed from Earth, Mars is exactly opposite the Sun in the sky, e.g., Mars rises in the east when the Sun sets in the west. A conjunction occurs when the Sun and Mars are at the same spot in the sky.

Timing the departure is critical; the spacecraft must depart Earth so that after its approximately 9-month cruise it arrives at the Mars orbit at the same time as the planet does. A Hohmann minimum energy transfer to Mars is shown in Figure 12.20. This is called an conjunction class mission because at departure Earth is located on the side of the Sun exactly opposite from where Mars will be at arrival.

If more energy is available or if the payload is reduced, the spacecraft accelerates to a higher speed for a shorter flight time; the departure timing would then be different from a Hohmann transfer.

The Mars orbit is more elliptical than the Earth orbit so the distance between the planets is different at each opposition. This distance is shortest if the conjunction occurs when Mars is at perihelion as shown in Figure 12.21.

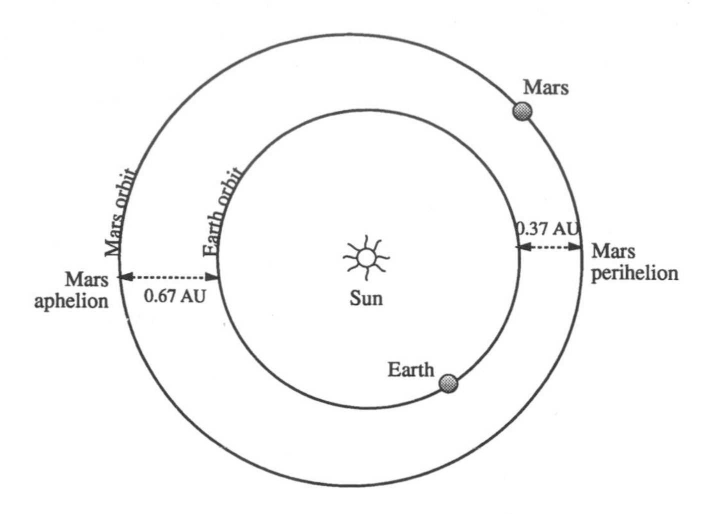

Figure 12.21 Mars orbit is more elliptical than Earth orbit. The distance between the two planets is much less when Mars is near perihelion.

A complicated mission involves taking the spacecraft past Venus for a gravity assist on the way to Mars The gravity of Venus accelerates the spacecraft so it "cracks the whip" around that planet and heads for Mars. Less energy is required to launch a larger payload but the trip takes more time, a total mission of 22 months.

The First Settlement

At first, a permanent colony on Mars would not be much different from a permanent base on the Moon. The need for breathable air, food, and other life support functions are about the same. Mars would be easier to colonize because water is available, the thin carbon dioxide atmosphere provides some protection from space radiation and can be used for fuel production, and the gravity is a bit higher.

The first problem that comes to mind is, of course, the atmosphere. It would be a major accomplishment if we could modify the Martian atmosphere so it could support our human needs. *Terraforming* is the process of modifying a planet to make it more Earth-like so humans can live a normal existence. It will be discussed later in this chapter. The first colonists, however, will have to live in closed, sealed habitats as they would on a space station or lunar colony. Work outside the habitat will be done in space suits. Figure 12.22 is an artist's concept of a Mars base.

The first habitat modules and supplies of food, air, and water will be imported along with machinery for mining ore, extracting usable minerals, and fabricating construction materials. If a lunar colony is established first, it could provide the oxygen for the trip and some of the necessities for establishing the Mars base at a lower energy cost than supplying it from Earth. Continuing dependence on Earth for fresh supplies would be far too expensive, so eventually colonists will have to learn to live off the land as our ancestors did when they migrated to new territory. In the case of Mars, the land is barren, but the technology to make it productive is available.

Buildings constructed on permafrost and heated on the inside could have structural problems if the permafrost melts beneath the building. On Earth, such buildings are built on piers above the ground, insulated from the ground. But on Mars it is necessary to protect the inhabitants from solar and galactic cosmic rays. As on the Moon, the solution is to dig a trench, put a habitat module in it, and cover it with several feet of soil. Storm cellars will be needed, also. After ore is dug from the mines, the tunnels can be sealed off and used as habitats.

Power for the colony could come from solar cells, concentrating solar electric generators, methane manufactured from hydrogen and carbon dioxide, or nuclear power plants. Because Mars is one-and-a-half times farther from the Sun than Earth is, solar energy is only 43 percent as intense as it is at Earth. Therefore, arrays of solar cells will have to be twice the size of arrays near Earth to produce the same amount of electricity.

Nuclear power generators could be installed in the polar ice cap to generate electricity and their waste heat could melt the ice to provide water for the colony located in a warmer climate near the equator. Then Mars might finally have its canals!

It is unlikely that any one nation alone will establish a base on either the Moon or Mars. Indeed, it appears unlikely that the first human expeditions to Mars will be carried on by a single nation. Like the International Space Station, human presence on the Moon and beyond will probably be the result of international cooperative efforts.

Indeed, the first Martians are probably alive on Earth today! Ultimately we may have self-sufficient cities on Mars where people live as normally as they do on Earth, going about their business of growing food, manufacturing products, doing scientific research, and enjoying life.

Cycling to Mars

Once a community is established on the red planet, it will be necessary to make regular passenger and cargo runs between Earth and Mars. Flights as described above are very expensive. However, Buzz Aldrin, the second man on the Moon,investigated the idea of a "bus" service between Earth and Mars by placing large rotating spacecraft into an orbit around the Sun that intersect the orbits of the two planets. At each encounter with Earth, a gravity assist boost transfers a spacecraft to an orbit that reaches the Martian orbit at the point where Mars will be the next time around. The gravity boost would move the spacecraft onto a shorter trajectory to Mars so the trip would take only 4 months. This cycling bus would continue out well beyond Mars orbit, then return to Earth for another gravity boost to meet Mars again. By continuously cycling, the need for accelerating and decelerating the large vehicles at each end of the trip is eliminated.

Figure 12.22 Artist's concept of a Mars base. Partially buried habitat is at the center, greenhouses to the right, and mine tunnels to the left rear. The vertical structure near the tunnel entrances is a well drilling rig. A "wagon train" vehicle made in segments climbs the hill in the foreground while a rocket lifts off at the right rear of the scene. Dish antennas on the mesa are for off-planet communications; the mast antenna is for communicating with roving exploration vehicles. Because Mars has an atmosphere, a very thin one, remotely piloted aircraft with very large wings can be used for transportation and exploration. In the lower right a geologist examines a newly discovered fossil. *Courtesy of Eagle Engineering; artist Pat Rawlings.*

Passengers bound for Mars leave Earth in a small passenger craft with a first stop at the space station in low Earth orbit. From there they ride a small shuttle craft to rendezvous with one of the cycling ships already equipped with an atmosphere and provisions supplied from the Moon or Mars. The refueled shuttle craft is hangared in the bus for use at the other end of the trip. At Mars the shuttle craft takes the passengers from the cycling bus orbit to a space station in low Mars orbit. A Mars lander carries the passengers to the surface.

Once the bus line is established, no fuel would be required to keep the buses running in their orbits. Power would be needed only for the life support equipment. Shuttle craft fuel requirements would be comparatively small.

Terraforming

There are many indications that Mars was once warmer and wetter. Can we make it that way again? Will human Martians someday stand by the side of a flowing stream, gaze at the blue sky above and the flowers below and think how much it reminds them of pictures they have seen of Earth? We have been discussing building comfortable human habitats in colonies on other worlds. Is it possible to change an entire world into an Earth-like human habitat? Technically, yes. But it requires time—decades or centuries.

Earth is warm and comfortable primarily because of its distance from the Sun and because of the composition of its atmosphere. The Moon is the same distance from the

Sun but, without an atmosphere, it is cold, dry, and bombarded with lethal radiation. Mars, farther from the Sun and with only a thin atmosphere, is also cold and bleak. On the other side, Venus, closer to the Sun and with a dense carbon dioxide atmosphere, is a 900°F (480°C) inferno.

The air blanket around Earth is transparent to the visible wavelengths of sunlight, the region of the spectrum which contains most of the solar energy. Visible light can penetrate almost undiminished through the air to warm the ground. On the other hand, heat from the Earth does not completely escape back to space because water vapor and carbon dioxide, both minor constituents of the atmosphere, strongly absorb the infrared wavelengths emitted by the warm ground and reradiate a significant portion back to the Earth. You are probably aware that deserts cool rapidly after sunset because the absence of water vapor in the air allows the heat to escape. Similarly on a clear, cloudless night, the temperature drops quickly for the same reason. Thus, these trace gases in the atmosphere act as a blanket, keeping the heat in.

To make Mars more Earth-like, then, the first step is to increase the density of the atmosphere and change its composition. Oxygen is needed by humans, but that may be difficult to find and it will not help warm the planet. So let us begin with water vapor and carbon dioxide, two gases which absorb infrared radiation and hold in the heat. Both are found on Mars, although the quantity of each is uncertain. We can begin by melting the ice caps. Orbiting mirrors can concentrate solar energy into the polar regions; white ice reflects most of the sunlight, so a coating of dark material spread over the ice first would absorb rather than reflect the energy. Where do we get the dark material? Phobos and Deimos are both made of dark matter and mass drivers can easily propel the stuff off the moons onto the planet. These moons probably contain water also, which would evaporate and aid in building the Martian atmosphere. The two little moons could give their all for the greening of Mars.

The carbon dioxide dry ice evaporates first, at a temperature of −108°F (−78°C). As the carbon dioxide content of the atmosphere increases, the heat loss to space decreases and the planet gradually warms. As the temperature goes up, still more carbon dioxide is released, further warming the planet and releasing still more carbon dioxide from the regolith. This positive feedback, referred to as a runaway greenhouse effect, continues until all the carbon dioxide is liberated into the atmosphere. It has been calculated that a temperature rise of only 8°F (4°C) at the south pole would start the process going and the polar dry ice would evaporate very quickly, perhaps in only 10 years. How much the temperature rises depends on how much carbon dioxide is on Mars, a largely unknown quantity at this time.

Introduction of other greenhouse gases, such as chlorofluorocarbons, to accelerate the process has also been suggested. Chlorofluorocarbons are the gases used in refrigerators and spray cans that are being accused of destroying Earth's ozone layer. They absorb different wavelengths of infrared than carbon dioxide does. One way to rid Earth of them and to speed up the warming of Mars may be to introduce them into the Martian atmosphere at a very early stage. They would initiate the runaway greenhouse effect and, after a few hundred years or so, would have broken down and dissipated.

Because Mars is farther from the Sun, an atmosphere twice as dense as Earth's would be needed to bring the average temperature above freezing, a temperature rise of 100°F (55°C). Once the temperature goes above freezing during the summer, water would again flow on Mars. Melting the water from the ice caps and permafrost is the second step in raising the planetary temperature. Water vapor in the air will contribute a great deal to the blanket. Mars could have a comfortable planetary temperature in perhaps 50 years.

People could not breathe the carbon dioxide/water vapor air, but they could go outside without pressurized space suits, carrying only oxygen bottles. As the atmospheric density increases and the temperature rises, we introduce primitive life, beginning with lichens and cyanobacteria (blue-green algae) which photosynthesize carbon dioxide from the air and nitrogen from the soil, building their own mass, releasing oxygen, and breaking rock down into fertile soil. Later, grasses, shrubs, trees, and food crops can be introduced to speed up the process. Slowly the oxygen content increases until the atmosphere is breathable by insects, then by birds and mammals. Because carbon dioxide acts as a blanket to keep the planet warm while oxygen does not, the quantity of carbon dioxide consuming plants will have to be carefully monitored.

This biological creation of a breathable atmosphere has happened before. The sequence of events described above is similar to the formation of Earth's atmosphere. How long would it take on Mars? Thousands of years, probably, but in that time the Martians may tire of living in closed habitats and space suits and will find some shortcuts. For example, genetic engineering of a particular life form suited to the harsh Martian climate could begin the oxygenation process at an earlier stage.

Other planets and moons could perhaps be terraformed, but with more difficulty than Mars. For a planet with plenty of oxygen but no water, hydrogen could be imported by tanker from Saturn. Or a water-bearing asteroid could be steered on a collision course. It would vaporize on impact yielding, at least temporarily, a water vapor atmosphere.

Ultimately the decision whether or not to terraform Mars will be made by the Martians themselves, that is, human Martians whose ancestors resided on Earth.

On to the Stars

Human travel beyond the Solar System is not at all feasible with our present science and technology. The distances involved are enormous. The next nearest star after the Sun, Alpha Centauri, is 4 light-years away; that is, light traveling at 186,000 miles per second (300,000 km per second) takes 4 years to travel that distance. The speed of light is a cosmic speed limit; nothing can travel faster. However, strange things begin to happen as one accelerates to high speeds. In particular, time slows down for the high-speed object relative to its surroundings. In his theory of relativity, Einstein postulated that this strange thing would occur and experiments have shown it to be true. It is not in the realm of science fiction; it really happens.

Some theoretical physicists studying relativity believe that Einstein's theories do not exclude travel at speeds faster than light under certain conditions. In space warped out of normal curvature by high mass objects as near black holes and worm holes, it may be possible to time travel and travel faster than light. The key is negative energy. Whether any practical application can be made of these revised theories remains to be seen. In the January 2000 *Scientific American,* Lawrence Ford and Thomas Roman discuss this in semipopular terms.

If some of the more exotic propulsion systems could be made to work to accelerate spaceships to speeds approaching the speed of light, or warping space for time travel and speeds faster than light, then it would be possible to send humans on interstellar flights. For example, by accelerating to 90 percent of the speed of light during the first half of the trip, then turning around and decelerating for the last half, it would take about 9 Earth-years to make the round trip to Alpha Centauri. But on the spaceship only a few months would pass. The crew would return from the trip a few months older while everyone on Earth would have aged 9 years. A round trip to the next large galaxy, in the constellation Andromeda, would require 40 or 50 years ship time, depending on how fast it accelerated and decelerated. But 5 million years would pass on Earth. Not many of their friends would be around to greet the crew when they returned! Perhaps in a century or two, the capability of travel at near light speed will be commonplace. But now, at the beginning of the twenty-first century, there is no propulsion system capable of accelerating to that speed.

Planets Around Other Stars

Even if we had a spaceship traveling near the speed of light, we have no idea in which direction to head to find a habitable planet, or whether habitable planets even exist around other stars. Stars are so far away that they appear as just points of light even through the largest telescopes on Earth. It is not possible to magnify the image large enough to see the disk of a star as we can see the disk of the Sun. A star shines from the energy generated by the nuclear furnace in its core, but planets just reflect the light of their parent star. Thus, we cannot directly see planets around other stars for two reasons: they are too small to resolve in a telescope and they are so close to their parent star that they cannot be seen in the glaring brilliance of the starlight. Looking for planets around other stars is especially difficult from Earth because the atmosphere scatters starlight, increasing the glare around a star's image. It may be possible for a telescope in space to separate the starlight from the planet light so that the two can be examined separately, revealing the composition of each. Such an experiment will be tried soon.

Indirect evidence of other planetary systems has come from the Infrared Astronomical Satellite which has shown the presence of cool dust clouds around several stars. These observations are consistent with the theory that planetary systems form when a star is born from a collapsing cloud of gas and dust. Most of the gas goes into the star while, over a period of millions of years, the remaining dust particles slowly accumulate into planets. Dust clouds and disks around other stars are a little easier to observe than individual planets. They may extend far from their parent star, and because they are heated by the radiation from their star, the particles making up the dust cloud are warmer than their surroundings and can be detected with an infrared telescope. Indeed, several such dust clouds and disks have been discovered, and where there is dust there may be planets. Some look like rings around the star, the explanation being that the inner part of the disk has been swept clear by the formation of planets. Figure 7.46 is a visual image of dust clouds around a star in the Orion nebula, made by the Hubble Space Telescope.

There are other ways to locate a planet. Like a satellite in orbit around Earth, the gravitational force between a planet and its parent star acts in both directions. Each pulls on the other. To a viewer a long distance away, the star would seem to move back and forth as the planet revolves around it. Astronomers are able to measure this wobble if the planet has a large enough mass. For example, Jupiter, one-thousandth the mass of the Sun, could probably be detected this way, provided our instruments could measure angles as small as a thousandth of a second of arc. (A second of arc is 1/3600 of a degree.) The technique obviously favors finding very large planets. Earth-size planets would not create a large enough wobble to measure.

Another way to detect a wobble is to measure the velocity of the star. The speed of a star moving toward or away from an observer can be measured by noting the shift in the wavelength of the light it emits. If it moves toward the observer, the wavelength decreases; if it is moving away, the wavelength increases. This phenomenon, called the *Doppler effect,* is described in MATHBOX 4.2. A periodic change in a star's direction—first toward us, then away from us—would indicate that it is in an orbit. From the magnitude of the shift, it is possible to determine the orbital period and the minimum size of the planet that causes the wobble.

By early 1999, planets had been detected about 19 sun-like stars by this method. The wobble of one of the stars appears to have several periodic variations superimposed on one another which could be explained by the existence of three planets in somewhat elliptical orbits with periods of 4.6 days, 242 days, and 3.5 years.

The light output from one of these stars was observed more closely by research astronomers. A large planet passing in front of the star will block some of the light and decrease the intensity, which should be measurable from Earth. Astronomers call this an *occultation*. Of course, the planet's orbit must be edgewise as viewed from the Earth; otherwise it would not occult the star. Several groups of researchers watched the same wobbling star and all reported the expected drop in light intensity. The length of time that the intensity decreased indicates how large the planet is and how fast it is moving. The results of these measurements are consistent with the measurement of the wobble. This is probably the most positive evidence to date of the existence of other planets.

Even though we find a planet around another star, we do not now have the means of determining whether it is habitable by human beings. To be habitable, the planet should have surface gravity similar to Earth gravity. Less than Earth gravity may be satisfactory, but it would be difficult to adapt to much more than one g. A breathable atmosphere and suitable temperatures would certainly be desirable although we can build comfortable habitats and go out only when necessary. Water and oxygen are the two essentials. It would be difficult if not impossible for Earth to provide continuing support to an extraterrestrial colony of human beings.

Any native inhabitants would hopefully be friendly. If intelligent, they would have to be willing to accept our colonists. If not intelligent, they must not be hostile. If only microbes, they must not cause incurable disease. Colonizing the Moon and Mars will provide experience as to the endurance of the human body and mind.

People who are headed for the stars must be equipped with the best total education possible. Everyone will have to be a worker bee. A group of settlers should include not only specialist scientists, engineers and test pilots, but people with a broad knowledge of the humanities and the arts as well as the sciences. We are, first of all, human and we must take our humanity with us. One curmudgeon said that we will probably take the two worst possible things into space: germs and management.

DISCUSSION QUESTIONS

1. Why can Earth hold dense atmosphere, while Mars has only a thin one and the Moon has none?
2. Compare and contrast a colony on Mars, a settlement on the Moon, a big wheel space station at L5, and the bases in Antarctica.
3. How would you select the first group of settlers for a colony on the Moon or on Mars? What skills? What ratio of women to men? What age groups? What health standards? Any mental health screening?
4. In terms of Kepler's second law, why is the southern hemisphere summer on Mars shorter than the northern hemisphere summer? Is the southern hemisphere winter also shorter than the northern hemisphere winter?
5. In terms of Kepler's third law, why is the Martian year nearly two Earth years long?
6. If you were going to Mars, what one personal item would you take with you besides the basic necessities of life? Why?
7. How would you govern a permanent settlement on Mars? Should it become an independent state? At what point in time?
8. Have humans had to adapt to inhospitable environments before? Where and how?
9. Design a calendar for Mars considering the length of the day, year, and seasons. Choose some logical length for something like a month.
10. How could you terraform Venus?
11. Would athletes who grew up on the Moon perform as well as those who came from Earth? Why?
12. When do you think the first manned flight will land on Mars? Would you like to go on the first flight? Why?

ADDITIONAL READING

Boston, Penelope J., ed. *The Case for Mars*. Univelt for American Astronautical Society, 1984. Conference papers.

Ford, Lawrence H., and Thomas A. Roman. "Negative Energy, Wormholes, and Warp Drive." *Scientific American,* January 2000. Possibility of traveling in time and faster than light. Difficult reading.

Golombek, Matthew P. "The Mars Pathfinder Mission." *Scientific American,* July 1998. Results of the Mars lander and rover which landed July 4, 1997.

Haberle, Robert M. "The Climate of Mars." *Scientific American,* May 1986.

Hartmann, William K., Ron Miller, and Pamela Lee. *Out of the Cradle: Exploring the Frontiers Beyond Earth.* Workman Publishers, 1984. Beautifully illustrated, popular reading.

Joels, Kerry M. *The Mars One Crew Manual.* Ballentine Books, 1985. Imaginatively prepared, technically accurate, readable.

Kargel, Jeffrey S., and Robert G. Strom. "Global Climatic Change on Mars." *Scientific American,* November 1996. The red planet as a water world.

Kasting, James F., et al. "How Climate Evolved on the Terrestrial Planets." *Scientific American,* February 1988. Factors which produce a comfortable climate.

Long, Michael E. "Mars on Earth." *National Geographic,* July 1999. A research station in northern Canada with an environment very similar to Mars.

Mauldin , John H. *Prospects for Interstellar Travel.* American Astronautical Society, Univelt, Inc., 1992. Detailed introduction to the problems of interstellar travel.

McKay, Christopher, ed. *The Case for Mars II.* Univelt for American Astronautical Society, 1985. Conference papers.

National Commission on Space. *Pioneering the Space Frontier.* Bantam Books, 1986. Presidential commission report on future goals in space.

Oberg, James E. *New Earths, Restructuring Earth and Other Planets.* New American Library, 1981. Popular treatise on terraforming.

Oberg, James E. *Mission to Mars.* New American Library, 1982. Popular book based on first Case for Mars conference.

Pieri, David. "Mars in the Millennium: A Brave New World?" *Launchspace,* May/June 1999. Discusses the many unmanned probes now exploring Mars.

Reiber, Duke B., ed. *The NASA Mars Conference.* Univelt for American Astronautical Society, 1988. Conference papers.

Ride, Sally K., et al. *Leadership and America's Future in Space.* NASA, 1987. Recommendations to NASA administrator on future national goals in space.

Schmitt, Harrison, and Gerald Kulcincki. "Helium-3: The Space Connection," Proceedings of the Ninth National Space Symposium, 1993. Summary paper with additional references.

Smith, I. Steve, Jr., and James A Cutts. "Floating in Space." *Scientific American,* November 1999. Using balloons to study the atmospheres of planets.

Stafford, Thomas, and the Synthesis Group. *America at the Threshold, America's Space Exploration Initiative.* Government Printing Office, 1991. Report to the president charting a course for future exploration of space, particularly the Moon and Mars..

Stoker, Carol, ed. *Case for Mars III: Strategies for Exploration.* American Astronautical Publication, Volume 74 and 75, Science and Technology Series. Univelt, Inc., 1989. A two-volume collection of papers presented at the third Case for Mars Conference, July 18–22, 1987.

Stoker, Carol R., and Carter Emmart, eds. *Strategies for Mars: A Guide to Human Exploration.* American Astronautical Publication, Volume 86, Science and Technology Series. Univelt, Inc., 1996. A collection of 26 papers covering all aspects of the subject.

Time-Life Books. *Voyage Through the Universe, Starbound.* Time, Inc. Book Co., 1991. One of a series of popular books about the Universe; addresses interplanetary and interstellar travel.

Zubrin, Robert M., and Christopher P. McKay. "Pioneering Mars." *Ad Astra,* September/October 1992. Original and unique ideas about exploring, settling, and terraforming Mars.

Zubrin, Robert. *The Case For Mars.* Free Press, 1996. The plan to settle the red planet and why we must.

PERIODICALS

Ad Astra. National Space Society, 922 Pennsylvania Ave., SE, Washington, DC 20003-2140. Bimonthly magazine of the National Space Society. Popular reading.

Final Frontier, The Magazine of Space Exploration. Final Frontier Publishing Co., Minneapolis, MN 55408. Bimonthly magazine, easy reading.

Mars Underground News. The Planetary Society, 65 N. Catalina Ave., Pasadena, CA 91106. A quarterly newsletter.

The Planetary Report. The Planetary Society, 65 N. Catalina Ave., Pasadena, CA 91106. Bimonthly magazine.

SSI Update, The High Frontier Newsletter. Space Studies Institute, P.O. Box 82, Princeton, NJ 08542. A bimonthly newsletter describing research sponsored by the Space Studies Institute founded by Gerard O'Neill.

NOTES

Chapter 13

Life in the Universe

Do living beings exist elsewhere in the universe? The answer is either yes or no. Right now we do not know which is correct. Either one would be astounding. How have we looked for extraterrestrial life? Is the probability of finding life elsewhere in the universe so low that searching would be a waste of time and money? There are two ways to search: by going and looking, or by remote sensing.

Astronauts and robot spacecraft have gone and looked throughout the Solar System. Astronauts have brought rocks and soil back from the Moon for examination; spacecraft have examined the other planets and moons; robot laboratories have tested the soil of Mars. A rock, which was blown off from Mars by an asteroid impact, landed on Earth and has been studied for signs of primitive life; results are inconclusive. So far, nothing living has been found anywhere except on Earth.

Space travel is time consuming. The fastest spacecraft, Voyager, launched in 1977 took 1½ years to reach Jupiter, over 2 years to Saturn, 9 years to Uranus, and 12 to Neptune. At that speed, Voyager would take some 50,000 years to reach the next nearest star. Exploration of the Solar System may be practical in a lifetime, but until faster spacecraft are built, alternative methods are being developed to look beyond.

Remote sensing offers a faster and cheaper way to look for signs of life on other planets and moons in our Solar System. Electromagnetic waves carry information about their source. We examine the electromagnetic energy coming from these planets to determine their temperatures, analyze the composition of their atmospheres, and try to deduce whether living organisms could possibly endure. We also listen for radio signals, radar pulses, and other electromagnetic waves, which may be artificially produced.

Both deliberately and unintentionally we have been signaling our presence to the universe. Daily we transmit radio, radar, and television signals which could be detected by sensitive receivers many light years away. If some intelligent, hopefully friendly aliens pick them up, they will learn about Earth and its inhabitants. Perhaps they are also sending out similar signals. By listening intensely to radio noises from space, we are searching for signals that might proclaim an intelligent source. That type of search is, of course, limited to intelligent extraterrestrials who have discovered radio communications. None has been found as yet.

We have also transmitted deliberate radio messages and attached audio and video recordings onto several spacecraft in the hope that alien beings may someday intercept them and learn about us.

The Chemistry of Life

Before we can search for life elsewhere, we need to define what it is we are looking for and to understand how living organisms manage to flourish on Earth.

What do all living things have in common? First, they all have the ability to reproduce, to replicate themselves cell by cell. Instructions for reproducing a cell are contained in the *DNA molecule,* which is an extremely long chain of billions of atoms. An integral part of each cell, DNA molecules reproduce by splitting lengthwise, unzipping, so to speak, each half joining with other atoms and molecules until two identical DNA replicas are made. Each one builds a duplicate cell, and so the growth of new cells and replacement of old dying cells continue.

Second, living organisms take in nourishment, manufacture cells, and excrete waste materials: they interact with their environment. On Earth, animals inhale oxygen and exhale carbon dioxide. They eat food and eliminate solid, liquid, and gaseous waste. Plants perform basically the same processes as animals. But, additionally, plants take in carbon dioxide from the air and nutrient-laden water from the soil to manufacture carbohydrates. During this process, energized by light falling on their leaves, they return more oxygen to the air than they consume.

Earth life is carbon-based. That is, all living organisms are composed of long chain molecules made up of carbon, oxygen, hydrogen, nitrogen, phosphorus, sulfur, and other elements with carbon as the "backbone." Carbon atoms combine with each other and with other elements in a multitude of ways to form large complex molecules that remain stable at Earth temperatures. It is not essential that life be carbon-based, but the chemical characteristics of carbon make it the most likely building block. In contrast, silicon, sometimes proposed as a base element for alien life, cannot form stable long-chain molecules. They tend to fall apart into useless fragments.

Atoms of elements bind together through the electrical force of attraction to form molecules. The nucleus of an atom carries a positive electrical charge and is surrounded by an electron cloud bearing an equal but negative charge. Thus, an atom is electrically neutral. If some of the negative charge is missing, however, the atom becomes a *positive ion* or if an excess of negative charge is attached, the atom becomes a *negative ion*. Because particles bearing opposite electrical charges attract one another, positive and negative ions move toward each other and bind together to form ionic compounds. For example, a positive sodium ion and a negative chlorine ion bind together to form sodium chloride, common table salt.

Ions of sodium are devoid of one electron. They therefore carry a single positive charge and are said to have a valence of $+1$. The *valence* indicates how many and what kind of excess charges the ion carries. Oxygen has a valence of -2 while hydrogen has a valence of $+1$. Thus, two hydrogen ions will bind with one oxygen ion to form a molecule of water; two hydrogen ions may also bind with two oxygen ions to form a molecule of hydrogen peroxide. Salt molecules and hydrogen peroxide molecules are the "end of the line"; they are the largest molecules that can be formed from those elements. Carbon is unique, however, in that it has a valence of -4 or $+4$ and can combine with other carbon ions as well as with ions of other elements to form very large molecules. Molecules built on carbon are called *organic* molecules and their study is called organic chemistry or biochemistry.

Molecules of water and salt may be represented by diagrams as follows:

$$\mathrm{H-O-H} \qquad \mathrm{Na-Cl}$$

H is the symbol for hydrogen, O for oxygen, Na for sodium, and Cl for chlorine. These diagrams, called structural formulas, show how the atoms are "hooked together." Some simple organic molecules are represented as follows:

$$\begin{array}{ccccc} & & \mathrm{H} & & \\ & & | & & \\ \mathrm{H} & - & \mathrm{C} & - & \mathrm{H} \\ & & | & & \\ & & \mathrm{H} & & \end{array} \qquad \qquad \begin{array}{ccccc} & & \mathrm{Cl} & & \\ & & | & & \\ \mathrm{Cl} & - & \mathrm{C} & - & \mathrm{H} \\ & & | & & \\ & & \mathrm{Cl} & & \end{array}$$

methane chloroform

These two compounds involve only one carbon atom each, but carbon atoms can also join with other carbon atoms. A simple sugar looks like this:

$$\begin{array}{ccccccccccccccc} & & \mathrm{H} & & \mathrm{H} & & \mathrm{H} & & \mathrm{H} & & \mathrm{H} & & \mathrm{H} & & \\ & & | & & | & & | & & | & & | & & | & & \\ \mathrm{H} & - & \mathrm{C} & - & \mathrm{C} & - & \mathrm{C} & - & \mathrm{C} & - & \mathrm{C} & - & \mathrm{C} & = & \mathrm{O} \\ & & | & & | & & | & & | & & | & & & & \\ & & \mathrm{OH} & & \mathrm{OH} & & \mathrm{OH} & & \mathrm{OH} & & \mathrm{OH} & & & & \end{array}$$

OH is the hydroxyl radical; it consists of the union of an oxygen atom, valence -2, and a hydrogen atom, valence $+1$. Thus the hydroxyl radical, taken as a unit, has a valence of -1. The double line connecting the O to the C indicates oxygen's valence of -2. A simple sugar, then, is a chain of six carbon atoms combined with hydrogen atoms, hydroxyl radicals, and an oxygen. More complex carbohydrates are formed by linking together simple sugar molecules.

A glycerin molecule is similar, but starts with a chain of three carbons and contains no oxygen:

$$\begin{array}{ccccccccc} & & \mathrm{H} & & \mathrm{H} & & \mathrm{H} & & \\ & & | & & | & & | & & \\ \mathrm{H} & - & \mathrm{C} & - & \mathrm{C} & - & \mathrm{C} & - & \mathrm{H} \\ & & | & & | & & | & & \\ & & \mathrm{OH} & & \mathrm{OH} & & \mathrm{OH} & & \end{array}$$

Simple fatty acids are structured like this:

$$\begin{array}{ccccccccccccccc} & & \mathrm{H} & & \mathrm{H} & & \mathrm{H} & & \mathrm{H} & & \mathrm{H} & & & & \\ & & | & & | & & | & & | & & | & & & & \\ \mathrm{H} & - & \mathrm{C} & - & \mathrm{C} & - & \mathrm{C} & - & \mathrm{C} & - & \mathrm{C} & - & \mathrm{C} & = & \mathrm{O} \\ & & | & & | & & | & & | & & | & & | & & \\ & & \mathrm{H} & & \mathrm{H} & & \mathrm{H} & & \mathrm{H} & & \mathrm{H} & & \mathrm{OH} & & \end{array}$$

and simple amino acids are structured:

$$\begin{array}{ccccccccccccccc} & & & & & & & & & & \mathrm{H} & & & & \\ & & & & & & & & & & | & & & & \\ & & \mathrm{H} & & \mathrm{H} & & \mathrm{H} & & \mathrm{H} & & \mathrm{N} & - & \mathrm{H} & & \\ & & | & & | & & | & & | & & | & & & & \\ \mathrm{H} & - & \mathrm{C} & - & \mathrm{C} & - & \mathrm{C} & - & \mathrm{C} & - & \mathrm{C} & - & \mathrm{C} & = & \mathrm{O} \\ & & | & & | & & | & & | & & | & & | & & \\ & & \mathrm{H} & & \mathrm{H} & & \mathrm{H} & & \mathrm{H} & & \mathrm{H} & & \mathrm{OH} & & \end{array}$$

N stands for nitrogen, and the combination of N and two H is an amino group. The combination of a C, an O, and a hydroxyl is a carboxyl group. Complex fats and proteins are made of these simpler molecules linked together into long chains. Only 20 amino acids are known. In various combinations they make up all the proteins of Earth.

A DNA molecule may contain billions of atoms. It takes the shape of a double helix, two spirals twisted around each other.

Sources of Organic Molecules

The significance of this discussion is that all life on Earth, all life that we know about, is composed of these organic molecules. Every plant, every animal that has ever been studied, from simple bacteria to complex human beings, all are made of the same 27 different building blocks: carbohydrates, fats, and proteins. There is, of course, a limit to the length of a molecular chain that can be constructed in this way. Temperature is a factor. In the optimum environment provided by typical Earth temperatures, some organic molecules may consist of hundreds of atoms. At higher temperatures, they simply fall apart. Boiling water or pasteurizing milk destroys bacteria because the heat breaks up their molecules.

Organic molecules are easily manufactured by nature. At the University of Chicago in the early 1950s, Harold Urey, a Nobel Prize–winning chemist, and Stanley Miller, a graduate student, mixed together methane (natural gas), ammonia, water vapor, and hydrogen, and ran an electric spark through the chamber. After several days, a brown coating of amino acids, formic acid, acetic acid, glycerin, and other organic compounds was found on the inside. The experiment is easy to perform; it is now done in undergraduate biology labs. But the conditions created for the experiment are also found commonly in natural environments. The gas mixture was thought to be similar to the original atmosphere of the Earth. Similar atmospheres are found on Jupiter, Saturn, and Titan. Electric sparks, lightning, were observed by Voyager as it passed by Jupiter. The experiment also succeeds if ultraviolet light is used instead of electric sparks. Ultraviolet is found in sunlight and starlight.

Organic molecules have been observed in interstellar dust clouds by listening with radio telescopes to the radio noise they produce. Although these organic molecules are not living things, they do show that organic chemistry takes place in other parts of the Galaxy, even in the cold, near-vacuum of interstellar clouds. They are known to be associated with ice crystals attached to grains of dust. At NASA Ames Research Center, recent laboratory experiments, which simulate conditions in interstellar space, give a clue to the origin of these organic molecules. In these experiments, simple gas molecules of water, methanol, and ammonia are sprayed onto a cold surface in a vacuum chamber. The gases freeze onto the surface and ultraviolet light is directed onto the crystals of mixed solid gases. The ultraviolet radiation breaks the molecular bonds and the fragments recombine to create more complex organic molecules. This process could take place in a turbulent, cold cloud of mixed gases and dust particles in interstellar space; a nearby star would provide the necessary ultraviolet light.

These experiments support the notion that if life exists elsewhere, its chemistry is most likely carbon based. Carbon is the only known element that can form complex but stable molecules needed for life and carbon is plentiful in our Galaxy. Presumably the phenomena, observed by Earth's scientists, happen everywhere in the universe. The presence of organic molecules on a planet does not necessarily mean that living organisms are present, however. The step from organic molecules to a living organism is a giant one, not at all well understood. Even the simplest single-celled organisms, bacteria, fungi, and protozoa, are marvelously complex living machines, which have never been created in a laboratory from simpler molecules.

Life on Earth

Earth is certainly a beautiful place: green plants, blue oxygen-nitrogen atmosphere, colorful birds and flowers, oceans of water and ice. There is none other like it, at least not in our Solar System. All these things are related to and interdependent with living things.

When we consider life on Earth we may think only of the prolific life on the surface. However, life thrives in extreme environments. Living organisms have been found in the most unlikely places: hot deserts, the frozen permafrost of Siberia and Antarctica, the tops of the tallest mountains, even high in the atmosphere. Bacteria have been found encased in salt crystals and pockets of brine trapped in salt deposits that formed when their ocean home evaporated hundreds of million years ago. Some survive massive doses of radiation and long periods without water. Creatures live in undersea volcanic vents at temperatures up to 230°F (110°C), above the boiling temperature of water at sea level. Bacteria colonies survive nearly 2 miles (3.2 km) underground at temperatures up to nearly 170°F (75°C) in the spaces between grains of sand in sandstone rocks and cracks in igneous rocks. How did they get into the rocks? Either they were present in the sediment that formed the rock or were carried by ground water seeping into the cracks. We must remember that life can survive in these extreme environments on Earth when we look for life on other planets.

What set of circumstances allows life to flourish on Earth? Earth was not always as it is today. Earth's original atmosphere was mostly hydrogen, methane (CH_4), sulfur oxides, ammonia (NH_3), carbon dioxide (CO_2), and water vapor (H_2O)—gases which were expelled from the interior of the planet through volcanic vents. The essential elements were there; they needed to be ordered and combined into the stuff of life. Earth's original environment was appropriate for the first primitive living things, *blue-green algae,* which, in turn, altered the environment of the entire planet, terraforming on a huge scale. At first there was little or no free oxygen. That came much later, from the metabolism of the blue-green algae that took carbon dioxide from the air and manufactured organic compounds with oxygen as a by-product, a process we call *photosynthesis*. In fact, free oxygen would have been lethal to early life forms. If a planet is to support the present life forms that we are familiar with, its atmosphere must contain elements found in organic molecules: carbon, oxygen, nitrogen, and hydrogen.

Even now, organic molecules from space are being deposited on Earth and probably on other suitable host planets by meteorites and passing comets. The Solar System formed about four billion years ago from a swirling interstellar gas cloud. Like the interstellar gas clouds we now see in the Galaxy, it would have been rich in organic compounds manufactured in space by the processes described in the previous section. The turbulence and heat when the Sun, planets, and moons formed would have destroyed organic molecules. However, the leftovers on the outer periphery of the Solar System would have retained their organic material. These leftovers subsequently have been entering the inner part of the Solar System as comets and meteorites. Organic compounds have been observed in comets and found in meteorites. When a comet enters the inner Solar System and approaches the warm Sun, two tails form, one containing gas and the other consisting of small particles sloughed off from the comet's head. Earth is constantly bathed in a rain of dust particles from passing comets, hundreds of tons per day. Because Earth's surface is about 75 percent water, most of the dust falls into the oceans where further organic chemistry can take place.

The importance of water or some other fluid environment to life as we understand it cannot be overemphasized. In a fluid, molecules are mobile and can meet and join together into cells and cells can join together into colonies. On Earth, the necessary fluid is water. Equally important, water absorbs ultraviolet rays, radiation which could disintegrate organic molecules. Earth's oceans provide protection from solar UV, allowing complex molecules to form and remain stable. The early atmosphere, without oxygen, afforded no such protection. Later when oxygen molecules were released into the atmosphere, they were split into atoms by the UV, and then recombined into three-atom-molecules of ozone. Ozone is a strong absorber of UV and provides the needed shield for living things to survive out of water on dry land.

Probably the most important circumstance conducive to nurturing life was Earth's location just the proper distance from the Sun so that its temperature was just right for liquid water to exist. Recent study indicates that the zone of satisfactory temperature around a star may be very narrow. A little farther from the Sun, a little colder, and Earth would have been frozen in the grips of a perpetual ice age. The ice and snow would reflect incoming solar energy, cooling Earth even more, eventually freezing all the oceans. A little closer to the Sun and oceans would warm up, increasing evaporation, and forming higher level clouds. This cloud blanket would further increase the temperature and evaporation, creating a runaway greenhouse effect. At still higher temperatures, the water could escape Earth's gravity and drift off into space.

As discussed earlier, temperature also determines the length of a molecular chain that can be constructed from the basic sugar, fatty acid, and amino acid molecules. At typical Earth temperatures these organic molecules may consist of hundreds or thousands of atoms. At higher temperatures they disintegrate. Other planets may not have such an appropriate temperature.

Another important factor is the size of the Earth, which determines the gravitational attraction for its oceans and atmosphere. On small planets or moons with low gravity, atmospheric gases warmed by the Sun would reach escape velocity and fly off never to return. Then, any water present would quickly evaporate into space. Mercury and the Moon have that problem. The atmosphere of Mercury, close to the Sun, was heated to high temperature by its proximity to the Sun and reached escape velocity.

Searches for Living Organisms in Our Solar System

If we find life somewhere besides Earth, we are most likely to find microorganisms. Microbes were the first and only occupants of Earth for billions of years. More complex organisms resided only in the last 600 million years or so. How then should we proceed to search other possible habitats for living organisms? The search for life thus far has taken several forms: performing on-site experiments on Mars, studying Martian meteorites, and examining data from probes to Jupiter and Saturn.

The Mars Viking Experiments

Influenced by Percival Lowell's observations of Mars at the beginning of this century, many people were convinced that Martians truly lived. Notions of what a Martian would look like led to many descriptions and drawings. If Mars had low gravity then the inhabitants would probably have long, spindly legs. A thin atmosphere meant large lung capacity. Sound would not propagate well in a thin atmosphere so large ears would be needed. And because of the planet's dryness, a long nose would be needed to filter out the dust.

In 1976, two Viking spacecraft carried robot laboratories to the surface of Mars to search for something that could be considered alive. They landed during the northern hemisphere summer, one at 23° north latitude, the other at 48° north, on opposite sides of the planet. A mechanical arm (Figure 13.1) scooped up Martian soil (Figure 13.2) and did various tests to see if it contained living organisms.

The Mars experiments were designed on the assumption that any life forms encountered would be carbon based. Experiments done prior to the Viking missions had showed that

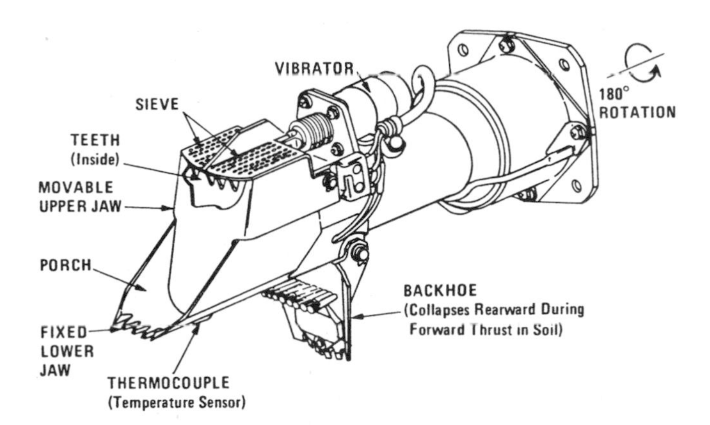

Figure 13.1 Viking scoop for sampling Martian soil. The upper jaw opens and the scoop, on the end of a mechanical arm, digs into the soil. With the upper jaw closed, the scoop is moved to the instrument, rotated upside down, and soil is sifted into the instrument's intake. A backhoe on the bottom enables the scoop to dig deeper into the soil. *Courtesy of NASA.*

organic chemistry could take place in Martian conditions simulated in Earth laboratories. When the experiments were carried out on the surface of Mars, reactions were observed, some of them almost violent reactions, but the results were ambiguous, inconclusive, and could be explained as simple chemical reactions.

In the *labeled release experiment* (Figure 13.3) a nutrient "soup" made of glucose, glycene, and other amino acids along with formic acid containing radioactive carbon-14 atoms was mixed with Martian soil. The idea was that if living organisms were in the soil they would eat the soup and exhale or excrete gases made up of molecules which included the carbon-14. Sensors would then look for the radioactive carbon. A large quantity of carbon-14, more than expected, was released. The problem was that the reaction rose to a high level almost immediately, then stopped and did not start again when more soup was added. If the gases were biologically produced, they should have continued to increase as the organisms metabolized, grew, and reproduced. The observed results seemed to indicate that the reaction was chemical rather than biological. Something on Mars with a high oxygen content fizzed and bubbled up when the soup was added. Whatever produced the reaction, perhaps an iron oxide from the soil and hydrogen peroxide from the atmosphere, was used up at the first introduction of the moist nutrient and therefore when more was added, nothing happened. To be sure that any positive result was not simply a chemical reaction, other samples of soil were used as controls. They were first heated to temperatures up to 320°F (160°C), then mixed with the soup, and the results were compared with the unheated soil. The heated samples showed less activity, indicating that the reaction may be biological. Heat destroys microorganisms so heated samples would show less action.

The second experiment, the *gas exchange experiment* (Figure 13.4), looked for signs of respiration. The chamber was filled with an inert gas that does not react chemically or biologically, then the soil and nutrient were introduced. If anything was exhaling carbon dioxide, nitrogen, oxygen, or other gases, it could be detected. At first, only a small amount of nutrient was injected into the chamber just to increase the humidity without saturating the soil. Carbon dioxide and oxygen were released immediately and rapidly, then ceased abruptly. Martian soil is probably full of carbon dioxide absorbed from the atmosphere and the moisture drove it out in this experiment. The production of oxygen is more difficult to understand, but was probably a chemical reaction like that seen in the labeled release experiment. Later, enough nutrient solution was added to saturate the soil and the mixture was allowed to incubate for nearly 7 months. Frequent measurements were made, but nothing further of significance happened.

The *pyrolytic release experiment* (Figure 13.5) searched for photosynthesis, the biological activity of Earth plants. Soil was sealed in a chamber along with a simulated Martian atmosphere of carbon dioxide and carbon monoxide that contained carbon-14. Artificial light that simulated the sunlight at Mars illuminated the chamber for 5 days. The gas was then removed and the soil sample baked at high temperature to break any organic molecules that formed. If the radioactive gas had been consumed, the molecule fragments released by the baking would contain carbon-14. Some was de-

Figure 13.2 Trenches dug into the Martian soil. The boom at the center holds weather-observing sensors. Trenches to the right of the boom are up to 12 inches deep. The boom's shadow can be seen just to the left. *Courtesy of NASA.*

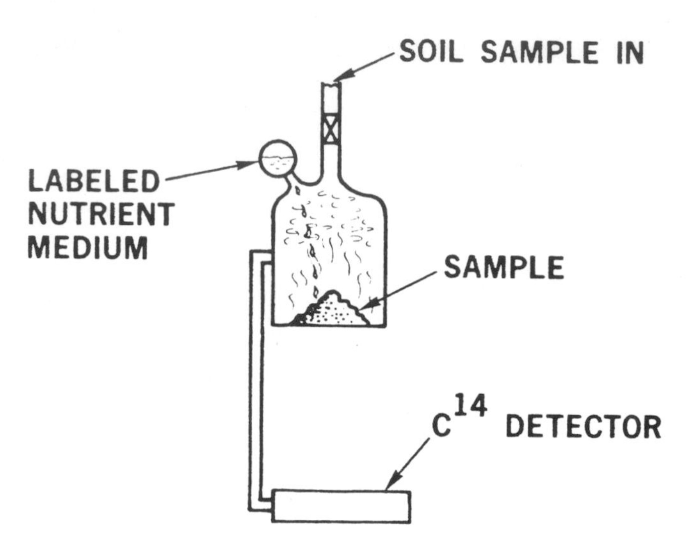

Figure 13.3 Labeled release experiment. *Courtesy of NASA.*

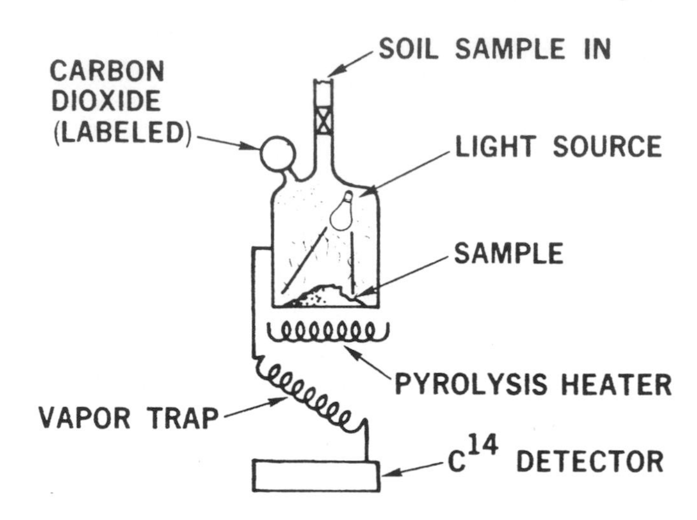

Figure 13.5 Pyrolytic release experiment. *Courtesy of NASA.*

tected in the experiment, but it may have been simply absorbed by the soil rather than metabolized by living organisms.

The Viking television pictures were, of course, examined in minute detail for any sign of life. Although the desert landscape would have been a perfect backdrop for a camel or a prospector with his mule, no hint of life, large or small, was seen (unless Martians are shaped like rocks and sit very quietly).

Also, a mass spectrometer analyzed soil samples to determine what molecules were present. Absolutely no organic molecules were found at either site. The apparatus was sensitive enough to detect organic molecules in concentrations as low as a few parts per billion. It was thought that the apparatus would detect at least a few organic molecules, some from meteorites if nothing else. It is now thought that, because of the thin atmosphere, the solar ultraviolet waves reach the ground with enough intensity to destroy all exposed organic molecules.

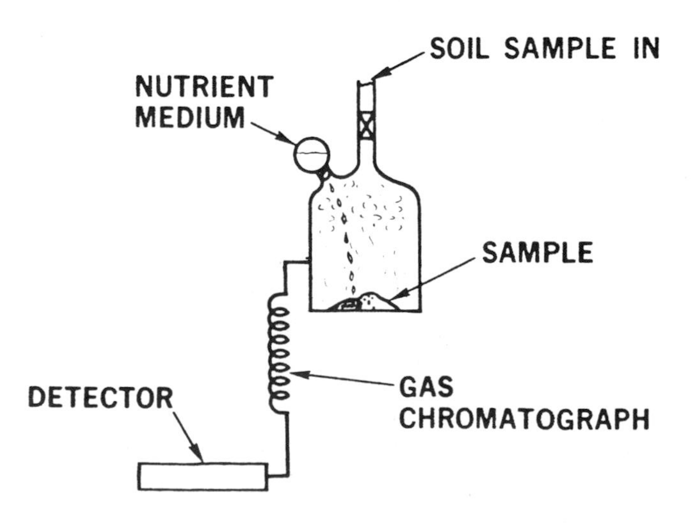

Figure 13.4 Gas exchange experiment. *Courtesy of NASA.*

The Martian Meteorite

In Chapter 4 we discussed how meteorites falling to Earth can accumulate in Antarctica. One of them, called ALH84001.0, about the size of a potato and weighing 4.2 pounds (1.9 kilograms), appears to contain fossilized microorganisms. It is shown in Figure 13.6 (Plate 32).

An asteroid striking Mars 16 million years ago ejected a large mass of Martian rocks and soil into space. Some of it reached escape speed and flew into orbits around the Sun. About 13,000 years ago one piece came close enough to Earth for gravity to pull it in through the atmosphere and land it on the Antarctic ice sheet. There it lay, sliding downhill with the snow and ice until it reached the base of a mountain. Eventually wind blew away its snow covering and left it exposed on the surface. There it was found in 1984, but its origin on Mars was not recognized until 1993. The chemical composition of its minerals matches that found by Viking and the isotope ratio of trapped gases matches the Martian atmosphere. Twelve similar meteorites have been identified as coming from Mars.

ALH84001.0 has been dated to 4.5 billion years ago, the same age as many other rocks found elsewhere in the Solar System, the time when the Solar System was formed. Between 3.6 and 4 billion years ago, Mars and other planets of the inner Solar System were being severely bombarded by asteroids and meteoroids. The craters we see on Mars, the Moon, and Mercury date from that era. Water saturated with dissolved carbon dioxide from the Martian atmosphere, carbonic acid, seeped into the cracks and fractures in the rocks. This acid, like that contained in soda water and soft drinks, reacted with the rock and formed carbonate minerals.

Some of these carbonate minerals and some unusual looking egg-shaped and tubular structures are visible in

ALH84001.0 under a high-resolution scanning electron microscope with great magnification. One of them is shown in Figure 13.7. They are very tiny, about 1/1000 the diameter of a human hair. Similar structures were recently identified in Earth rocks as *nanobacteria*. The nano-prefix indicates their tiny size. Organic molecules called polycyclic aromatic hydrocarbons (PAH) were detected in the carbonates as well. In addition, grains of magnetite, a magnetic mineral with shapes similar to those produced by bacteria on Earth, were found. These molecules and minerals could have been formed individually by purely chemical processes. However, those processes require different environmental conditions that cannot occur simultaneously. That is, they are not normally found together except when they are produced by bacteria. For example, if the carbonates formed at high temperature, then the magnetite would have lost its magnetism. These findings prompted the researchers to announce that they had discovered indications of primitive life on Mars.

The findings are under intense critical scrutiny in the scientific community. Four things in particular are questioned:

1. The structures are smaller than any found on Earth, too small to contain the large molecules needed by a living organism.
2. These carbonates form in hot water at temperatures too high for living organisms to survive.
3. The organic molecules could be contamination that entered the meteorite while it was in Antarctica.
4. The amino acids found in the meteorite are the kinds found on Earth, but not in other meteorites.

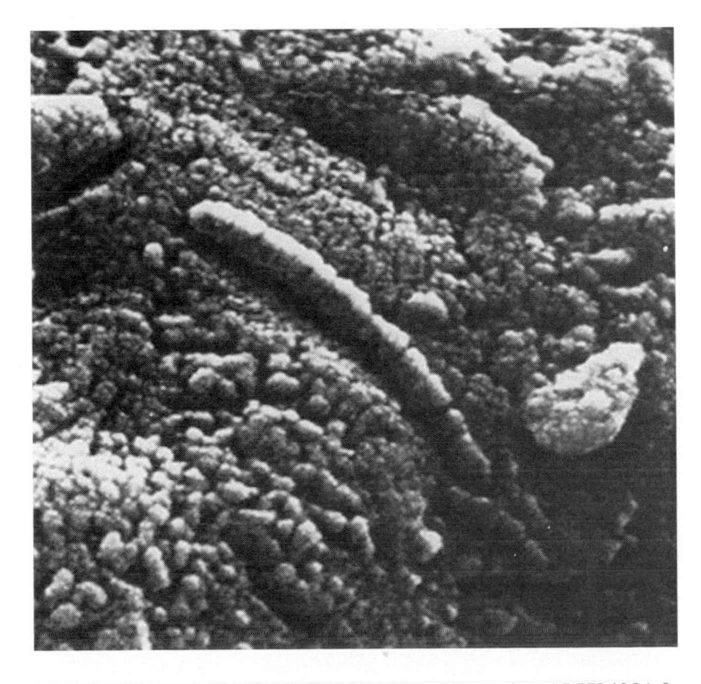

Figure 13.7 A structure inside the Mars meteorite, ALH84001.0, tentatively identified as nanobacteria. *Courtesy of NASA.*

These concerns do not invalidate the proposition that ALH84001.0 contains signs of primitive Martian life; they point out the complexity of the situation and the need for additional research. Such are the workings of science.

Future Mars Experiments

The Viking experiments all showed some unexpected reactions taking place, different from any observed on Earth or with lunar soil, but none could definitely be associated with biological activity. This does not say that there is nothing living on Mars; it only says that there was no firm indication in the results at these two locations. There may be more favorable environments elsewhere, perhaps in deep canyons, at the polar caps, or deep in the soil. Scientists designed the experiments based on what they knew about life on Earth. Perhaps there are living organisms on Mars, but they just didn't like the soup we sent. Their metabolism may be quite different from Earth organisms.

Possibly there was once life on Mars, but it became extinct as the atmosphere dwindled, water dried up, and climate changed. The tiny Sojourner rover that landed on Mars in the summer of 1997 examined Martian rocks and found evidence of a warmer and wetter climate. A denser atmosphere and warmer temperature may have existed for as long as a billion years, enough time for primitive life to take hold. Our first crew to land on Mars should include a paleontologist to search for fossils in dried-up lake beds, ice caps, former hot springs, areas around craters where debris blown out from the impact now lies on the surface, and places where water once flowed.

The extinct volcanoes on Mars also give evidence of a denser atmosphere and a warmer climate. Active volcanoes have hot interiors which spew out large volumes of gas. Perhaps warmer temperatures remain underground where water could still exist in the liquid form.

If evidence of life, either past or present, is found on Mars, an important question is whether or not it is related to us. Is it made of the same amino acids, proteins, and the same DNA? The Mars meteorites found in Antarctica are pieces of the red planet, blown off by comet or asteroid impacts. Such impacts were frequent in the early days of the Solar System. Perhaps Earth material was sent to Mars in the same way. So, were living organisms originally transported from Mars to Earth or vice versa? Or were the first organisms transplanted from some other place to both planets via comets or asteroids? If Martian organisms have the same genetic code, which is shared by all life on Earth, then we may have come from the same origin. If not, then there may have been a second genesis.

We must take care in sending our spacecraft to other planets that we do not send Earth life with them, not even microorganisms. Not only could they contaminate the spacecraft's experiments or future experiments, but also they could conceivably grow, spread, and compete with the planet's indigenous life. Component parts of the Viking spacecraft were cleaned and sterilized and were assembled in a clean room. The entire spacecraft then underwent an elaborate and expensive heat-sterilization process. These procedures added about 10 percent to the cost of the mission.

Europa

Photographs of Jupiter's moon, Europa, by the Voyager spacecraft showed a smooth surface with a multitude of dark-colored lines. Scientists speculated that the surface was fractured ice. Data and images from the Galileo spacecraft touring through the Jovian system confirmed the idea that Europa probably has a liquid water ocean beneath its ice-covered surface and that the cracks were frozen dirty water upwelling from beneath. Figure 7.27 (Plate 15) is an image of Europa from Galileo.

Liquid water on a moon five times as far from the Sun as Earth is seems impossible. Why doesn't it freeze? The source of the energy is the gravitational tugging at Europa by Jupiter and the other three Galilean moons: Io, Callisto, and Ganymede. Just as Earth's ocean tides rise and fall because of the gravitational attraction of the Sun and the Moon, so also does Europa feel a tidal tug as it moves in its elliptical orbit closer to and farther from Jupiter or when it passes one of the other moons or when one of the other moons passes it. This tidal action warms the water and keeps it from freezing.

Also, early in Europa's history, volcanic vents perhaps spewed out hot water which would have been ideal places for living organisms to thrive. Europa may be the most promising place in the Solar System to search for life.

Titan

Both water and carbon dioxide lie on the frozen surface of Saturn's largest moon, Titan, so it is unlikely that anything is alive there. However, the atmosphere of Titan appears to be very much like the early atmosphere of Earth and can give us some insight into the processes by which our atmosphere evolved.

The Drake Equation

Finding *microbial life* on other planets would be interesting and would help us place ourselves in the universe. Finding *intelligent life* on another planet would be profound, perhaps the most important discovery of all time. Frank Drake of Cornell University has been asking questions about life in the universe since the 1950s. At one point he organized his thinking by setting the unknowns into an equation to calculate the number of advanced civilizations in our Milky Way Galaxy that would be capable of communicating with us. The equation and some estimated answers to the question are shown in MATHBOX 13.1. Let us examine the ideas Drake incorporated into his equation.

Planets Around Other Stars

The second factor in the Drake equation is the number of stars that have planets. Many astronomers think that the creation of planets and moons always accompanies the birth of a star, that they all form simultaneously out of the same cloud of dust and gas. Since there are at least 100 billion stars in our Milky Way Galaxy and some 200 billion galaxies, there may be many planetary systems. However, about a quarter of the stars are very old, born when the Galaxy was just forming out of the primordial hydrogen. They could have no planets because no heavy elements existed at that time for the creation of planets. That still leaves about 25 billion candidate stars to consider in our Galaxy alone.

The successful search for planets around other stars was discussed in Chapter 12. These findings lead some astronomers to estimate that about 2 to 5 percent of the Sun-like stars in our Galaxy have detectable planets. However, many may have small, undetectable planets. Considering the short time we have been able to detect planets and the prevalence of dust disks, other astronomers think as many as half the stars have planets.

Life on Other Planets

The third factor in the Drake equation is the number of planets per star that are in the proper location to have an environment that will support living organisms. With the possibility of billions of planets in the Milky Way Galaxy, it seems that living creatures should thrive in many places, but there are other factors to consider.

To support life, a planet must have a suitable temperature and gravity to hold an atmosphere and liquid water. About half the stars in the Galaxy have companions, two or three or more stars in orbit around a common center of mass. Planets in such star systems would undergo extremes of temperatures as they moved around in the star group and would probably be unsuitable for life as we know it.

The parent star must have a long enough lifetime to allow life to fully develop. On Earth, simple forms of life date back about 3 billion years. At 5.5 billion years of age, our Sun is about halfway through its lifetime. But many stars burn at a furious rate and live only a few million years. Life may not have sufficient time to develop on their planets.

Our Solar System has nine planets and about 50 moons. Most of these objects are too cold, too hot, too small, or too something else to support living organisms. We know for sure that higher life forms exist only on one, our Earth. Given the extreme environments in which life thrives on Earth, other planets may have hospitable niches to support primitive life forms. Mars's ice caps are not that much different from Earth's. Titan's atmosphere, the ice-covered seas of Europa, or the clouds of Jupiter may harbor simple organisms. Some scientists believe that given a suitable environ-

MATHBOX 13.1

The Drake Equation

The Drake equation calculates N, the number of advanced civilizations now existing in our Galaxy, which would be capable of transmitting and receiving radio signals.

$$N = R * P * E * L * I * C * T$$

R is the number of stars formed per year in the Galaxy. There are about 200 billion stars in the Galaxy and, since the Galaxy is about 10 billion years old, the average rate of production of stars would be 20 per year over the lifetime of the Galaxy. Astronomers estimate that they are now forming at the rate of about 1 per year but at a much higher rate in the past. A reasonable number for R may be 5.

P is the fraction of stars that have planets. If all stars have planets, then P equals one; if half have planets, then $P = 0.5$. Considering recent research, let us take $P = 0.25$.

E is the number of planets per star that have suitable conditions to support life. In our Solar System, the only one we know anything about, E could be about 5, including moons.

L is the fraction of those planets on which life actually does develop. $L = 1$ if life develops inevitably in any suitable environment. If life is a very fragile and unlikely occurrence, then L is very small. Given that organic molecules are found "everywhere," that microorganisms can live in extreme environments, and that they existed on Earth very quickly after Earth formed, $L = 1$ may be a good choice.

I is the fraction of those life forms that develop intelligence. This is the most controversial factor of all.

C is the fraction of intelligent species that become technically developed and discover radio communications.

T is the time in years that a technically developed civilization manages to survive.

Values for these last three factors are much more speculative. Any numbers could be used depending on your viewpoint. There is not yet any scientific basis or experience for making a choice.

To find how many advanced civilizations exist in the Galaxy, multiply all these factors together.

Let us, then, enter these values into the Drake equation: $R=5$; $P=0.25$; $E=5$; $L=1$; $I=0.1$ (1 in 10); $C=1$; and $T=5{,}000$ years. Multiplying these together gives $N=50{,}000$. Fifty thousand intelligent civilizations! But if we take more conservative numbers of, say, $L=0.1$ (1 in 10), $C=0.1$ (1 in 10) and $T=100$ years, then N is only 1: us.

If we want to consider the possibility of any type of life, intelligent or not, including bacteria, algae, or other primitive forms, we omit I and C from the equation. Then T becomes the time that any kind of life form manages to survive. On Earth, simple forms of life date back about 3 billion years. Using the conservative numbers from the previous example and 3 billion years for T gives an N of 15 billion places where any type of life now exists in the galaxy.

The value of the Drake equation lies not so much in the answer you get for N; you can get any result you want depending on your choice of numbers. The equation is important because it provides a starting point for considering the factors involved in the search for life in the universe.

ment, it is inevitable that life will develop. Others believe life to be a fragile and unlikely occurrence.

Future experiments are planned to determine whether the planets around other stars have signs of life. Animal life inhales oxygen and exhales carbon dioxide while plant life takes in carbon dioxide and releases oxygen. Water is an essential ingredient of life and methane is another by-product. The four "greenhouse gases"—ozone, carbon dioxide, water vapor, and methane—absorb sunlight and emit infrared waves. Oxygen does not strongly emit infrared, but ozone does and is a sensitive indicator of the presence of oxygen. Detecting the infrared signatures of these four gases coming from a planet would be strong evidence that living organisms doing organic chemistry are probably present.

As mentioned before, seeing a planet in the glare of its parent star presents a difficult problem. However, stars emit much less infrared than visible light while planets emit more infrared than reflected starlight.

Intelligent Life

No one knows for sure how intelligence arises. Is it a natural development of all life? Scientists are strongly divided on the question. Some would say that we humans are indeed the only intelligent beings in the Galaxy. Others believe that life is a natural part of the physics and chemistry of the universe and that the Galaxy must be teeming with life. To a biologist, an event that occurs once in the life of a planet is rare; to an astronomer, an event that occurs in the lifetime of a planet has a high probability of occurring again, given the lifetime of planets. A famous biologist, Ernst Mayr of Harvard, points out that "Physicists think more deterministically than biologists. They tend to say that if life has originated somewhere, it will also develop intelligence in due time. The biologist, on the other hand, is impressed by the improbability of such a development."

Presumably, if a species is intelligent and has an appropriate physical structure such as sensors of its surroundings, fingers, and motion it will discover the natural laws of the universe and become technologically capable. There is less controversy over this factor. Of course, a species could be intelligent and yet not be able to build radio transmitters. We consider dolphins and gorillas intelligent and we attempt to communicate with them, but they could not communicate across interstellar space.

How long an intelligent species can survive is the last factor. Paleontologists have found remains of dinosaurs that were about 5 feet (1.5 meters) tall and weighted about 100 pounds (60 kilograms). They had hands and bird feet and their brain cavities were about the size of human brains. Given a few million years they might have showed signs of intelligence, but an asteroid impact 65 million years ago killed them off.

Humans have survived plague, pestilence, climate changes, meteoroid impacts, and other natural disasters. Can we survive our own technology in the form of nuclear weapons, environmental destruction, and pollution? Perhaps technical civilizations promptly destroy themselves within a hundred years or so after discovering nuclear power. But perhaps those species that develop technologically are also smart enough to figure out how to survive for millions of years.

If we search and find another advanced civilization, it will be perhaps the most important discovery ever made. On the other hand, if we come to the realization that we are alone, that too will be a most profound conclusion.

Search for Extraterrestrial Intelligence (SETI)

To look for living organisms in general, we must either go there and look or we must infer their presence from their biological activity or their organic chemistry. But to search for intelligent creatures elsewhere in the universe we have alternate methods. So far, the search has involved transmitting a radio message, sending videodisks on spacecraft, and listening for intelligent radio signals.

An advanced civilization will almost certainly discover the electromagnetic spectrum and ways to produce electromagnetic waves: light, radio, infrared, and so on. In doing so, they give themselves away. The waves travel at the fastest speed possible, the speed of light, spreading throughout the Galaxy to be intercepted by anyone with a sensitive enough detector and the patience to search carefully and thoroughly.

Earthlings have been generating radio waves since the early 1900s, radar pulses since World War II, and television since the 1940s. The earliest radio broadcasts were weak and were of such a low frequency that they did not penetrate the ionosphere into space. However, for at least 60 years broadcasts from Earth have been traveling through space at the speed of light. The first signals to penetrate the ionosphere are now 60 light years from Earth and have passed at least 3000 stars. Has anyone been listening out there? Does anyone know we are here? Perhaps someone on a planet 60 light years from us has just received one of those old programs and is sending a response in our direction. Sixty years from now it will reach us. Two-way communications across interstellar distances is a very slow process indeed!

If we hope to communicate with intelligent aliens, they must reach their intelligent activity level at the same time as we do, or with a time lag to allow for messages to travel at the speed of light. Perhaps there are civilizations that have been transmitting radio waves for a longer time. If we do receive a transmission from a planet say a thousand light years away, they certainly will not be as backward as we are. Their message would have originated while we were still in the Dark Ages. In those intervening thousand years they may have progressed to something beyond radio or television, something we have not yet discovered and cannot imagine. Maybe they are no longer listening. They could have beamed a message toward us for several thousand years and given up in say 1850, before we learned about radio waves. There may be only a narrow window in time when interstellar communications are possible between two intelligent civilizations.

It would seem easy enough to conduct a search of the heavens for intelligent signals, but it is not as simple as it first appears. The problem does not lie in the sensitivity of the receiver. The Pioneer 10 spacecraft has a 1-watt transmitter and a SETI receiver found its signal when it was 3.3 billion miles (5.3 billion kilometers) away. The problem is the need to cover millions of frequencies. Where do we point our antenna? Where do we tune the dial? What kind of signal do we look for? The electromagnetic spectrum is a gigantic place. For example, the FM radio band (Figure 13.8) covers 20 MHz, from 88 to 108 MHz on the dial. A *hertz* (Hz) is one oscillation per second, and a *megahertz* (MHz) is one million hertz. There are 20 million 1-hertz channels in the FM radio band alone. Each television channel has a bandwidth of 6 MHz and there are 82 such channels in the television bands. In addition there are the police, fire, and emergency bands, the satellite communications bands, and the microwave bands. There are a total of 100 billion (100,000 million) 1-hertz channels in the entire radio band. So, where do we begin?

We certainly do not want to start in the FM radio band. It would be difficult to pick out an extraterrestrial through the rock music. The television bands are out, too, for a similar reason. By international agreement, however, a number of frequency bands have been set aside as quiet bands in which no human transmissions are made. They are used primarily by radio astronomers for scientific research. Their selection has been somewhat arbitrary, but one in particular is at the natural resonant frequency of the hydrogen atom, near 1420 MHz, a frequency used by radio telescopes to locate clouds of hydrogen gas in interstellar space. If intelligent extraterrestrials think like human scientists, they may think of the hydrogen resonance frequency as an interstellar hailing frequency. Radio astronomers everywhere would have sensitive equipment tuned to that band. If they wanted to make contact with other radio astronomers it would be logical to transmit a signal on or near that frequency. But, perhaps those far superior extraterrestrials, if they exist, tried that thousands of years ago and gave up. Nonetheless, the hydrogen frequency appears to be a good place to start a search.

Figure 13.8 FM radio band from 88 to 108 megahertz. The AM radio band is between 540 and 1600 kilohertz.

The entire microwave band runs from 1000 MHz (1 GHz) to 10,000 MHz (10 GHz). In that frequency range cosmic background noise is low and Earth's atmosphere is most transparent. Microwaves are selectively absorbed by water vapor in the atmosphere which prevents them from reaching the ground, but the absorption is at a minimum between 1400 MHz and 1720 MHz and that band is used for most SETI projects. There are 320 million 1-Hz channels to examine in that band! We have narrowed the problem from 100 billion to 230 million, still a sizable problem.

Where do we look? There are at least 100 billion stars in our galaxy. How do we choose? Young stars may be a poor choice because it takes millions, perhaps billions of years for advanced life forms to evolve. Old stars from the origin of the universe would also be a poor choice because in the beginning the universe consisted of only hydrogen and helium. There were no heavy elements such as oxygen, nitrogen, iron, silicon, and all the others from which a planet could form. Best choice seems to be Sun-like stars at the middle or toward the end of their life cycle.

What kind of signal do we look for? In Chapter 6, we noted that electromagnetic waves are emitted by everything at a temperature above absolute zero. The universe is filled with such waves from natural sources, but they are random and without pattern, like the sizzling noise you hear as you tune the FM radio between stations. If an extraterrestrial intelligent being produces electromagnetic waves, they will contain a pattern, a recognizable message. Like searching for a needle in a haystack, you will know when you find it.

Two types of signals would stand out against the cosmic background noise: radio, television, and radar type transmissions which unintentionally spill out into the universe, and transmissions specially designed for interstellar contacts. Signals would be distinguishable from natural noise initially from their greater signal strength, then by their modulation, the rhythmic shifting of frequency or changing strength. The first detection will probably not be good enough to get the intelligence out of the signal. It may require receiving it and studying it for a long period of time before being able to decode and find the message, but there would be no doubt that it was from an intelligent source.

SETI Surveys

The first SETI project, Ozma, was by Frank Drake in 1960 using an 85-foot (26-meter) dish antenna at Green Bank, West Virginia. Drake spent 400 hours looking at two nearby Sun-type stars in a 400,000 hertz band around 1420 MHz.

Receiving equipment was not automated, so it was a tedious task to scan the possible frequencies.

Since then there have been over 70 attempts, some for only a few hours and some continuing year after year. The former Soviet Union and a half dozen other countries have made serious efforts.

In the early 1980s, Paul Horowitz of Harvard University designed and built a computerized receiver-signal processor that listened to 128,000 frequencies simultaneously and tested it on the big dish antenna at Arecibo, Puerto Rico, in 1982, probing 250 stars in only 75 hours. The equipment was so successful that several iterations followed, each with greater capability than the previous one.

The newest and largest of the SETI projects, BETA (Billion-channel Extra Terrestrial Assay), uses an 84-foot (25-meter) dish antenna in Massachusetts and an analyzer that can examine 250-million channels simultaneously. Three antennas are connected to the system. East and west pointing antennas are feedhorns from the big dish; a broad beam antenna listens to Earth. The system listens to a 40 MHz wide band in 0.5 Hz increments, i.e. 80 million channels times three antenna feeds equals 240 million channels. The dish is stationary set at a certain elevation angle. As the Earth rotates, it carries the big dish with it, sweeping out a half-degree strip across the sky. Thus, an extraterrestrial signal would first be detected on the east pointing feedhorn, then as the Earth rotates, the west feedhorn would pick it up. If the Earth-pointing antenna also receives it, the signal is automatically disregarded as being human made.

BETA automatically discarded all but about a million signals per month, which needed further examination. These ultimately fell into four categories: noise, earthly radio interference, equipment malfunction, and others.

Horowitz's equipment designs were originally sponsored by NASA. Funding by Congress was intermittent and the project was often ridiculed by congressmen, the media, and opponents saying that searching for ET was a silly waste of money. Congress cut off funds in 1993, but the search continues, sponsored by The Planetary Society and funded with private donations, including $100,000 worth of computer chips donated by Micron Technology and a very large contribution from Stephen Spielberg.

Another SETI project, called SETI@home, has a free ride on the Arecibo big dish antenna in Puerto Rico. Any time a research astronomer is using the antenna, SETI equipment taps in to search for extraterrestrial signals coming from wherever the astronomer happens to be looking. The data thus collected are sent to a network of over a million volunteers for analysis on their personal computers. Each volunteer receives a batch of data via the Internet. A screen saver turns on and does the analyses whenever the PC is sitting idle. Depending on the speed of the PC, it may take from 8 to 50 hours to analyze 110 seconds of data over a 9700 Hz band. On completion of a batch, the computer automatically calls the data center at Berkley, transmits the results, and downloads another batch. Some participants leave their computers running continuously. Anyone with a capable PC can participate. URL is http://setiathome.ssl.berkley.edu.

Optical SETI

On Earth, we are just beginning to transmit telephone, television, Internet, and other information on light waves at high speed from place to place using fiber-optic cables. Intelligent extraterrestrial beings may be using laser light beams for communicating from planet to planet. If Earth happens to be along the line of sight of the laser beam, it may be possible to listen in on them.

In the past, several experiments have searched for short laser pulses and for the characteristic colors of lasers. The equipment now used to search for planets around other stars can detect pulses as short as a few billionths of a second and is precisely what is needed to search the heavens for optical communications. A project now in progress is studying existing data collected in the search for planets; other experimenters are taking new data. They are looking for short, bright pulses that may be coming from a laser operated by some intelligent being on a distant planet. Pulses that repeat in a definite pattern would be particularly significant. A continuous beam of light at a laser wavelength would also be important; it could be a beacon. No such signals have been found yet.

What Do You Say to an Extraterrestrial?

During the last century, before radio was invented, people had no doubt that other celestial bodies were inhabited, that the Moon and Mars were populated by intelligent beings. Radio was unknown as a means of communications, but several scholars proposed other means of letting the extraterrestrials know we are here. One suggested planting a forest of pine trees in Siberia in the shape of a right triangle large enough to be visible from the Moon. Another suggested draping a black cloth over a large white surface of Earth and moving it back and forth. From Mars it would seem to blink. Yet another said to dig a 20-mile (32-kilometer) wide trench in some geometric shape across a desert, fill it with kerosene, and ignite it.

Now radio is the obvious means of communication. But what should we say? Extraterrestrials would not understand any Earth language, may not have 10 fingers on which to base a decimal counting system, and probably have a completely different biology. But, an advanced civilization would have one language in common with us. Mathematics and the sciences of chemistry, physics, and astronomy are universal.

A message intended to initiate communications with another world would be designed specifically so it could be eas-

```
000000101010100000000000000101000000101000000001
001000100010001001011001010101010101010101000010
010000000000000000000000000000000000000000011000
000000000000000000011010000000000000000000011010
000000000000000000010101000000000000000000001111
100000000000000000000000000000000000110000011100
011000011000100000000000000001100100001101000011
000110000011010111110111110111110111110000000000
000000000000000000010000000000000000000010000000
000000000000000000000001000000000000000000001111
110000000000000001111100000000000000000000000001
100001100001110001100010000000010000000000001000
011010000110001110011010111110111110111110111
111000000000000000000000000000000100000001100000
000100000000000001100000000000000000010000011000
000000111111000000110000001111100000000000001100
000000000000100000000010000000001000001000000011
000000010000000011000011000000010000000000001100
010000011000000000000000000110011000000000000011
000100000110000000000110000011000000010000000100
000010000000001000000100000000110000000001000100
000000110000000001000100000000000100000000100000
100000001000000010000000100000000000000110000000
000011000000000110000000000010001110101100000000
000010000000010000000000000000010000001111100000
000000100001011101001011011000000010011100100
111111101110000111000000110111000000000000101000
001110110010000000101000000111111001000000001010
000011000000010000001101100000000000000000000000
000000000000001110000001000000000000000000111010
000101010101010011100000000001010101000000000
000000010100000000000000001111100000000000000000
011111111100000000000000111000000000111000000000
110000000000001100000000110100000000000010110000
110011000000001100110000100010100000010100010
001000100100010010001000000000100010100010000
```

Figure 13.9 The Arecibo message.

ily decoded. A simple message might consist of the first ten *prime numbers,* 1, 2, 3, 5, 7, 11, 13, 17, 19, 23, transmitted in sequence using on-off pulses: one pulse, pause, two pulses, pause, three pulses, pause, etc., to the end, long pause, and begin again. Such a sequence would not occur randomly in a natural radio noise source, and would be immediately recognized by an extraterrestrial mathematician. It could serve as a beacon, in effect saying, "We are here."

A much more complex message was transmitted from the radio telescope antenna at Arecibo, Puerto Rico, in November 1974. The antenna, a section of a sphere, is hung over a bowl shaped valley in the hill country, far from man-made noise sources. Shown in Figure 12.6, it has an area of 20 acres (8 hectares), more than all other radio telescopes put together. The message consisted of a series of on-off pulses, 1679 of them, displayed in Figure 13.9, which, when arranged in the proper pattern, produce a picture. A smart extraterrestrial would know how to arrange them by noticing that 1679 is the product of two prime numbers, 23 times 73. If they are arranged in 23 rows of 73 pulses each, they do not produce anything of obvious intelligence (Figure 13.10), but ordering them into 73 rows of 23 pulses each creates the pictogram as shown in Figure 13.11. (We have used the symbol * for ones and a blank for zeros so the pattern shows up more clearly.) Notice the human figure at lines 46 to 55. The top of the pictogram, lines 1 to 4 reading from right to left, starts with the binary numbers from one to ten, followed by the atomic numbers for the elements of life and formulas for the organic molecules of life, all using binary numbers. At the bottom is the shape and size of the Arecibo radio telescope. In between are a diagram of the Solar System and the shape of a DNA helix.

This message was transmitted toward M13, a great globular cluster of about half a million stars in the constellation Hercules some 24,000 light years away. If beings are intelligent enough to pick up the message, they may be intelligent enough to decode it. If they send an answer back, travel time will be another 24,000 years, to be received by our descen-

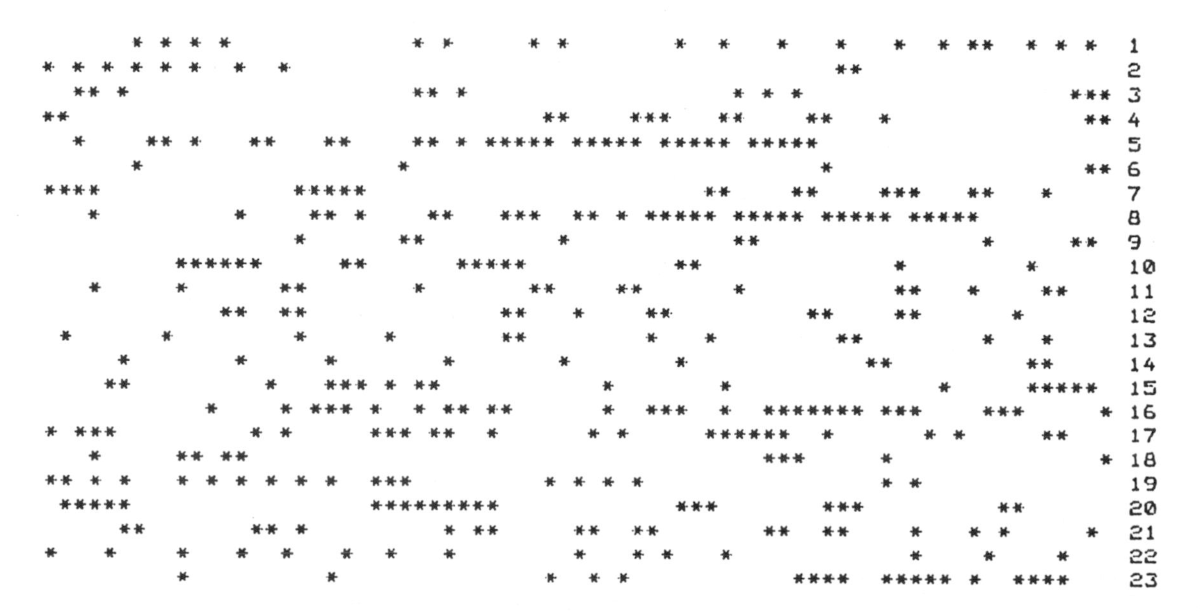

Figure 13.10 The Arecibo message incorrectly decoded.

dants in the year 49,974! Who can tell where the human species will be at that time?

The Arecibo transmission was not so much an attempt to contact another civilization as it was an exploration of the possibilities in interstellar communication, getting scientists to think about what is involved and how to go about it.

Pioneer 10 and Pioneer 11, leaving the Solar System, carry metal plates with an engraved message from Earth (Figure 13.12). Each plate shows a human couple in front of an outline of the spacecraft, drawn to scale to show the size of humans. To the left is a diagram locating the Sun with respect to 14 pulsars and at the bottom is a sketch of the Solar System showing the path of Pioneer.

The Voyager spacecraft, also on their way out of the Solar System, carry recordings of pictures, voices, and a diagram message from Earth to any creature intelligent enough to intercept it and translate it. The recording is something like a CD with music (Bach, Mozart, Beethoven, Chuck Berry, primitive drums), sounds of Earth (baby crying, singing, waves), pictures of Earth, and for international acceptance of the project, the names of leaders of all countries of the world. Instructions for playing the record are on the cover (Figure 13.13) along with diagrams similar to those on Pioneer. If extraterrestrial space travelers catch Voyager and figure out how to make it run, their first response back to Earth will probably be, "Send more Mozart!"

Scientists worldwide have agreed on a set of procedures to follow if an extraterrestrial message is detected. The discoverer attempts to verify the signal and notifies other SETI scientists and research organizations to verify that it is extraterrestrial. The International Astronomical Union (IAU) sends word to other scientists worldwide for independent verification. All parties studying the signal are to record it and store it permanently for future research. When the message is verified, the IAU notifies the Secretary General of the United Nations, the Institute for Space Law, the International Telecommunications Union (ITU), and other interested organizations. Then the discoverer tells the rest of the world. Meanwhile, the ITU does what is necessary to clear the frequency of other transmissions so the alien signal can be heard more clearly and others can listen.

The final question: if we do get a message from a highly advanced extraterrestrial civilization, should we respond? By international agreement, no response will be sent until an international consultation takes place. If they are like us, but further advanced with the capability for interstellar travel, maybe it would be unwise to let them know of our presence. Human beings have spread over the entire Earth, the stronger oppressing the weaker as they migrated. Would the extraterrestrials have the same goal? They may already know we are here, but do not want to have anything to do with us. Or, if their population is growing exponentially, if they seek greener pastures, if their motivation is simply to conquer,

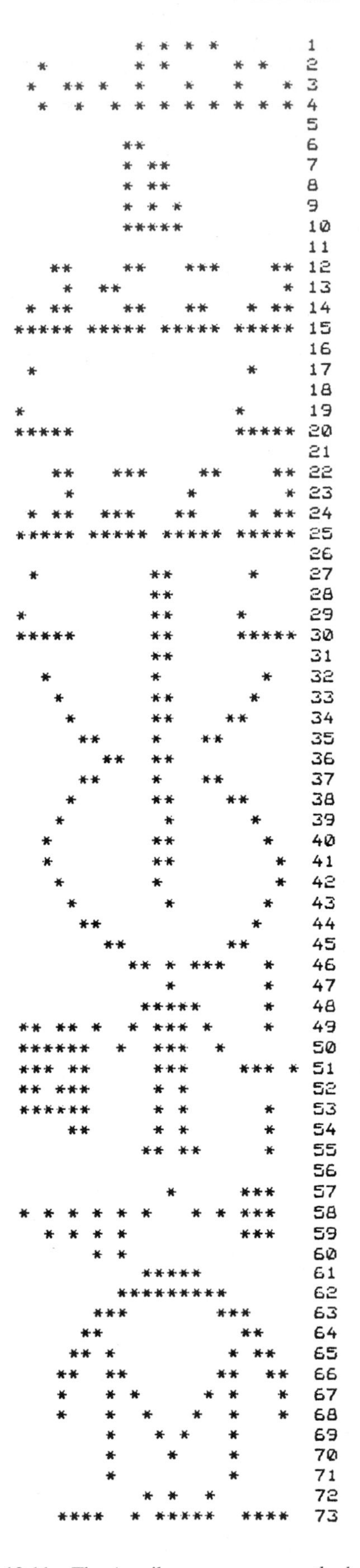

Figure 13.11 The Arecibo message correctly decoded.

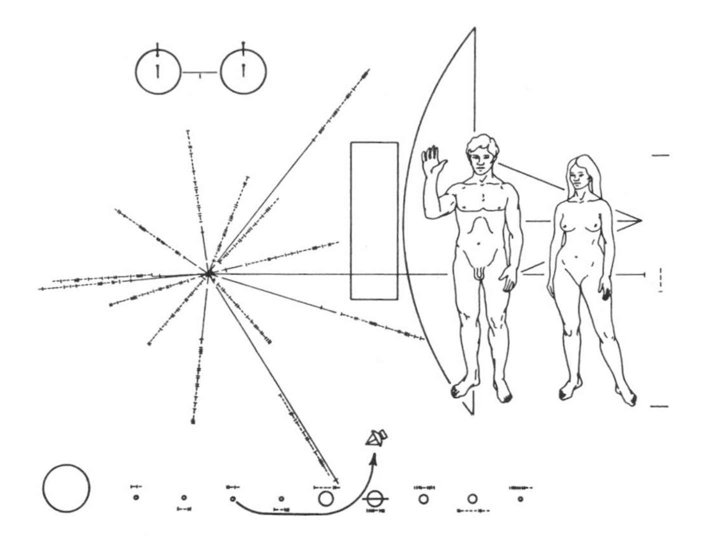

Figure 13.12 Plaque carried out of the Solar System on Pioneer 10 and 11 spacecraft. Short vertical lines and dashes are binary numbers. An extraterrestrial smart enough to catch the spacecraft should be smart enough to understand the numbers. *Courtesy of NASA.*

Figure 13.13 Record attached to outside of Voyager spacecraft. It is protected by a cover carrying instructions on how to play it along with the pulsar map from Pioneer 10 and 11. In addition, a small amount of pure uranium-238 is attached; an extraterrestrial will be able to determine approximately when the spacecraft was launched using radiometric dating techniques. *Courtesy of NASA.*

then perhaps we should keep to ourselves. More likely, if their society has survived technological advancement without self-destruction, then they have probably learned to be peaceful and benevolent. We would find a great deal to learn from such a society.

Physicist Enrico Fermi once asked his colleagues: If life is common in the Universe, why haven't they shown up here? There are many stars much older than ours where intelligent life would have had plenty of time to develop high tech means of travel through the universe. An intelligent species with a will to do so could colonize a large region of the Galaxy in a few million years. This is called Fermi's Paradox. There are several possible responses. Perhaps no one has learned how to deal with the enormous distances and long time required for interstellar travel. Maybe they know how but do not have the resources to accomplish it. Maybe they have been here but have a "Prime Directive" prohibiting interference with a developing civilization. Or maybe, indeed, we are the first to reach this level of technical know-how.

DISCUSSION QUESTIONS

1. What effect would the discovery of the existence of intelligent life elsewhere in the universe have on human society?
2. If intelligent beings were found on some other planet, in what ways would they be like humans? In what way would they be different physically and mentally?
3. If we find anything living on Mars, should we terraform the planet, even if it means death to the native organisms? What if the native life is only bacteria-like?
4. Make up a message to send to extraterrestrials that is different from the Pioneer, Voyager, and Arecibo messages. Be sure it is decipherable by nonhumans.
5. Watch a TV program with the sound turned down as an extraterrestrial might see it and try to interpret it as an extraterrestrial might.
6. Why is it true that when we look out into space we look back in time?
7. Given the results of the Viking search for life on Mars, devise a follow on experiment that might be more conclusive.

ADDITIONAL READING

Angel, J. Roger P., and Neville J. Woolf, "Searching for Life on Other Planets." *Scientific American,* April 1996. Proposed space-based telescope system.

Bernstein, Max P., Scott A. Sandford, and Louis J. Allamandola. "Life's Far-flung Raw Materials." *Scientific American,* July 1999. Sources of complex organic molecules.

Black, David C., "Worlds around Other Stars." *Scientific American,* January 1991. Astronomical searches for planets outside our Solar System.

Davies, Paul. "Did Earthlife Come From Mars?" *Star Date,* September/October 1999. Readable article for general public.

Drake, Frank, *Is Anyone Out There?* 1992.

Feinberg, G., and R. Shapiro. *Life Beyond Earth.* Morrow and Co., 1980.

Frederickson, James K., and Tullis C. Orstott. "Microbes Deep Inside the Earth." *Scientific American,* October 1996. Technical and detailed, but readable.

Gibson, Everett K., Jr., et al. "The Case for Relic Life on Mars." *Scientific American,* December 1997. The researchers who investigated the Martian meteorite tell their story.

Horowitz, Norman. "The Search for Life on Mars." *Scientific American,* November 1977. Viking experiments and results.

Huang, Su-Shu. "Life Outside the Solar System." *Scientific American,* April 1960. Fundamentals.

Kuznik, Frank. "Aliens in the Basement." *Air & Space,* September 1992. About UFOs and alien bodies hidden by the U.S. Air Force.

McDonald, Gene. "The Stuff of Life: Must Life Be Carbon-based?" *The Planetary Report,* March/April 1998. About the chemistry of life.

McInnis, Doug. "Wanted: Life-bearing Planets." *Astronomy,* April 1998. Readable popular article.

NASA. *SETI.* NASA NP-114. U.S., Government Printing Office, revised June 1990. Picture book with basic information.

Pappalardo, Robert T., et al. "The Hidden Ocean of Europa." *Scientific American,* October 1999. The possibility of living organisms on Europa.

Ponnamperuma, Cyril, and A. G. W. Cameron. *Interstellar Communication: Scientific Perspectives.* Houghton Mifflin Co., 1974. Complete, technical, readable.

Poynter, Margaret, and Michael J. Klein. *Cosmic Quest: Searching for Intelligent Life Among the Stars.* Athenium, 1984. Simple, easy reading.

Sagan, Carl, and Frank Drake. "The Search for Extraterrestrial Intelligence." *Scientific American,* May 1975. Readable summary of SETI to 1975.

Sagan, Carl, et al. *Murmurs of Earth, The Voyager Interstellar Record.* Random House, 1978. Details of images and music on Voyager.

PERIODICALS

Bioastronomy News. Published quarterly by The Planetary Society, Pasadena, CA.

Planetary Report. Bimonthly publication of The Planetary Society, Pasadena, CA.

SETI Quest, The Magazine of SETI and Bioastronomy. Published quarterly by Helmers Publishing Co., Peterborough, NH.

NOTES

Appendix

Space Shuttle Flights

Flight	Launch	Duration (Days)	Orbiter	Primary Mission
STS-1	Apr 12, 1981	2	*Columbia*	First test flight
STS-2	Nov 12, 1981	2	*Columbia*	Test remote manipulator system; Earth survey instruments
STS-3	Mar 22, 1982	8	*Columbia*	Test manipulator arm; materials processing
STS-4	Jun 27, 1982	7	*Columbia*	Test manipulator arm with scientific payload; materials processing; first DOD payload
STS-5	Nov 11, 1982	5	*Columbia*	Launch two communications satellites
STS-6	Apr 4, 1983	5	*Challenger*	First EVA spacewalk; deploy NASA TDRS satellite
STS-7	Jun 18, 1983	6	*Challenger*	Launch two comm satellites; release and recapture tests
STS-8	Aug 30, 1983	7	*Challenger*	Deploy comm satellite; test remote manipulator arm
STS-9	Nov 28, 1983	10	*Columbia*	First Spacelab
41-B	Feb 3, 1984	11	*Challenger*	First test of manned maneuvering unit, backpack propulsion system; launch two comm satellites (boosters failed to send them to synchronous altitude)
41-C	Apr 6, 1984	13	*Challenger*	Retrieve, repair, redeploy Solar Maximum Mission using MMU; deploy Long Duration Exposure Facility
41-D	Aug 30, 1984	6	*Discovery*	Launch three comm satellites; test large solar array.
41-G	Oct 5, 1984	13	*Challenger*	Launch Earth Radiation Budget Explorer; imaging radar
51-A	Nov 8, 1984	16	*Discovery*	Launch two comm satellites; retrieve two failed satellites from 41-B
51-C	Jan 24, 1985	3	*Discovery*	DOD payload
51-D	Apr 12, 1985	7	*Discovery*	Deploy two commercial satellites (one booster failed)
51-B	Apr 29, 1985	7	*Challenger*	Spacelab; materials processing
51-G	June 17, 1985	7	*Discovery*	Deploy three communications satellites; Spartan experiment
51-F	Jul 29, 1985	8	*Challenger*	Spacelab, solar physics, astrophysics, plasma physics
51-I	Aug 27, 1985	7	*Discovery*	Deploy three comm satellites; retrieve, repair, redeploy failed satellite from 51-D
51-J	Oct 3, 1985	7	*Atlantis*	DOD mission
61-A	Oct 30, 1985	7	*Challenger*	Spacelab, materials processing
61-B	Nov 26, 1985	7	*Atlantis*	Deploy three comm satellites; assemble large structures
61-C	Jan 12, 1986	6	*Columbia*	Deploy comm satellite; IR imaging; image Comet Halley
51-L	Jan 28, 1986	0	*Challenger*	The accident
NO FLIGHTS FROM JANUARY 28, 1986 TO SEPTEMBER 29, 1988				
STS-26	Sep 29, 1988	4	*Discovery*	Launch TDRS comm satellite
STS-27	Dec 2, 1988	4	*Atlantis*	DOD payload
STS-29	Mar 13, 1989	5	*Discovery*	Launch TDRS comm satellite
STS-30	May 4, 1989	4	*Atlantis*	Launch Magellan to Venus
STS-28	Aug 8, 1989	5	*Columbia*	DOD payload
STS-34	Oct 18, 1989	5	*Atlantis*	Launch Galileo to Jupiter
STS-33	Nov 22, 1989	5	*Discovery*	DOD payload

Flight	Launch	Duration (Days)	Orbiter	Primary Mission
STS-32	Jan 9, 1990	11	*Columbia*	Launch Syncom satellite; Retrieve LDEF
STS-36	Fed 28, 1990	5	*Atlantis*	DOD payload
STS-31	Apr 24, 1990	5	*Discovery*	Deploy Hubble Space Telescope
STS-41	Oct 6, 1990	4	*Discovery*	Launch Ulysses to Sun
STS-38	Nov 15, 1990	5	*Atlantis*	DOD payload
STS-35	Dec 2, 1990	9	*Columbia*	Astrophysics experiments
STS-37	Apr 5, 1991	6	*Atlantis*	Deploy Gamma Ray Observatory
STS-39	Apr 28, 1991	8	*Discovery*	Space science
STS-40	Jun 5, 1991	9	*Columbia*	Space science, microgravity experiments
STS-43	Aug 2, 1991	9	*Atlantis*	Launch TDRS comm satellite
STS-48	Sep 12, 1991	6	*Discovery*	Deploy UARS atmospheric research satellite
STS-44	Nov 24, 1991	7	*Atlantis*	Launch DSP early warning satellite
STS-42	Jan 22, 1992	8	*Discovery*	Microgravity research
STS-45	Mar 24, 1992	9	*Atlantis*	Space environment research lab
STS-49	May 7, 1992	9	*Endeavour*	Retrieve, repair, release Intelsat
STS-50	Jun 25, 1992	14	*Columbia*	Microgravity laboratory
STS-46	Jul 31, 1992	8	*Atlantis*	European retrievable satellite; tethered satellite
STS-47	Sep 12, 1992	8	*Endeavour*	Spacelab materials and life sciences
STS-52	Oct 22, 1992	10	*Columbia*	Deploy laser satellite; microgravity lab
STS-53	Dec 2, 1992	7	*Discovery*	DOD payload
STS-54	Jan 13, 1993	6	*Endeavour*	Launch TDRS comm satellite
STS-56	Apr 8, 1993	9	*Discovery*	Space environment research lab
STS-55	Apr 26, 1993	10	*Columbia*	Spacelab life sciences
STS-57	Jun 21, 1993	10	*Endeavour*	Retrieve EURECA satellite; Spacehab
STS-51	Sep 12, 1993	10	*Discovery*	Deploy research satellite
STS-58	Oct 18, 1993	14	*Columbia*	Spacehab life sciences
STS-61	Dec 2, 1993	10	*Endeavour*	Repair Hubble Space Telescope
STS-60	Feb 3, 1994	8	*Discovery*	Wake Shield; Spacelab microgravity experiments
STS-62	Mar 4, 1994	14	*Columbia*	Microgravity research
STS-59	Apr 9, 1994	10	*Endeavour*	Space Radar Laboratory
STS-65	Jul 8, 1994	15	*Columbia*	International Microgravity Lab
STS-64	Sep 9, 1994	11	*Discovery*	Spaceflight research
STS-68	Sep 30, 1994	11	*Endeavour*	Space Radar Lab
STS-63	Feb 3, 1995	9	*Discovery*	Shuttle-*Mir* rendezvous and fly around; Spacelab
STS-67	Mar 2, 1995	15	*Endeavour*	ASTRO 2 Spacelab; Ultraviolet telescopes
STS-71	Jun 27, 1995	10	*Atlantis*	First Shuttle-*Mir* docking
STS-70	Jul 13, 1995	9	*Discovery*	Launch TDRS-G
STS-69	Sep 7, 1995	11	*Endeavour*	Wake Shield; SPARTAN
STS-73	Oct 20, 1995	16	*Columbia*	Microgravity Science Laboratory (MSL)
STS-74	Nov 12, 1995	8	*Atlantis*	Second Shuttle-*Mir* docking; IMAX
STS-72	Jan 11, 1996	9	*Endeavour*	Space Flyer units; UV experiments
STS-75	Feb 22, 1996	16	*Columbia*	Tethered satellite; Microgravity payload
STS-76	Mar 22, 1996	9	*Atlantis*	Third Shuttle-*Mir* docking; Spacehab
STS-77	May 19, 1996	10	*Endeavour*	SPARTAN; Spacehab
STS-78	Jun 20, 1996	17	*Columbia*	Life and microgravity Spacelab
STS-79	Sep 14, 1996	12	*Atlantis*	Fourth Shuttle-*Mir* docking; SPACEHAB; IMAX
STS-80	Nov 19, 1996	18	*Columbia*	UV spectrograph; Wake Shield experiment
STS-81	Jan 12, 1997	10	*Atlantis*	Fifth Shuttle-*Mir* docking
STS-82	Feb 11, 1997	10	*Discovery*	Second Hubble Space Telescope (HST) servicing
STS-83	Apr 4, 1997	4	*Columbia*	Early return—fuel cell problem; Microgravity Science Laboratory-1
STS-84	May 15, 1997	9	*Atlantis*	Sixth Shuttle-*Mir* docking
STS-94	Jul 1, 1997	16	*Columbia*	Microgravity Science Laboratory-1 (MSL-1 reflight)
STS-85	Aug 7, 1997	12	*Discovery*	IR Spectrometers; Atmosphere-Shuttle Pallet Satellite

Flight	Launch	Duration (Days)	Orbiter	Primary Mission
STS-86	Sep 25, 1997	11	*Atlantis*	Seventh Shuttle-*Mir* docking
STS-87	Nov 19, 1997	16	*Columbia*	U.S. Microgravity Payload; EVA
STS-89	Jan 22, 1998	9	*Endeavour*	Eighth Shuttle-*Mir* docking
STS-90	Apr 17, 1998	17	*Columbia*	Final Spacelab mission
STS-91	Jun 2, 1998	11	*Discovery*	Ninth and final Shuttle-*Mir* docking
STS-95	Oct 29, 1998	9	*Discovery*	John Glenn's flight; SPACEHAB
STS-88	Dec 4, 1998	12	*Endeavour*	1st International Space Station flight
STS-96	May 27, 1999	10	*Discovery*	1st International Space Station docking
STS-93	Jul 22, 1999	5	*Columbia*	Chandra X-Ray Observatory
STS-103	Dec 2, 1999	10	*Discovery*	Hubble Space Telescope servicing
STS-99	Feb 11, 2000	11	*Endeavour*	Shuttle Radar Topography Mission
STS-101	May 19, 2000	10	*Atlantis*	ISS assembly flight

All 2000–2001 flights will be to the ISS except 1 research mission.

Glossary

ABM Antiballistic missile, as in ABM weapon or ABM Treaty.

acceleration A change in velocity per unit time; a change in either the speed or direction of a moving object over a certain time interval.

aerobraking A maneuver in which a spacecraft accomplishes an orbital change by dropping into the upper atmosphere of a planet to reduce its energy instead of firing retrorockets.

airlock An airtight chamber used to move between spaces of different pressure, from the orbiter to the cargo bay, for example.

alpha particle The nucleus of a helium atom consisting of two protons and two neutrons.

amino acid One of the fundamental organic molecules which includes groups containing two atoms of hydrogen bonded to one atom of nitrogen; the structural unit of all proteins.

aphelion The point which is farthest from the Sun on an elliptical orbit around the Sun.

apoapsis The point of an orbit which is farthest from its parent body.

apogee The point which is farthest from Earth on an elliptical orbit around Earth.

apogee kick Firing rockets at apogee to decrease the eccentricity of the orbit, to make it more nearly circular.

apolune The point which is farthest from the Moon on an elliptical orbit around the Moon.

apsides The two extreme ends of the major axis of an elliptical orbit; *see also* line of apsides.

asteroid A piece of rock, metal, or frozen gases in space, larger than a meteoroid but smaller than a planet.

atom The smallest particle of an element, composed of electrons, protons, and neutrons.

attitude The orientation or position of a vehicle.

ballistic Referring to the path followed by an object which is affected only by the forces of gravity and atmospheric drag.

barcan dunes Crescent-shaped sand dunes.

bends *See* decompression sickness.

blue-green algae One of the earliest lower forms of plant life.

booster rocket A rocket used to provide large initial thrust for launch, usually dropped off after it has served its purpose.

burn Rocket engine firing.

cardiovascular deconditioning Weakening of the heart and blood vessels.

centripetal force The force required to keep an object moving in a curved path; it is directed toward the center of the curve.

chromosphere The lower atmosphere of the Sun.

circadian rhythm A natural periodicity in human activity or bodily function, such as the 24-hour sleep-wake cycle.

cislunar space The region of space between Earth and the Moon.

clevis A fitting with a U-shaped end.

combustion The chemical process in which a fuel oxidizes, that is, combines with oxygen, usually accompanied by heat and fire.

comet A ball of dust and frozen gas traveling in an orbit of high eccentricity around the Sun; as it approaches the Sun, solar heat and the solar wind cause particles to evaporate, break off, and trail into space producing a visible tail.

convection The process by which heat is transferred in a fluid by the bulk motion of warmer fluid of lesser density rising in cooler surroundings of greater density.

core Referring to the Sun, the central region in which nuclear fusion takes place.

corona The outermost part of the solar atmosphere which extends out into the Solar System past the Earth.

coronagraph A telescope designed for viewing the Sun's corona.

cosmic ray Nucleus of an atom traveling at very high speed through space; protons coming from the Sun are called solar cosmic rays while heavier nuclei coming from outside the Solar System are called galactic cosmic rays.

cryogenic Involving very low temperatures.

decompression sickness Illness produced by a too-rapid decrease in air pressure surrounding a body, causing bubbles of nitrogen in the blood to expand painfully.

dehydrate Remove the water from a substance, particularly for the preservation of food.

delta-V (Δv) A change in velocity.

deorbit Fire a rocket engine to reduce the energy of a vehicle so it leaves orbit and returns to Earth.

directed energy weapon An antimissile weapon which uses an intense beam of electromagnetic energy, such as a laser beam, or of small particles, such as hydrogen atoms, to burn through the shell of a missile or damage its electronics.

diurnal Referring to a daily occurrence.

DNA The organic molecule which is part of all living cells, containing the genetic instructions for the cell to reproduce.

docking Joining together two orbiting vehicles.

drag Frictional force which decelerates a vehicle, particularly atmospheric friction.

dysbarism Decompression sickness; the bends.

eccentricity A number which tells the elongation of an ellipse or an orbit.

electrolysis Separating the positive ions from the negative ions in a substance by passing an electric current through it; in the case of water, hydrogen collects at the negative electrode (cathode) and oxygen collects at the positive electrode (anode).

electromagnetic radiation The energy carried by electromagnetic waves.

electromagnetic spectrum The range of wavelengths of electromagnetic waves including, in order of decreasing wavelength, radio, microwaves, infrared, light, ultraviolet, x-rays, and gamma rays.

electromagnetic waves The means of transmitting energy at the speed of light, produced by the oscillation or acceleration of electrically charged particles.

electron A small particle of matter carrying a negative electrical charge, one of the three constituent particles of an atom.

electrophoresis A process by which different materials suspended in a fluid are separated from one another by applying an electrical field.

element A basic substance consisting of only one kind of atom.

ELINT Electronic intelligence gathering.

ellipse A closed oval-shaped curve.

end effector Robotics term for a grasping device.

ephemeris A set of numbers which specify the location of a celestial body or satellite in space at various times.

ergonomics The study of human factors in space operations.

escape velocity The minimum velocity an object must have to completely escape the gravitational attraction of a planet (or other celestial body) and neither fall back nor enter an orbit around it.

ET The Space Shuttle's external tank.

EVA Extravehicular activity, working outside a spacecraft.

external tank Referring to the Space Shuttle, the propellant tank which carries liquid hydrogen and liquid oxygen for the orbiter main engines.

extraterrestrial Originating or existing beyond the Earth's atmosphere.

filaments Referring to the Sun, dark streaks on the Sun which are actually prominences, masses of hot gas suspended in the atmosphere, which appear dark against the bright disk of the Sun.

flare A sudden, explosive emission of electromagnetic radiation in the solar chromosphere, sometimes accompanied by the ejection of high-speed protons, called solar cosmic rays.

force A push or a pull which accelerates an object.

fossil fuels Coal, oil, and natural gas, the fuels formed from organisms which lived millions of years ago and became buried in sediments.

fuel A substance which burns in the presence of oxygen.

fuel cell A device which produces electricity by combining hydrogen and oxygen into water.

fusion A nuclear process by which smaller atomic nuclei fuse together to form larger nuclei, such as the fusion of hydrogen into helium.

g Symbol for the acceleration of gravity at the surface of the Earth.

galaxy A basic structure of the universe consisting of billions of stars, dust, and gas, held together by their mutual gravitational attraction, usually in a spherical, elliptical, or spiral shape.

gamma rays Electromagnetic waves with wavelength shorter than x-rays.

geostationary orbit *See* geosynchronous orbit.

geosynchronous orbit An orbit approximately 22,300 miles above the equator with a period of one day, the same as the rotational period of the Earth; a satellite in geosynchronous orbit appears to remain stationary in the sky as viewed from the Earth.

gimbal A device on which to mount a rocket engine to provide angular movement to steer the vehicle.

grain The segment of solid propellant enclosed in a rocket engine.

grapple A fixture attached to a satellite or other payload so the Space Shuttle remote manipulator arm can grasp it.

gravitation The natural force of attraction between any two objects.

H-alpha A particular wavelength of red light produced by glowing hydrogen gas; solar flares are readily seen by their H-alpha light.

heliopause The edge of the Solar System; the boundary between the solar wind and the interplanetary medium.

heliosphere The region of space occupied by the Solar System where the solar wind dominates.
Hohmann transfer Transfer trajectory between two orbits which requires the least amount of energy.
hyperbola An open curve, not closed as an ellipse or circle.
hypergolic Referring to the spontaneous ignition of a fuel and oxidizer when they come into contact.
hyperspectral Referring to remote sensing with sensors covering hundreds of discrete spectral bands.
hypoxia Illness produced by lack of oxygen.
ICBM Intercontinental ballistic missile.
igneous rock Rocks formed by solidification of molten material called magma.
inclination Referring to an orbit, the angle between the orbital plane and the plane of the Earth's equator measured at the point where a spacecraft crosses from the southern hemisphere to the northern hemisphere.
inertial upper stage (IUS) A rocket booster attached to a satellite carried in the cargo bay of the Space Shuttle orbiter to boost the satellite to a higher orbit. *See also* payload assist module.
infrared Electromagnetic waves with wavelengths in the range 1 to 1000 micrometers, longer than light but shorter than radio waves.
infrastructure The basic facilities, installations, and equipment needed for an organization or a system to operate.
inhibitor A coating over the areas of a solid propellant grain that are not supposed to burn.
injector A mechanical device that forces or sprays fuel and oxidizer into the rocket engine combustion chamber.
intelligence Referring to the military, gathering information about a potential adversary.
interstellar medium The thin gas and dust mixture that lies between the stars.
ion An atom or combination of atoms which carries an electrical charge due to the addition or removal of electrons.
ionize To produce an ion by adding or removing electrons from a neutral atom or molecule.
ionosphere Region of a planet's atmosphere in which the atoms are ionized by solar x-ray and ultraviolet radiation.
IRAS Infrared Astronomical Satellite.
irradiate Expose to radiation, particularly to destroy the bacteria in food to preserve it.
IUS *See* inertial upper stage.
kinetic energy The energy of a moving object.
kinetic energy weapon An antimissile weapon that fires a projectile at the missile to destroy it by collision.
Lagrangian point *See* libration point.
laser A device which produces an intense beam of electromagnetic waves of a single wavelength, most commonly light waves of a single color.
LEM *See* LM.
LEO Low Earth orbit.
libration point A location in space where the gravitational attraction of two or more celestial bodies adds to zero, that is, a neutral point in the gravitational field; also called Lagrangian point after the mathematician who first investigated this.
light Electromagnetic waves to which the human eye is sensitive, with wavelengths of approximately 0.5 micrometer.
light year A measure of interstellar distance; the distance light travels in a year, about 6 trillion miles (10 trillion km).
lignin A substance produced by plants which helps give them their structural strength so they stand upright.
limb Referring to the Sun, the edge of the solar disk.
line of apsides Major axis of an elliptical orbit.
lithify To turn into rock.
LM Lunar module, the part of the Apollo spacecraft that landed on the Moon.
LOX Liquid oxygen.
Mach 1 The speed of sound; Mach 2 is twice the speed of sound and so on.
maglev *See* magnetic levitation.
magnetic field A visual image of the lines of force exerted by a magnet.
magnetic levitation A method of accelerating a vehicle along a track using electromagnetic fields.
magnetosphere Region around a planet in which the planet's magnetic field is sufficiently strong to prevent or impede the solar wind particles from entering.
manned maneuvering unit Device equipped with thrusters which an astronaut in a space suit can attach to and then fly freely, untethered away from the orbiter.
mass Most simply, the amount of matter contained in an object. Mass varies with velocity, increasing significantly as it approaches the speed of light. More precisely, a measure of a body's resistance to acceleration.
mass driver An electrically powered device which accelerates projectiles to very high speed, perhaps to escape velocity.
MECO Main engine cutoff, referring to a flight of the Space Shuttle.
metabolism The chemical processes that take place in an organism to maintain life.
meteor Bright streak of light in the night sky produced by a meteoroid falling toward Earth and burning up from the heat of friction with the atmosphere.
meteorite A meteoroid that has survived the fall through the atmosphere and lands on the surface of the Earth.
meteoroid A piece of rock, metal, or frozen gas in space, varying in size from microscopic to asteroid size.
meter Metric unit of length equal to about a yard, 39.37 inches to be precise.
MHz Megahertz; 1 million oscillations per second.
microfossils Microscopic remains of plant or animal life found in rocks.

micrometeoroid A microscopically small meteoroid.
micrometer One-millionth of a meter, also called a micron.
micron One-millionth of a meter; 1 micrometer.
microwaves Electromagnetic waves with wavelengths in the range 1 to 30 centimeters (0.4 inch to 1 foot).
Milky Way The spiral Galaxy in which we live.
missile An object that is fired or thrown at a target; a projectile.
MMU *See* manned maneuvering unit.
module A self-contained, standardized component of a spacecraft, space station, or habitat, designed for ease of assembly with other components.
muscular deconditioning Weakening of the muscle tissue due to lack of use while weightless.
nanobacteria Very tiny bacteria.
NASA National Aeronautics and Space Administration.
neutron A small particle of matter with no electrical charge, one of the three constituent particles of an atom.
node On a space station, a connecting module to which other modules are attached, like a hallway in a building with doors to rooms and to the outside.
NORAD North American Aerospace Defense Command.
nozzle The exhaust duct of a rocket combustion chamber through which the combustion gases are accelerated to higher velocity.
OMS *See* orbital maneuvering system.
OMV *See* orbital maneuvering vehicle.
orbit Closed path followed by a satellite.
orbital injection Firing rocket engines to transfer a spacecraft from a ballistic trajectory into a closed orbit.
orbital maneuvering system A pair of rocket engines located in pods next to the tail of the Space Shuttle orbiter, used for orbital insertion, changing orbit, and deorbiting.
orbital maneuvering vehicle A remotely operated spacecraft which delivers satellites from a space station to higher orbits and returns them for repair and maintenance.
orbital mechanics The science and mathematics of orbits.
orbiter The Space Shuttle's rocket-powered aircraft-spacecraft.
organic Referring to a substance containing carbon as the basic atom of its molecular structure.
O-ring A rubber-like seal between segments of a solid rocket engine.
oxidizer A substance which provides oxygen to support the combustion of a fuel, including oxygen itself.
paleontologist A scientist who specializes in the study of fossils and ancient life forms.
PAM *See* payload assist module.
panchromatic A band covering the visible part of the electromagnetic spectrum and producing a grey-scale (black-and-white) image.
payload assist module (PAM) A solid rocket booster attached to a payload in the Shuttle orbiter's cargo bay for the purpose of boosting the payload to a higher orbit. *See also* inertial upper stage.
perigee The point which is closest to Earth on an elliptical orbit around Earth.
perigee kick Firing rockets at perigee to increase the eccentricity of the orbit.
perihelion The point which is closest to the Sun on an elliptical orbit around the Sun.
period Referring to an orbit, the time required for a spacecraft to make one complete orbit as viewed from space.
photon Bundle of energy that acts like a massless particle.
photosphere The visible surface of the Sun.
photosynthesis The process by which plants convert carbon dioxide from the air, light from the Sun, and water from the Earth into carbohydrates with oxygen as a by-product.
photovoltaic cell A semiconductor device which converts light directly into electricity; called a solar cell when used in sunlight.
phytoplankton Tiny algae plants which float in the ocean; the basic organism of the oceanic food chain.
pitch Up and down rotational motion of the nose and tail of an aircraft or spacecraft. *See also* roll; yaw.
pixel Contraction of "picture element"; one bit of an image.
plage On the Sun, a bright region of hot gas in the chromosphere.
plasma A gas composed of a mixture of neutral and ionized atoms.
polarimeter An instrument for measuring the polarization of light waves.
prime number A number which is evenly divisible only by itself and one.
progressive burn The burn of a solid propellant grain in which the thrust increases with time.
prominence Referring to the Sun, a cloud of hot gas in the solar atmosphere appearing bright when seen on the edge of the disk against the dark background of space, and appearing as a dark filament when seen against the bright surface of the Sun.
propellants The fuels and oxidizers burned in rocket engines to produce thrust.
proton A small particle of matter carrying a positive electrical charge; one of the three constituent particles of an atom.
proton event Referring to the Sun, an outburst of high energy protons during a flare which reach the vicinity of the Earth.
rad A unit of absorption of radiation in a living organism; 600 rads in a day would be a lethal dose for most humans.
radar A device which transmits a microwave pulse, receives the returned pulse which is reflected from an object, and calculates the distance and direction to the object.
radiation The transfer of energy from place to place by means of electromagnetic waves; also, a somewhat am-

biguous term applied to high energy waves and particles coming from the Sun, interstellar space, or radioactive substances.

radiation sickness Illness caused by exposure to large doses of radiation, in severe cases involving nausea, diarrhea, vomiting, dehydration, destruction of blood cells, and perhaps death.

ramjet A reaction engine in which only the air entering the front intake at high speed prevents exhaust gases from leaving in that direction.

RCS The Space Shuttle orbiter's reaction control system, 44 thrusters which control its orientation in space.

reaction engine An engine that produces thrust by the directed expulsion of mass, usually hot gases.

reconnaissance Gathering images and other data about an area of interest, usually applied to military intelligence.

reentry The return of a spacecraft into the Earth's atmosphere.

refraction Bending of electromagnetic waves from a straight line path as they pass from one medium into another.

regolith Broken and powdered rock on the surface of a planet; on the Moon regolith is produced by constant bombardment of the surface by meteoroids, large and small.

regression In orbital mechanics, the rotation of the major axis of an elliptical orbit due to variations of gravity in different parts of the parent body.

regressive burn The burn of a solid propellant grain during which the thrust decreases with time.

rehydrate Add water to a dehydrated food to prepare it for eating.

remote sensing Learning something about an object from the electromagnetic waves emitted and reflected from it at a distance without physical contact.

rendezvous Come together at a specified time and place.

resolution Referring to remote sensing, the size of the smallest object that can be distinguished from its surroundings.

retrofire To fire a rocket in the direction of motion in order to reduce the energy of the orbit

retrorocket A rocket that fires in the direction of motion.

rocket A reaction engine that carries both fuel and oxidizer so it can operate in the absence of air.

roll Rotation of an aircraft or spacecraft around the axis through the nose and tail. *See also* pitch; yaw.

satellite An object in orbit around another object; an object in orbit around Earth, usually one which has some useful purpose; moons are satellites of planets; planets are satellites of the Sun.

scramjet Supersonic ram jet.

SDI Strategic Defense Initiative, the U.S. program for development of a defense against ballistic missiles.

sedimentary rock Rocks formed from the accumulation of sediment in water or windblown sediments.

sensor A device which responds to electromagnetic waves impinging on it, usually by producing or modifying an electrical current.

SETI Search for extraterrestrial intelligence.

signature Referring to remote sensing, the characteristic intensities of various wavelengths of electromagnetic radiation coming from an object which uniquely identify that object.

silicate A compound containing silicon and oxygen.

sinter Compress and fuse together by heat.

solar cell *See* photovoltaic cell.

solar wind A steady flow of particles, mostly protons and electrons, from the Sun into interplanetary space, an extension of the solar atmosphere.

solid rocket booster A booster rocket using solid propellants.

space adaptation syndrome Illness similar to seasickness, experienced by many astronauts during the first day or two of weightlessness.

spacecraft Any vehicle designed to operate in space; an unmanned spacecraft in a closed orbit around Earth is usually called a satellite.

Space Shuttle A four-part, mostly reuseable vehicle consisting of an orbiter-spacecraft-aircraft, two solid rocket boosters, and an external fuel tank.

space station A habitat-workshop-laboratory permanently in orbit, with crews and supplies brought from Earth as needed.

specific impulse A number which indicates the effectiveness of a fuel-oxidizer combination; the time it takes to burn 1 pound of fuel while it is producing 1 pound of thrust.

spectral bands A range of wavelengths that a particular sensor responds to or that has some particular meaning for interpreting the image.

spectrum *See* electromagnetic spectrum.

SRB The Space Shuttle's solid rocket booster.

sunspot A "cool" dark region in the photosphere of the Sun where magnetic fields retard the flow of heat from the interior to the surface.

Sun-synchronous An orbit around Earth which crosses the equator at precisely the same local time on each pass.

tang The part of a fitting that fits into a clevis.

teleoperate Operate equipment or instruments by remote control using a television monitor and radio link.

termination shock The point at which the solar wind drops from supersonic to subsonic as it approaches the heliopause.

terminator On a planet, the line which divides daylight from darkness.

terraform Modify the environment of a planet to make it more Earth-like and habitable by humans.

terrestrial Pertaining to Earth and its life forms.

tetrahedron A solid object with four equilateral triangular shaped sides.

thermostabilize Cook food at temperatures to preserve it by destroying the bacteria.

thrust Force produced by a rocket engine.

thruster A small rocket engine or gas jet used to control the attitude and, to a lesser extent, the orbital speed of a spacecraft.

topography Referring to the varying altitude of the land surface.

trajectory Path followed by a vehicle.

transfer orbit Segment of an orbit which a spacecraft follows to move from one orbit to another.

troposphere The lowest 6 miles (10 km), approximately, of the Earth's atmosphere, the region in which all weather occurs.

turbojet A reaction engine in which air is taken in at the front and compressed, fuel is mixed and burned, and the exhaust gases are ejected out the back.

ultraviolet Electromagnetic waves with wavelengths shorter than visible light but longer than x-rays.

valence An integer number which tells how one element combines chemically with another; refers to the number of outermost electrons in an atom.

Van Allen belts Regions in the Earth's magnetosphere where solar electrons and protons become trapped by the magnetic field.

velocity The motion of an object described by its speed and direction.

vernier engine A small rocket or thruster used to make fine adjustments in speed or attitude.

vestibular apparatus Organ in the inner ear that gives a sense of balance and motion.

wavelength The distance between two adjacent crests or troughs of a wave.

weight The force of gravity acting on a mass.

weightlessness Condition of free fall or zero-g in which objects in a spacecraft are weightless.

x-rays Electromagnetic waves with wavelengths in the range from one-hundredth to one-millionth of a micrometer.

yaw Left and right rotational motion of the nose and tail of an aircraft or spacecraft. *See also* pitch; roll.

zero g A misleading term implying the absence of gravity, but meaning the absence of weight, i.e., weightlessness.

Index

Note: The letter P denotes a color plate number.